Matthias Gerdts

Optimal Control of ODEs and DAEs

Also of Interest

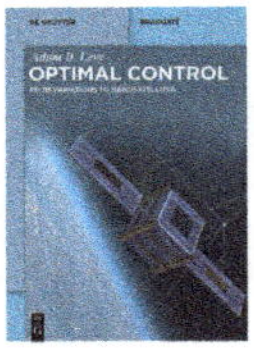

Optimal Control
From Variations to Nanosatellites
Adam B. Levy, 2023
ISBN 978-3-11-128983-0, e-ISBN (PDF) 978-3-11-129015-7

Computational Physics
With Worked Out Examples in FORTRAN® and MATLAB®
Michael Bestehorn, 2023
ISBN 978-3-11-078236-3, e-ISBN (PDF) 978-3-11-078252-3

Optimization and Control for Partial Differential Equations
Uncertainty quantification, open and closed-loop control, and shape optimization
Edited by Roland Herzog, Matthias Heinkenschloss, Dante Kalise, Georg Stadler, Emmanuel Trélat, 2022
ISBN 978-3-11-069596-0, e-ISBN (PDF) 978-3-11-069598-4
in: Radon Series on Computational and Applied Mathematics
ISSN 1865-3707

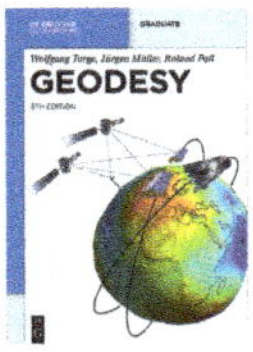

Geodesy
Wolfgang Torge, Jürgen Müller, Roland Pail, 2023
ISBN 978-3-11-072329-8, e-ISBN (PDF) 978-3-11-072330-4

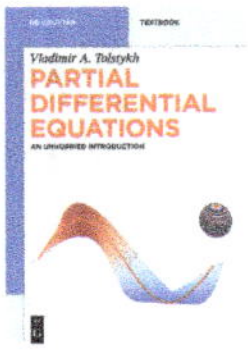

Partial Differential Equations
An Unhurried Introduction
Vladimir A. Tolstykh, 2020
ISBN 978-3-11-067724-9, e-ISBN (PDF) 978-3-11-067725-6

Matthias Gerdts

Optimal Control of ODEs and DAEs

2nd edition

DE GRUYTER
OLDENBOURG

Mathematics Subject Classification 2020
49K15, 49M05, 49M15, 49M25, 49M37, 49N90, 49L20, 90C31, 90C11, 93B52, 34B15, 34A09

Author
Prof. Dr. Matthias Gerdts
Universität der Bundeswehr München
Institut für Angewandte Mathematik und
Wissenschaftliches Rechnen
Fakultät für Luft- und Raumfahrttechnik
Werner-Heisenberg-Weg 39
85577 Neubiberg
Germany
matthias.gerdts@unibw.de
http://www.unibw.de/ingmathe

ISBN 978-3-11-079769-5
e-ISBN (PDF) 978-3-11-079789-3
e-ISBN (EPUB) 978-3-11-079793-0

Library of Congress Control Number: 2023942990

Bibliographic information published by the Deutsche Nationalbibliothek
The Deutsche Nationalbibliothek lists this publication in the Deutsche Nationalbibliografie; detailed bibliographic data are available on the Internet at http://dnb.dnb.de.

Cover image: Prof. Matthias Gerdts
Typesetting: VTeX UAB, Lithuania
Printing and binding: CPI books GmbH, Leck

www.degruyter.com

Preface

This book contains material developed and collected over the last twenty years in my research and teaching activities. It has been used in lecture series on optimal control for master students at the University of Hamburg, the University of Würzburg, and the University of the Bundeswehr Munich.

The book primarily addresses master and PhD students as well as researchers in applied mathematics, but also engineers or scientists with a good background in mathematics and an interest in optimal control.

The intention of the book is to provide both the theoretical and computational tools that are necessary to investigate and solve optimal control problems with ordinary differential equations and differential-algebraic equations. An emphasis is placed on the interplay between the continuous optimal control problem, which typically is defined and analyzed in a Banach space setting, and optimal control problems in discrete time, which are obtained by discretization and lead to structured finite-dimensional optimization problems. The theoretical parts of the book require some knowledge of functional analysis, the numerically oriented parts require knowledge from linear algebra and numerical analysis. Examples are provided for illustration purposes.

Optimal control is a huge field, and not all topics and aspects can be covered in this book. Hence, the contents reflect my personal preferences in favor of necessary optimality conditions, direct and indirect solution methods, mixed-integer optimal control, and real-time optimal control. Topics like the existence of solutions, pseudospectral methods, and sufficient optimality conditions are not covered in detail.

Parts of the book were presented at a summer school on discrete dynamic programming held by Prof. Frank Lempio and myself in Borovets and Sofia, Bulgaria, in 2003. This summer school was funded by the Bulgarian Academy of Sciences and DAAD within the "Center of Excellence for Applications of Mathematics". The second edition of the book contains – next to minor corrections and an update of literature – further exercises and examples, especially in Chapter 5. Moreover, the presentation of model-predictive control in Chapter 6 was revised, amongst others, with material from a postgraduate course on Applied Optimal Control, which I gave during a visiting professorship in the Department of Mechanics, Mathematics, and Management at Politecnico di Bari in 2022. Details on the exploitation of sparsity structures in fully discretized optimal control problems were added in Chapter 5. I hope these adaptions improve the presentation and close some gaps in the first edition.

Last but not least, I would like to express my sincere gratitude to all current and former members of my Engineering Mathematics working group at the University of the Bundeswehr Munich and to my former colleagues in Bayreuth, Hamburg, Birmingham, and Würzburg for the very pleasant and enjoyable time and all the valuable support I received.

https://doi.org/10.1515/9783110797893-201

I hope this book on optimal control problems with ordinary differential equations and differential-algebraic equations is helpful to the reader. Corrections and suggestions are welcome!

Neubiberg, September 2023 Matthias Gerdts

Contents

1 Introduction

Historically, optimal control problems evolved from variational problems. Variational problems have been investigated more thoroughly since 1696, although the first variational problems, such as Queen Dido's problem, were formulated in the ancient world already. In 1696, Johann Bernoulli (1667–1748) posed the Brachistochrone problem to other famous contemporary mathematicians like Sir Isaac Newton (1643–1727), Gottfried Wilhelm Leibniz (1646–1716), Jacob Bernoulli (1654–1705), Guillaume François Antoine Marquis de L'Hôspital (1661–1704), and Ehrenfried Walter von Tschirnhaus (1651–1708). Each of these distinguished mathematicians was able to solve the problem. An interesting description of the Brachistochrone problem with many additional historical remarks, as well as the solution approach of Johann Bernoulli exploiting Fermat's principle, can be found in [264].

Optimal control problems generalize variational problems by separating control and state variables and admitting control constraints. To this end, the *dynamic behavior* of the *state* of a technical, economical, or biological system is described by dynamic equations. For instance, we might be interested in the development of the population size of a specific species during a certain time period, or we want to describe the dynamical behavior of chemical processes or mechanical systems, or the development of the profit of a company during the next five years, say.

The dynamic behavior of a given system typically can be influenced by the choice of *control variables*. For instance, the breeding of rabbits can be influenced by the incorporation of diseases or natural predators. A car can be controlled by the steering wheel, the accelerator pedal, and the brakes. A chemical process can be controlled, for instance, by increasing or decreasing the temperature. The profit of a company is influenced, for instance, by the prices of its products or the number of employees.

Most often, the state or control variables cannot assume any value, but are subject to *control* or *state constraints*. These constraints may result from safety regulations, economical restrictions, or physical limitations, such as the temperature in a reactor has to be lower than a specific threshold, only a certain budget is available, the altitude of an airplane should be larger than ground level, and the steering angle of a car is limited by a maximum steering angle.

Finally, we are particularly interested in those state and control variables that fulfill all constraints and, moreover, minimize or maximize a given *objective function*. For example, if the objective of a company is to maximize profit or to minimize operational costs, a race driver intends to minimize lap time.

In summary, optimal control problems have the following ingredients in common: *State and control variables, dynamic equations in terms of differential equations, state or control constraints, and an objective function to be minimized or maximized.*

The following well-known examples illustrate the ingredients of optimal control problems.

https://doi.org/10.1515/9783110797893-001

Example 1.0.1 (Spline problem, Minimum energy problem, see [41]). Consider a rod in the (t, x)-plane, which is fixed at the positions $(0, 0)$ and $(1, 0)$ such that it assumes the angle α to the t-axis, see Figure 1.1.

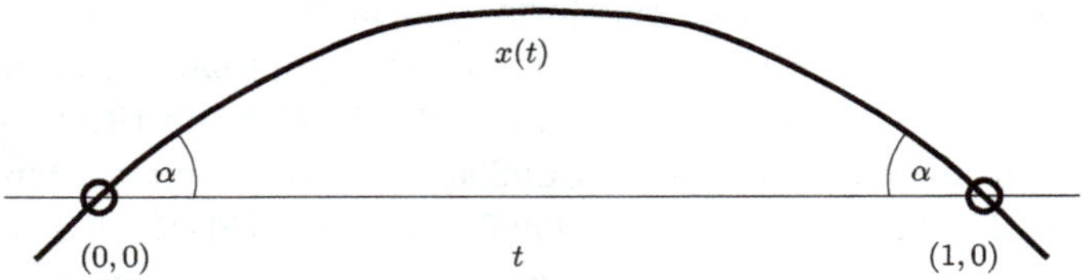

Figure 1.1: The spline problem (minimum energy problem): minimal bending energy of a fixed rod.

We are interested in the shape $x(t)$ of the rod. The rod will assume a shape that minimizes the bending energy, which is given by the curve integral

$$E = c \int_0^L \kappa(s)^2 ds, \tag{1.1}$$

where c is a material dependent constant, and κ denotes curvature. The shape function $x(t)$ is described by the curve

$$\gamma : [0,1] \longrightarrow \mathbb{R}^2, \quad t \mapsto \gamma(t) = \begin{pmatrix} t \\ x(t) \end{pmatrix}.$$

The arc length $s(t)$ is given by

$$s(t) = \int_0^t \|\gamma'(\tau)\| d\tau = \int_0^t \sqrt{1 + x'(\tau)^2}\, d\tau.$$

The length L of the curve between the points $(0, 0)$ and $(1, 0)$ is

$$L = s(1) - s(0) = \int_0^1 \|\gamma'(t)\| dt = \int_0^1 \sqrt{1 + x'(t)^2}\, dt.$$

The function $s(\cdot)$ is a continuously differentiable parameter transformation (injective, s^{-1} continuously differentiable), and the curve γ can be parameterized with respect to the arc length s:

$$\tilde{\gamma} : [0, L] \longrightarrow \mathbb{R}^2, \quad \ell \mapsto \tilde{\gamma}(\ell) = \gamma(s^{-1}(\ell)).$$

In particular, it holds (as $\gamma' \neq 0$)

$$\tilde{\gamma}'(s(t)) = \frac{\gamma'(t)}{\|\gamma'(t)\|}.$$

The curvature κ is defined by

$$\kappa(\ell) := \|\tilde{y}''(\ell)\|.$$

After some computations, we find

$$\kappa(s(t)) = \frac{x''(t)}{\sqrt{1+x'(t)^2}^3}.$$

With the substitution $s \mapsto s(t)$ in (1.1) the bending energy reads as

$$E = c\int_0^L \kappa(s)^2 ds = c\int_0^1 \kappa(s(t))^2 \sqrt{1+x'(t)^2}\, dt = c\int_0^1 \frac{x''(t)^2}{\sqrt{1+x'(t)^2}^5}\, dt.$$

For $|x'(t)| \ll 1$, it simplifies to

$$E \approx c\int_0^1 x''(t)^2 dt.$$

With $x_1(t) := x(t)$, $x_2(t) := x'(t)$, $u(t) := x''(t)$, we obtain an optimal control problem, which is known as *spline problem* or *minimum energy problem*, see [41, p. 120, Sec. 3.11, Ex. 2]:

Minimize

$$\int_0^1 u(t)^2 dt$$

subject to

$$x_1'(t) = x_2(t), \quad x_1(0) = x_1(1) = 0,$$
$$x_2'(t) = u(t), \quad x_2(0) = -x_2(1) = \tan\alpha.$$

The functions x_1 and x_2 denote the *state* of the problem, and the function u is the *control*. The problem can be further augmented by a state constraint $x_1(t) \le x_{\max}$. The solution of the optimal control problem then describes the shape of a fixed rod subject to a load:

Minimize

$$\int_0^1 u(t)^2 dt$$

subject to the differential equations

$$x_1'(t) = x_2(t),$$
$$x_2'(t) = u(t),$$

the boundary conditions

$$x_1(0) = x_1(1) = 0, \quad x_2(0) = -x_2(1) = \tan \alpha,$$

and the state constraint

$$x_1(t) - x_{\max} \leq 0.$$

□

Example 1.0.2 (Goddard problem, see [144]). A rocket of mass m lifts off vertically at time $t = 0$ with (normalized) altitude $h(0) = 1$ and velocity $v(0) = 0$, compare Figure 1.2. The task is to choose the thrust $u(t)$ and the final time t_f such that the altitude $h(t_f)$ at final time t_f is maximized:

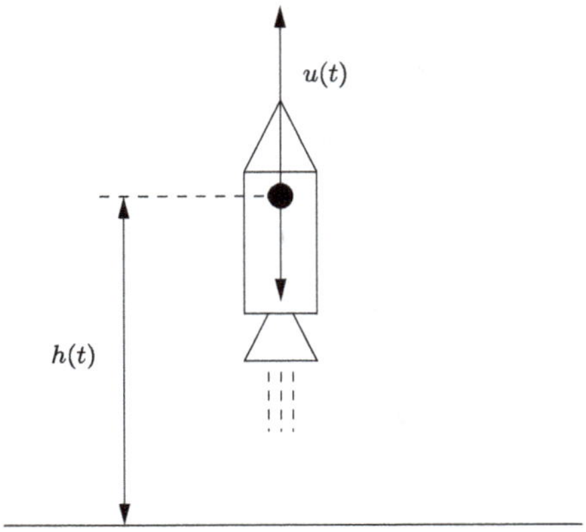

Figure 1.2: Goddard problem: Vertical ascent of a rocket.

Minimize $-h(t_f)$ *subject to the differential equations*

$$\dot{h}(t) = v(t),$$
$$\dot{v}(t) = \frac{1}{m(t)}\Big(u(t) - D\big(v(t), h(t)\big)\Big) - \frac{1}{h(t)^2}, \tag{1.2}$$
$$\dot{m}(t) = -\frac{u(t)}{c} \tag{1.3}$$

with initial conditions

$$h(0) = 1, \quad v(0) = 0, \quad m(0) = 1,$$

terminal condition $m(t_f) = 0.6$, *and control constraint*

$$0 \leq u(t) \leq 3.5.$$

Herein, the fuel consumption is assumed to be proportional to the thrust, which leads to (1.3) with $c = 0.5$. $D(v,h)$ in (1.2) denotes the drag defined by

$$D(v,h) = \frac{C_D F_D}{2m_0 g}\rho(h)v^2$$

with the air density $\rho(h) = \rho_0 \exp(500(1-h))$ and $\frac{\rho_0 C_D F_D}{m_0 g} = 620$.

The term $-1/h^2$ in (1.2) models the (normalized) acceleration due to gravity. □

The following example leads to an optimal control problem with a discrete control set – a so-called integer optimal control problem.

Example 1.0.3 (Lotka–Volterra fishing problem, see [289]). The following problem models an optimal fishing strategy on the fixed time horizon $[0,12]$ with the aim to control the bio-masses of a predator population and a prey fish population to the prescribed steady state $(1,1)$:

Minimize

$$\int_0^{12} (x_1(t)-1)^2 + (x_2(t)-1)^2 \, dt$$

subject to

$$\begin{aligned}
\dot{x}_1(t) &= x_1(t) - x_1(t)x_2(t) - 0.4x_1(t)u(t), \\
\dot{x}_2(t) &= -x_2(t) + x_1(t)x_2(t) - 0.2x_2(t)u(t), \\
u(t) &\in \{0,1\}, \\
x_1(0) &= 0.5, \\
x_2(0) &= 0.7.
\end{aligned}$$

The control u may only assume the two values 0 and 1. Hence, the control set is *discrete.* □

The dynamics are given in terms of ordinary differential equations (ODEs) in the previous examples. However, many processes in natural sciences are modeled by partial differential equations (PDEs) leading to PDE constrained optimal control problems. A typical sample problem is described next, and it will be revisited later on.

Example 1.0.4 (Distributed control of 2D Stokes equation). The instationary 2D Stokes equation

$$\begin{aligned}
v_t &= \Delta v - \nabla p + u, \\
0 &= \operatorname{div}(v), \\
v(0,x,y) &= 0, \quad (x,y) \in \Omega, \\
v(t,x,y) &= 0, \quad (t,x,y) \in (0,T) \times \partial\Omega
\end{aligned}$$

is a model for the motion of a laminar flow of a Newtonian fluid in two space dimensions x and y. Herein, v denotes the velocity vector, p the pressure, and u a distributed control in the domain $Q := (0,T) \times \Omega$ with $\Omega = (0,1) \times (0,1)$ and $T > 0$.

Let a desired velocity field v_d be given on the domain Q. The task is to minimize the distance of the controlled velocity field v to the desired velocity field v_d, see Figure 1.3 for a numerical solution.

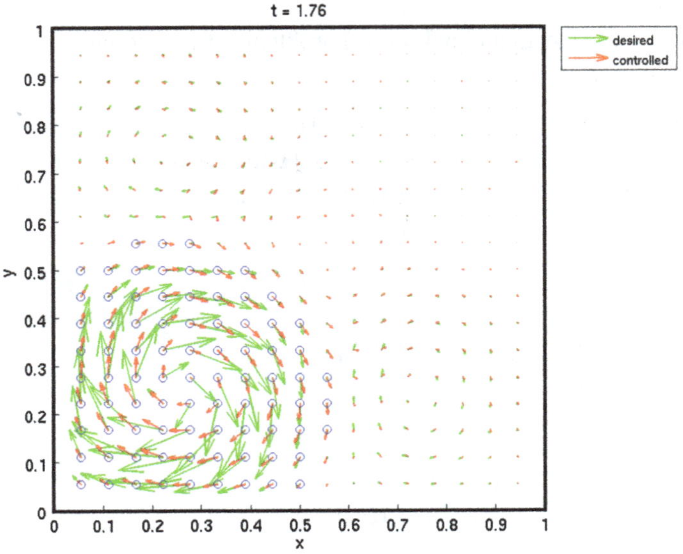

Figure 1.3: Actual velocity field from the 2D Stokes equation and desired velocity field.

This leads to an optimal control problem with a partial differential equation:

Minimize

$$\frac{1}{2}\int_Q \left\|v(t,x,y) - v_d(t,x,y)\right\|^2 dxdydt + \frac{\alpha}{2}\int_Q \left\|u(t,x,y)\right\|^2 dxdydt \tag{1.4}$$

subject to

$$\begin{aligned} v_t &= \Delta v - \nabla p + u, && \text{in } Q,\\ 0 &= \operatorname{div}(v), && \text{in } Q,\\ v(0,x,y) &= 0, && (x,y) \in \Omega,\\ v(t,x,y) &= 0, && (t,x,y) \in (0,T) \times \partial\Omega \end{aligned}$$

and

$$u_{\min} \le u \le u_{\max} \quad \text{in } Q.$$

□

Numerous additional applications of optimal control problems can be found in natural sciences, engineering sciences, economy, and other disciplines. Strongly motivated by military applications, optimal control theory and solution methods have evolved rapidly since 1950. The decisive breakthrough was achieved by the Russian mathematician Lev S. Pontryagin (1908–1988) and his coworkers V. G. Boltyanskii, R. V. Gamkrelidze, and E. F. Mishchenko in proving the maximum principle, which provides necessary optimality conditions for optimal control problems, see [274]. Almost at the same time, Magnus R. Hestenes [162] also proved a similar theorem. Since then, many contributions in view of necessary conditions, e. g., [171, 139, 181, 256, 179, 169, 238, 239, 159], sufficient conditions, e. g., [240, 330, 217, 236, 235, 219], sensitivity analysis and real-time optimal control, e. g., [16, 49, 50, 220, 221, 222, 242, 243, 244, 241, 265, 266, 267, 268], and numerical methods and analysis, e. g., [42, 70, 74, 71, 41, 79, 61, 29, 184, 259, 258, 158, 316, 269, 25, 163, 325, 223, 45, 81, 80], have been released.

Since the 1970s, differential-algebraic equations (DAEs) have been considered more intensively due to their importance in mechanics, process engineering, and electronics. DAEs are composite systems of differential equations and algebraic equations. In an industrial environment such systems are often generated automatically by software packages like SIMPACK. Although DAEs are related to ODEs, they possess different properties and require tailored methods regarding analysis and numerical approaches. The main differences to ODEs can be found with regard to *stability properties*, choice of *consistent initial values*, and *smoothness of solutions depending on input functions*. In this book, we investigate certain classes of DAE optimal control problems with regard to *necessary optimality conditions*, *direct and indirect solution methods*, *sensitivity analysis*, and *real-time optimization*.

!

Throughout this book, Θ_X denotes the zero element of some vector space X. If no confusion is possible, we will use Θ. In the special cases, $X = \mathbb{R}^n$ and $X = \mathbb{R}$ the zero elements are denoted by $0_{\mathbb{R}^n}$ and 0, respectively. The unit matrix of dimension n is denoted by $I_n \in \mathbb{R}^{n\times n}$. The Euclidian norm on $\mathbb{R}^n$ is denoted by $\|\cdot\| = \|\cdot\|_2$.

We use the convention that the dimension of a real-valued vector x is denoted by $n_x \in \mathbb{N}$, that is $x \in \mathbb{R}^{n_x}$.

In most cases, time-dependent processes are investigated, and hence the independent variable of a mapping $z : \mathcal{I} \longrightarrow \mathbb{R}^{n_z}$ with $\mathcal{I} \subseteq \mathbb{R}$ is denoted by t and it is associated with time, although it can have a different meaning in specific applications. The derivative of z with respect to time at time t is denoted by $\dot{z}(t) = \frac{d}{dt}z(t) = z'(t)$.

To simplify notation, we often use the abbreviation $f[t]$ for a function of type $f(t, z(t))$, which depends on time and on one or more time-dependent functions. It will be clear from the context, at what argument f is evaluated, typically this will be an optimal solution of an optimal control problem.

The partial derivatives of a function $f : \mathbb{R}^{n_x} \times \mathbb{R}^{n_y} \longrightarrow \mathbb{R}^{n_f}$, $(x, y) \mapsto f(x, y)$, at (x, y) are denoted by $f'_x(x, y)$ and $f'_y(x, y)$.

1.1 DAE optimal control problems

DAEs are composite systems of differential equations and algebraic equations and often are viewed as differential equations on manifolds. The most general form of a DAE reads as follows:

Definition 1.1.1 (General DAE). Let $\mathcal{I} := [t_0, t_f] \subseteq \mathbb{R}$, $t_0 < t_f$, be a compact time interval, $F : \mathcal{I} \times \mathbb{R}^{n_z} \times \mathbb{R}^{n_z} \times \mathbb{R}^{n_u} \longrightarrow \mathbb{R}^{n_z}$ a sufficiently smooth mapping. The function $u : \mathcal{I} \longrightarrow \mathbb{R}^{n_u}$ is considered an external input, and it is referred to as *control* (*control function*, *control variable*).

The implicit differential equation

$$F(t, z(t), \dot{z}(t), u(t)) = 0_{\mathbb{R}^{n_z}}, \quad t \in \mathcal{I}, \tag{1.5}$$

is called *differential-algebraic equation (DAE)*. The function z in Equation (1.5) is called *state* (*state function, state variable*) of the DAE. □

If the partial derivative $F'_{\dot{z}}(\cdot)$ in Equation (1.5) happens to be non-singular, then Equation (1.5) is just an ODE in implicit form, and the implicit function theorem allows to solve (1.5) for $\dot{z}$ in order to obtain an explicit ODE of type $\dot{z}(t) = f(t, z(t), u(t))$. Hence, explicit ODEs are special cases of DAEs. The more interesting case occurs if the partial derivative $F'_{\dot{z}}(\cdot)$ is *singular*. In this case, Equation (1.5) cannot be solved directly for $\dot{z}$, and it includes differential equations and algebraic equations at the same time.

Example 1.1.2. The DAE

$$F(z(t), \dot{z}(t), u(t)) := \begin{pmatrix} \dot{z}_1(t) - z_2(t) \\ \dot{z}_2(t) - z_3(t) - u_1(t) \\ z_2(t) - u_2(t) \end{pmatrix} = 0_{\mathbb{R}^3}$$

for $z = (z_1, z_2, z_3)^\top$ and $u = (u_1, u_2)^\top$ contains differential equations for the components z_1 and z_2 and an algebraic equation relating z_2 and u_2. Note that the component z_3 is not determined explicitly. Differentiating the algebraic equation with respect to time and exploiting the differential equation for $\dot{z}_2$ yields

$$0 = \dot{z}_2(t) - \dot{u}_2(t) = z_3(t) + u_1(t) - \dot{u}_2(t)$$

if u_2 is differentiable, and thus

$$z_3(t) = -u_1(t) + \dot{u}_2(t).$$

Hence, the algebraic constraint implicitly defines the component z_3. □

DAEs have been discussed intensively since the early 1970s. Solution properties and structural properties of DAEs have been in the main focus since then, see, e. g., [271, 84, 145, 10, 52, 53, 54, 55, 231, 232, 230].

Though, at a first glance, DAEs seem to be very similar to ODEs, they possess different solution properties as it was pointed out in [271, 35]. Particularly, DAEs and ODEs possess different stability properties (see the *perturbation index* in Definition 1.1.12), and initial values $z(t_0) = z_0$ have to be defined properly to guarantee at least locally unique solutions (see *consistent initial values* in Definition 1.1.16). Moreover, in the context of control problems, the control input u needs to be *sufficiently smooth* to obtain a meaningful solution.

These difficulties are illustrated in the following examples.

Example 1.1.3. Consider the linear DAE with constant coefficients

$$\begin{pmatrix} 0 & 0 \\ 1 & 1 \end{pmatrix} \begin{pmatrix} \dot{z}_1(t) \\ \dot{z}_2(t) \end{pmatrix} = \begin{pmatrix} -1 & 1 \\ 0 & 0 \end{pmatrix} \begin{pmatrix} z_1(t) \\ z_2(t) \end{pmatrix} + \begin{pmatrix} u_1(t) \\ u_2(t) \end{pmatrix}, \quad t \in [t_0, t_f],$$

where $u = (u_1, u_2)^\top$ is a sufficiently smooth control input. The first equation yields

$$z_1(t) = z_2(t) + u_1(t), \tag{1.6}$$

and by differentiation, we obtain

$$\dot{z}_1(t) = \dot{z}_2(t) + \dot{u}_1(t).$$

Introducing this relation into the second equation yields

$$u_2(t) = \dot{z}_1(t) + \dot{z}_2(t) = 2\dot{z}_2(t) + \dot{u}_1(t),$$

and thus

$$\dot{z}_2(t) = \frac{1}{2}(u_2(t) - \dot{u}_1(t)).$$

The latter differential equation possesses a unique solution for every initial value $z_2(t_0)$, whenever u_2 and $\dot{u}_1$ are integrable functions. The component z_1 and its initial value (!) are then uniquely determined by Equation (1.6). □

Example 1.1.3 is a special case of the more general setting in

Example 1.1.4 (Canonical form of Weierstraß). Let $A, B \in \mathbb{R}^{n_z \times n_z}$ be given matrices and $u : [t_0, t_f] \longrightarrow \mathbb{R}^{n_z}$ a given control function. Consider the linear DAE

$$A\dot{z}(t) = Bz(t) + u(t) \tag{1.7}$$

for $t \in [t_0, t_f]$. The matrix pair (A, B) is called *regular matrix pair*, if there exists λ with $\det(\lambda A - B) \neq 0$.

For a regular matrix pair (A, B) the so-called *Weierstraß canonical form* exists, i. e. there exist non-singular matrices P and Q with

$$PAQ = \begin{pmatrix} I & \Theta \\ \Theta & N \end{pmatrix}, \quad PBQ = \begin{pmatrix} J & \Theta \\ \Theta & I \end{pmatrix},$$

where Θ denotes a zero matrix of appropriate dimension, J and N are Jordan matrices of appropriate dimension, and N is nilpotent, compare [193, Theorem 2.7, p. 16].

Using the matrices P and Q, Equation (1.7) can be rewritten equivalently as

$$PAQQ^{-1}\dot{z}(t) = PBQQ^{-1}z(t) + Pu(t),$$

and since Q does not depend on time,

$$PAQ\frac{d}{dt}(Q^{-1}z(t)) = PBQQ^{-1}z(t) + Pu(t).$$

The Weierstraß canonical form allows to decompose the DAE (1.7) equivalently into

$$I\dot{\tilde{z}}_1(t) = J\tilde{z}_1(t) + \tilde{u}_1(t), \tag{1.8}$$

$$N\dot{\tilde{z}}_2(t) = I\tilde{z}_2(t) + \tilde{u}_2(t), \tag{1.9}$$

where $\tilde{z} := (\tilde{z}_1, \tilde{z}_2)^\top := Q^{-1}z$ and $\tilde{u} := (\tilde{u}_1, \tilde{u}_2)^\top := Pu$.

Equation (1.8) is a standard linear ODE that possesses a unique solution for any integrable control input $\tilde{u}_1$ and any initial value.

Equation (1.9) is a DAE again, but in a canonical form. Let $p \in \mathbb{N}$ denote the *degree of nilpotency* of the matrix N, that is, it holds $N^p = \Theta$ and $N^k \neq \Theta$ for $k = 0, \ldots, p-1$. Then it is easy to verify that

$$\tilde{z}_2(t) := -\sum_{k=0}^{p-1} N^k \tilde{u}_2^{(k)}(t) \tag{1.10}$$

is a solution of Equation (1.9) provided that $\tilde{u}_2$ is sufficiently smooth, compare [193, Lemma 2.8, p. 17].

The figure p is called the *index* of the matrix pair (A, B) and the DAE (1.7), respectively. It can be shown that the index is independent of the way the Weierstraß canonical form is calculated, see [193, Lemma 2.10, p. 18].

Equation (1.10) reveals three important observations that are characteristic for DAEs:

(a) $\tilde{u}_2$ needs to be sufficiently smooth to allow for a meaningful solution of (1.10).

(b) The initial value $\tilde{z}_2(t_0)$ is uniquely determined by (1.10) and *cannot* be chosen arbitrarily. The initial value has to be *consistent*.

(c) If $\tilde{u}_2$ is interpreted as a sufficiently smooth perturbation in (1.7), then the derivatives up to order $p-1$ of the perturbation will influence the solution $\tilde{z}_2$. This motivates the definition of the *perturbation index* in Definition 1.1.12 below.

Canonical forms for time-dependent matrices A and B using the *strangeness index* can be found in [193, Theorem 3.17, p. 74]. □

The above examples indicate that the DAE (1.5) in its full complexity is too general and therefore too challenging for theoretical and numerical investigations, especially since canonical forms like the Weierstraß canonical form in Example 1.1.4 cannot be extended to general nonlinear DAEs without imposing additional structural restrictions.

Whenever necessary, we restrict the problem to more simple problems. Most often, we discuss so-called *semi-explicit DAEs*.

Definition 1.1.5 (Semi-explicit DAE). Let $\mathcal{I} = [t_0, t_f] \subset \mathbb{R}$, $t_0 < t_f$, be a compact time interval. Let the state $z = (x, y)^\top$ in (1.5) be decomposed into components $x : \mathcal{I} \longrightarrow \mathbb{R}^{n_x}$ and $y : \mathcal{I} \longrightarrow \mathbb{R}^{n_y}$. Let $u : \mathcal{I} \longrightarrow \mathbb{R}^{n_u}$ be a given control function.

A DAE of type

$$\dot{x}(t) = f(t, x(t), y(t), u(t)), \tag{1.11}$$

$$0_{\mathbb{R}^{n_y}} = g(t, x(t), y(t), u(t)), \tag{1.12}$$

is called *semi-explicit DAE* on $\mathcal{I}$. Herein, $x(\cdot)$ is referred to as *differential variable*, and $y(\cdot)$ is called *algebraic variable*. Correspondingly, (1.11) is called *differential equation* and (1.12) *algebraic equation*. □

The implicit DAE (1.5) can be transformed formally into a semi-explicit DAE by introducing an artificial algebraic variable y as follows:

$$\dot{z}(t) = y(t), \tag{1.13}$$

$$0_{\mathbb{R}^{n_z}} = F(t, z(t), y(t), u(t)). \tag{1.14}$$

The transformation inherits a potential pitfall, though. According to the differential equation (1.13), one might expect the solution z of (1.13)–(1.14) to be absolutely continuous for an essentially bounded input y, which is the standard setting in optimal control. The following example shows that this does not necessarily lead to a meaningful solution, which satisfies the algebraic equation (1.14) as well.

Example 1.1.6. Consider the nonlinear DAE

$$\begin{pmatrix} 0 & 0 \\ u(t) & 1 \end{pmatrix} \begin{pmatrix} \dot{z}_1(t) \\ \dot{z}_2(t) \end{pmatrix} = \begin{pmatrix} -1 & u(t) \\ 0 & 0 \end{pmatrix} \begin{pmatrix} z_1(t) \\ z_2(t) \end{pmatrix} + \begin{pmatrix} 0 \\ u(t) \end{pmatrix} \tag{1.15}$$

for $t \in [0, 2]$. Let the initial values $z_1(0) = 0$ and $z_2(0) = 1$ be given, and let the control u be piecewise defined by

$$u(t) = \begin{cases} 0, & \text{if } 0 \le t < 1, \\ 1, & \text{if } 1 \le t \le 2. \end{cases}$$

Piecewise evaluation of the DAE yields the following:

For $0 \le t < 1$, we find the solution

$$z_1(t) = 0, \quad z_2(t) = 1.$$

For $1 \le t \le 2$, we obtain the solution

$$z_1(t) = u(t)z_2(t) = z_2(t) = \frac{1+t}{2}.$$

The component z_1 is discontinuous at $t = 1$.

The transformation (1.13)–(1.14) yields

$$\begin{pmatrix} \dot{z}_1(t) \\ \dot{z}_2(t) \end{pmatrix} = \begin{pmatrix} y_1(t) \\ y_2(t) \end{pmatrix},$$
$$\begin{pmatrix} 0 & 0 \\ u(t) & 1 \end{pmatrix} \begin{pmatrix} y_1(t) \\ y_2(t) \end{pmatrix} = \begin{pmatrix} -1 & u(t) \\ 0 & 0 \end{pmatrix} \begin{pmatrix} z_1(t) \\ z_2(t) \end{pmatrix} + \begin{pmatrix} 0 \\ u(t) \end{pmatrix}$$

with initial values $z_1(0) = 0$ and $z_2(0) = 1$.

For $0 \le t < 1$, we find

$$z_1(t) = 0, \quad y_1(t) = 0, \quad y_2(t) = 0, \quad z_2(t) = 1.$$

For $1 \le t \le 2$, we obtain

$$z_1(t) = u(t)z_2(t) = z_2(t), \quad y_1(t) + y_2(t) = 1.$$

Differentiating $z_1(t) = z_2(t)$, it follows $y_1(t) = \dot{z}_1(t) = \dot{z}_2(t) = y_2(t)$ and thus $y_1(t) = y_2(t) = \frac{1}{2}$, since $y_1(t) + y_2(t) = 1$. Hence, we found

$$y_1(t) = y_2(t) = \begin{cases} 0, & \text{if } 0 \le t < 1, \\ \frac{1}{2}, & \text{if } 1 \le t \le 2. \end{cases}$$

Integration of the differential equations and obeying the initial values yields

$$z_1(t) = \begin{cases} 0, & \text{if } 0 \le t < 1, \\ \frac{t-1}{2}, & \text{if } 1 \le t \le 2, \end{cases} \qquad z_2(t) = \begin{cases} 1, & \text{if } 0 \le t < 1, \\ \frac{t+1}{2}, & \text{if } 1 \le t \le 2. \end{cases}$$

However, these functions do not satisfy the algebraic constraint $z_1(t) = z_2(t)$ for $1 \le t \le 2$. On the other hand, the equation $z_1(t) = z_2(t)$ for $t \in [1, 2]$ can only be satisfied if z_1 (or z_2) jumps at $t = 1$. Hence, the original DAE (1.15) and the transformed DAE only have the same solution if the latter permits jumps in the differential state. □

Example 1.1.6 shows that in order to have the same meaningful solution as in (1.5), one needs to permit impulsive solutions in the transformed system (1.13)–(1.14) allowing the differential variable z to jump. This would complicate the following optimal control problems considerably, and for that reason, we exclude such situations in the sequel by the following definition of a solution of the semi-explicit DAE (1.11)–(1.12).

Definition 1.1.7. Consider the semi-explicit DAE in Definition 1.1.5. (x, y) is called a *solution* of (1.11)–(1.12) for a given control input u, if (1.12), and

$$x(t) = x(t_0) + \int_{t_0}^{t} f(\tau, x(\tau), y(\tau), u(\tau))d\tau$$

are satisfied for $t \in \mathcal{I}$. □

Consider

Problem 1.1.8 (DAE optimal control problem). Let $\mathcal{I} = [t_0, t_f] \subset \mathbb{R}$, $t_0 < t_f$, be a compact time interval,

$$\begin{aligned}
\varphi &: \mathbb{R} \times \mathbb{R} \times \mathbb{R}^{n_x} \times \mathbb{R}^{n_x} \longrightarrow \mathbb{R},\\
f_0 &: \mathcal{I} \times \mathbb{R}^{n_x} \times \mathbb{R}^{n_y} \times \mathbb{R}^{n_u} \longrightarrow \mathbb{R},\\
f &: \mathcal{I} \times \mathbb{R}^{n_x} \times \mathbb{R}^{n_y} \times \mathbb{R}^{n_u} \longrightarrow \mathbb{R}^{n_x},\\
g &: \mathcal{I} \times \mathbb{R}^{n_x} \times \mathbb{R}^{n_y} \times \mathbb{R}^{n_u} \longrightarrow \mathbb{R}^{n_y},\\
c &: \mathcal{I} \times \mathbb{R}^{n_x} \times \mathbb{R}^{n_y} \times \mathbb{R}^{n_u} \longrightarrow \mathbb{R}^{n_c},\\
s &: \mathcal{I} \times \mathbb{R}^{n_x} \longrightarrow \mathbb{R}^{n_s},\\
\psi &: \mathbb{R} \times \mathbb{R} \times \mathbb{R}^{n_x} \times \mathbb{R}^{n_x} \longrightarrow \mathbb{R}^{n_\psi}
\end{aligned}$$

sufficiently smooth functions, and $\mathcal{U} \subseteq \mathbb{R}^{n_u}$ a non-empty set.

Minimize the objective function

$$\varphi(t_0, t_f, x(t_0), x(t_f)) + \int_{t_0}^{t_f} f_0(t, x(t), y(t), u(t))dt$$

with respect to

$$x : \mathcal{I} \longrightarrow \mathbb{R}^{n_x}, \quad y : \mathcal{I} \longrightarrow \mathbb{R}^{n_y}, \quad u : \mathcal{I} \longrightarrow \mathbb{R}^{n_u}$$

subject to the semi-explicit DAE

$$\begin{aligned}
\dot{x}(t) &= f(t, x(t), y(t), u(t)),\\
0_{\mathbb{R}^{n_y}} &= g(t, x(t), y(t), u(t)),
\end{aligned}$$

the mixed control-state constraint

$$c(t, x(t), y(t), u(t)) \leq 0_{\mathbb{R}^{n_c}}, \tag{1.16}$$

the pure state constraint

$$s(t, x(t)) \leq 0_{\mathbb{R}^{n_s}}, \tag{1.17}$$

the boundary condition

$$\psi(t_0, t_f, x(t_0), x(t_f)) = 0_{\mathbb{R}^{n_\psi}}, \tag{1.18}$$

and the control set constraint

$$u(t) \in \mathcal{U}. \tag{1.19}$$

The time points t_0 or t_f in Problem 1.1.8 can be fixed or unknown in which case t_0 or t_f are additional optimization variables. If t_0 is not fixed, then t_0 is called *free initial time*, and Problem 1.1.8 is a *problem with free initial time*. If t_f is not fixed, then t_f is called *free final time*, and Problem 1.1.8 is a *problem with free final time*.

Problem 1.1.8 is called *autonomous* if the functions φ, f_0, f, g, c, s, ψ do not explicitly depend on the time; otherwise, it is called *non-autonomous*.

In Problem 1.1.8, x and y are considered to define the *state* of the process, while u is the *control*. In fact, this assignment is somehow arbitrary from the optimal control theoretic point of view, since both, u and y, can be viewed as control variables.

This becomes clearer in Chapter 3 on local minimum principles for DAE optimal control problems, where it can be seen that both functions, algebraic variable y and control u, have very similar properties. Actually, this is the reason why the functions φ, ψ, and s only depend on x and not on y and u. However, recall that the DAE, in reality, is a model for, e. g., a robot, a car, an electric circuit, or a power plant. Hence, the DAE has a meaning for itself independent of whether it occurs in an optimal control problem or not. In this context, it is necessary to distinguish between control u and algebraic variable y. The essential difference is that an operator can choose the control u, whereas the algebraic variable y typically cannot be controlled directly since it results from the state component x and the input u. For instance, in the context of mechanical multi-body systems, the algebraic variable y corresponds physically to a constraint force.

A simple example of a DAE optimal control problem is given by the pendulum problem. Despite its simplicity, it already exhibits the main difficulties when dealing with DAEs.

Example 1.1.9 (Pendulum problem). Consider a rigid pendulum of length $\ell > 0$ and mass $m > 0$ that is mounted to a motor. The motion of the pendulum under influence of the acceleration due to gravity $g = 9.81\,[\mathrm{m/s^2}]$ can be controlled by applying a torque $\ell u(t)$ through the motor, see Figure 1.4.

Minimize

$$\int_0^{t_f} u(t)^2 dt$$

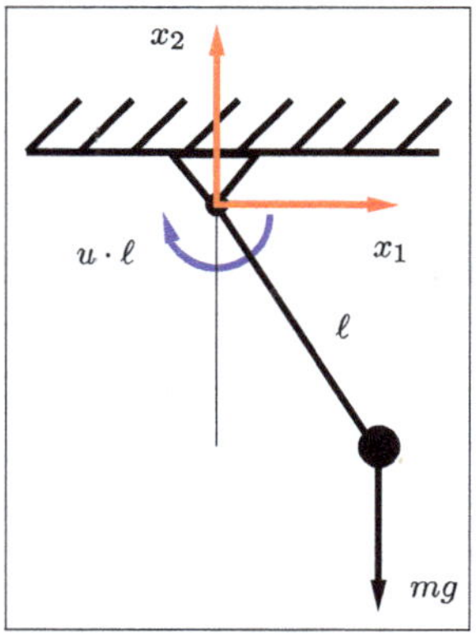

Figure 1.4: Control of a rigid pendulum.

with respect to

$$x = (x_1, \ldots, x_4)^\top : [0, t_f] \longrightarrow \mathbb{R}^4, \quad y : [0, t_f] \longrightarrow \mathbb{R}, \quad u : [0, t_f] \longrightarrow \mathbb{R}$$

subject to the equations of motion of the mathematical pendulum given by

$$\dot{x}_1(t) = x_3(t), \tag{1.20}$$

$$\dot{x}_2(t) = x_4(t), \tag{1.21}$$

$$m\dot{x}_3(t) = \quad - 2x_1(t)y(t) + \frac{u(t)x_2(t)}{\ell}, \tag{1.22}$$

$$m\dot{x}_4(t) = -mg - 2x_2(t)y(t) - \frac{u(t)x_1(t)}{\ell}, \tag{1.23}$$

$$0 = x_1(t)^2 + x_2(t)^2 - \ell^2, \tag{1.24}$$

and the boundary conditions

$$x_1(0) = \ell, \quad x_2(0) = x_3(0) = x_4(0) = x_1(t_f) = x_3(t_f) = 0. \tag{1.25}$$

With $\mathcal{U} := \mathbb{R}, f_0(u) := u^2, g(x) := x_1^2 + x_2^2 - \ell^2$, and

$$f(x, y, u) := \begin{pmatrix} x_3, \\ x_4, \\ -\frac{2}{m} x_1 y + \frac{u x_2}{m\ell} \\ -g - \frac{2}{m} x_2 y - \frac{u x_1}{m\ell} \end{pmatrix}, \quad \psi(x(0), x(t_f)) := \begin{pmatrix} x_1(0) - \ell \\ x_2(0) \\ x_3(0) \\ x_4(0) \\ x_1(t_f) \\ x_3(t_f) \end{pmatrix}$$

the problem fits into Problem 1.1.8.

Note that the algebraic variable y is not explicitly defined by (1.20)–(1.24). However, if we differentiate the algebraic constraint (1.24) twice with respect to time, we obtain the equations

$$0 = 2(x_1(t)\dot{x}_1(t) + x_2(t)\dot{x}_2(t)),$$

$$0 = 2(\dot{x}_1(t)^2 + \dot{x}_2(t)^2 + x_1(t)\ddot{x}_1(t) + x_2(t)\ddot{x}_2(t)).$$

Introducing (1.22)–(1.23) into the latter and solving for y yields

$$y(t) = \frac{m}{2\ell^2}(\dot{x}_1(t)^2 + \dot{x}_2(t)^2 - x_2(t)g).$$

In general, the algebraic variable y is defined implicitly either by the algebraic constraint or by its derivatives. In this example, three differentiations of the algebraic constraint with respect to time are necessary to eventually obtain a differential equation for the algebraic variable y, and the DAE is said to have *differentiation index* three.

Moreover, initial values $x(0)$ and $y(0)$ have to satisfy not only the algebraic constraint (1.24), i. e. $g(x(0)) = 0$, but also its derivatives at the initial time, i. e.

$$\begin{aligned} 0 &= 2(x_1(0)\dot{x}_1(0) + x_2(0)\dot{x}_2(0)), \\ y(0) &= \frac{m}{2\ell^2}(\dot{x}_1(0)^2 + \dot{x}_2(0)^2 - x_2(0)g). \end{aligned}$$

Consequently, initial values cannot be chosen arbitrarily as it would have been possible for ordinary differential equations. Instead, initial values have to be chosen *consistent* with the DAE. □

Often, algebraic equations are added to a problem formulation in order to force a dynamic system to follow a prescribed path. We illustrate this for a problem from flight path optimization.

Example 1.1.10 (Flight path optimization). Consider the following scenario: During the ascent phase of a hypersonic flight, system a malfunction forces the ascent to be stopped. It is assumed that the flight system is still able to maneuver with the restriction that the propulsion system is damaged, and hence the thrust is zero. For security reasons, an emergency landing trajectory of the maximum range is sought, compare [246, 47, 48].

This scenario leads to the following optimal control problem on the time interval $t \in [0, t_f]$ with free final time t_f:

Minimize the negative range with respect to the initial position

$$\Phi(C_L, \mu, t_f) = -\left(\frac{\Lambda(t_f) - \Lambda(0)}{\Lambda(0)}\right)^2 - \left(\frac{\Theta(t_f) - \Theta(0)}{\Theta(0)}\right)^2$$

subject to the differential equations for the velocity v, the inclination γ, the azimuth angle χ, the altitude h, the latitude Λ, and the longitude Θ,

$$\dot{v} = -D(v, h; C_L)\frac{1}{m} - g(h)\sin\gamma + \omega^2\cos\Lambda(\sin\gamma\cos\Lambda - \cos\gamma\sin\chi\sin\Lambda)R(h), \tag{1.26}$$

$$\begin{aligned} \dot{\gamma} = {} & L(v, h; C_L)\frac{\cos\mu}{mv} - \left(\frac{g(h)}{v} - \frac{v}{R(h)}\right)\cos\gamma + 2\omega\cos\chi\cos\Lambda \\ & + \omega^2\cos\Lambda(\sin\gamma\sin\chi\sin\Lambda + \cos\gamma\cos\Lambda)\frac{R(h)}{v}, \end{aligned} \tag{1.27}$$

$$\dot{\chi} = L(v,h;C_L)\frac{\sin\mu}{mv\cos\gamma} - \cos\gamma\cos\chi\tan\Lambda\frac{v}{R(h)}$$
$$+ 2\omega(\sin\chi\cos\Lambda\tan\gamma - \sin\Lambda) - \omega^2\cos\Lambda\sin\Lambda\cos\chi\frac{R(h)}{v\cos\gamma}, \quad (1.28)$$
$$\dot{h} = v\sin\gamma, \quad (1.29)$$
$$\dot{\Lambda} = \cos\gamma\sin\chi\frac{v}{R(h)}, \quad (1.30)$$
$$\dot{\Theta} = \cos\gamma\cos\chi\frac{v}{R(h)\cos\Lambda}. \quad (1.31)$$

Box constraints for the controls C_L (lift coefficient) and μ (angle of bank) are given by

$$0.01 \le C_L \le 0.18326,$$
$$-\frac{\pi}{2} \le \mu \le \frac{\pi}{2}.$$

The initial state corresponds to a position above Bayreuth (Germany):

$$\begin{pmatrix} v(0) \\ \gamma(0) \\ \chi(0) \\ h(0) \\ \Lambda(0) \\ \Theta(0) \end{pmatrix} = \begin{pmatrix} 2150.5452900 \\ 0.1520181770 \\ 2.2689279889 \\ 33900.000000 \\ 0.8651597102 \\ 0.1980948701 \end{pmatrix}.$$

A terminal condition is

$$h(t_f) = 500.$$

Finally, the dynamic pressure constraint

$$q(v,h) \le q_{\max}$$

with $q_{\max} = 60000\ [N/m^2]$ has to be obeyed.

Herein, lift L, air density ρ, drag D, radius R, drag coefficient C_D, acceleration due to gravity g, and dynamic pressure q are defined by

$$\begin{aligned} L(v,h,C_L) &= q(v,h)\,F\,C_L, & \rho(h) &= \rho_0\exp(-\beta h), \\ D(v,h,C_L) &= q(v,h)\,F\,C_D(C_L), & R(h) &= r_0 + h, \\ C_D(C_L) &= C_{D_0} + k\,C_L^2, & g(h) &= g_0(r_0/R(h))^2, \\ q(v,h) &= \frac{1}{2}\rho(h)v^2, \end{aligned}$$

with constants

$$\begin{aligned} F &= 305, & r_0 &= 6.371\cdot 10^6, & C_{D_0} &= 0.017, \\ k &= 2, & \rho_0 &= 1.249512, & \beta &= 1/6900, \\ g_0 &= 9.80665, & \omega &= 7.27\cdot 10^{-5}, & m &= 115000. \end{aligned}$$

The numerical solution of the emergency maneuver with and without dynamic pressure constraint using the techniques in Chapter 5 is depicted in Figure 1.5.

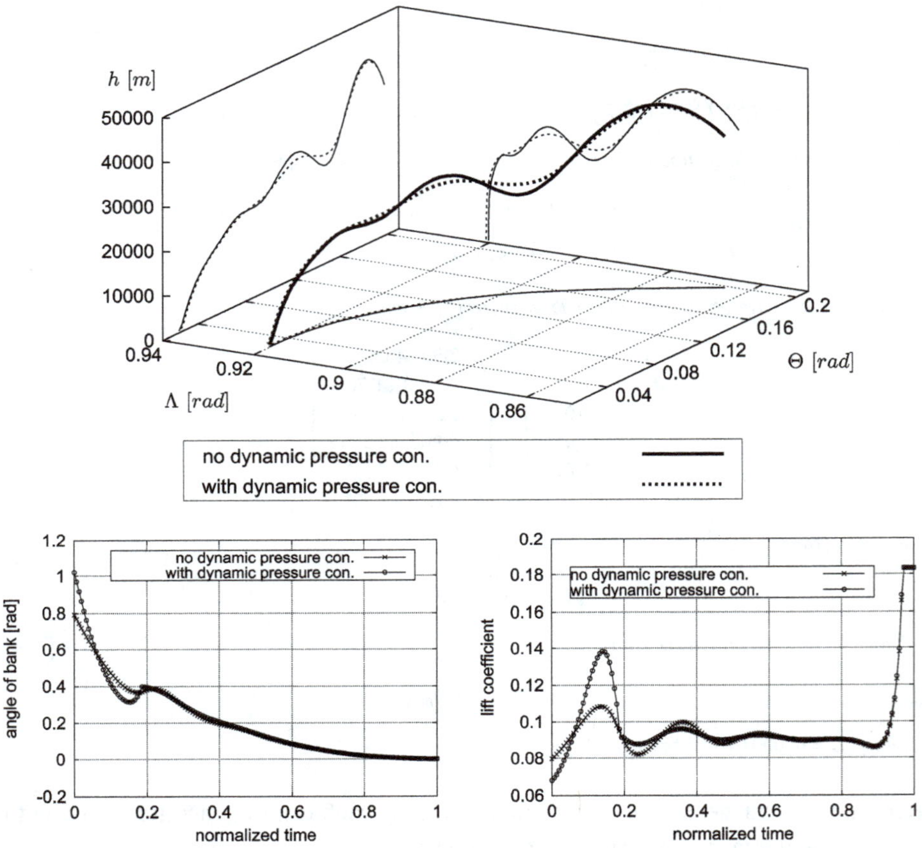

Figure 1.5: Comparison of the solution with and without dynamic pressure constraint: flight trajectory (top), controls μ (angle of bank), and C_L (lift coefficient) (bottom, normalized time).

Flight path optimization with a prescribed path

A modification is obtained if a prescribed path needs to be followed during the emergency landing maneuver. The path to be followed is expressed in terms of the longitude and latitude and is given for simplicity by the straight line

$$\Lambda(t) + \Theta(t) = \Lambda(0) + \Theta(0) \quad \text{for } t \in [0, t_f].$$

Adding this algebraic equation to the dynamics (1.26)–(1.31) yields a DAE if one of the two controls, C_L or μ, is considered an algebraic variable. In this example, we choose $\mu(t)$ as an algebraic variable. Analyzing the algebraic constraint reveals that the differentiation index of the DAE is three, i. e. the algebraic constraint needs to be differentiated three times with respect to time to derive a differential equation for the algebraic variable μ. Differentiating the algebraic constraint leads to

$$\dot{\Lambda}(t) + \dot{\Theta}(t) = 0 \quad \text{for } t \in [0, t_f].$$

Introducing the differential equations for Λ and Θ yields the algebraic constraint

$$0 = \frac{v(t)\cos\gamma(t)}{R(h(t))}\left(\sin\chi(t) + \frac{\cos\chi(t)}{\cos\Lambda(t)}\right) \quad \text{for } t \in [0, t_f].$$

This imposes an additional constraint for the initial conditions. For the initial value,

$$\begin{pmatrix} v(0) \\ \gamma(0) \\ \chi(0) \\ h(0) \\ \Lambda(0) \\ \Theta(0) \end{pmatrix} = \begin{pmatrix} 2150.5452900 \\ 0.1520181770 \\ \text{free} \\ 33900.000000 \\ 0.8651597102 \\ 0.1980948701 \end{pmatrix}$$

consistent values for χ and μ are given by $\chi(0) \approx -0.9955$ and $\mu(0) \approx 0.3695$.

Figure 1.6 shows the numerical solution with dynamic pressure constraint and a prescribed path. □

DAEs also occur in discretized PDE constrained optimal control problems.

Example 1.1.11 (Discretized 2D Stokes equation). Let us revisit Example 1.0.4 and the 2D instationary Stokes equation

$$\begin{aligned} v_t &= \Delta v - \nabla p + u, \\ 0 &= \operatorname{div}(v), \\ v(0,x,y) &= 0, \quad (x,y) \in \Omega, \\ v(t,x,y) &= 0, \quad (t,x,y) \in (0,T) \times \partial\Omega \end{aligned}$$

with $\Omega := (0,1) \times (0,1)$. We apply the method of lines in order to discretize the Stokes equation in space. For simplicity, an equidistant grid with step-size

$$h := \frac{1}{N}, \quad N \in \mathbb{N},$$

is used to discretize Ω by the grid

$$\Omega_h := \{(x_i, y_j) \in \mathbb{R}^2 \mid x_i = ih,\ y_j = jh,\ i,j = 0,\dots,N\}.$$

For $i,j = 0,\dots,N$ consider approximations

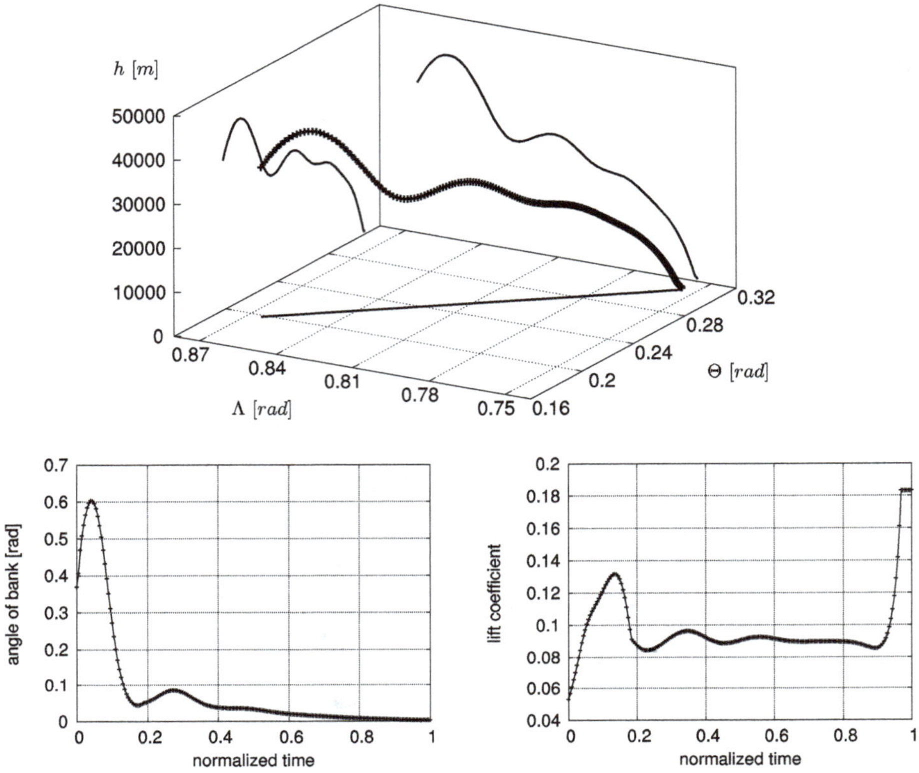

Figure 1.6: Flight trajectory (top), algebraic variable μ (angle of bank), and control C_L (lift coefficient) (bottom, normalized time) for the emergency landing maneuver on a prescribed path.

$$v_{ij}(t) \approx v(t, x_i, y_j), \quad p_{ij}(t) \approx p(t, x_i, y_j), \quad u_{ij}(t) \approx u(t, x_i, y_j).$$

The operators Δv, ∇p, and $\operatorname{div}(v)$ are replaced by finite difference approximations

$$\begin{aligned}
\Delta v(t, x_i, y_j) &\approx \frac{1}{h^2}\big(v_{i+1,j}(t) + v_{i-1,j}(t) + v_{i,j+1}(t) + v_{i,j-1}(t) - 4v_{i,j}(t)\big),\\
\nabla p(t, x_i, y_j) &\approx \frac{1}{h}\big(p_{i+1,j}(t) - p_{i,j}(t), p_{i,j+1}(t) - p_{i,j}(t)\big)^\top,\\
\operatorname{div}(v)(t, x_i, y_j) &\approx \frac{1}{h}\big(v^1_{i,j}(t) - v^1_{i-1,j}(t) + v^2_{i,j}(t) - v^2_{i,j-1}(t)\big),
\end{aligned}$$

for $i, j = 1, \dots, N-1$. Herein, v^1 and v^2 denote the components of $v = (v^1, v^2)^\top$, respectively. The undefined pressure components $p_{i,j}$ with $i = N$ or $j = N$ are set to zero.

Introducing these approximations into the optimal control problem in Example 1.0.4 yields an optimal control problem with a differential-algebraic equation:

Minimize

$$\frac{1}{2}\int_0^T \|v_h(t) - v_{d,h}(t)\|^2 dt + \frac{\alpha}{2}\int_0^T \|u_h(t)\|^2 dt$$

subject to

$$\begin{aligned} \dot{v}_h(t) &= A_h v_h(t) + B_h p_h(t) + u_h(t), \quad v_h(0) = \Theta, \\ \Theta &= B_h^\top v_h(t), \end{aligned}$$

and

$$u_{\min} \le u_h(t) \le u_{\max}.$$

Herein,

$$\begin{aligned} v_h &= (v_{1,1},\dots,v_{N-1,1},v_{1,2},\dots,v_{N-1,2},\dots,v_{1,N-1},\dots,v_{N-1,N-1})^\top, \\ p_h &= (p_{1,1},\dots,p_{N-1,1},p_{1,2},\dots,p_{N-1,2},\dots,p_{1,N-1},\dots,p_{N-1,N-1})^\top, \\ u_h &= (u_{1,1},\dots,u_{N-1,1},u_{1,2},\dots,u_{N-1,2},\dots,u_{1,N-1},\dots,u_{N-1,N-1})^\top, \end{aligned}$$

and $A_h \in \mathbb{R}^{2(N-1)^2\times 2(N-1)^2}$, $M \in \mathbb{R}^{2(N-1)\times 2(N-1)}$, $B_h \in \mathbb{R}^{2(N-1)^2\times(N-1)^2}$, $N, Q \in \mathbb{R}^{2(N-1)\times(N-1)}$ with

$$A_h = \frac{1}{h^2}\begin{pmatrix} M & I & & & \\ I & M & I & & \\ & \ddots & \ddots & \ddots & \\ & & I & M & I \\ & & & I & M \end{pmatrix}, \quad B_h = \frac{1}{h}\begin{pmatrix} N & Q & & \\ & \ddots & \ddots & \\ & & N & Q \\ & & & N \end{pmatrix},$$

and

$$Q = \begin{pmatrix} 0 & & & \\ -1 & & & \\ & 0 & & \\ & -1 & & \\ & & \ddots & \\ & & & 0 \\ & & & -1 \end{pmatrix},$$

and

$$M = \begin{pmatrix} -4 & & 1 & & & \\ & -4 & & 1 & & \\ 1 & & \ddots & & \ddots & \\ & \ddots & & \ddots & & 1 \\ & & 1 & & -4 & \\ & & & 1 & & -4 \end{pmatrix}, \quad N = \begin{pmatrix} 1 & -1 & & & & \\ 1 & 0 & & & & \\ & 1 & -1 & & & \\ & 1 & 0 & & & \\ & & \ddots & \ddots & & \\ & & & & 1 & -1 \\ & & & & 1 & 0 \\ & & & & & 1 \\ & & & & & 1 \end{pmatrix}.$$

Consider the DAE

$$\begin{aligned} \dot{v}_h(t) &= A_h v_h(t) + B_h p_h(t) + u_h(t), \quad v_h(0) = \Theta, \\ \Theta &= B_h^\top v_h(t) \end{aligned}$$

with differential state v_h, algebraic variable p_h, and control u_h. Differentiation of the algebraic constraint $0 = B_h^\top v_h(t)$ yields

$$\Theta = B_h^\top (A_h v_h(t) + B_h p_h(t) + u_h(t)). \tag{1.32}$$

The matrix $B_h^\top B_h$ turns out to be positive definite, and thus, it is non-singular. Hence, the above equation can be solved for the discretized pressure p_h:

$$p_h = -(B_h^\top B_h)^{-1} B_h^\top (A_h v_h(t) + u_h(t)). \tag{1.33}$$

The discretized Stokes equation is said to be of *index two*, because differentiating equation (1.32) again with respect to time would lead to a differential equation for p_h. Note, however, that the control u_h appears in equation (1.32). As u_h is typically an element of the space L_∞ (Banach space of essentially bounded functions, see Chapter 2), a further differentiation with respect to time is not permitted without additional smoothness assumptions. Consequently, the algebraic variable p_h in (1.33) possesses basically the same smoothness properties as u_h, that is, p_h has also to be considered a function in the space L_∞. □

The fact that y is defined by x and u is mathematically taken into account by imposing additional regularity assumptions. The degree of regularity of the DAE (1.11)–(1.12) is measured by a quantity called *index*. For specially structured DAEs, e. g., Hessenberg DAEs, the various index definitions coincide, whereas for general systems, several different index definitions exist, among them are the *differentiation index*, the *perturbation index*, the *strangeness index*, the *tractability index*, and the *structural index*, see [84, 106, 107, 53, 155, 193, 231, 233, 232]. We only review the perturbation index in more detail as this index illustrates one of the major difficulties related to DAEs: The influence of perturbations on solutions.

1.1.1 Perturbation index

The perturbation index indicates the influence of perturbations and their derivatives on the solution and therefore addresses the *stability of DAEs*.

Definition 1.1.12 (Perturbation index, see [155]). The DAE (1.5) has *perturbation index* $p \in \mathbb{N}$ along a solution z on $[t_0, t_f]$ if $p \in \mathbb{N}$ is the smallest number such that for all functions $\tilde{z}$ satisfying the perturbed DAE

$$F(t, \tilde{z}(t), \dot{\tilde{z}}(t), u(t)) = \delta(t), \tag{1.34}$$

there exists a constant S depending on F, u, and $t_f - t_0$ with

$$\|z(t) - \tilde{z}(t)\| \leq S\Big(\|z(t_0) - \tilde{z}(t_0)\| + \max_{t_0 \leq \tau \leq t}\|\delta(\tau)\| + \cdots + \max_{t_0 \leq \tau \leq t}\|\delta^{(p-1)}(\tau)\|\Big) \tag{1.35}$$

for all $t \in [t_0, t_f]$, whenever the expression on the right is less than or equal to a given bound.

The *perturbation index* is $p = 0$ if the estimate

$$\|z(t) - \tilde{z}(t)\| \leq S\left(\|z(t_0) - \tilde{z}(t_0)\| + \max_{t_0 \leq \tau \leq t_f}\left\|\int_{t_0}^{\tau} \delta(s)ds\right\|\right) \tag{1.36}$$

holds.

The DAE is said to be of *higher index* if $p \geq 2$. □

We intend to investigate the perturbation index for some DAEs and make use of the subsequent lemma.

Lemma 1.1.13 (Gronwall). *Let $w, z : [t_0, t_f] \longrightarrow \mathbb{R}$ be integrable functions and $L \geq 0$ a constant with*

$$w(t) \leq L \int_{t_0}^{t} w(\tau)d\tau + z(t)$$

for almost every $t \in [t_0, t_f]$. Then it holds

$$w(t) \leq z(t) + L \int_{t_0}^{t} \exp(L(t - \tau))z(\tau)d\tau$$

for almost every $t \in [t_0, t_f]$. If z in addition is (essentially) bounded, then it holds

$$w(t) \leq \|z(\cdot)\|_{\infty} \exp(L(t - t_0))$$

for almost every $t \in [t_0, t_f]$.

Proof. According to the assumption, we may write

$$w(t) = a(t) + z(t) + \delta(t)$$

with the absolutely continuous function

$$a(t) := L \int_{t_0}^{t} w(\tau)d\tau$$

and a nonpositive integrable function $\delta(\cdot) \le 0$. Introducing the expression for w in a yields

$$a(t) = L \int_{t_0}^{t} a(\tau)d\tau + L \int_{t_0}^{t} (z(\tau) + \delta(\tau))d\tau.$$

Hence, a solves the inhomogeneous linear differential equation

$$\dot{a}(t) = La(t) + L\big(z(t) + \delta(t)\big)$$

for almost every $t \in [t_0, t_f]$ with initial value $a(t_0) = 0$. The well-known solution formula for linear differential equations yields

$$a(t) = L \int_{t_0}^{t} \exp(L(t-\tau))(z(\tau) + \delta(\tau))d\tau$$

and

$$w(t) = L \int_{t_0}^{t} \exp(L(t-\tau))(z(\tau) + \delta(\tau))d\tau + z(t) + \delta(t).$$

Since $\delta(t) \le 0$, the first assertion holds. If z is even (essentially) bounded, we find

$$w(t) \le \|z(\cdot)\|_\infty \left(1 + L \int_{t_0}^{t} \exp(L(t-\tau))d\tau\right) = \|z(\cdot)\|_\infty \exp(L(t - t_0)).$$ □

The last assertion of Gronwall's lemma obviously remains true if we apply the norm $\|\cdot\|_\infty$ only to the interval $[t_0, t]$ instead of the whole interval $[t_0, t_f]$.

The ODE case

To illustrate the perturbation index, we start with the initial value problem (IVP)

$$\dot{x}(t) = f(t, x(t), u(t)), \quad x(t_0) = x_0,$$

where f is assumed to be Lipschitz continuous with respect to x uniformly with respect to t and u.

The (absolutely continuous) solution x can be written in integral form:

$$x(t) = x_0 + \int_{t_0}^{t} f(\tau, x(\tau), u(\tau))d\tau.$$

Now, consider the perturbed IVP

$$\dot{\tilde{x}}(t) = f(t, \tilde{x}(t), u(t)) + \delta(t), \quad \tilde{x}(t_0) = \tilde{x}_0$$

with an integrable perturbation $\delta : [t_0, t_f] \longrightarrow \mathbb{R}^{n_x}$ and its solution

$$\tilde{x}(t) = \tilde{x}_0 + \int_{t_0}^{t} (f(\tau, \tilde{x}(\tau), u(\tau)) + \delta(\tau))d\tau.$$

It holds

$$\begin{aligned}\|x(t) - \tilde{x}(t)\| &\leq \|x_0 - \tilde{x}_0\| + \int_{t_0}^{t} \|f(\tau, x(\tau), u(\tau)) - f(\tau, \tilde{x}(\tau), u(\tau))\| d\tau + \left\| \int_{t_0}^{t} \delta(\tau)d\tau \right\| \\ &\leq \|x_0 - \tilde{x}_0\| + L\int_{t_0}^{t} \|x(\tau) - \tilde{x}(\tau)\| d\tau + \left\| \int_{t_0}^{t} \delta(\tau)d\tau \right\|.\end{aligned}$$

Application of Gronwall's lemma yields

$$\begin{aligned}\|x(t) - \tilde{x}(t)\| &\leq \left(\|x_0 - \tilde{x}_0\| + \max_{t_0 \leq \tau \leq t} \left\| \int_{t_0}^{\tau} \delta(s)ds \right\| \right) \exp(L(t - t_0)) \\ &\leq \left(\|x_0 - \tilde{x}_0\| + \max_{t_0 \leq \tau \leq t_f} \left\| \int_{t_0}^{\tau} \delta(s)ds \right\| \right) \exp(L(t_f - t_0)).\end{aligned}$$

Hence, the ODE has perturbation index $p = 0$.

The index-1 case

Consider the DAE (1.11)–(1.12) with solution (x, y) and initial value $x(t_0) = x_0$ and the perturbed IVP

$$\dot{\tilde{x}}(t) = f(t, \tilde{x}(t), \tilde{y}(t), u(t)) + \delta_f(t), \quad \tilde{x}(t_0) = \tilde{x}_0, \tag{1.37}$$

$$0_{\mathbb{R}^{n_y}} = g(t, \tilde{x}(t), \tilde{y}(t), u(t)) + \delta_g(t) \tag{1.38}$$

in $[t_0, t_f]$. Assume that

(a) f is Lipschitz continuous with respect to x and y with Lipschitz constant L_f uniformly with respect to t and u.

(b) g is continuously differentiable, and $g_y'(t, x, y, u)$ is non-singular and bounded for all $(t, x, y, u) \in [t_0, t_f] \times \mathbb{R}^{n_x} \times \mathbb{R}^{n_y} \times \mathcal{U}$.

According to (b), the equation

$$0_{\mathbb{R}^{n_y}} = g(t, x, y, u) + \delta_g$$

can be solved for $y \in \mathbb{R}^{n_y}$ for any $t \in [t_0, t_f]$, $x \in \mathbb{R}^{n_x}$, $u \in \mathcal{U}$ by the implicit function theorem:

$$\tilde{y} = Y(t, x, u, \delta_g), \quad g(t, x, Y(t, x, u, \delta_g), u) + \delta_g = 0_{\mathbb{R}^{n_y}}.$$

Furthermore, Y is locally Lipschitz continuous with respect to t, x, u, δ_g with Lipschitz constant L_Y. Let $(x(\cdot), y(\cdot))$ denote the unperturbed solution of (1.37)–(1.38) for $\delta_f \equiv 0_{\mathbb{R}^{n_x}}$, $\delta_g \equiv 0_{\mathbb{R}^{n_y}}$ and $(\tilde{x}(\cdot), \tilde{y}(\cdot))$ the perturbed solution of (1.37)–(1.38). For the algebraic variables, we get the estimate

$$\begin{aligned}\|y(t) - \tilde{y}(t)\| &= \|Y(t, x(t), u(t), 0_{\mathbb{R}^{n_y}}) - Y(t, \tilde{x}(t), u(t), \delta_g(t))\| \\ &\leq L_Y(\|x(t) - \tilde{x}(t)\| + \|\delta_g(t)\|).\end{aligned}$$

With (a) we obtain

$$\begin{aligned}\|x(t) - \tilde{x}(t)\| &\leq \|x_0 - \tilde{x}_0\| + \int_{t_0}^{t} \|f(\tau, x(\tau), y(\tau), u(\tau)) - f(\tau, \tilde{x}(\tau), \tilde{y}(\tau), u(\tau))\| d\tau + \left\| \int_{t_0}^{t} \delta_f(\tau) d\tau \right\| \\ &\leq \|x_0 - \tilde{x}_0\| + L_f \int_{t_0}^{t} \|x(\tau) - \tilde{x}(\tau)\| + \|y(\tau) - \tilde{y}(\tau)\| d\tau + \left\| \int_{t_0}^{t} \delta_f(\tau) d\tau \right\| \\ &\leq \|x_0 - \tilde{x}_0\| + L_f(1 + L_Y) \int_{t_0}^{t} \|x(\tau) - \tilde{x}(\tau)\| d\tau + L_f \int_{t_0}^{t} \|\delta_g(\tau)\| d\tau + \left\| \int_{t_0}^{t} \delta_f(\tau) d\tau \right\|.\end{aligned}$$

Using the same arguments as in the ODE case, we end up in the estimate

$$\|x(t) - \tilde{x}(t)\| \leq \Big(\|x_0 - \tilde{x}_0\| + T\Big(L_f \max_{t_0 \leq \tau \leq t} \|\delta_g(\tau)\| + \max_{t_0 \leq \tau \leq t} \|\delta_f(\tau)\| \Big) \Big) \cdot \exp(L_f(1 + L_Y)(t - t_0)),$$

where $T := t_f - t_0$. Hence, if assumptions (a) and (b) are valid, then the DAE (1.11)–(1.12) has perturbation index $p = 1$.

The index-k case

The above procedure will work for a class of DAEs called Hessenberg DAEs.

Definition 1.1.14 (Hessenberg DAE). Let $k \geq 2$. The semi-explicit DAE

$$\begin{aligned}
\dot{x}_1(t) &= f_1(t, y(t), x_1(t), x_2(t), \dots, x_{k-2}(t), x_{k-1}(t), u(t)),\\
\dot{x}_2(t) &= f_2(t, x_1(t), x_2(t), \dots, x_{k-2}(t), x_{k-1}(t), u(t)),\\
&\vdots \qquad \ddots\\
\dot{x}_{k-1}(t) &= f_{k-1}(t, x_{k-2}(t), x_{k-1}(t), u(t)),\\
0_{n_y} &= g(t, x_{k-1}(t), u(t))
\end{aligned} \tag{1.39}$$

is called *Hessenberg DAE of order k*. Herein, the algebraic variable is y, and the differential variable is $x = (x_1, \dots, x_{k-1})^\top$. □

For $i = 1, \dots, k-1$ let f_i be i times continuously differentiable, and let u be sufficiently smooth. By $(k-1)$-fold differentiation of the algebraic constraint, application of the implicit function theorem, and repeating the arguments for ODEs and index-one DAEs, it is easy to see that the Hessenberg DAE of order k has perturbation index k if the matrix

$$M(\cdot) := g'_{x_{k-1}}(\cdot) \cdot f'_{k-1,x_{k-2}}(\cdot) \cdots f'_{2,x_1}(\cdot) \cdot f'_{1,y}(\cdot) \tag{1.40}$$

is non-singular.

Another index concept corresponds to the definition of the order of active state constraints in optimal control problems, see [239]. Notice that the algebraic constraint (1.12) can be interpreted as a state respectively mixed control-state constraint in an appropriate optimal control problem, which is active on $[t_0, t_f]$. The idea is to differentiate the algebraic constraint (1.12) with respect to time and to replace the derivative $\dot{x}$ according to (1.11), until the resulting equations can be solved for the derivative $\dot{y}$ by the implicit function theorem. The smallest number of differentiations needed is called *differentiation index*.

Definition 1.1.15 (Differentiation index, compare [154, p. 455]). The DAE (1.5) with u sufficiently smooth has *differentiation index* $p \in \mathbb{N}_0$ if $p \in \mathbb{N}_0$ is the smallest number such that the equations

$$\begin{aligned}
F(t, z(t), \dot{z}(t), u(t)) &= 0_{\mathbb{R}^{n_z}},\\
\frac{d}{dt} F(t, z(t), \dot{z}(t), u(t)) &= 0_{\mathbb{R}^{n_z}},\\
&\vdots\\
\frac{d^p}{dt^p} F(t, z(t), \dot{z}(t), u(t)) &= 0_{\mathbb{R}^{n_z}}
\end{aligned}$$

allow deriving by algebraic manipulations an explicit ODE

$$\dot{z}(t) = \tilde{f}(t, z(t), u(t), \ldots, u^{(p)}(t)).$$

This ODE is called *underlying ODE*. □

We illustrate the differentiation index for (1.11)–(1.12) and assume

(a) g and u are continuously differentiable.

(b) $g_y'(t,x,y,u)$ is non-singular, and $g_y'(t,x,y,u)^{-1}$ is bounded for all $(t,x,y,u) \in [t_0,t_f] \times \mathbb{R}^{n_x} \times \mathbb{R}^{n_y} \times \mathcal{U}$.

Differentiation of the algebraic constraint (1.12) with respect to time yields

$$\begin{aligned} 0_{\mathbb{R}^{n_y}} &= g_t'[t] + g_x'[t] \cdot \dot{x}(t) + g_y'[t] \cdot \dot{y}(t) + g_u'[t] \cdot \dot{u}(t) \\ &= g_t'[t] + g_x'[t] \cdot f[t] + g_y'[t] \cdot \dot{y}(t) + g_u'[t] \cdot \dot{u}(t), \end{aligned}$$

where, e. g., $g_t'[t]$ is an abbreviation for $g_t'(t, x(t), y(t), u(t))$. Since $g_y'[t]^{-1}$ exists and is bounded, we can solve this equality for $\dot{y}$ and obtain the differential equation

$$\dot{y}(t) = -(g_y'[t])^{-1}(g_t'[t] + g_x'[t] \cdot f[t] + g_u'[t] \cdot \dot{u}(t)). \tag{1.41}$$

Hence, it was necessary to differentiate the algebraic equation once in order to obtain a differential equation for the algebraic variable y. Consequently, the differentiation index is one. Equations (1.11) and (1.41) define the *underlying ODE* of the DAE (1.11)–(1.12). Often ODEs are viewed as index-zero DAEs since the underlying ODE is identical with the ODE itself, and differentiations are not needed.

In the same way, it can be shown that the Hessenberg DAE of order $k \geq 2$ has differentiation index k if the matrix M in (1.40) is non-singular.

Consequently, the differentiation index and the perturbation index coincide for Hessenberg DAEs, see also [53]. For more general DAEs, the difference between perturbation index and differentiation index can be arbitrarily high, see [154, p. 461].

---

i Notice that the index calculations are questionable in the presence of a control variable u. We assumed that u is sufficiently smooth, which is not the case for many optimal control problems. Nevertheless, the optimal control u is often at least piecewise smooth, so the interpretation remains valid locally.

1.1.2 Consistent initial values

DAEs not only differ in their stability behavior from explicit ODEs. Another difference is that initial values have to be defined properly, that is, they have to be *consistent*. We restrict the discussion to Hessenberg DAEs of order k. A general definition of consistency can be found in [35].

Define

$$g^{(0)}(t, x_{k-1}(t), u(t)) := g(t, x_{k-1}(t), u(t)). \tag{1.42}$$

An initial value for x_{k-1} at $t = t_0$ has to satisfy this equality, of course. However, this is not sufficient because time derivatives of (1.42) also impose additional restrictions on initial values. This can be seen as follows. Differentiation of $g^{(0)}$ with respect to time and substitution of $\dot{x}_{k-1}(t) = f_{k-1}(t, x_{k-2}(t), x_{k-1}(t), u(t))$ leads to the equation

$$\begin{aligned} 0_{\mathbb{R}^{n_y}} &= g'_t[t] + g'_{x_{k-1}}[t] \cdot f_{k-1}(t, x_{k-2}(t), x_{k-1}(t), u(t)) + g'_u[t] \cdot \dot{u}(t) \\ &=: g^{(1)}(t, x_{k-2}(t), x_{k-1}(t), u(t), \dot{u}(t)), \end{aligned}$$

which has also to be satisfied. Recursive application of this differentiation and substitution process leads to the equations

$$0_{\mathbb{R}^{n_y}} = g^{(j)}(t, x_{k-1-j}(t), \dots, x_{k-1}(t), u(t), \dot{u}(t), \dots, u^{(j)}(t)) \tag{1.43}$$

for $j = 1, 2, \dots, k-2$ and

$$0_{\mathbb{R}^{n_y}} = g^{(k-1)}(t, y(t), x_1(t), \dots, x_{k-1}(t), u(t), \dot{u}(t), \dots, u^{(k-1)}(t)). \tag{1.44}$$

Since the equations (1.43)–(1.44) do not occur explicitly in the original system (1.39), these equations are called *hidden constraints* of the Hessenberg DAE. With this notation, consistency is defined as follows.

Definition 1.1.16 (Consistent initial value). Let $x(t_0)$ and $y(t_0)$ be given, and let u be sufficiently smooth. The differential variable $x(t_0)$ is called *consistent* for the Hessenberg DAE of order k if equations (1.42)–(1.43) are satisfied for $j = 1, 2, \dots, k-2$. $y(t_0)$ is called consistent if $y(t_0)$ satisfies the equation (1.44) at $t = t_0$. □

The following optimal control problem shows that control components may occur in the algebraic constraint.

Example 1.1.17 (Communicated by O. Kostyukova). Consider the following optimal control problem:

Minimize

$$\int_0^3 2x_1(t) + \frac{1}{2}y(t)^2 dt$$

subject to the DAE

$$\dot{x}_1(t) = x_2(t), \qquad x_1(0) = 1,$$
$$\dot{x}_2(t) = y(t) + u_1(t), \qquad x_2(0) = \frac{2}{3},$$
$$\dot{x}_3(t) = u_2(t), \qquad x_3(0) = 1,$$
$$0 = x_2(t) - u_2(t),$$

the control constraint

$$-1 \le u_1(t) \le 1,$$

and the state constraint

$$0 \le x_3(t) \le 200.$$

As the algebraic variable does not appear in the algebraic constraint, this problem has at least index two. Differentiating the algebraic constraint is questionable as u_2 is not necessarily differentiable. However, as x_2 is an absolutely continuous function, so is u_2 by the algebraic constraint. Hence, the optimal control component u_2 is smoother as expected.

An alternative way to avoid the appearance of control variables in the algebraic equations is to smoothen the controls. In this case, an artificial differential equation

$$\dot{u}_2(t) = u_3(t), \quad u_2(0) = x_2(0) = \frac{2}{3}$$

with new control u_3 can be introduced, where u_2 is treated as a state variable. □

The previous example shows that the dependence on derivatives of the control in (1.43) and (1.44) can be avoided by introducing additional state variables $\xi_j, j = 0, \ldots, k-2$, by

$$\xi_0 := u, \ \xi_1 := \dot{u}, \ldots, \xi_{k-2} := u^{(k-2)}$$

satisfying the differential equations

$$\dot{\xi}_0 = \xi_1, \ldots, \dot{\xi}_{k-3} = \xi_{k-2}, \ \dot{\xi}_{k-2} = \tilde{u},$$

and to consider $\tilde{u} := u^{(k-1)}$ as the new control, see [252]. Clearly, this approach is nothing else than constructing a sufficiently smooth control u for the original problem. The resulting problem is not equivalent to the original problem anymore. Nevertheless, this strategy is very useful from a practical point of view.

1.1.3 Index reduction and stabilization

The perturbation index measures the stability of the DAE. With increasing perturbation index, the DAE suffers from increasing ill-conditioning since not only the perturbation

δ occurs in (1.35) but also derivatives thereof. Thus, even for small perturbations (in the supremum norm), large deviations in the respective solutions are possible. Higher index DAEs occur, e. g., in mechanics, see [102, 306, 10], process system engineering, see [88], and electrical engineering, see [234, 151].

To avoid the problems of severe ill-conditioning, it is possible to reduce the index by replacing the algebraic constraint in (1.39) by its derivative $g^{(1)}$ defined in (1.43). It is easy to see that the resulting DAE has perturbation index $k-1$. More generally, the algebraic constraint

$$g^{(0)}(t, x_{k-1}(t), u(t)) = 0_{\mathbb{R}^{n_y}}$$

in (1.42) can be replaced by any of the constraints

$$g^{(j)}(t, x_{k-1-j}(t), \dots, x_{k-1}(t), u(t), \dot{u}(t), \dots, u^{(j)}(t)) = 0_{\mathbb{R}^{n_y}}$$

in (1.43) with $1 \le j \le k-1$. Then the resulting DAE has perturbation index $k-j$. The drawback of this index reduction approach is that numerical integration methods suffer from the so-called *drift-off effect*. Since a numerical integration scheme, if applied to an index reduced DAE, only obeys hidden constraints up to a certain level, the numerical solution will not satisfy the higher level constraints, and it can be observed that the magnitude of violation increases with time t. Hence, the numerical solution drifts off the neglected constraints, even if the initial value was consistent. Especially for long time intervals, the numerical solution may deviate substantially from the solution of the original DAE. One possibility to avoid this drawback is to perform a projection step onto the neglected constraints after each successful integration step for the index reduced system, see [14, 86].

Another approach is to *stabilize* the index reduced problem by adding the neglected constraints to the index reduced DAE. The resulting system is an overdetermined DAE, and numerical integration methods have to be adapted, see [102, 103]. The stabilization approach, if applied to mechanical multi-body systems, is equivalent with the Gear–Gupta–Leimkuhler stabilization (GGL-stabilization) in [108].

The above outlined procedures are illustrated for a very important field of applications – mechanical multi-body systems.

Example 1.1.18 (Mechanical multi-body systems). Let a system of n rigid bodies be given. Every body of mass m_i has three translational and three rotational degrees of freedom. The position of body i in a fixed reference system is given by its position $r_i = (x_i, y_i, z_i)^\top$. The orientation of the bodies coordinate system with respect to the reference coordinate system is given by the angles α_i, β_i, and γ_i. Hence, body i is characterized by the coordinates

$$q_i = (x_i, y_i, z_i, \alpha_i, \beta_i, \gamma_i)^\top \in \mathbb{R}^6$$

and the whole multi-body system is described by

$$q = (q_1, \dots, q_n)^\top \in \mathbb{R}^{6n}.$$

In general, the motion of the n bodies is restricted by holonomic constraints

$$0_{\mathbb{R}^{n_g}} = g(q).$$

The kinetic energy of the multi-body system is given by

$$T(q, \dot{q}) = \frac{1}{2} \sum_{i=1}^{n} (m_i \dot{r}_i(q, \dot{q})^\top \dot{r}_i(q, \dot{q}) + w_i(q, \dot{q})^\top J_i w_i(q, \dot{q})),$$

where $\dot{r}_i$ denotes the velocity of body i, w_i denotes the angular velocity of body i with respect to the reference system, and J_i is the moment of inertia of body i. The Euler-Lagrangian equations of the first kind are given by

$$\begin{aligned} \frac{d}{dt}(T'_{\dot{q}}(q, \dot{q}))^\top - (T'_q(q, \dot{q}))^\top &= F(q, \dot{q}, u) - g'(q)^\top \lambda, \\ 0_{\mathbb{R}^{n_g}} &= g(q), \end{aligned}$$

where F is the vector of applied forces and torques, which may depend on the control u. Explicit calculation of the derivatives leads to the *descriptor form* of mechanical multi-body systems:

$$\begin{aligned} \dot{q}(t) &= v(t), \\ M(q(t))\dot{v}(t) &= f(q(t), v(t), u(t)) - g'(q(t))^\top \lambda(t), \\ 0_{\mathbb{R}^{n_g}} &= g(q(t)), \end{aligned}$$

where M is the symmetric and positive definite mass matrix, and f includes the applied forces and torques and the Coriolis forces. Multiplication of the second equation by M^{-1} yields a Hessenberg DAE of order 3.

Let g be twice continuously differentiable. The constraint $g(q(t)) = 0_{\mathbb{R}^{n_g}}$ is called *constraint on position level*. Differentiation with respect to time of this algebraic constraint yields the *constraint on velocity level*

$$g'(q(t)) \cdot v(t) = 0_{\mathbb{R}^{n_g}}$$

and the *constraint on acceleration level*

$$g'(q(t)) \cdot \dot{v}(t) + g''_{qq}(q(t))(v(t), v(t)) = 0_{\mathbb{R}^{n_g}}.$$

Replacing $\dot{v}$ by

$$\dot{v}(t) = M(q(t))^{-1}(f(q(t), v(t), u(t)) - g'(q(t))^\top \lambda(t))$$

yields

$$0_{\mathbb{R}^{n_g}} = g'(q(t))M(q(t))^{-1}(f(q(t), v(t), u(t)) - g'(q(t))^\top \lambda(t)) + g''_{qq}(q(t))(v(t), v(t)).$$

If $\operatorname{rank}(g'(q)) = n_g$, then the matrix $g'(q)M(q)^{-1}g'(q)^\top$ is non-singular, and the latter equation can be solved for the algebraic variable λ. By the same reasoning as before, the descriptor form has index three. Consistent initial values (q_0, v_0, λ_0) have to satisfy the constraints on position, velocity, and acceleration level.

For numerical methods, it is advisable to perform an index reduction, i. e. the constraint on position level is replaced by the constraint on velocity or acceleration level. An even better idea is to use the *Gear–Gupta–Leimkuhler stabilization (GGL-stabilization)*

$$\dot{q}(t) = v(t) - g'(q(t))^\top \mu(t), \tag{1.45}$$

$$M(q(t))\dot{v}(t) = f(q(t), v(t), u(t)) - g'(q(t))^\top \lambda(t), \tag{1.46}$$

$$0_{\mathbb{R}^{n_g}} = g(q(t)), \tag{1.47}$$

$$0_{\mathbb{R}^{n_g}} = g'(q(t)) \cdot v(t). \tag{1.48}$$

This DAE is equivalent to an index-two Hessenberg DAE, if the second equation is multiplied by M^{-1}. Furthermore, differentiation of the first algebraic equation yields

$$0_{\mathbb{R}^{n_g}} = g'(q(t)) \cdot (v(t) - g'(q(t))^\top \mu(t)) = -g'(q(t))g'(q(t))^\top \mu(t).$$

If $\operatorname{rank}(g'(q)) = n_g$, then $g'(q)g'(q)^\top$ is non-singular, and it follows $\mu \equiv 0_{\mathbb{R}^{n_g}}$. Hence, the GGL-stabilization (1.45)–(1.48) is equivalent to the overdetermined (stabilized) descriptor form

$$\begin{aligned}
\dot{q}(t) &= v(t),\\
M(q(t))\dot{v}(t) &= f(q(t), v(t), u(t)) - g'(q(t))^\top \lambda(t),\\
0_{\mathbb{R}^{n_g}} &= g(q(t)),\\
0_{\mathbb{R}^{n_g}} &= g'(q(t)) \cdot v(t).
\end{aligned}$$

□

Example 1.1.19 (Pendulum problem, revisited). The equations of motion in GGL formulation for the pendulum problem in Example 1.1.9 read as follows:

$$\begin{aligned}
\dot{x}_1(t) &= x_3(t) - 2x_1(t)\mu(t),\\
\dot{x}_2(t) &= x_4(t) - 2x_2(t)\mu(t),\\
m\dot{x}_3(t) &= \qquad - 2x_1(t)\lambda(t) + \frac{u(t)x_2(t)}{\ell},\\
m\dot{x}_4(t) &= -mg - 2x_2(t)\lambda(t) - \frac{u(t)x_1(t)}{\ell},
\end{aligned}$$

$$
\begin{aligned}
0 &= x_1(t)^2 + x_2(t)^2 - \ell^2,\\
0 &= x_1(t)x_3(t) + x_2(t)x_4(t),
\end{aligned}
$$

where μ denotes the additional multiplier. Indeed, differentiation of the first algebraic constraint yields

$$
\begin{aligned}
0 &= 2x_1(t)\dot{x}_1(t) + 2x_2(t)\dot{x}_2(t)\\
&= 2x_1(t)(x_3(t) - 2x_1(t)\mu(t)) + 2x_2(t)(x_4(t) - 2x_2(t)\mu(t))\\
&= -4(x_1(t)^2 + x_2(t)^2)\mu(t)\\
&= -4\ell^2\mu(t)
\end{aligned}
$$

and hence $\mu \equiv 0$ in every solution of the DAE. □

A formal procedure of stabilizing general DAEs of type

$$
\begin{aligned}
\dot{z}(t) &= v(t),\\
F(t, z(t), v(t), u(t)) &= 0_{\mathbb{R}^{n_z}}
\end{aligned}
$$

works as follows. Let G be a function with

$$G(t, z(t), v(t), u(t)) = 0_{\mathbb{R}^{n_G}}$$

for almost all t, and let G_z' have full rank. Moreover, assume that G is invariant, i. e.

$$\frac{d}{dt}G[t] = G_t'[t] + G_z'[t]\cdot v(t) + G_v'[t]\cdot\dot{v}(t) + G_u'[t]\dot{u}(t) \equiv 0_{\mathbb{R}^{n_G}}.$$

Stabilization in the sense of Gear–Gupta–Leimkuhler yields the system

$$
\begin{aligned}
\dot{z}(t) &= v(t) - (G_z'[t])^\top \cdot \mu(t),\\
F(t, z(t), v(t), u(t)) &= 0_{\mathbb{R}^{n_z}},\\
G(t, z(t), v(t), u(t)) &= 0_{\mathbb{R}^{n_G}}.
\end{aligned}
$$

The invariance of G yields

$$
\begin{aligned}
0_{\mathbb{R}^{n_G}} &= \frac{d}{dt}G[t]\\
&= G_t'[t] + G_z'[t]\cdot v(t) - G_z'[t]\cdot G_z'[t]^\top\cdot\mu(t) + G_v'[t]\cdot\dot{v}(t) + G_u'[t]\cdot\dot{u}(t)\\
&= -G_z'[t]\cdot G_z'[t]^\top\cdot\mu(t).
\end{aligned}
$$

Since $G_z'[t]$ has full rank, the matrix $G_z'[t]G_z'[t]^\top$ is non-singular. This implies $\mu(t) \equiv 0_{\mathbb{R}^{n_G}}$. Hence, both systems are equivalent.

We close the section with an example of a switched system, which illustrates that the index is not necessarily constant over time.

Example 1.1.20 (Docking problem). We consider a simple docking problem of two rigid bodies of masses m_1 and m_2, respectively, in the (x, y)-plane. Let us assume that the bodies are disconnected in the time interval $[0, t_c)$ and move freely until an adhesive contact occurs at time t_c. Let the bodies be firmly coupled after the contact for $t \geq t_c$.

What happens to the motion of the bodies after docking?

Let the first body move according to the equations of motion

$$
\begin{aligned}
m_1 \ddot{x}_1(t) &= -\frac{m_1}{r^2} x_1(t) v_1(t)^2, \\
m_1 \ddot{y}_1(t) &= -\frac{m_1}{r^2} y_1(t) v_1(t)^2
\end{aligned}
$$

with velocity $v_1(t) = \sqrt{\dot{x}_1(t)^2 + \dot{y}_1(t)^2}$ in the time interval $[0, t_c)$ with initial value $(x_1(0), y_1(0), \dot{x}_1(0), \dot{y}_1(0)) = (r, 0, 0, \bar{v})$. In fact, this system describes the motion on a circle of radius $r > 0$ around the origin of the coordinate system with velocity $\bar{v}$.

Let the second body move on a straight line before docking according to the equations of motion

$$
\begin{aligned}
m_2 \ddot{x}_2(t) &= 0, \\
m_2 \ddot{y}_2(t) &= 0
\end{aligned}
$$

with initial values $(x_2(0), y_2(0), \dot{x}_2(0), \dot{y}_2(0)) = (r, -2\pi, 0, \bar{v})$. Please note that the initial conditions are chosen such that a docking occurs at the point $(r, 0)$ with relative velocity equal to zero at time $t_c = 2\pi/\bar{v}$.

After docking the two bodies are coupled by the algebraic equations (docking conditions)

$$
0 = x_1(t) - x_2(t) \quad \text{and} \quad 0 = y_1(t) - y_2(t),
$$

and the motion of the coupled two-body system can be described by the DAE

$$
\begin{aligned}
m_1 \ddot{x}_1(t) &= -\frac{m_1}{r^2} x_1(t) v_1(t)^2 - \lambda_1(t), && (1.49) \\
m_1 \ddot{y}_1(t) &= -\frac{m_1}{r^2} y_1(t) v_1(t)^2 - \lambda_2(t), && (1.50) \\
m_2 \ddot{x}_2(t) &= \lambda_1(t), && (1.51) \\
m_2 \ddot{y}_2(t) &= \lambda_2(t), && (1.52) \\
0 &= x_1(t) - x_2(t) \\
0 &= y_1(t) - y_2(t)
\end{aligned}
$$

for $t \geq t_c$ with $v_1(t) = \sqrt{\dot{x}_1(t)^2 + \dot{y}_1(t)^2}$, compare Example 1.1.18. This is a Hessenberg DAE of index three. At t_c, consistent differential and algebraic variables are required, that is,

$$\begin{aligned} x_1(t_c) &= x_2(t_c), & \dot{x}_1(t_c) &= \dot{x}_2(t_c), & \ddot{x}_1(t_c) &= \ddot{x}_2(t_c), \\ y_1(t_c) &= y_2(t_c), & \dot{y}_1(t_c) &= \dot{y}_2(t_c), & \ddot{y}_1(t_c) &= \ddot{y}_2(t_c). \end{aligned}$$

In our example, these consistency conditions are satisfied at $t_c = 2\pi/\bar{v}$ with

$$\lambda_1(t_c) = -\frac{m_1 m_2}{(m_1 + m_2) r^2} x_1(t_c) v_1(t_c)^2, \quad \lambda_2(t_c) = -\frac{m_1 m_2}{(m_1 + m_2) r^2} y_1(t_c) v_1(t_c)^2,$$

and we obtain a "smooth docking". Figure 1.7 shows snapshots of the docking maneuver.

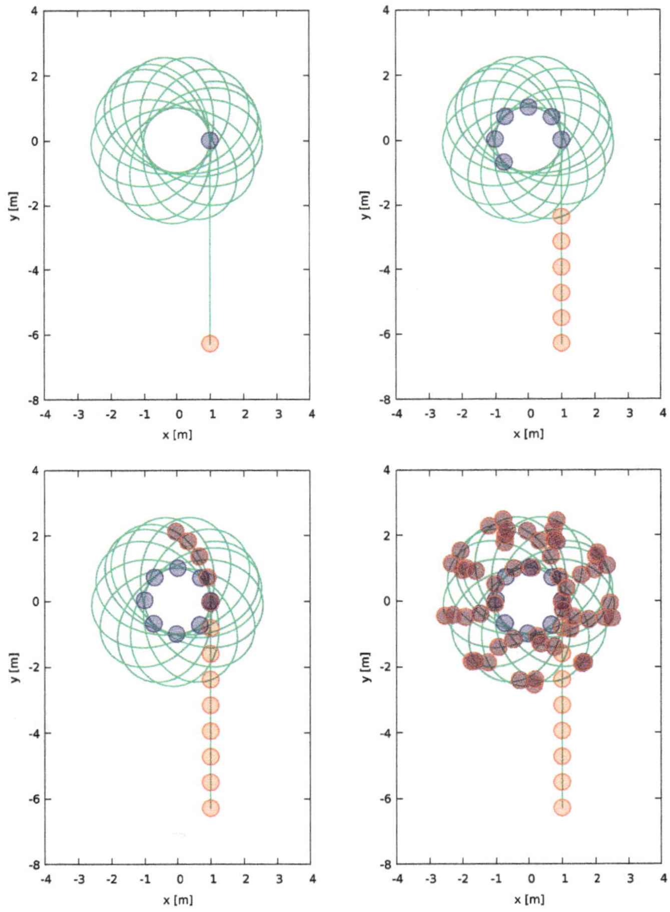

Figure 1.7: Snapshots of the docking maneuver: Initial constellation (top left), shortly before docking (top right), shortly after docking (bottom left), and beyond docking (bottom right).

Please note that the motion before docking can be described equivalently by (1.49)–(1.52) with the algebraic constraints $\lambda_1(t) = 0$ and $\lambda_2(t) = 0$, which yields an index-one DAE. This example thus illustrates that the index can change over time, which is a natural effect especially for piecewise defined dynamics. □

1.2 Transformation techniques

Several useful transformation techniques are discussed that allow the transformation of fairly general optimal control problems to standard form. However, it should be mentioned that the size of the transformed problem may increase or additional nonlinearities may be introduced by the transformations, and thus, tailored methods for the original problem formulation can be more efficient.

1.2.1 Transformation of a Bolza problem to a Mayer problem

Depending on the form of the objective function

$$\varphi(t_0, t_f, x(t_0), x(t_f)) + \int_{t_0}^{t_f} f_0(t, x(t), y(t), u(t))dt$$

in the optimal control problem 1.1.8, we call it a

Bolza problem, if $\varphi \not\equiv 0$ and $f_0 \not\equiv 0$ or
Mayer problem, if $\varphi \not\equiv 0$ and $f_0 \equiv 0$ or
Lagrange problem, if $\varphi \equiv 0$ and $f_0 \not\equiv 0$.

Bolza or Lagrange problems can be transformed equivalently to a Mayer problem. To this end, we introduce an additional state x_0 with

$$\dot{x}_0(t) = f_0(t, x(t), y(t), u(t)), \quad x_0(t_0) = 0.$$

Then,

$$x_0(t_f) = x_0(t_0) + \int_{t_0}^{t_f} f_0(t, x(t), y(t), u(t))dt = \int_{t_0}^{t_f} f_0(t, x(t), y(t), u(t))dt$$

and

$$\varphi(t_0, t_f, x(t_0), x(t_f)) + \int_{t_0}^{t_f} f_0(t, x(t), y(t), u(t))dt = \varphi(t_0, t_f, x(t_0), x(t_f)) + x_0(t_f).$$

The latter yields a Mayer problem.

1.2.2 Transformation to fixed time interval

Problem 1.1.8 with free initial or final time can be transformed into an equivalent problem with fixed initial and final times. This is achieved by the linear time transformation

$$t(\tau) := t_0 + \tau(t_f - t_0), \quad \tau \in [0, 1]. \tag{1.53}$$

Define

$$\begin{aligned}\bar{x}(\tau) &:= x(t(\tau)) = x(t_0 + \tau(t_f - t_0)),\\ \bar{y}(\tau) &:= y(t(\tau)) = y(t_0 + \tau(t_f - t_0)),\\ \bar{u}(\tau) &:= u(t(\tau)) = u(t_0 + \tau(t_f - t_0)).\end{aligned}$$

Then,

$$\begin{aligned}\frac{d}{d\tau}\bar{x}(\tau) &= \dot{x}(t(\tau)) \cdot t'(\tau)\\ &= (t_f - t_0)f(t(\tau), x(t(\tau)), y(t(\tau)), u(t(\tau)))\\ &= (t_f - t_0)f(t(\tau), \bar{x}(\tau), \bar{y}(\tau), \bar{u}(\tau)).\end{aligned}$$

The quantities t_0 and/or t_f can be viewed as constant states with

$$\begin{aligned}\frac{d}{d\tau}t_0(\tau) &= 0, \quad t_0(0) \text{ free},\\ \frac{d}{d\tau}t_f(\tau) &= 0, \quad t_f(0) \text{ free},\end{aligned}$$

or more efficiently as additional scalar optimization parameters. We obtain the equivalent transformed problem (with t_0 and t_f as additional states):

Minimize

$$\varphi(t_0(0), t_f(1), \bar{x}(0), \bar{x}(1)) + \int_0^1 (t_f(\tau) - t_0(\tau))f_0(t(\tau), \bar{x}(\tau), \bar{y}(\tau), \bar{u}(\tau))d\tau$$

subject to

$$\frac{d}{d\tau}\bar{x}(\tau) = (t_f(\tau) - t_0(\tau))f(t(\tau), \bar{x}(\tau), \bar{y}(\tau), \bar{u}(\tau)),$$

$$
\begin{aligned}
0_{\mathbb{R}^{n_y}} &= g\big(t(\tau), \bar{x}(\tau), \bar{y}(\tau), \bar{u}(\tau)\big),\\
\frac{d}{d\tau} t_0(\tau) &= 0,\\
\frac{d}{d\tau} t_f(\tau) &= 0,\\
s\big(t(\tau), \bar{x}(\tau)\big) &\le 0_{\mathbb{R}^{n_s}},\\
c\big(t(\tau), \bar{x}(\tau), \bar{y}(\tau), \bar{u}(\tau)\big) &\le 0_{\mathbb{R}^{n_c}},\\
\psi\big(t_0(0), t_f(1), \bar{x}(0), \bar{x}(1)\big) &= 0_{\mathbb{R}^{n_\psi}},\\
\bar{u}(\tau) &\in \mathcal{U}.
\end{aligned}
$$

1.2.3 Transformation to autonomous problem

A non-autonomous Problem 1.1.8 can be transformed into an autonomous problem by introducing an additional state according to the differential equation

$$\dot{T}(t) = 1, \quad T(t_0) = t_0. \tag{1.54}$$

The equivalent autonomous problem reads as follows:

Minimize

$$\varphi\big(T(t_0), T(t_f), x(t_0), x(t_f)\big) + \int_{t_0}^{t_f} f_0\big(T(t), x(t), y(t), u(t)\big)dt$$

subject to

$$
\begin{aligned}
\dot{x}(t) &= f\big(T(t), x(t), y(t), u(t)\big),\\
0_{\mathbb{R}^{n_y}} &= g\big(T(t), x(t), y(t), u(t)\big),\\
\dot{T}(t) &= 1,\\
s\big(T(t), x(t)\big) &\le 0_{\mathbb{R}^{n_s}},\\
c\big(T(t), x(t), y(t), u(t)\big) &\le 0_{\mathbb{R}^{n_c}},\\
\psi\big(T(t_0), T(t_f), x(t_0), x(t_f)\big) &= 0_{\mathbb{R}^{n_\psi}},\\
T(t_0) &= t_0,\\
u(t) &\in \mathcal{U}.
\end{aligned}
$$

1.2.4 Transformation of Chebyshev problems

A Chebyshev problem aims at minimizing the objective function

$$\max_{t\in[t_0,t_f]} h\big(t, x(t), y(t), u(t)\big)$$

subject to the standard constraints of the DAE optimal control problem 1.1.8. Define

$$\alpha := \max_{t\in[t_0,t_f]} h(t, x(t), y(t), u(t)).$$

Then,

$$h(t, x(t), y(t), u(t)) \leq \alpha \quad \text{in } [t_0, t_f]. \tag{1.55}$$

This is an additional mixed control-state constraint, and the Chebyshev problem is equivalent to minimizing α subject to the standard constraints and the mixed control-state constraint (1.55).

The parameter α can be treated as a real optimization variable or a way to add an artificial constant state variable and the differential equation

$$\dot{\alpha}(t) = 0, \quad \alpha(t_0) \text{ free.}$$

1.2.5 Transformation of L_1-minimization problems

An L_1-minimization problem aims at minimizing an objective function that contains the term

$$\|u\|_1 = \int_{t_0}^{t_f} |u(t)|dt,$$

where $u : [t_0, t_f] \longrightarrow \mathbb{R}$ is supposed to be single-valued for notational convenience.

Such an objective function is not differentiable, and the corresponding optimal control problem is non-smooth. The problem can be transformed into an equivalent smooth problem by introducing two artificial controls $u^+(t) := \max\{u(t), 0\} \geq 0$ and $u^-(t) := -\min\{u(t), 0\} \geq 0$. Then it holds

$$u(t) = u^+(t) - u^-(t)$$

and

$$|u(t)| = u^+(t) + u^-(t).$$

An equivalent smooth optimal control problem is obtained by replacing $\|u\|_1$ by

$$\|u\|_1 = \int_{t_0}^{t_f} u^+(t) + u^-(t)dt, \tag{1.56}$$

by adding the control constraints $u^+(t) \geq 0$ and $u^-(t) \geq 0$, and by replacing all remaining occurrences of u by $u^+ - u^-$.

Note that in an optimal solution one of the values $u^+(t)$ or $u^-(t)$ will be zero almost everywhere depending on the sign of $u(t)$ since otherwise the expression in (1.56) would not be minimal.

1.2.6 Transformation of interior-point constraints

Problems with intermediate point constraints of type

$$\psi(x(t_0), x(t_1), \dots, x(t_N)) = 0_{\mathbb{R}^{n_\psi}}$$

and objective function

$$\varphi(x(t_0), x(t_1), \dots, x(t_N)) + \sum_{i=0}^{N-1} \int_{t_i}^{t_{i+1}} f_{0,i}(t, x(t), y(t), u(t))dt$$

with $t_N = t_f$ can be transformed into Problem 1.1.8 using linear time transformations

$$t^i(\tau) := t_i + \tau(t_{i+1} - t_i), \quad \tau \in [0,1], \; i = 0, \dots, N-1,$$

that map each interval $[t_i, t_{i+1}]$ onto $[0,1]$. In addition, for $\tau \in [0,1]$ and $i = 0, \dots, N-1$, new states $x^i(\tau) := x(t^i(\tau))$, $y^i(\tau) := y(t^i(\tau))$ and new controls $u^i(\tau) := u(t^i(\tau))$ are introduced with additional continuity conditions

$$x^i(1) - x^{i+1}(0) = 0_{\mathbb{R}^{n_x}}, \quad i = 0, \dots, N-2.$$

The transformed problem then reads as follows:

Minimize

$$\varphi\big(x^0(0), x^1(0), \dots, x^{N-1}(0), x^{N-1}(1)\big) + \sum_{i=0}^{N-1} \int_0^1 (t_{i+1} - t_i) f_{0,i}\big(t^i(\tau), x^i(\tau), y^i(\tau), u^i(\tau)\big) d\tau$$

subject to

$$\begin{aligned}
\dot{x}^i(\tau) - (t_{i+1} - t_i) f\big(t^i(\tau), x^i(\tau), y^i(\tau), u^i(\tau)\big) &= 0_{\mathbb{R}^{n_x}}, & i &= 0, \dots, N-1, \\
g\big(t^i(\tau), x^i(\tau), y^i(\tau), u^i(\tau)\big) &= 0_{\mathbb{R}^{n_y}}, & i &= 0, \dots, N-1, \\
s\big(t^i(\tau), x^i(\tau)\big) &\leq 0_{\mathbb{R}^{n_s}}, & i &= 0, \dots, N-1, \\
c\big(t^i(\tau), x^i(\tau), y^i(\tau), u^i(\tau)\big) &\leq 0_{\mathbb{R}^{n_c}}, & i &= 0, \dots, N-1, \\
\psi\big(x^0(0), x^1(0), \dots, x^{N-1}(0), x^{N-1}(1)\big) &= 0_{\mathbb{R}^{n_\psi}},
\end{aligned}$$

$$x^i(1) - x^{i+1}(0) = 0_{\mathbb{R}^{n_x}}, \quad i = 0, \dots, N-2,$$
$$u^i(\tau) \in \mathcal{U}, \quad i = 0, \dots, N-1.$$

A problem with interior point constraints and free final time is discussed in the following example.

Example 1.2.1 (Interior point constraints, communicated by Roland Herzog). The task is to find a path with initial and terminal constraints that passes in minimal time through a defined ε-neighborhood of given via-points x_i, $i = 1, \dots, N$. This task occurs in the path planning of machines that have to perform a given task at prescribed points. It can be modeled as the following control problem:

Minimize t_f subject to

$$\begin{aligned} \dot{x}(t) &= v(t), \quad x(0) = x_0, \\ \dot{v}(t) &= a(t), \quad v(0) = v_0, v(t_f) = v_0, \\ \dot{a}(t) &= u(t), \quad a(0) = a_0, a(t_f) = a_0, \end{aligned}$$

and

$$\begin{aligned} u_{\min} &\le u(t) \le u_{\max}, \\ \|x(t_i) - x_i\| &\le \varepsilon, \quad i = 1, \dots, N, \\ v(t) &\le v_{\max}, \\ a(t) &\le a_{\max}. \end{aligned}$$

The final time t_f and the intermediate time points t_i, $i = 1, \dots, N$, are free, and the problem can be transformed into an equivalent problem on the fixed time interval $[0, 1]$ using the transformation $t(\tau) := \tau t_f$ with $\tau \in [0, 1]$. The transformed problem reads as follows:

Minimize t_f subject to

$$\begin{aligned} \dot{x}(\tau) &= t_f v(\tau), \quad x(0) = x_0, \\ \dot{v}(\tau) &= t_f a(\tau), \quad v(0) = v_0, v(1) = v_0, \\ \dot{a}(\tau) &= t_f u(\tau), \quad a(0) = a_0, a(1) = a_0, \\ u_{\min} &\le u(\tau) \le u_{\max}, \\ \|x(\tau_i t_f) - x_i\| &\le \varepsilon, \quad i = 1, \dots, N, \\ v(\tau) &\le v_{\max}, \\ a(\tau) &\le a_{\max}. \end{aligned}$$

The intermediate time points $\tau_i := t_i / t_f$ are still free. The variable time transformation $\tau(\tilde{\tau}) = \int_0^{\tilde{\tau}} w(s)ds$ with additional control $w(\cdot) \ge 0$, $\tau_i = \tau(\tilde{\tau}_i)$ and fixed (equidistant) $\tilde{\tau}_i$, $i = 1, \dots, N$, can be used to transform the problem into an equivalent problem with fixed intermediate time points. The transformed problem reads as follows:

Minimize t_f subject to

$$
\begin{aligned}
\dot{x}(\tilde{\tau}) &= t_f w(\tilde{\tau}) v(\tilde{\tau}), & x(0) &= x_0, \\
\dot{v}(\tilde{\tau}) &= t_f w(\tilde{\tau}) a(\tilde{\tau}), & v(0) &= v_0, v(1) = v_0, \\
\dot{a}(\tilde{\tau}) &= t_f w(\tilde{\tau}) u(\tilde{\tau}), & a(0) &= a_0, a(1) = a_0, \\
\dot{\tau}(\tilde{\tau}) &= w(\tilde{\tau}), & \tau(0) &= 0, \tau(1) = 1, \\
u_{\min} &\le u(\tilde{\tau}) \le u_{\max}, \\
\left\| x(\tilde{\tau}_i) - x_i \right\| &\le \varepsilon, \quad i = 1, \dots, N, \\
v(\tilde{\tau}) &\le v_{\max}, \\
a(\tilde{\tau}) &\le a_{\max}, \\
w(\tilde{\tau}) &\ge 0.
\end{aligned}
$$

Sample data: $N = 4$, $\tilde{\tau}_i = i/4$, $i = 0, \dots, 4$, $\varepsilon = 0.1$, $v_{\max} = a_{\max} = 1$,

$$
x_i = x(\tilde{\tau}_i) = \begin{pmatrix} i+1 \\ (-1)^i \varepsilon \end{pmatrix}, \quad i = 0, \dots, 3, \quad x_4 = x(\tilde{\tau}_4) = \begin{pmatrix} 5 \\ 0 \end{pmatrix},
$$

$$
v_0 = \begin{pmatrix} 0.4 \\ 0 \end{pmatrix}, \quad a_0 = \begin{pmatrix} 0 \\ 0 \end{pmatrix}, \quad u_{\min} = -u_{\max} = \begin{pmatrix} -1 \\ -1 \end{pmatrix}.
$$

The solution is depicted in Figure 1.8. The time points t_i, $i = 1, \dots, 4$, and the final time t_f are given by

$$
\begin{aligned}
t_1 &= 1.3751117339512176, \\
t_2 &= 2.3876857361783093, \\
t_3 &= 3.4636193112119735, \\
t_4 &= 4.7688769330936207, \\
t_f &= 4.768876933093622.
\end{aligned}
$$

□

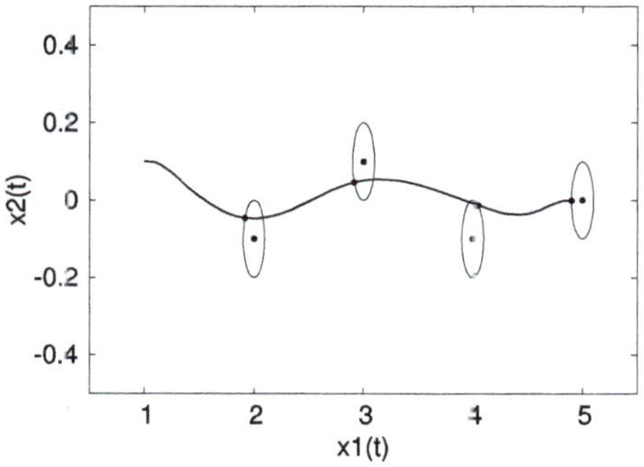

Figure 1.8: Minimum time path through an ε-neighborhood of given points x_i, $i = 1, \dots, 4$, with radius $\varepsilon = 0.1$. The neighborhoods are depicted as circles with center x_i. The solid points on the path at the boundary of the circles indicate the position $x(t_i)$ of the process at time points t_i, $i = 1, \dots, 4$.

1.3 Overview

Figure 1.9 attempts to classify different approaches towards Problem 1.1.8. It does not claim to provide a complete overview on methods, and, of course, alternative classifications are also possible.

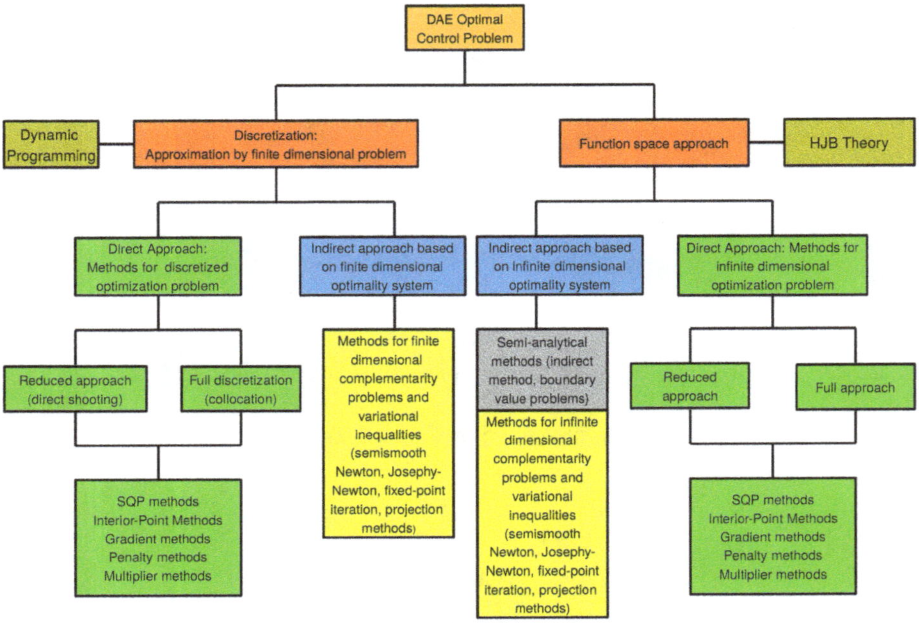

Figure 1.9: Classification of solution approaches for a DAE optimal control problem.

The main distinction, which is made in Figure 1.9, is the discretization approach versus the function space approach. The paradigm of the discretization method is to approximate the DAE optimal control problem by a finite dimensional optimization problem using a suitable discretization scheme. This approach is widely known as *'first discretize, then optimize approach'*. The discretization approach is discussed in Chapter 5. Suitable integration schemes for DAEs, which are fundamental to discretization methods, are briefly summarized in Chapter 4.

In contrast, the function space approach considers the DAE optimal control problem as an infinite dimensional optimization problem, typically in a suitable Banach space setting. This is known as *'first optimize, then discretize approach'*. Typical methods of the function space approach are described in Chapter 8. In the latter, discretization techniques are only used to approximate the single steps of the function space method.

The branches below this first main branch into direct discretization approach and function space approach basically summarize the same techniques, either to solve the optimization problem *directly* using suitable optimization algorithms or to solve it *indirectly* by solving the first-order necessary optimality conditions with suitable techniques for variational inequalities or complementarity problems. The only fundamental difference is that the methods in the left branch of Figure 1.9 are applied to finite-dimensional problems, while the methods in the right branch have to be suitable for infinite dimensional problems.

The classical *indirect approach* for optimal control problems plays an extra role in the function space branch and can be considered a semi-analytical method. Herein, necessary optimality conditions given in terms of the famous minimum principle are exploited in order to set up a nonlinear boundary value problem that a minimizer has to satisfy necessarily. This exploitation is usually not done automatically by an algorithm but by the user. The user needs to extract all information that is provided and sometimes hidden in first- or second-order necessary optimality conditions. Especially in the presence of state or control constraints, this is not an easy task as the switching structure of the problem, which is the sequence of singular sub-arcs and active and inactive parts, is usually unknown, and good initial guesses are mandatory. Hence, the user needs to have a sound knowledge of optimal control theory and insights into the problem-specific structure.

Dynamic Programming and Hamilton–Jacobi–Bellman (HJB) theory aim to characterize the *value function* of the respective optimal control problems in discrete or continuous time. Depending on the problem setting, either a fixed-point iteration, a recursion, or a first-order partial differential equation has to be solved to obtain the value function of the underlying optimal control problem. Once the value function is known, it is possible to extract a feedback control law from it.

The knowledge of first-order necessary optimality conditions is fundamental to all approaches. In Chapter 2, such optimality conditions are provided in terms of Fritz John conditions for infinite optimization problems. In Chapter 3, the necessary Fritz John conditions are evaluated for DAE optimal control problems by exploitation of the special structure. The resulting optimality conditions provide local minimum principles for certain classes of DAE optimal control problems.

Chapter 7 is devoted to mixed-integer optimal control problems. These problems include control variables that are restricted to discrete sets, for instance, a gear shift control in a car. The local minimum principles of Chapter 3 do not hold for such problems as they require convex control sets with a non-empty interior. By applying the Dubovitsky–Milyutin trick, see [139, p. 95], [169, p. 148], it is possible to obtain a *global* minimum principle for mixed-integer optimal control problems. This trick exploits a time transformation technique, which turns out to be of use for the derivation of numerical approaches for mixed-integer optimal control problems as well.

Chapter 6 addresses the problem of controlling a system in real-time. Next to the classic LQ feedback controller, an open loop controller is described based on a sensitivity analysis of the discretized optimal control problem. Finally, the discretization techniques in Chapter 5 are exploited to construct a (nonlinear) model-predictive feedback controller, which is capable of taking state or control constraints into account.

1.4 Exercises

Consider the flight of an aircraft of mass m with center of gravity $S(t) = (x(t), h(t))^\top$ and velocity $v(t)$ in two space dimensions x (range) and h (altitude).

The flight of the aircraft is subject to the weight force $W = mg$, where g denotes acceleration due to gravity, and the aerodynamic forces lift

$$L(v, h, C_L) := q(v, h) \cdot F \cdot C_L$$

and drag

$$D(v, h, C_L) := q(v, h) \cdot F \cdot C_D(C_L),$$

where $q(v, h) := \frac{1}{2} \cdot \rho(h) \cdot v^2$ denotes the dynamic pressure, $\rho(h)$ the air density, C_L the lift coefficient, $C_D(C_L) = C_{D_0} + kC_L^2$ the drag coefficient, C_{D_0} and k are constants, and F is the efficient surface.

The aircraft can be controlled by the thrust $T(t) \in [0, T_{\max}]$ in its longitudinal direction and by the lift coefficient $C_L(t) \in [C_{L,\min}, C_{L,\max}]$.

For simplicity, it is assumed that all forces apply at the center of gravity S, see figure.

(a) Use Newton's law

$$\text{force} = \text{mass} \times \text{acceleration}$$

to formulate differential equations for $x(t)$, $h(t)$, $v(t)$, and $\gamma(t)$, where γ denotes the pitch angle, see figure.

(b) Formulate an optimal control problem for the flight of the aircraft with the aim to maximize the flight range. The aircraft is supposed to start at time $t_0 = 0$ in a given initial position, and it has to reach at final time point $t_f > 0$ a given altitude $h(t_f) = h_f \geq 0$ with $\gamma(t_f) = 0$.

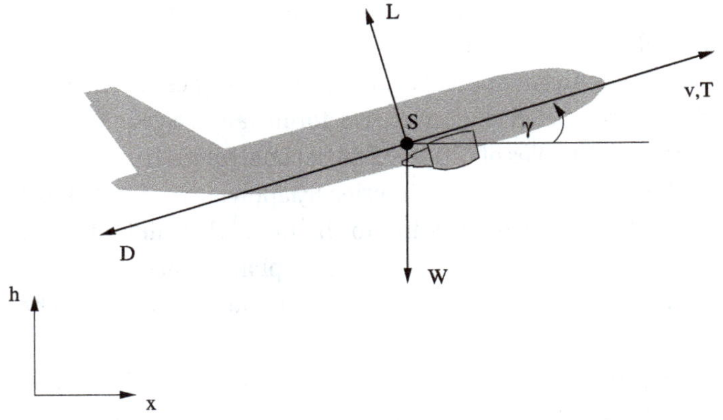

Derive the equations of motion of the following two mechanical multi-body systems, determine the differentiation index, write down the Gear–Gupta–Leimkuhler stabilization, and find consistent initial values:

(a) The quick return mechanism can be controlled by a torque u that applies at body two:

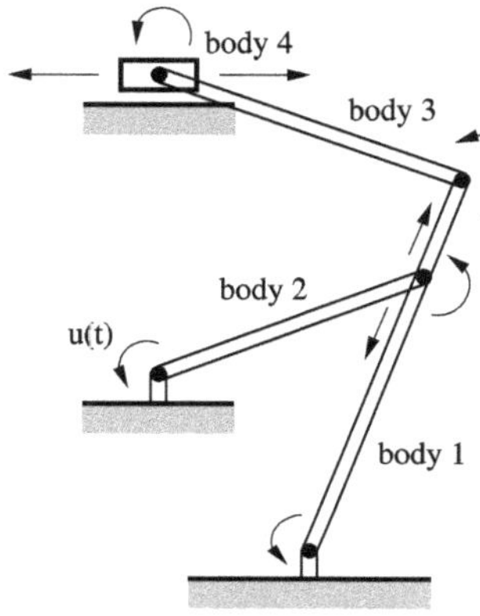

(b) The slider crank mechanism can be controlled by a torque u that applies at body one:

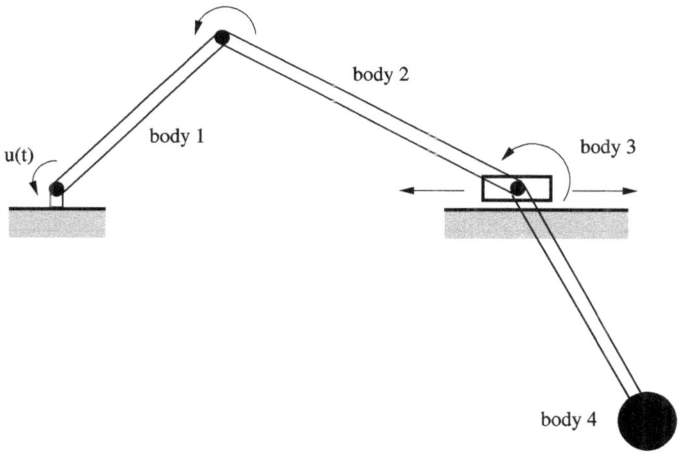

Consider a trolley of mass m moving on a bar at height z with a swinging load of mass m_2 attached, see figure.

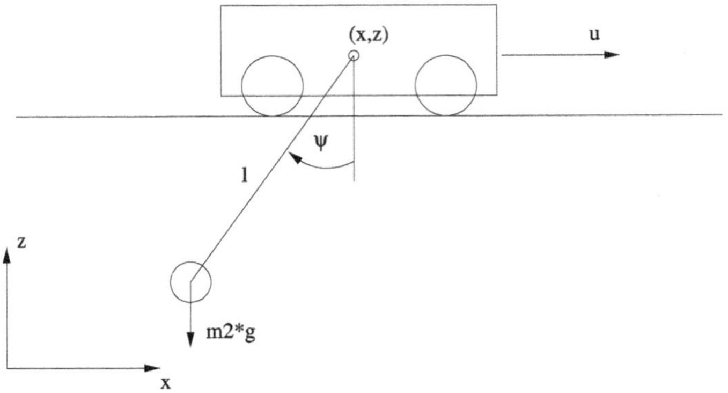

(a) Formulate a DAE that describes the motion of the trolley. Analyze the DAE, find its underlying ODE, and write down the Gear–Gupta–Leimkuhler stabilization for the problem.

(b) At time $t_0 = 0$, the trolley starts at the position $x(0) = 0$, and velocity $\dot{x}(0) = 0$ with the load at rest. At free final time t_f, the trolley is supposed to reach the position $x(t_f) = 1$ with velocity $\dot{x}(t_f) = 0$ and with the load at rest again.
Formulate suitable optimal control problems subject to the above restrictions for
(i) A time minimal motion of the trolley.
(ii) A motion with minimal swinging of the load.

Investigate the following DAE regarding solvability depending on the controls u_1 and u_2:

$$\begin{aligned}\dot{z}_1(t) + \dot{z}_2(t) + z_1(t) + z_2(t) + u_1(t) &= 0,\\ z_1(t) + z_2(t) + u_2(t) &= 0.\end{aligned}$$

Let $M \in \mathbb{R}^{n_z \times n_z}$ be an arbitrary matrix. Transform the quasilinear DAE

$$M\dot{z}(t) = f\big(t, z(t), u(t)\big)$$

into an equivalent semi-explicit DAE. *Hint:* Decompose the matrix M suitably.

Find the differentiation index and the perturbation index for the following nonlinear DAE, compare [53, 154]:

$$z_n(t)\begin{pmatrix} 0 & 1 & & \\ & \ddots & \ddots & \\ & & \ddots & 1 \\ & & & 0 \end{pmatrix}\begin{pmatrix} \dot{z}_1(t) \\ \dot{z}_2(t) \\ \vdots \\ \dot{z}_n(t) \end{pmatrix} + \begin{pmatrix} z_1(t) \\ z_2(t) \\ \vdots \\ z_n(t) \end{pmatrix} = \begin{pmatrix} 0 \\ 0 \\ \vdots \\ 0 \end{pmatrix}.$$

How large is the difference between the differentiation index and the perturbation index in terms of $n \in \mathbb{N}$?

Let the DAE

$$
\begin{aligned}
F_1\big(x(t), \dot{x}(t), y_1(t), y_2(t)\big) &= 0_{\mathbb{R}^{n_x}},\\
F_2\big(x(t), y_1(t), y_2(t)\big) &= 0_{\mathbb{R}^{n_{y_1}}},\\
F_3\big(x(t)\big) &= 0_{\mathbb{R}^{n_{y_2}}}
\end{aligned}
$$

with sufficiently smooth functions F_1, F_2, and F_3 be given. State conditions such that the DAE has differentiation index two.

Consider Example 1.1.20 and investigate non-smooth dockings, that is, docking problems with non-zero relative velocity at contact time t_c.

Hint: In general, velocities will be discontinuous in this case, and for finding physically meaningful consistent values, further relations, e. g., preservation of energy, need to be exploited.

Transform the following optimal control problem on the time interval $[0, t_f]$ with free final time $t_f > 0$ into an equivalent optimal control problem on a fixed time interval $[0, 1]$:

Minimize $-m(t_f)$ subject to the constraints

$$
\begin{aligned}
\dot{r}(t) &= \omega(t), & r(0) &= 1,\ r(t_f) = 1.525,\\
\dot{\omega}(t) &= \frac{v(t)^2}{r(t)} - \frac{1}{r(t)^2} + u(t)\frac{c}{m(t)}\sin\psi(t), & \omega(0) &= 0,\ \omega(t_f) = 0,\\
\dot{v}(t) &= -\frac{\omega(t)v(t)}{r(t)} + u(t)\frac{c}{m(t)}\cos\psi(t), & v(0) &= 1,\ v(t_f) = \frac{1}{\sqrt{1.525}},\\
\dot{m}(t) &= -u(t), & m(0) &= 1,\\
u(t) &\in [0, 0.075]. & &
\end{aligned}
$$

Let a vertically ascending rocket with time-dependent mass m and thrust $T = cu$ obey the equations of motion

$$
\begin{aligned}
\dot{h}(t) &= v(t),\\
\dot{v}(t) &= \frac{c}{m(t)}u - g,\\
\dot{m}(t) &= -u,
\end{aligned}
$$

where h denotes the altitude, v is the velocity, u is the constant thrust level, g is the acceleration due to gravity, and c is a constant.

(a) Solve the equations of motion given the initial values $h(0) = 0$, $v(0) = 0$, $m(0) = m_0 > 0$.

(b) Find u such that the altitude $h(t_f)$ at time $t_f > 0$ becomes maximal under the restriction

$$m_0 - m(t_f) = \Delta m,$$

where $\Delta m > 0$ is given.

(c) Sketch $m(t)$, $v(t)$, $h(t)$ for $u = u_{opt}$ from (b) and $u = u_{max} = 9.5$ [kg/s]. Use the data $m_0 = 215$ [kg], $\Delta m = 147$ [kg], $c = 2.06$ [km/s], $g = 0.00981$ [km/s^2].

(d) Now let u be time-dependent with

$$u(t) = \begin{cases} u_1, & 0 \le t \le t_f, \\ 0, & t > t_f. \end{cases}$$

Find $u_1 \in (0, u_{max}]$ and $T_f > t_f$ such that $h(T_f)$ becomes maximal. The time point t_f is again defined by the restriction $m_0 - m(t_f) = \Delta m$.

Can a Mayer problem be transformed into an equivalent Lagrange problem?

2 Infinite optimization problems

This chapter is devoted to the formulation of DAE optimal control problems and their interpretation as infinite dimensional optimization problems. For the readers' convenience, the functional analytic background needed for deriving optimality conditions is provided inasmuch as necessary. Although most of the material is standard, we found that some parts are hard to find in the literature, for instance, the more advanced properties of Stieltjes integrals and the results on variational equalities and inequalities involving functions of bounded variation and Stieltjes integrals in Chapter 3.

After the functional analytic background has been provided, we start with a DAE optimal control problem in standard form, which can be obtained by application of the transformation techniques in Section 1.2 to more general problem classes. The optimal control problem is considered an infinite-dimensional optimization problem with cone constraints. First-order necessary optimality conditions are provided for infinite optimization problems in general and finite-dimensional optimization problems as a special case. These optimality conditions are the basis for the local minimum principles in Chapter 3, the global minimum principle in Section 7.1, the Lagrange–Newton method in Section 8.2, and the discretization methods in Chapter 5.

For the sake of completeness, we provide a detailed presentation of the necessary optimality conditions, including most of the proofs. We found it useful to include the proof of first-order necessary Fritz John conditions since it sheds light on the many difficulties arising in infinite dimensions. Furthermore, the exposition attempts to combine several versions of necessary conditions from the literature, which differ slightly in their assumptions or the problem classes under consideration, and to adapt them to our purposes. Readers familiar with infinite and finite-dimensional optimization theory may skip large parts of this chapter. However, Chapter 2 is important for all upcoming theoretical and numerical approaches towards optimal control problems.

2.1 Function spaces

For the interpretation of the DAE optimal control problem 1.1.8 as an infinite dimensional optimization problem, it is necessary to specify the differentiability assumptions on the differential variable x, the algebraic variable y, and the control variable u. As already pointed out in Chapter 1, differential and algebraic states exhibit different smoothness properties. For the semi-explicit DAE (1.11)–(1.12), it is reasonable to choose x, y, and u from the function spaces

$$x \in W_{1,\infty}^{n_x}(\mathcal{I}),$$
$$y \in L_{\infty}^{n_y}(\mathcal{I}),$$
$$u \in L_{\infty}^{n_u}(\mathcal{I}),$$

https://doi.org/10.1515/9783110797893-002

which will be introduced in this section. To this end, some fundamental definitions and results from functional analysis are summarized in the following subsections.

2.1.1 Topological spaces, Banach spaces, and Hilbert spaces

Let $\mathbb{K} = \mathbb{R}$ or $\mathbb{K} = \mathbb{C}$, and let X be a set. Let two algebraic operations $+ : X \times X \longrightarrow X$ (addition) and $\cdot : \mathbb{K} \times X \longrightarrow X$ (scalar multiplication) be defined. Recall that $(X, +)$ is called an *Abelian group*, if

(a) $(x + y) + z = x + (y + z)$ holds for all $x, y, z \in X$ (associative law).
(b) There exists an element $\Theta_X \in X$ with $\Theta_X + x = x$ for all $x \in X$ (existence of zero element).
(c) For every $x \in X$, there exists $x' \in X$ with $x + x' = \Theta_X$ (existence of inverse).
(d) $x + y = y + x$ holds for all $x, y \in X$ (commutative law).

---

i Throughout, Θ_X denotes the zero element of $(X, +)$. If no confusion is possible, we will use simply Θ. In the special cases $X = \mathbb{R}^n$ and $X = \mathbb{R}$ the zero elements are denoted by $0_{\mathbb{R}^n}$ and 0, respectively.

Definition 2.1.1 (Vector space, linear space). $(X, +, \cdot)$ is called a *vector space* or *linear space* over $\mathbb{K}$, if $(X, +)$ is an Abelian group and if the following computational rules are satisfied:

(a) $(s \cdot t) \cdot x = s \cdot (t \cdot x)$ for all $s, t \in \mathbb{K}, x \in X$.
(b) $s \cdot (x + y) = s \cdot x + s \cdot y$ for all $s \in \mathbb{K}, x, y \in X$.
(c) $(s + t) \cdot x = s \cdot x + t \cdot x$ for all $s, t \in \mathbb{K}, x \in X$.
(d) $1 \cdot x = x$ for all $x \in X$. □

To define continuity of functions, we need to define neighborhoods of points.

Definition 2.1.2 (Topology, open set, closed set, topological vector space, continuity).

(a) A *topology* τ on a set X is a subset of the power set of X with the following properties:
  (i) $\emptyset, X$ belong to τ.
  (ii) The union of arbitrary many elements of τ belongs to τ.
  (iii) The intersection of finitely many elements of τ belongs to τ.
(b) The elements of a topology τ are called *open sets*.
(c) The complement of an open set is a *closed set*.
(d) The tuple (X, τ) is called a *topological vector space* over $\mathbb{K}$, if
  (i) X is a vector space over $\mathbb{K}$.
  (ii) X is endowed with a topology τ.
  (iii) Addition and multiplication by scalars are continuous functions in the given topology.

(e) Let (X, τ_X) and (Y, τ_Y) be topological vector spaces and $f : X \longrightarrow Y$ a mapping. f is called *continuous at* $x \in X$ if for all open sets $V \in \tau_Y$ containing $y = f(x) \in Y$, there exists an open set $U \in \tau_X$ in X with $x \in U$ and $f(U) \subseteq V$.
f is called *continuous on* X if f is continuous at every $x \in X$. □

To define a topology in a vector space, it is sufficient to specify a basis $\mathcal{A}$ of open sets around zero. A basis around zero is characterized by the property that for any neighborhood V of zero, there exists $U \in \mathcal{A}$ such that $U \subseteq V$. Herein, every open set containing x is called a *neighborhood of* x. Then, a set S is open if and only if for every $x \in S$, there exists an element $U \in \mathcal{A}$ such that $x + U \subseteq S$.

Definition 2.1.3 (Interior point, closure, boundary point, dense set). Let S be a subset of a topological vector space (X, τ).

(a) $x \in S$ is called an *interior point of* S if there is a neighborhood $U \in \tau$ of x with $U \subseteq S$. The set of all interior points of S is denoted by $\text{int}(S)$.
(b) The *closure* $\text{cl}(S)$ *of* S is the set of all points x satisfying $U \cap S \neq \emptyset$ for all neighborhoods U of x.
(c) x is a *boundary point of* S if $x \in \text{cl}(S)$ and $x \notin \text{int}(S)$.
(e) A subset S of a topological vector space (X, τ) is called *dense in* X if $cl(S) = X$ holds. □

S is open if and only if $S = \text{int}(S)$. S is closed if and only if $S = \text{cl}(S)$.

Some measure of distance is necessary to define convergence of sequences.

Definition 2.1.4 (Metric space, Cauchy sequence, convergence). (a) A *metric space* is a tuple (X, d), where X is a set, and $d : X \times X \longrightarrow \mathbb{R}$ is a mapping such that for every $x, y, z \in X$ it holds
(i) $d(x,y) \geq 0$ and $d(x,y) = 0$, if and only if $x = y$.
(ii) $d(x,y) = d(y,x)$.
(iii) $d(x,y) \leq d(x,z) + d(z,y)$.
d is called a *metric* on X.

(b) Let (X, d) be a metric space. The sequence $\{x_n\}_{n\in\mathbb{N}}$ is said to *converge* to $x \in X$, if $\lim_{n\to\infty} d(x_n, x) = 0$. We write

$$\lim_{n\to\infty} x_n = x \quad \text{or} \quad x_n \longrightarrow x \quad \text{as } n \longrightarrow \infty.$$

(c) A sequence $\{x_n\}_{n\in\mathbb{N}}$ in a metric space (X, d) is called a *Cauchy sequence* if for every $\varepsilon > 0$, there is an $N(\varepsilon) \in \mathbb{N}$ such that $d(x_n, x_m) < \varepsilon$ for all $n, m > N(\varepsilon)$.
(d) A metric space (X, d) is called *complete* if every Cauchy sequence from X has a limit in X. □

In a metric space every convergent sequence is a Cauchy sequence.
If a sequence converges in a metric space, then its limit is unique.

In what follows, normed spaces are considered most often. The norm measures the length of a vector.

Definition 2.1.5 (Norm). Let X be a vector space over $\mathbb{K} = \mathbb{R}$ or $\mathbb{K} = \mathbb{C}$. The tuple $(X, \|\cdot\|_X)$ is called a *normed vector space* if $\|\cdot\|_X : X \longrightarrow \mathbb{R}$ is a mapping such that for every $x, y \in X$ and every $\lambda \in \mathbb{K}$, it holds
(a) $\|x\|_X \geq 0$ and $\|x\|_X = 0$ if and only if $x = \Theta_X$.
(b) $\|\lambda x\|_X = |\lambda| \cdot \|x\|_X$.
(c) $\|x + y\|_X \leq \|x\|_X + \|y\|_X$.

The mapping $\|\cdot\|_X$ is called a *norm on X*. □

Since every norm defines a metric by $d(x,y) := \|x - y\|_X$, the terminologies 'convergence, completeness, Cauchy sequence, ...' can be translated directly to normed spaces $(X, \|\cdot\|_X)$. Most often, we will work in Banach spaces.

Definition 2.1.6 (Banach space). A complete normed vector space is called *a Banach space*. □

In the Banach space $(X, \|\cdot\|_X)$ for $x \in X$ and $r > 0$, let

$$B_r(x) := \{y \in X \mid \|y - x\|_X < r\},$$
$$\overline{B_r(x)} := \{y \in X \mid \|y - x\|_X \leq r\}$$

denote the open and closed balls around x with radius r, respectively. Then, $\mathcal{A} := \{B_r(\Theta_X) \mid r > 0\}$ defines a basis of open sets about zero. We say, the norm $\|\cdot\|_X$ induces the *strong topology on X*.

Finally, we introduce Hilbert spaces that permit to investigate orthogonality relations of vectors with respect to an inner product.

Definition 2.1.7 (Inner product, scalar product). Let X be a vector space over $\mathbb{K} = \mathbb{R}$ or $\mathbb{K} = \mathbb{C}$. The mapping $\langle\cdot,\cdot\rangle_X : X \times X \longrightarrow \mathbb{K}$ is called an *inner product* or *scalar product* if the following conditions hold for all $x, y, z \in X$ and all $\lambda \in \mathbb{K}$.
(a) $\langle x, y\rangle_X = \overline{\langle y, x\rangle_X}$.
(b) $\langle x + y, z\rangle_X = \langle x, z\rangle_X + \langle y, z\rangle_X$.
(c) $\langle x, \lambda y\rangle_X = \lambda\langle x, y\rangle_X$.
(d) $\langle x, x\rangle_X \geq 0$ and $\langle x, x\rangle_X = 0$ if and only if $x = \Theta_X$. □

Definition 2.1.8 (Pre-Hilbert space, Hilbert space). A *pre-Hilbert space* is a vector space endowed with an inner product. A complete pre-Hilbert space is called *a Hilbert space*. □

Notice that $\|x\|_X := \sqrt{\langle x, x\rangle_X}$ defines a norm on the pre-Hilbert space X.

From now on, unless otherwise specified, $(X, \|\cdot\|_X), (Y, \|\cdot\|_Y), (Z, \|\cdot\|_Z), \dots$ denote real Banach spaces over $\mathbb{K} = \mathbb{R}$ with norms $\|\cdot\|_X, \|\cdot\|_Y, \|\cdot\|_Z, \dots$. If no confusion is possible, we omit the norms, call $X, Y, Z, \dots$ Banach spaces, and assume that appropriate norms are defined.

2.1.2 Mappings and dual spaces

Basic notions of mappings are summarized in

Definition 2.1.9. Let $T : X \longrightarrow Y$ be a mapping from a Banach space $(X, \|\cdot\|_X)$ into a Banach space $(Y, \|\cdot\|_Y)$.

(a) The *image of* T is defined by

$$\operatorname{im}(T) := \{T(x) \mid x \in X\}.$$

The *kernel (null-space) of* T is defined by

$$\ker(T) := \{x \in X \mid T(x) = \Theta_Y\}.$$

Given a set $S \subseteq Y$, the *pre-image of* S *under* T is defined by

$$T^{-1}(S) := \{x \in X \mid T(x) \in S\}.$$

(b) T is called *linear* if

$$T(x_1 + x_2) = T(x_1) + T(x_2), \qquad T(\lambda x_1) = \lambda T(x_1)$$

holds for all $x_1, x_2 \in X, \lambda \in \mathbb{R}$.

(c) Let $D \subseteq X$ be open. A mapping $T : D \longrightarrow Y$ is called *locally Lipschitz continuous at* $x \in D$ *with constant* L if there exists $\varepsilon > 0$ such that

$$\|T(y) - T(z)\|_Y \le L\|y - z\|_X \quad \text{for every } y, z \in B_\varepsilon(x).$$

(d) T is called *bounded* if $\|T(x)\|_Y \le C \cdot \|x\|_X$ holds for all $x \in X$ and some constant $C \ge 0$.

(e) If $Y = \mathbb{R}$, then T is called a *functional*.

(f) The set of all linear continuous functionals on X endowed with the norm

$$\|T\|_{X^*} := \sup_{\|x\|_X \le 1} |T(x)|$$

is called *dual space of* X and is denoted by X^*. This norm defines the *strong topology on* X^*.

X is called *reflexive* if $(X^*)^* = X$ holds with respect to the strong topology.

(g) Let $T : X \longrightarrow Y$ be linear. The *adjoint operator* $T^* : Y^* \longrightarrow X^*$ is a linear operator defined by

$$T^*(y^*)(x) = y^*(T(x))$$

for all $y^* \in Y^*$ and all $x \in X$. □

The following statements can be found in textbooks on functional analysis, e. g., [212, 281, 1, 326]:

(a) $\ker(T)$ is a closed subspace of X if $T : X \longrightarrow Y$ is a linear and continuous mapping.
(b) A linear operator is continuous if and only if it is bounded.
(c) If the linear operator T is continuous, so is the adjoint operator T^*.
(d) The set $\mathcal{L}(X, Y)$ of all linear continuous mappings from X into Y endowed with the *operator norm*

$$\|T\|_{\mathcal{L}(X,Y)} := \sup_{x \neq \Theta_X} \frac{\|T(x)\|_Y}{\|x\|_X} = \sup_{\|x\|_X \leq 1} \|T(x)\|_Y = \sup_{\|x\|_X = 1} \|T(x)\|_Y$$

is a Banach space.
(e) The dual space of a Banach space is a Banach space.
(f) The product space

$$X = X_1 \times X_2 \times \cdots \times X_n$$

of n Banach spaces $(X_1, \|\cdot\|_{X_1}), \ldots, (X_n, \|\cdot\|_{X_n})$ equipped with one of the norms

$$\|x\|_X = \max_{1 \leq i \leq n} \|x_i\|_{X_i}, \quad \|x\|_X = \left(\sum_{i=1}^{n} \|x_i\|_{X_i}^p \right)^{1/p}, \quad p \in \mathbb{N},$$

is again a Banach space. The dual space of X is given by

$$X^* = \{x^* = (x_1^*, x_2^*, \ldots, x_n^*) \mid x_i^* \in X_i^*,\ i = 1, \ldots, n\},$$
$$x^*(x) = \sum_{i=1}^{n} x_i^*(x_i).$$

The following theorem allows to identify the dual space X^* of a Hilbert space X with X itself. In particular, Hilbert spaces are reflexive.

Theorem 2.1.10 (Riesz). *Let X be a Hilbert space and $x^* \in X^*$. Then, there exists a unique element $x \in X$ with $x^*(y) = \langle x, y\rangle_X$ for all $y \in X$ and $\|x^*\|_{X^*} = \|x\|_X$.* □

2.1.3 Derivatives, mean-value theorem, and implicit function theorem

The differentiability of a mapping $T : X \longrightarrow Y$, where X and Y are Banach spaces, is of central importance in analyzing functions and plays a key role in deriving necessary optimality conditions. The common definitions of directional differentiability and differentiability in terms of linear approximations for functions $T : \mathbb{R} \longrightarrow \mathbb{R}$ are extended to the Banach space setting. Throughout this book, we use directional derivatives and Fréchet-derivatives. The Fréchet-derivative coincides with the common derivative in the case $X = Y = \mathbb{R}$ and $T : \mathbb{R} \longrightarrow \mathbb{R}$. Further differentiability concepts like the Hadamard differentiability are not considered, although such concepts would allow proving, e. g., the Fritz John conditions 2.3.23 under weaker assumptions, see [210].

Definition 2.1.11 (Directional derivative, Gâteaux-derivative, Fréchet-derivative). Let X and Y be Banach spaces, $x \in X$, and $T : X \longrightarrow Y$ a mapping.

(a) T is called *directionally differentiable at x in direction $d \in X$* if the limit

$$T'(x;d) = \lim_{\alpha \downarrow 0} \frac{1}{\alpha}(T(x + \alpha d) - T(x))$$

exists. $T'(x;d)$ is called *directional derivative of T at x in direction d*.

(b) T is called *Gâteaux-differentiable at x* if there exists a continuous linear operator $\delta T(x) : X \longrightarrow Y$ with

$$\lim_{\alpha \downarrow 0} \frac{T(x + \alpha d) - T(x) - \alpha \delta T(x)(d)}{\alpha} = \Theta_Y$$

for all $d \in X$. The operator $\delta T(x)$ is called *Gâteaux-differential of T at x*.

(c) T is called *Fréchet-differentiable at x* if there exists a continuous linear operator $T'(x) : X \longrightarrow Y$ with

$$\lim_{\|d\|_X \longrightarrow 0} \frac{T(x + d) - T(x) - T'(x)(d)}{\|d\|_X} = \Theta_Y$$

for all $d \in X$. The operator $T'(x)$ is called *(Fréchet-)derivative of T at x*.
If the mapping $T' : X \longrightarrow \mathcal{L}(X, Y)$, $x \mapsto T'(x)$, is continuous in the strong topology of $\mathcal{L}(X, Y)$, then T is called *continuously differentiable*.

(d) Let $X = X_1 \times X_2$. $T : X_1 \times X_2 \longrightarrow Y$ is called *partially Fréchet-differentiable with respect to the component x_1 at $(\hat{x}_1, \hat{x}_2) \in X_1 \times X_2$*, if $T(\cdot, \hat{x}_2) : X_1 \longrightarrow Y$ is Fréchet-differentiable at $\hat{x}_1$.
The partial Fréchet-derivative of T with respect to x_1 at $(\hat{x}_1, \hat{x}_2)$ is denoted by $T'_{x_1}(\hat{x}_1, \hat{x}_2)$. A similar definition holds for the component x_2. □

The following statements can be found in [169] and are provided without proof. Basically, the same computational rules and properties known from the common derivative in Euclidean spaces still hold for the Fréchet-derivative. In particular, the chain rule applies. Moreover, if $T : X_1 \times X_2 \longrightarrow Y$ is Fréchet-differentiable at $(\hat{x}_1, \hat{x}_2)$, then T is also partially Fréchet-differentiable with respect to x_1 and x_2 and for all $(x_1, x_2) \in X_1 \times X_2$,

$$T'(\hat{x}_1, \hat{x}_2)(x_1, x_2) = T'_{x_1}(\hat{x}_1, \hat{x}_2)(x_1) + T'_{x_2}(\hat{x}_1, \hat{x}_2)(x_2).$$

If $T : X \longrightarrow Y$ is Gâteaux-differentiable at x, then T is directionally differentiable, and the directional derivative and Gâteaux-derivative coincide: $T'(x; d) = \delta T(x)(d)$.

If $T : X \longrightarrow Y$ is Fréchet-differentiable at x, then T is continuous, and Gâteaux-differentiable at x and the Fréchet-derivative and the Gâteaux-derivative coincide: $T'(x)(d) = \delta T(x)(d)$. Conversely, if T is continuous and continuously Gâteaux-differentiable in some neighborhood of $\hat{x} \in X$, then T is Fréchet-differentiable in this neighborhood with $T'(x) = \delta T(x)$.

Higher-order Fréchet-derivatives are defined recursively. Let $k \geq 2$. T is called k-times Fréchet-differentiable if T' is $(k-1)$-times Fréchet-differentiable. In particular, T is twice Fréchet-differentiable at $\hat{x}$ if the mapping

$$T'(\cdot) : X \longrightarrow \mathcal{L}(X, Y), \quad x \mapsto T'(x),$$

is Fréchet-differentiable at $\hat{x}$, i. e. $T''(\hat{x})$ is a continuous linear operator from X into $\mathcal{L}(X, Y)$:

$$\begin{aligned} T''(\hat{x}) &\in \mathcal{L}(X, \mathcal{L}(X, Y)), \\ T''(\cdot) &: X \longrightarrow \mathcal{L}(X, \mathcal{L}(X, Y)). \end{aligned}$$

Hence, for every $d_1, d_2 \in X$, we have

$$\begin{aligned} T''(\hat{x})(d_1)(\cdot) &\in \mathcal{L}(X, Y), \\ T''(\hat{x})(d_1)(d_2) &\in Y. \end{aligned}$$

Notice that $T''(\hat{x})(d_1)(\cdot)$ is linear for every $d_1 \in X$, and $T''(\hat{x})(\cdot)(d_2)$ is also linear for every $d_2 \in X$. Hence, $T''(\hat{x})$ is a *bilinear mapping*, and for notational simplicity, we write

$$T''(\hat{x})(\cdot, \cdot) \quad \text{respectively} \quad T''(\hat{x})(d_1, d_2).$$

In general, the kth Fréchet-derivative of T is a k-linear mapping:

$$T^{(k)}(\hat{x})(d_1, \dots, d_k).$$

Example 2.1.12. For $X = \mathbb{R}^n$ and $Y = \mathbb{R}$, the Fréchet-derivative is just the standard derivative of the function $T : \mathbb{R}^n \longrightarrow \mathbb{R}$.

The first derivative for $d = (d_1, \dots, d_n)^\top$ is given by

$$T'(\hat{x})(d) = \sum_{i=1}^{n} \frac{\partial T(\hat{x})}{\partial x_i} d_i = \nabla T(\hat{x})^\top d.$$

The second derivative for $d = (d_1, \dots, d_n)^\top$ and $h = (h_1, \dots, h_n)^\top$ is

$$T''(\hat{x})(d,h) = \sum_{i,j=1}^{n} \frac{\partial^2 T(\hat{x})}{\partial x_i \partial x_j} d_i h_j = d^\top \nabla^2 T(\hat{x}) h,$$

where $\nabla^2 T(\hat{x})$ is the Hessian matrix of T at $\hat{x}$. The kth derivative of T for $d_i = (d_{i,1}, \dots, d_{i,n})^\top$, $i = 1, \dots, k$, is

$$T^{(k)}(\hat{x})(d_1, \dots, d_k) = \sum_{i_1=1}^{n} \cdots \sum_{i_k=1}^{n} \frac{\partial^k T(\hat{x})}{\partial x_{i_1} \cdots \partial x_{i_k}} d_{1,i_1} \cdots d_{k,i_k}.$$

□

In addition, the mean-value theorem holds true under weak differentiability assumptions:

Theorem 2.1.13 (Mean-value theorem, see [169, p. 27]). *Let X and Y be linear topological spaces, $D \subseteq X$ be open, and $h \in X$ such that $x + \alpha h \in D$ for every $0 \le \alpha \le 1$. Let $T : D \longrightarrow Y$ be Gâteaux-differentiable for every point $x + \alpha h$, $0 \le \alpha \le 1$.*

(a) *Then*

$$T(x+h) - T(x) = \int_0^1 \delta T(x + \alpha h)(h) d\alpha,$$

provided the mapping $x \mapsto \delta T(x)(h)$ is continuous for all $x + \alpha h$, $0 \le \alpha \le 1$.

(b) *If X and Y are Banach spaces, then*

$$\|T(x+h) - T(x)\|_Y \le \sup_{0 \le \alpha \le 1} \|\delta T(x + \alpha h)\|_{\mathcal{L}(X,Y)} \cdot \|h\|_X.$$

Moreover, for any $\tilde{T} \in \mathcal{L}(X,Y)$, we have the estimate

$$\|T(x+h) - T(x) - \tilde{T}(h)\|_Y \le \sup_{0 \le \alpha \le 1} \|\delta T(x + \alpha h) - \tilde{T}\|_{\mathcal{L}(X,Y)} \cdot \|h\|_X.$$

□

Especially in the context of DAEs, the following implicit function theorem is an indispensable tool that we have already used in Chapter 1:

Theorem 2.1.14 (Implicit function theorem, see [169, p. 29]). *Let X, Y, and Z be Banach spaces, $D \subseteq X \times Y$ be some neighborhood of the point $(\hat{x}, \hat{y}) \in X \times Y$, and $T : D \longrightarrow Z$ be continuously Fréchet-differentiable. Let $T(\hat{x}, \hat{y}) = \Theta_Z$, and let the partial derivative $T'_y(\hat{x}, \hat{y})$ be linear, continuous, and bijective.*

Then there exist open balls $B_\varepsilon(\hat{x}) \subseteq X$ and $B_\delta(\hat{y}) \subseteq Y$ with $\varepsilon, \delta > 0$ and a mapping $y : B_\varepsilon(\hat{x}) \longrightarrow Y$ with $y(\hat{x}) = \hat{y}$ and

$$T(x, y(x)) = \Theta_Z \quad \textit{for all } (x, y(x)) \in B_\varepsilon(\hat{x}) \times B_\delta(\hat{y}).$$

Moreover, $y(\cdot)$ is Fréchet-differentiable in $B_\varepsilon(\hat{x})$ and

$$y'(x) = -T'_y(x, y(x))^{-1}(T'_x(x, y(x))) \quad \text{for all } (x, y(x)) \in B_\varepsilon(\hat{x}) \times B_\delta(\hat{y}).$$ □

The assumptions of the implicit function theorem imply that the inverse operator of the partial derivative $T'_y(\hat{x}, \hat{y})$ exists and $T'_y(\hat{x}, \hat{y})^{-1}$ is linear and continuous due to the following

Theorem 2.1.15 ([212, p. 109, Theorem 4]). *Let X and Y be Banach spaces and $T : X \longrightarrow Y$ a continuous, linear, surjective, and injective operator. Then the inverse operator T^{-1} exists, and it is linear and continuous.* □

A sufficient condition for the existence of the inverse is provided in the following theorem.

Theorem 2.1.16 ([212, p. 106, Theorem 4]). *Let X and Y be Banach spaces and $T : X \longrightarrow Y$ a surjective and continuous linear operator. Let there be some constant $C > 0$ with*

$$\|T(x)\|_Y \geq C\|x\|_X$$

for every $x \in X$. Then the inverse operator T^{-1} exists, and it is linear and continuous. □

2.1.4 L_p-spaces, $W_{q,p}$-spaces, absolutely continuous functions, functions of bounded variation

We introduce specific Banach spaces that are used to properly define the DAE optimal control problem 2.2.1 in Section 2.2.

Definition 2.1.17 (L_p-spaces). Let $\mathcal{I} := [t_0, t_f] \subset \mathbb{R}$ be a compact interval with $t_0 < t_f$.

(a) Let $1 \leq p < \infty$. The space $L_p(\mathcal{I})$ consists of all measurable functions $f : \mathcal{I} \longrightarrow \mathbb{R}$ with

$$\int_{t_0}^{t_f} |f(t)|^p dt < \infty,$$

where the integral denotes the Lebesgue integral.

(b) The space $L_\infty(\mathcal{I})$ consists of all measurable functions $f : \mathcal{I} \longrightarrow \mathbb{R}$, which are essentially bounded, i. e.

$$\operatorname*{ess\,sup}_{t \in \mathcal{I}} |f(t)| := \inf_{\substack{N \subset \mathcal{I} \\ N \text{ is set of measure zero}}} \sup_{t \in \mathcal{I} \setminus N} |f(t)| < \infty.$$

(c) For $1 \leq p \leq \infty$ the space, $L_p^n(\mathcal{I})$ is defined to be the product space

$$L_p^n(\mathcal{I}) := L_p(\mathcal{I}) \times \cdots \times L_p(\mathcal{I}),$$

where each element f of $L_p^n(\mathcal{I})$ is a mapping from $\mathcal{I}$ into $\mathbb{R}^n$. □

Functions that differ only on a set of measure zero, i. e. functions that are equal almost everywhere, are considered to be the same in Definition 2.1.17. In this sense, the elements of the spaces $L_p(\mathcal{I})$, $1 \le p \le \infty$, are equivalence classes.

The spaces $L_p(\mathcal{I})$, $1 \le p < \infty$, endowed with the norm

$$\|f\|_p := \left(\int_{t_0}^{t_f} |f(t)|^p dt \right)^{1/p},$$

and the space $L_\infty(\mathcal{I})$, endowed with the norm

$$\|f\|_\infty := \operatorname*{ess\,sup}_{t \in \mathcal{I}} |f(t)|,$$

are Banach spaces, see [186, Theorems 2.8.2, 2.11.7].

For $1 < p < \infty$, the dual space of $L_p(\mathcal{I})$ is given by $L_q(\mathcal{I})$, where $1/p + 1/q = 1$, i. e. for $f^* \in L_p(\mathcal{I})^*$, there exists a unique function $g \in L_q(\mathcal{I})$ such that

$$f^*(f) = \int_{t_0}^{t_f} f(t)g(t)dt \quad \text{for all } f \in L_p(\mathcal{I}),$$

and $\|f^*\| = \|g\|_q$, see [186, Theorem 2.9.5]. For $1 < p < \infty$, the spaces L_p are reflexive, see [186, Theorem 2.10.1].

The dual space of $L_1(\mathcal{I})$ is given by $L_\infty(\mathcal{I})$, i. e. for $f^* \in L_1(\mathcal{I})^*$, there exists a unique function $g \in L_\infty(\mathcal{I})$ such that

$$f^*(f) = \int_{t_0}^{t_f} f(t)g(t)dt \quad \text{for all } f \in L_1(\mathcal{I}),$$

see [186, Theorem 2.11.8]. The spaces $L_1(\mathcal{I})$ and $L_\infty(\mathcal{I})$ are not reflexive, see [186, Theorems 2.11.10, 2.11.11].

The dual space of $L_\infty(\mathcal{I})$ does not have a nice structure. According to [186, Remark 2.17.2] the dual space $L_\infty(\mathcal{I})$ is isometrically isomorph with the space of all finitely additive measures on the family of measurable subsets of $\mathcal{I}$, which are absolutely continuous with respect to the Lebesgue measure.

$L_2(\mathcal{I})$ is a Hilbert space with the inner product

$$\langle f, g \rangle_{L_2} := \int_{t_0}^{t_f} f(t)g(t)dt.$$

Definition 2.1.18 (Absolutely continuous function). Let $\mathcal{I} := [t_0, t_f]$ be a compact interval with $t_0 < t_f$. A function $f : \mathcal{I} \longrightarrow \mathbb{R}$ is said to be *absolutely continuous* if for every $\varepsilon > 0$, there exists $\delta(\varepsilon) > 0$ such that

$$\sum_{i=1}^{m} |b_i - a_i| < \delta(\varepsilon) \quad \Longrightarrow \quad \sum_{i=1}^{m} |f(b_i) - f(a_i)| < \varepsilon,$$

where $m \in \mathbb{N}$ is arbitrary and $(a_i, b_i) \subseteq \mathcal{I}$, $i = 1, \dots, m$, are disjoint intervals. The set of all absolutely continuous functions $f : \mathcal{I} \longrightarrow \mathbb{R}$ is denoted by $AC(\mathcal{I})$. □

The following properties of absolutely continuous functions can be found in [253]:

Theorem 2.1.19 (Properties of absolutely continuous functions). *Let $\mathcal{I} := [t_0, t_f]$ be a compact interval with $t_0 < t_f$.*

(a) *If $f \in L_1(\mathcal{I})$, then the undetermined integral*

$$F(t) := C + \int_{t_0}^{t} f(\tau)d\tau$$

is absolutely continuous, and it holds

$$F'(t) = f(t) \quad \text{almost everywhere in } \mathcal{I}.$$

(b) *If $f' \in L_1(\mathcal{I})$ exists everywhere and it is finite, then it holds*

$$f(t) = f(a) + \int_{t_0}^{t} f'(\tau)d\tau. \tag{2.1}$$

Furthermore, the partial integration

$$\int_{t_0}^{t_f} f(t)g'(t)dt + \int_{t_0}^{t_f} g(t)f'(t)dt = [f(t)g(t)]_{t_0}^{t_f}$$

holds for absolutely continuous functions f and g on $\mathcal{I}$.

(c) *An absolutely continuous function on $\mathcal{I}$ is continuous, of bounded variation, f' exists almost everywhere in $\mathcal{I}$ and satisfies (2.1) for $t \in \mathcal{I}$.*
If f' is zero almost everywhere in $[a, b]$, then f is constant.
If the derivatives f' and g' of absolutely continuous functions f and g are equal almost everywhere, then the difference $f - g$ is constant. □

The following spaces are natural spaces for solutions of differential equations. The space $W_{1,\infty}(\mathcal{I})$ is a natural choice for the components of the differential state x in Problem 1.1.8.

Definition 2.1.20 ($W_{q,p}$-spaces). Let $\mathcal{I} := [t_0, t_f] \subset \mathbb{R}$ be a compact interval with $t_0 < t_f$ and $1 \le q, p \le \infty$.

(a) The space $W_{q,p}(\mathcal{I})$ consists of all absolutely continuous functions $f : \mathcal{I} \longrightarrow \mathbb{R}$ with absolutely continuous derivatives up to order $q - 1$ and

$$\|f\|_{q,p} < \infty,$$

where the norm is given by

$$\|f\|_{q,p} := \left(\sum_{i=0}^{q} \|f^{(i)}\|_p^p \right)^{1/p}, \quad 1 \le p < \infty,$$
$$\|f\|_{q,\infty} := \max_{0 \le i \le q} \|f^{(i)}\|_\infty.$$

(b) For $1 \le q, p \le \infty$, the space $W_{q,p}^n(\mathcal{I})$ is defined as the product space

$$W_{q,p}^n(\mathcal{I}) := W_{q,p}(\mathcal{I}) \times \cdots \times W_{q,p}(\mathcal{I}),$$

where each element f of $W_{q,p}^n(\mathcal{I})$ is a mapping from $\mathcal{I}$ into $\mathbb{R}^n$. □

For $1 \le q, p \le \infty$, the space $W_{q,p}(\mathcal{I})$ endowed with the norm $\|\cdot\|_{q,p}$ is a Banach space. The spaces $W_{q,2}(\mathcal{I})$, $1 \le q \le \infty$, are Hilbert spaces with the inner product

$$\langle f, g \rangle_{W_{q,2}} := \sum_{i=0}^{q} \int_a^b f^{(i)}(t) g^{(i)}(t) dt.$$

Definition 2.1.21 (k-times continuously differentiable functions). Let $\mathcal{I} := [t_0, t_f] \subset \mathbb{R}$ be a compact interval with $t_0 < t_f$ and $n, k \in \mathbb{N}$.

The space of *continuous functions* $f : \mathcal{I} \longrightarrow \mathbb{R}^n$ is denoted by $\mathcal{C}^n(\mathcal{I}) = \mathcal{C}_0^n(\mathcal{I})$.

The space of *k-times continuously differentiable functions* $f : \mathcal{I} \longrightarrow \mathbb{R}^n$ is denoted by $\mathcal{C}_k^n(\mathcal{I})$. □

The elements of the space $W_{1,\infty}(\mathcal{I})$ are continuous functions, but not all of them are continuously differentiable, and hence $\mathcal{C}_1(\mathcal{I}) \subset W_{1,\infty}(\mathcal{I}) \subset \mathcal{C}(\mathcal{I})$.

The space $\mathcal{C}^n(\mathcal{I})$ of continuous vector-valued functions $f : \mathcal{I} \longrightarrow \mathbb{R}^n$ endowed with the norm

$$\|f\|_\infty = \max_{t \in \mathcal{I}} \|f(t)\|,$$

where $\|\cdot\|$ is any vector norm on $\mathbb{R}^n$, is a Banach space.

If pure state constraints (1.17) are present in the DAE optimal control problem 1.1.8, which are continuous functions, then the Lagrange multipliers associated with the pure state constraints in the Fritz John conditions 2.3.23 are elements of the dual space of the space of continuous functions. Hence, explicit representations for the elements of the dual space are required in order to represent the Lagrange multipliers. It turns out that the elements of the dual space of $\mathcal{C}(\mathcal{I})$ can be represented by *Riemann–Stieltjes integrals* (see Definition 2.1.24 below), which are defined by *functions of bounded variation*, see Definition 2.1.23 below. It holds

Theorem 2.1.22 (Riesz' representation theorem, see [253, p. 266]). *Let $\Phi : \mathcal{C}(\mathcal{I}) \longrightarrow \mathbb{R}$ be a continuous linear functional. Then there exists a function μ of bounded variation on $\mathcal{I}$ such that for every $f \in \mathcal{C}(\mathcal{I})$, it holds*

$$\Phi(f) = \int_{t_0}^{t_f} f(t)d\mu(t).$$

□

In Theorem 2.1.22, the function of bounded variation μ is defined almost everywhere in $\mathcal{I}$ with exception of an additive constant, see [143, p. 59], [215, p. 113].

Definition 2.1.23 (Function of bounded variation). Let $\mathcal{I} := [t_0, t_f] \subset \mathbb{R}$ be a compact interval with $t_0 < t_f$.

(a) $\mu : \mathcal{I} \longrightarrow \mathbb{R}$ is said to be of *bounded variation* if there exists a constant K such that for any partition

$$\mathbb{G}_m := \{t_0 < t_1 < \cdots < t_m = t_f\}$$

of $[t_0, t_f]$, it holds

$$\sum_{i=1}^{m} |\mu(t_i) - \mu(t_{i-1})| \leq K.$$

The *total variation of* μ is

$$TV(\mu, t_0, t_f) := \sup_{\text{all } \mathbb{G}_m} \sum_{i=1}^{m} |\mu(t_i) - \mu(t_{i-1})|.$$

The space of all functions of bounded variation on $\mathcal{I}$ is called $BV(\mathcal{I})$.

(b) The space $BV^n(\mathcal{I})$ is defined as the product space

$$BV^n(\mathcal{I}) := BV(\mathcal{I}) \times \cdots \times BV(\mathcal{I}),$$

where each element μ of $BV^n(\mathcal{I})$ is a mapping from $\mathcal{I}$ into $\mathbb{R}^n$.

(c) The *space of normalized functions of bounded variations* $NBV^n(\mathcal{I})$ consists of all functions $\mu \in BV^n(\mathcal{I})$, which are continuous from the right on (t_0, t_f) and satisfy $\mu(t_0) = 0_{\mathbb{R}^n}$.

□

A norm on $NBV(\mathcal{I})$ is given by $\|\mu\|_{NBV} := TV(\mu, t_0, t_f)$. With this norm, the correspondence between the dual space of $\mathcal{C}(\mathcal{I})$ and $NBV(\mathcal{I})$ in Theorem 2.1.22 is unique. Hence, $NBV(\mathcal{I})$ is the dual space of $\mathcal{C}(\mathcal{I})$.

We summarize some facts about functions of bounded variation, see [253]:

(a) The derivative $\dot{\mu}$ of a function $\mu \in BV(\mathcal{I})$ exists almost everywhere in $\mathcal{I}$, and the integral $\int_{t_0}^{t_f} \dot{\mu}(t)dt$ exists.

(b) $\mu \in BV(\mathcal{I})$ possesses only countably many jumps. For every jump point of μ, the left- and right-sided limits exist.

(c) Every function μ of bounded variation can be represented as

$$\mu(t) = \mu_a(t) + \mu_d(t) + \mu_s(t),$$

where μ_a is absolutely continuous, μ_d is the jump function of μ given by

$$\begin{aligned}\mu_d(t_0) &= 0,\\ \mu_d(t) &= (\mu(t_0+) - \mu(t_0)) + \sum_{t_i<t}(\mu(t_i+) - \mu(t_i-)) + (\mu(t) - \mu(t-))\end{aligned}$$

for $t \in \mathcal{I}$ with left-sided limits $\mu(t_i-)$, $\mu(t-)$ and right-sided limits $\mu(t_i+)$ at the jump points t_i of μ, respectively, and μ_s is singular. Herein, μ_s is called *singular* if it is a non-constant continuous function of bounded variation, whose derivative is zero almost everywhere. Note that a singular function is not absolutely continuous since then it would be constant.

If μ is continuous, then μ_d is zero. If μ is continuous and monotonically increasing, then μ_a and μ_s are also monotonically increasing.

The Stieltjes integral in Theorem 2.1.22 generalizes the Riemann integral.

Definition 2.1.24 (Riemann–Stieltjes integral). Let $f, \mu : \mathcal{I} \longrightarrow \mathbb{R}$ be two functions. For $m \in \mathbb{N}$, let a partition

$$\mathbb{G}_m := \{t_0 < t_1 < \cdots < t_m = t_f\}$$

of $\mathcal{I}$ be given. Let $\xi_i \in [t_{i-1}, t_i]$ be arbitrary and

$$S_m(f, \mu) := \sum_{i=1}^{m} f(\xi_i)(\mu(t_i) - \mu(t_{i-1})).$$

$S(f, \mu)$ is called *a Riemann–Stieltjes integral of f with respect to μ*, and we write

$$S(f, \mu) = \int_{t_0}^{t_f} f(t)d\mu(t),$$

if for every $\varepsilon > 0$, there exists $\delta > 0$ such that for every partition $\mathbb{G}_m$, $m \in \mathbb{N}$, with

$$\max_{0 \le i \le m-1} \{t_{i+1} - t_i\} < \delta,$$

it holds

$$|S_m(f, \mu) - S(f, \mu)| < \varepsilon.$$

In case of vector-valued functions $f, \mu : \mathcal{I} \longrightarrow \mathbb{R}^n$, we use the abbreviation

$$\int_{t_0}^{t_f} f(t)^\top d\mu(t) := \sum_{i=1}^{n} \int_{t_0}^{t_f} f_i(t) d\mu_i(t).$$

Note that the convergence of the Riemann–Stieltjes integral in Definition 2.1.24 is supposed to be independent of the choice of the points ξ_i.

The existence of the Riemann–Stieltjes integral is guaranteed if f is continuous and μ is of bounded variation on $\mathcal{I}$. Notice that the Riemann integral is a special case of the Riemann–Stieltjes integral for $\mu(t) = t$. Some properties of the Riemann–Stieltjes integral are summarized below, see [253, 327].

(a) If one of the integrals $\int_{t_0}^{t_f} f(t) d\mu(t)$ or $\int_{t_0}^{t_f} \mu(t) df(t)$ exists, so does the other, and it holds

$$\int_{t_0}^{t_f} f(t) d\mu(t) + \int_{t_0}^{t_f} \mu(t) df(t) = [f(t)\mu(t)]_{t_0}^{t_f}.$$

This is the *partial integration rule.*

(b) It holds

$$\int_{t_0}^{t_f} d\mu(t) = \mu(t_f) - \mu(t_0).$$

If μ is constant, then

$$\int_{t_0}^{t_f} f(t) d\mu(t) = 0.$$

(c) If f is continuous and μ is of bounded variation in $\mathcal{I}$, then

$$F(t) = \int_{t_0}^{t} f(\tau) d\mu(\tau), \quad t \in \mathcal{I},$$

is of bounded variation. Moreover,

$$F(t+) - F(t) = f(t)(\mu(t+) - \mu(t)), \quad t_0 \le t < t_f, \tag{2.2}$$

$$F(t) - F(t-) = f(t)(\mu(t) - \mu(t-)), \quad t_0 < t \le t_f. \tag{2.3}$$

(d) If f is continuous, $g \in L_1(\mathcal{I})$, and

$$\mu(t) = \int_c^t g(\tau)d\tau, \quad t_0 \le c \le t_f, \; t_0 \le t \le t_f,$$

then μ is of bounded variation on (t_0, t_f) and

$$\int_{t_0}^{t_f} f(t)d\mu(t) = \int_{t_0}^{t_f} f(t)g(t)dt = \int_{t_0}^{t_f} f(t)\dot{\mu}(t)dt.$$

The latter integral is the Lebesgue integral.
If g is continuous, h is of bounded variation in $\mathcal{I}$, and

$$\mu(t) = \int_c^t g(\tau)dh(\tau), \quad t_0 \le c \le t_f, \; t_0 \le t \le t_f,$$

then

$$\int_{t_0}^{t_f} f(t)d\mu(t) = \int_{t_0}^{t_f} f(t)g(t)dh(t).$$

(e) If f is of bounded variation and μ is absolutely continuous on $\mathcal{I}$, then

$$\int_{t_0}^{t_f} f(t)d\mu(t) = \int_{t_0}^{t_f} f(t)\dot{\mu}(t)dt,$$

where the integral on the right is the Lebesgue integral.

(e) If f is continuous and μ is monotonically increasing, then there exists $\xi \in \mathcal{I}$ such that

$$\int_{t_0}^{t_f} f(t)d\mu(t) = f(\xi)(\mu(b) - \mu(a)).$$

This is a *mean-value theorem.*

Lemma 2.1.25. *Let f be continuous and μ monotonically increasing on $\mathcal{I}$. Let μ be differentiable at t. Then it holds*

$$\frac{d}{dt}\int_{t_0}^{t} f(s)d\mu(s) = f(t)\frac{d}{dt}\mu(t).$$

Proof. Let μ be differentiable at t. Define

$$F(t) := \int_{t_0}^{t} f(s)d\mu(s).$$

The mean-value theorem yields

$$F(t+h) - F(t) = \int_{t}^{t+h} f(s)d\mu(s) = f(\xi_h)(\mu(t+h) - \mu(t)), \quad t \le \xi_h \le t+h.$$

Hence,

$$\lim_{h\to 0}\frac{F(t+h) - F(t)}{h} = \lim_{h\to 0} f(\xi_h)\cdot\frac{\mu(t+h) - \mu(t)}{h} = f(t)\cdot\frac{d}{dt}\mu(t).$$

Observe that $t \le \xi_h \le t+h \longrightarrow t$ holds. □

2.2 The DAE optimal control problem as an infinite optimization problem

This section aims to provide the connection between DAE optimal control problems and infinite optimization problems. In fact, the DAE optimal control problem is an infinite optimization problem with a special structure. For the upcoming theoretical investigations, it is convenient to restrict the general DAE optimal control problem 1.1.8 to standard form. To this end, the discussion is restricted to DAE optimal control problems on a fixed time interval $[t_0, t_f]$. In Section 1.2, transformation techniques are summarized, which can be used to transform various problem settings, including Problem 1.1.8, into the following standard form:

Problem 2.2.1 (DAE optimal control problem in standard form). Let $\mathcal{I} := [t_0, t_f] \subset \mathbb{R}$ be a non-empty compact time interval with $t_0 < t_f$ fixed. Let

$$\begin{aligned}
\varphi &: \mathbb{R}^{n_x} \times \mathbb{R}^{n_x} \longrightarrow \mathbb{R},\\
f_0 &: \mathcal{I} \times \mathbb{R}^{n_x} \times \mathbb{R}^{n_y} \times \mathbb{R}^{n_u} \longrightarrow \mathbb{R},\\
f &: \mathcal{I} \times \mathbb{R}^{n_x} \times \mathbb{R}^{n_y} \times \mathbb{R}^{n_u} \longrightarrow \mathbb{R}^{n_x},
\end{aligned}$$

$$g : \mathcal{I} \times \mathbb{R}^{n_x} \times \mathbb{R}^{n_y} \times \mathbb{R}^{n_u} \longrightarrow \mathbb{R}^{n_y},$$
$$c : \mathcal{I} \times \mathbb{R}^{n_x} \times \mathbb{R}^{n_y} \times \mathbb{R}^{n_u} \longrightarrow \mathbb{R}^{n_c},$$
$$s : \mathcal{I} \times \mathbb{R}^{n_x} \longrightarrow \mathbb{R}^{n_s},$$
$$\psi : \mathbb{R}^{n_x} \times \mathbb{R}^{n_x} \longrightarrow \mathbb{R}^{n_\psi}$$

be sufficiently smooth functions and $\mathcal{U} \subseteq \mathbb{R}^{n_u}$ be a set.

Minimize the objective function

$$\varphi(x(t_0), x(t_f)) + \int_{t_0}^{t_f} f_0(t, x(t), y(t), u(t))dt \tag{2.4}$$

with respect to

$$x \in W_{1,\infty}^{n_x}(\mathcal{I}), \quad y \in L_{\infty}^{n_y}(\mathcal{I}), \quad u \in L_{\infty}^{n_u}(\mathcal{I}),$$

subject to the semi-explicit DAE

$$\dot{x}(t) = f(t, x(t), y(t), u(t)), \tag{2.5}$$
$$0_{\mathbb{R}^{n_y}} = g(t, x(t), y(t), u(t)), \tag{2.6}$$

the mixed control-state constraint

$$c(t, x(t), y(t), u(t)) \le 0_{\mathbb{R}^{n_c}}, \tag{2.7}$$

the pure state constraint

$$s(t, x(t)) \le 0_{\mathbb{R}^{n_s}}, \tag{2.8}$$

the boundary condition

$$\psi(x(t_0), x(t_f)) = 0_{\mathbb{R}^{n_\psi}}, \tag{2.9}$$

and the set constraint

$$u(t) \in \mathcal{U}. \tag{2.10}$$

Suppose the DAE is not given in semi-explicit form, but in the general form (1.5). In that case, it is a nontrivial task to identify the differential variables and algebraic variables in the vector z, to assign appropriate smoothness properties and the corresponding function spaces. This assignment may even change with time, depending on the current state z and the control u. A nice idea to define appropriate function spaces was used in [17] for the quasi-linear DAE

$$B\dot{z}(t) = f(t, z(t), u(t)),$$

where B is a fixed $n_z \times n_z$-matrix. In this case, one may use the space

$$\hat{W}_{1,\infty}^{n_z}(\mathcal{I}) := \{z \in L_{\infty}^{n_z}(\mathcal{I}) \mid Bz \in W_{1,\infty}^{n_z}(\mathcal{I})\}$$

for the state vector z.

For the derivation of first-order necessary optimality conditions, it is convenient to rewrite the DAE optimal control problem 2.2.1 as an infinite optimization problem in appropriate Banach spaces. Necessary optimality conditions of Fritz John type are then applied to the latter. The exploitation of the special structure of the Fritz John conditions eventually leads to a local minimum principle for the DAE optimal control problem in Chapter 3. To this end, the vector $z := (x, y, u)$ of optimization variables in Problem 2.2.1 is an element of the Banach space $(Z, \|\cdot\|_Z)$ defined by

$$Z := W_{1,\infty}^{n_x}(\mathcal{I}) \times L_{\infty}^{n_y}(\mathcal{I}) \times L_{\infty}^{n_u}(\mathcal{I})$$
$$\|(x, y, u)\|_Z := \max\{\|x\|_{1,\infty}, \|y\|_\infty, \|u\|_\infty\}.$$

The objective function in (2.4) defines the mapping $J : Z \longrightarrow \mathbb{R}$ with

$$J(x, y, u) := \varphi(x(t_0), x(t_f)) + \int_{t_0}^{t_f} f_0(t, x(t), y(t), u(t))dt.$$

The equality constraints (2.5), (2.6), (2.9) of the optimal control problem define the operator equation

$$H(x, y, u) = \Theta_V,$$

where $H = (H_1, H_2, H_3) : Z \longrightarrow V$ is given by

$$H_1(x, y, u) := f(\cdot, x(\cdot), y(\cdot), u(\cdot)) - \dot{x}(\cdot),$$
$$H_2(x, y, u) := g(\cdot, x(\cdot), y(\cdot), u(\cdot)),$$
$$H_3(x, y, u) := -\psi(x(t_0), x(t_f)),$$

and the Banach space $(V, \|\cdot\|_V)$ is defined by

$$V := L_{\infty}^{n_x}(\mathcal{I}) \times L_{\infty}^{n_y}(\mathcal{I}) \times \mathbb{R}^{n_\psi},$$
$$\|(v_1, v_2, v_3)\|_V := \max\{\|v_1\|_\infty, \|v_2\|_\infty, \|v_3\|_2\}.$$

---

i It has to be pointed out that the choice of the image space of the component H_2 is crucial and needs to be restricted further depending on the actual structure of the DAE (2.5), (2.6). The space $L_{\infty}^{n_y}(\mathcal{I})$ is a convenient choice for the general setting in Problem 2.2.1. In Chapter 3, however, we require the derivative operator (H_1', H_2') to be surjective. In general, surjectivity cannot be obtained for the above choice. For instance, if the DAE is an index-two Hessenberg system with u and y not appearing in g, then surjectivity can be shown only if H_2 is considered as a mapping into the space $W_{1,\infty}^{n_y}(\mathcal{I})$, compare Lemma 3.1.2.

Definition 2.2.2 (Cone, dual cone).

(a) A subset K of a vector space X is called *cone with vertex at* Θ_X if $k \in K$ implies $\alpha k \in K$ for all scalars $\alpha \geq 0$.

(b) Let K be a cone with vertex at Θ_X and $x_0 \in X$. $x_0 + K$ is called *cone with vertex at* x_0.

(c) Let $K \subseteq X$ be a subset of a Banach space X. The *positive dual cone of* K (or *positive polar cone of* K or *positive conjugate cone of* K) is defined as

$$K^+ := \{x^* \in X^* \mid x^*(k) \geq 0 \text{ for all } k \in K\}.$$

The *negative dual cone of* K (or *negative polar cone of* K or *negative conjugate cone of* K) is defined as

$$K^- := \{x^* \in X^* \mid x^*(k) \leq 0 \text{ for all } k \in K\}.$$

Functionals in K^+ are called positive on K, and functionals in K^- are called negative on K. □

The dual cones K^+ and K^- are non-empty closed convex cones. **i**

Example 2.2.3. Let $X = \mathbb{R}^2$. Then the sets

$$\begin{aligned}
K_1 &:= \{(x_1, x_2)^\top \in \mathbb{R}^2 \mid x_1 \geq 0,\ x_2 \geq 0\},\\
K_2 &:= \{(x_1, x_2)^\top \in \mathbb{R}^2 \mid x_1 \geq 0 \text{ or } x_2 \geq 0\},\\
K_3 &:= \{(x_1, x_2)^\top \in \mathbb{R}^2 \mid x_1 \geq 0\},\\
K_4 &:= \{(x_1, x_2)^\top \in \mathbb{R}^2 \mid x_1 + x_2 = 0\},\\
K_5 &:= \{(x_1, x_2)^\top \in \mathbb{R}^2 \mid x_1 > 0,\ x_2 > 0\} \cup \{0_2\}
\end{aligned}$$

are cones. K_1, K_3, K_4, K_5 are convex, K_2 is not convex. K_1, K_2, K_3, K_4 are closed, K_5 is not closed. With exception of K_4, all cones have interior points.

The following sets are closed convex cones in L_∞ and L_2, respectively:

$$\begin{aligned}
K_6 &:= \{x \in L_\infty(\mathcal{I}) \mid x(t) \geq 0 \text{ almost everywhere in } \mathcal{I}\},\\
K_7 &:= \{x \in L_2(\mathcal{I}) \mid x(t) \geq 0 \text{ almost everywhere in } \mathcal{I}\}.
\end{aligned}$$

However, only K_6 has interior points, while K_7 has no interior points! □

Define

$$\begin{aligned}
K_1 &:= \{k \in L_\infty^{n_c}(\mathcal{I}) \mid k(t) \geq 0_{\mathbb{R}^{n_c}} \text{ almost everywhere in } \mathcal{I}\},\\
K_2 &:= \{k \in \mathcal{C}^{n_s}(\mathcal{I}) \mid k(t) \geq 0_{\mathbb{R}^{n_s}} \text{ in } \mathcal{I}\}.
\end{aligned}$$

Then, $K := K_1 \times K_2$ is a convex cone with non-empty interior in the Banach space $(W, \|\cdot\|_W)$, which is defined by

$$W := L_\infty^{n_c}(\mathcal{I}) \times \mathcal{C}^{n_s}(\mathcal{I}),$$

$$\|(w_1, w_2)\|_W := \max\{\|w_1\|_\infty, \|w_2\|_\infty\}.$$

The inequality constraints (2.7), (2.8) of the optimal control problem define the cone constraint

$$G(x, y, u) \in K,$$

where $G = (G_1, G_2) : Z \longrightarrow W$ is given by

$$G_1(x, y, u) := -c(\cdot, x(\cdot), y(\cdot), u(\cdot)),$$
$$G_2(x, y, u) := -s(\cdot, x(\cdot)).$$

Finally, define the set $S \subseteq Z$ by

$$S := W_{1,\infty}^{n_x}(\mathcal{I}) \times L_\infty^{n_y}(\mathcal{I}) \times U_{ad},$$

where

$$U_{ad} := \{u \in L_\infty^{n_u}(\mathcal{I}) \mid u(t) \in \mathcal{U} \text{ almost everywhere in } \mathcal{I}\}$$

denotes the set of admissible controls. S is closed and convex with non-empty interior if $\mathcal{U}$ is closed and convex with non-empty interior.

Summarizing, the DAE optimal control problem 2.2.1 is equivalent to

Problem 2.2.4 (Infinite optimization problem).
Minimize $J(z)$ with respect to $z = (x, y, u) \in Z$ subject to the constraints

$$G(z) \in K, \quad H(z) = \Theta_V, \quad z \in S.$$

Finally, we investigate differentiability properties of the mappings J, H, and G. To this end, our standing smoothness assumptions for the functions in Problem 2.2.1 are given in Assumption 2.2.5, compare [169, Examples 9, 11]. Notice that we do not assume continuity with respect to the time component, but only measurability. This will be important in Section 7.1.

Assumption 2.2.5. Let the functions $\varphi, \psi, f_0, f, g, c, s$ in Problem 2.2.1 satisfy the following smoothness conditions:

(a) φ and ψ are continuously differentiable with respect to all arguments.

(b) Let $\hat{x} \in W_{1,\infty}^{n_x}(\mathcal{I})$, $\hat{y} \in L_\infty^{n_y}(\mathcal{I})$, $\hat{u} \in L_\infty^{n_u}(\mathcal{I})$ be given, and let M be a sufficiently large convex compact neighborhood of

$$\{(\hat{x}(t), \hat{y}(t), \hat{u}(t)) \in \mathbb{R}^{n_x+n_y+n_u} \mid t \in \mathcal{I}\}.$$

(i) The mappings $t \mapsto s(t,x)$ and

$$t \mapsto f_0(t,x,y,u), \quad t \mapsto f(t,x,y,u),$$
$$t \mapsto g(t,x,y,u), \quad t \mapsto c(t,x,y,u)$$

are measurable for every $(x,y,u) \in M$.

(ii) The mappings $x \mapsto s(t,x)$ and

$$(x,y,u) \mapsto f_0(t,x,y,u), \quad (x,y,u) \mapsto f(t,x,y,u),$$
$$(x,y,u) \mapsto g(t,x,y,u), \quad (x,y,u) \mapsto c(t,x,y,u)$$

are continuously differentiable in M uniformly for $t \in \mathcal{I}$.

(iii) The derivatives

$$f'_{0,(x,y,u)}, \quad f'_{(x,y,u)}, \quad g'_{(x,y,u)}, \quad c'_{(x,y,u)}$$

are bounded in $\mathcal{I} \times M$. □

Under Assumption 2.2.5, Fréchet-differentiability can be obtained for J, G, and H in a similar way as illustrated in the following result for a simplified setting. It also indicates a possible choice of M.

Theorem 2.2.6. *Let $\hat{x} \in W^{n_x}_{1,\infty}(\mathcal{I})$ be given, and let $f : \mathcal{I} \times \mathbb{R}^{n_x} \longrightarrow \mathbb{R}^{n_x}$, $(t,x) \mapsto f(t,x)$, be a function satisfying the conditions in Assumption 2.2.5 with*

$$M := \{x \in \mathbb{R}^{n_x} \mid \text{there exists } t \in \mathcal{I} \text{ with } \|x - \hat{x}(t)\| \le r\}, \quad r > 0.$$

Then the mapping $T : W^{n_x}_{1,\infty}(\mathcal{I}) \longrightarrow L^{n_x}_{\infty}(\mathcal{I})$ defined by

$$T(x(\cdot)) := \dot{x}(\cdot) - f(\cdot, x(\cdot))$$

is continuously Fréchet-differentiable in $\hat{x}$ with

$$T'(\hat{x})(x) = \dot{x}(\cdot) - f'_x(\cdot, \hat{x}(\cdot))x(\cdot).$$

Proof. (a) Apparently $T'(\hat{x})(\cdot)$ is linear. It remains to show the continuity of the operator.

For almost every $t \in \mathcal{I}$ and $x \in W^{n_x}_{1,\infty}(\mathcal{I})$, it holds

$$\begin{aligned}\|T'(\hat{x})(x)(t)\| &\le \|\dot{x}(t)\| + \|f'_x(t, \hat{x}(t))\| \cdot \|x(t)\| \\ &\le \|\dot{x}\|_\infty + C \cdot \|x\|_\infty \\ &\le (1 + C) \cdot \|x\|_{1,\infty}.\end{aligned}$$

Hence, $T'(\hat{x})(\cdot)$ is continuous. As f_x' is supposed to be continuous with respect to x and bounded in M uniformly with respect to $t \in \mathcal{I}$, $T'(\cdot)$ is continuous in M.

(b) It remains to show

$$\lim_{\|x\|_{1,\infty}\to 0} \frac{\|T(\hat{x}+x) - T(\hat{x}) - T'(\hat{x})(x)\|_\infty}{\|x\|_{1,\infty}} = 0.$$

For almost every $t \in \mathcal{I}$, it holds by the mean-value theorem

$$\begin{aligned}\Delta(x)(t) &:= (T(\hat{x}+x) - T(\hat{x}) - T'(\hat{x})(x))(t)\\ &= -(f(t,\hat{x}(t)+x(t)) - f(t,\hat{x}(t))) + f_x'(t,\hat{x}(t))x(t)\\ &= -\int_0^1 (f_x'(t,\hat{x}(t)+\tau x(t)) - f_x'(t,\hat{x}(t)))x(t)d\tau\end{aligned}$$

and thus

$$\|\Delta(x)(t)\| \le \sup_{\tau\in[0,1]} \|f_x'(t,\hat{x}(t)+\tau x(t)) - f_x'(t,\hat{x}(t))\| \cdot \|x\|_{1,\infty}.$$

According to (b), (ii) in Assumption 2.2.5, $f_x'(t,\cdot)$ is uniformly continuous on the compact set M. Hence, for every $\varepsilon > 0$, there exists $\delta \in (0,r]$ with

$$\|f_x'(t,x_1) - f_x'(t,x_2)\| \le \varepsilon \quad \text{for all } x_1,x_2 \in M,\ t \in \mathcal{I},\ \|x_1 - x_2\| \le \delta.$$

Let $\|x\|_{1,\infty} \le \delta$ and $\tau \in [0,1]$. Then,

$$\hat{x}(t), \hat{x}(t) + \tau x(t) \in M \quad \text{for all } t \in \mathcal{I}$$

and

$$\|\hat{x}(t) + \tau x(t) - \hat{x}(t)\| = \tau\|x(t)\| \le \|x\|_{1,\infty} \le \delta.$$

The choice of δ implies

$$\|\Delta(x)\|_\infty \le \varepsilon\|x\|_{1,\infty} \quad \text{for all } \|x\|_{1,\infty} \le \delta,$$

and hence

$$\lim_{\|x\|_{1,\infty}\to 0} \frac{\|T(\hat{x}+x) - T(\hat{x}) - T'(\hat{x})(x)\|_\infty}{\|x\|_{1,\infty}} = 0.$$

□

Similar arguments can be used to prove Fréchet-differentiability of J, G, and H under Assumption 2.2.5. The derivatives are then given by

$$J'(\hat{x},\hat{y},\hat{u})(x,y,u) = \varphi'_{x_0}x(t_0) + \varphi'_{x_f}x(t_f) + \int_{t_0}^{t_f} f'_{0,x}[t]x(t) + f'_{0,y}[t]y(t) + f'_{0,u}[t]u(t)dt, \quad (2.11)$$

and

$$H'_1(\hat{x},\hat{y},\hat{u})(x,y,u) = f'_x[\cdot]x(\cdot) + f'_y[\cdot]y(\cdot) + f'_u[\cdot]u(\cdot) - \dot{x}(\cdot), \quad (2.12)$$
$$H'_2(\hat{x},\hat{y},\hat{u})(x,y,u) = g'_x[\cdot]x(\cdot), \quad (2.13)$$
$$H'_3(\hat{x},\hat{y},\hat{u})(x,y,u) = -\psi'_{x_0}x(t_0) - \psi'_{x_f}x(t_f), \quad (2.14)$$

and

$$G'_1(\hat{x},\hat{y},\hat{u})(x,y,u) = -c'_x[\cdot]x(\cdot) - c'_y[\cdot]y(\cdot) - c'_u[\cdot]u(\cdot), \quad (2.15)$$
$$G'_2(\hat{x},\hat{y},\hat{u})(x,y,u) = -s'_x[\cdot]x(\cdot), \quad (2.16)$$

compare also [179, pp. 94–95], where continuity with respect to the argument t was assumed.

For notational convenience, we used the abbreviations

$$\varphi'_{x_0} := \varphi'_{x_0}(\hat{x}(t_0), \hat{x}(t_f)), \quad f'_x[t] := f'_x(t, \hat{x}(t), \hat{y}(t), \hat{u}(t)),$$

and in a similar way $\varphi'_{x_f}, f'_{0,x}[t], f'_{0,y}[t], f'_{0,u}[t], c'_x[t], c'_y[t], c'_u[t], s'_x[t], f'_y[t], f'_u[t], g'_x[t], \psi'_{x_0}, \psi'_{x_f}$ for the respective derivatives.

2.3 Necessary conditions for infinite optimization problems

Problem 2.2.4 is a special case of the following general infinite optimization problem:

Problem 2.3.1 (General optimization problem). Let Z be a Banach space, $\Sigma \subseteq Z$ a non-empty set, and $J : Z \longrightarrow \mathbb{R}$ a functional.

Minimize $J(z)$ with respect to $z \in \Sigma$.

The following terminology is used:

(a) J is called *objective function*.

(b) A vector z is called *admissible* or *feasible* for Problem 2.3.1 if $z \in \Sigma$. Σ is called *admissible set* or *feasible set* for Problem 2.3.1.

(c) $\hat{z} \in \Sigma$ is called *global minimum* of Problem 2.3.1 if

$$J(\hat{z}) \leq J(z) \quad \text{for all } z \in \Sigma. \quad (2.17)$$

$\hat{z} \in Z$ is called *strict global minimum* of Problem 2.3.1 if '<' holds in (2.17) for all $z \in \Sigma$, $z \neq \hat{z}$.

(d) $\hat{z} \in \Sigma$ is called *local minimum* of Problem 2.3.1 if there exists $\varepsilon > 0$ such that

$$J(\hat{z}) \leq J(z) \quad \text{for all } z \in \Sigma \cap B_\varepsilon(\hat{z}). \tag{2.18}$$

$\hat{z} \in \Sigma$ is called *strict local minimum* of Problem 2.3.1 if '<' holds in (2.18) for all $z \in \Sigma \cap B_\varepsilon(\hat{z})$, $z \neq \hat{z}$.

(e) Problem 2.3.1 is called *unconstrained* if $\Sigma = Z$ holds.

(f) Problem 2.3.1 is called *convex* if J and Σ are convex.

For a convex optimization problem, every local minimum is a global one. This can be seen as follows. Assume $\hat{z}$ is a local minimum, but not a global one. Then there exists $z \in \Sigma$ with $J(z) < J(\hat{z})$. The convexity of J yields

$$J\big(\alpha z + (1-\alpha)\hat{z}\big) \leq \alpha J(z) + (1-\alpha)J(\hat{z}) < J(\hat{z}) \quad \text{for all } 0 < \alpha \leq 1,$$

which contradicts the local minimality of $\hat{z}$.

As it was shown in Section 2.2, the structure in Problem 2.2.4 is of special interest for the investigation of optimal control problems in view of necessary conditions and numerical algorithms. The admissible set for Problem 2.2.4 is given by

$$\Sigma = S \cap G^{-1}(K) \cap H^{-1}(\Theta_V),$$

where $G^{-1}(K) := \{z \in Z \mid G(z) \in K\}$ denotes the pre-image of K under G, and $H^{-1}(\Theta_V) := \{z \in Z \mid H(z) = \Theta_V\}$ is the pre-image of Θ_V under H. Problem 2.2.4 is referred to as standard infinite optimization problem:

Problem 2.3.2 (Standard infinite optimization problem). Let Banach spaces $(Z, \|\cdot\|_Z)$, $(V, \|\cdot\|_V)$, $(W, \|\cdot\|_W)$ be given. Let $J : Z \longrightarrow \mathbb{R}$ be a functional, $G : Z \longrightarrow W$, $H : Z \longrightarrow V$ operators, $S \subseteq Z$ a closed convex set, and $K \subseteq W$ a closed convex cone with vertex at Θ_W.

Minimize $J(z)$ with respect to $z \in S$ subject to the constraints

$$G(z) \in K, \quad H(z) = \Theta_V.$$

Standard nonlinear programs in finite-dimensional Euclidean spaces are special instances of Problem 2.3.2 with the particular Banach spaces

$$Z := \mathbb{R}^{n_z}, \quad W := \mathbb{R}^{n_G}, \quad V := \mathbb{R}^{n_H},$$

and the closed convex cone $K \subseteq \mathbb{R}^{n_G}$ with vertex at $0_{\mathbb{R}^{n_G}}$ and non-empty interior $\text{int}(K) \neq \emptyset$ defined by

$$K := \{k \in \mathbb{R}^{n_G} \mid k \leq 0_{\mathbb{R}^{n_G}}\}.$$

The constraint $G(z) \in K$ in Problem 2.3.2 is then equivalent to the *inequality constraint* $G(z) \leq 0_{\mathbb{R}^{n_G}}$. Due to this relation, the cone constraint $G(z) \in K$ is also referred to as *inequality constraint*.

Direct discretization methods for DAE optimal control problems immediately lead to finite nonlinear optimization problems. For this reason, we will investigate standard finite nonlinear optimization problems in detail.

Problem 2.3.3 (Standard finite optimization problem). Let

$$\begin{aligned} J &: \mathbb{R}^{n_z} \longrightarrow \mathbb{R}, \\ G = (G_1, \dots, G_{n_G})^\top &: \mathbb{R}^{n_z} \longrightarrow \mathbb{R}^{n_G}, \\ H = (H_1, \dots, H_{n_H})^\top &: \mathbb{R}^{n_z} \longrightarrow \mathbb{R}^{n_H} \end{aligned}$$

be continuously differentiable functions and $S \subseteq \mathbb{R}^{n_z}$ a closed convex set.

Minimize $J(z)$ with respect to $z \in S$ subject to the constraints

$$\begin{aligned} G_i(z) &\leq 0, \quad i = 1, \dots, n_G, \\ H_i(z) &= 0, \quad i = 1, \dots, n_H. \end{aligned}$$

The admissible set of Problem 2.3.3 is given by

$$\Sigma = \{z \in S \mid G(z) \leq 0_{\mathbb{R}^{n_G}}, H(z) = 0_{\mathbb{R}^{n_H}}\}.$$

The active constraints are particularly important.

Definition 2.3.4 (Index set of active inequality constraints). Let $\hat{z} \in \mathbb{R}^{n_z}$ be admissible for Problem 2.3.3. The index set

$$A(\hat{z}) := \{i \in \{1, \dots, n_G\} \mid G_i(\hat{z}) = 0\}$$

is called *index set of active inequality constraints at $\hat{z}$*. □

In order to derive the necessary conditions for a local minimum of Problem 2.3.2, the set $S \subseteq Z$ usually cannot be an arbitrary set but has to fulfill additional conditions, e. g., S has to be closed and convex with non-empty interior. Nevertheless, there are also practically important problems, e. g., optimal control problems or mixed integer optimization problems, where S is a discrete set. Fortunately, it is also possible to derive necessary conditions for such problems, but among other things, additional convexity or differentiability conditions have to be imposed on the remaining constraints and the objective function. Application of these necessary conditions to optimal control problems will result in the famous global minimum principle for optimal control problems in Chapter 7.

2.3.1 Existence of a solution

The existence of a solution for Problem 2.3.1, where $\Sigma \subseteq Z$ is a compact set and J is lower semi-continuous, is ensured by the famous Weierstraß Theorem. Herein, Σ is called *compact* if, for every sequence $\{z_n\}_{n\in\mathbb{N}}$ in Σ, there exists a subsequence $\{z_{n_k}\}_{k\in\mathbb{N}}$ converging to an element $z \in \Sigma$. In finite-dimensional spaces, compactness is equivalent with closedness and boundedness. In infinite-dimensional spaces, this is not true. For instance, it is shown in [215, p. 40] that the unit sphere, in general, is not compact.

Definition 2.3.5 (Lower semi-continuity, upper semi-continuity). A functional $J : Z \longrightarrow \mathbb{R}$ is called *upper semi-continuous at* z if, for every sequence $\{z_n\}_{n\in\mathbb{N}}$ with $z_n \longrightarrow z$ as $n \longrightarrow \infty$, it holds

$$\limsup_{n\to\infty} J(z_n) \leq J(z).$$

A functional $J : Z \longrightarrow \mathbb{R}$ is called *lower semi-continuous at* z if, for every sequence $\{z_n\}_{n\in\mathbb{N}}$ with $z_n \longrightarrow z$ as $n \longrightarrow \infty$, it holds

$$J(z) \leq \liminf_{n\to\infty} J(z_n).$$

□

Theorem 2.3.6 (Weierstraß). *Let Σ be a compact subset of a normed vector space Z, and let $J : \Sigma \longrightarrow \mathbb{R}$ be lower semi-continuous. Then J achieves its minimum on Σ.*

Proof. (See [215, p. 40]). Assume that J is not bounded from below on Σ. Then there is a sequence $\{z_n\}_{n\in\mathbb{N}}$ in Σ with $J(z_n) \leq -n$. Since Σ is compact, there exists a convergent subsequence

$$\lim_{k\to\infty} z_{n_k} = \hat{z}$$

with $J(z_{n_k}) \leq -n_k$ for all $k \in \mathbb{N}$. Since J is lower semi-continuous, it follows

$$J(\hat{z}) \leq \liminf_{k\to\infty} J(z_{n_k}).$$

Hence, $\{J(z_{n_k})\}_{k\in\mathbb{N}}$ is bounded from below by $J(\hat{z}) \in \mathbb{R}$. This contradicts

$$J(z_{n_k}) \leq -n_k \longrightarrow -\infty.$$

Hence, J is bounded from below on Σ.

This, in turn, implies that

$$\hat{J} := \inf_{z\in\Sigma} J(z)$$

is a real number, and for any $n \in \mathbb{N}$, there exists $z_n \in \Sigma$ with $J(z_n) \leq \hat{J} + \frac{1}{n}$. Since Σ is compact, there exists a convergent subsequence $z_{n_k} \longrightarrow \hat{z}$ with

$$J(z_{n_k}) \leq \hat{J} + \frac{1}{n_k} \quad \text{for all } k \in \mathbb{N}.$$

Since J is lower semi-continuous, it follows

$$\hat{J} \leq J(\hat{z}) \leq \liminf_{k \to \infty} J(z_{n_k}) \leq \hat{J}.$$

Hence, J assumes its minimum on Σ. □

A generalization can be found in [6]:

Theorem 2.3.7. *Let $\Sigma \subseteq Z$, and let $J : \Sigma \longrightarrow \mathbb{R}$ be a lower semi-continuous function on Σ. Let the set*

$$lev(J, J(\tilde{z})) \cap \Sigma = \{z \in \Sigma \mid J(z) \leq J(\tilde{z})\}$$

be non-empty and compact for some $\tilde{z} \in \Sigma$. Then J achieves its minimum on Σ.

Proof. According to the Theorem of Weierstraß, there exists $\hat{z} \in lev(J, J(\tilde{z})) \cap \Sigma$ with $J(\hat{z}) \leq J(z)$ for all $z \in lev(J, J(\tilde{z})) \cap \Sigma$. For

$$z \in \Sigma \setminus (lev(J, J(\tilde{z})) \cap \Sigma) = \Sigma \setminus lev(J, J(\tilde{z})),$$

it holds

$$J(z) > J(\tilde{z}) \geq J(\hat{z}).$$

Hence, $\hat{z}$ is a minimum of J on Σ. □

Admittedly, the assumptions of the above existence results are difficult to verify for a given optimization problem like the optimal control problem 2.2.1. More specific existence results for certain optimal control problems can be found in [263] and the literature cited therein.

2.3.2 Conic approximation of sets

Conic approximations to sets play an important role in the formulation of necessary conditions for constrained optimization problems. We will summarize some important cones. To this end, it is convenient to define the arithmetic operations '+', '−', and '·' for given subsets A and B of a vector space X and a scalar $\lambda \in \mathbb{R}$ as follows:

$$A \pm B := \{a \pm b \mid a \in A,\ b \in B\}, \quad \lambda A := \{\lambda a \mid a \in A\}.$$

Note: If A and B are convex, then $A + B$ and λA are also convex.

Definition 2.3.8 (Conic hull). Let $A \subseteq X$ be a subset of a vector space X and $x \in A$. The set

$$\operatorname{cone}(A,x) := \{\alpha(a - x) \mid \alpha \geq 0,\ a \in A\} = \mathbb{R}_+ \cdot (A - \{x\})$$

is called *conic hull of* $A - \{x\}$. □

The definition of a conic hull immediately implies that the conic hull of $A - \{x\}$ can be written as

$$\operatorname{cone}(A,x) = \{a - \alpha x \mid \alpha \geq 0,\ a \in A\} = A - \{\alpha x \mid \alpha \geq 0\} = A - \mathbb{R}_+ \cdot \{x\},$$

if A is a cone with vertex at Θ_X. Moreover, if A is a convex cone with vertex at Θ_X, then it holds

$$\operatorname{cone}(A,x) = A + \{\alpha x \mid \alpha \in \mathbb{R}\} = A + \mathbb{R} \cdot \{x\}.$$

To see this, note that $\operatorname{cone}(A,x) \subseteq A + \mathbb{R} \cdot \{x\}$. Hence, it suffices to show that given an element $y \in A + \mathbb{R} \cdot \{x\}$ it can be written as $a - \alpha x$ with $a \in A$ and $\alpha \geq 0$. So, let $y = a_1 + \alpha_1 x$, $a_1 \in A$, $\alpha_1 \in \mathbb{R}$. If $\alpha_1 \leq 0$, we are done. If $\alpha_1 > 0$, then from

$$y = a_1 + \alpha_1 x + x - x = a_1 + \underbrace{(1 + \alpha_1)x}_{=:a_2 \in A} - x = 2\underbrace{\left(\frac{1}{2}a_1 + \frac{1}{2}a_2\right)}_{\in A} - x$$

it follows $y \in \operatorname{cone}(A,x)$.

The following cone is motivated by the idea of approximating the admissible set of an optimization problem by tangential directions, see Figure 2.1.

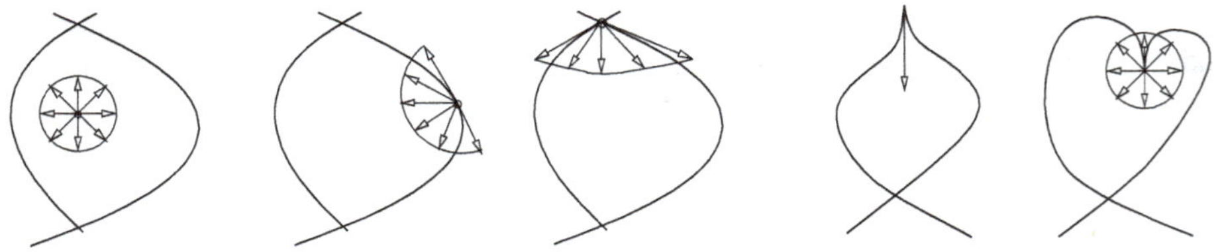

Figure 2.1: Tangent cones to different sets.

Definition 2.3.9 (Tangent cone). Let Σ be a non-empty subset of the Banach space Z. The *(sequential) tangent cone to* Σ *at* $\hat{z} \in \Sigma$ is defined by

$$T(\Sigma,\hat{z}) := \left\{ d \in Z \;\middle|\; \begin{array}{l} \text{there exist sequences } \{\alpha_k\}_{k\in\mathbb{N}},\ \alpha_k \downarrow 0, \text{ and } \{z_k\}_{k\in\mathbb{N}},\ z_k \in \\ \Sigma, \text{ with } \lim_{k\to\infty} z_k = \hat{z}, \text{ such that } \lim_{k\to\infty}(z_k - \hat{z})/\alpha_k = d \\ \text{holds.} \end{array} \right\}. \tag{2.19}$$

□

The tangent cone is a closed cone with vertex at zero. If $\hat{z}$ happens to be an interior point of Σ, then $T(\Sigma, \hat{z}) = Z$. If Σ is a convex set, so the tangent cone is convex and can be written as

$$\begin{aligned} T(\Sigma, \hat{z}) &= \{d \in Z \mid \exists \alpha_k \downarrow 0,\ d_k \longrightarrow d \ :\ \hat{z} + \alpha_k d_k \in \Sigma\} \\ &= \overline{\{d \in Z \mid \exists \alpha > 0 \ :\ \hat{z} + \alpha d \in \Sigma\}}. \end{aligned}$$

It is easy to show, see [179, p. 31], that the tangent cone can be written as

$$T(\Sigma, \hat{z}) = \left\{ d \in Z \,\middle|\, \begin{array}{l} \text{there exist } \sigma > 0 \text{ and a mapping } r : (0, \sigma] \longrightarrow Z \text{ with } \lim_{\varepsilon \downarrow 0} \frac{r(\varepsilon)}{\varepsilon} = \\ \Theta_Z \text{ and a sequence } \{\alpha_k\}_{k \in \mathbb{N}},\ \alpha_k \downarrow 0 \text{ with } \hat{z} + \alpha_k d + r(\alpha_k) \in \Sigma \text{ for all} \\ k \in \mathbb{N}. \end{array} \right\}.$$

The following theorem addresses operator equations as they appear in Problem 2.3.2.

Theorem 2.3.10 (Ljusternik, see [179, p. 40]). *Let Z and V be Banach spaces, $H : Z \longrightarrow V$ a mapping, and $\hat{z} \in M := \{z \in Z \mid H(z) = \Theta_V\}$. Let H be continuous in a neighborhood of $\hat{z}$ and continuously Gâteaux-differentiable at $\hat{z}$. Let the Gâteaux-derivative $\delta H(\hat{z})$ be surjective. Let $\hat{d} \in Z$ be given with $\delta H(\hat{z})(\hat{d}) = \Theta_V$. Then there exist $\varepsilon_0 > 0$ and a mapping*

$$r : (0, \varepsilon_0] \longrightarrow Z, \quad \lim_{\varepsilon \downarrow 0} \frac{r(\varepsilon)}{\varepsilon} = \Theta_Z$$

such that

$$H(\hat{z} + \varepsilon \hat{d} + r(\varepsilon)) = \Theta_V$$

holds for every $\varepsilon \in (0, \varepsilon_0]$. In particular, it holds

$$\{d \in Z \mid \delta H(\hat{z})(d) = \Theta_V\} = T(M, \hat{z}).$$

□

The upcoming necessary conditions involve a conic approximation of the set S and a linearization of the constraints $G(z) \in K$ and $H(z) = \Theta_V$ in Problem 2.3.2.

Definition 2.3.11 (Linearizing cone). Let Z, V, and W be Banach spaces. Let $G : Z \longrightarrow W$ and $H : Z \longrightarrow V$ be Fréchet-differentiable. The *linearizing cone of K and S at $\hat{z}$* is given by

$$T_{\text{lin}}(K, S, \hat{z}) := \{d \in \text{cone}(S, \hat{z}) \mid G'(\hat{z})(d) \in \text{cone}(K, G(\hat{z})),\ H'(\hat{z})(d) = \Theta_V\}.$$

□

Example 2.3.12 (Finite dimensional case). The linearizing cone for Problem 2.3.3 is given by

$$T_{\text{lin}}(K, S, \hat{z}) = \{d \in \text{cone}(S, \hat{z}) \mid G_i'(\hat{z})(d) \le 0,\ i \in A(\hat{z}),\ H_i'(\hat{z})(d) = 0,\ i = 1, \ldots, n_H\},$$

where $A(\hat{z})$ denotes the index set of active inequality constraints at $\hat{z}$.

□

A relation between tangent cone and linearizing cone is given by

Corollary 2.3.13. *Let G and H be Fréchet-differentiable. Let* $\text{cone}(S,\hat{z})$ *and* $\text{cone}(K,G(\hat{z}))$ *be closed. Then it holds* $T(\Sigma,\hat{z}) \subseteq T_{\text{lin}}(K,S,\hat{z})$.

Proof. Let $d \in T(\Sigma,\hat{z})$. Then there are sequences $\alpha_k \downarrow 0$ and $z_k \longrightarrow \hat{z}$ with $z_k \in S$, $G(z_k) \in K, H(z_k) = \Theta_V$ and $(z_k - \hat{z})/\alpha_k \longrightarrow d$. Since $(z_k - \hat{z})/\alpha_k \in \text{cone}(S,\hat{z})$ and $\text{cone}(S,\hat{z})$ is closed, it holds $d \in \text{cone}(S,\hat{z})$. The continuity of $H'(\hat{z})(\cdot)$ implies

$$\Theta_V = \lim_{k\to\infty} H(\hat{z})((z_k - \hat{z})/\alpha_k) = H(\hat{z})(d).$$

Furthermore, since G is Fréchet-differentiable, it holds

$$G(z_k) = G(\hat{z}) + G'(\hat{z})(z_k - \hat{z}) + \varepsilon_k \|z_k - \hat{z}\|_Z$$

with some sequence $\{\varepsilon_k\}_{k\in\mathbb{N}} \subseteq W$, $\varepsilon_k \longrightarrow \Theta_W$. Hence,

$$\underbrace{\frac{G(z_k) - G(\hat{z})}{\alpha_k}}_{\in \text{cone}(K,G(\hat{z}))} = G'(\hat{z})\left(\frac{z_k - \hat{z}}{\alpha_k}\right) + \varepsilon_k \left\|\frac{z_k - \hat{z}}{\alpha_k}\right\|_Z.$$

Since $\text{cone}(K, G(\hat{z}))$ is closed, the term on the left converges to an element in $\text{cone}(K,G(\hat{z}))$, while the term on the right converges to $G'(\hat{z})(d)$. Thus, $G'(\hat{z})(d) \in \text{cone}(K,G(\hat{z}))$. Together with $d \in \text{cone}(S,\hat{z})$ and $H'(\hat{z})(d) = \Theta_V$, this shows $d \in T_{\text{lin}}(K,S,\hat{z})$. □

A geometrically motivated first-order necessary condition for Problem 2.3.1 involves the tangent cone.

Theorem 2.3.14. *Let* $J : Z \longrightarrow \mathbb{R}$ *be Fréchet-differentiable at* $\hat{z}$*, and let* $\hat{z}$ *be a local minimum of Problem* 2.3.1. *Then,*

$$J'(\hat{z})(d) \geq 0 \quad \textit{for all } d \in T(\Sigma,\hat{z}).$$

Proof. Let $d \in T(\Sigma,\hat{z})$. Then there are sequences $\alpha_k \downarrow 0$, $z_k \longrightarrow \hat{z}$, $z_k \in \Sigma$ with $d = \lim_{k\to\infty}(z_k - \hat{z})/\alpha_k$. Since $\hat{z}$ is a local minimum and J is Fréchet-differentiable at $\hat{z}$, it follows

$$0 \leq J(z_k) - J(\hat{z}) = J'(\hat{z})(z_k - \hat{z}) + o(\|z_k - \hat{z}\|_Z).$$

Division by $\alpha_k > 0$ yields

$$0 \leq J'(\hat{z})\underbrace{\left(\frac{z_k - \hat{z}}{\alpha_k}\right)}_{\to d} + \underbrace{\left\|\frac{z_k - \hat{z}}{\alpha_k}\right\|_Z}_{\to\|d\|_Z} \cdot \underbrace{\frac{o(\|z_k - \hat{z}\|_Z)}{\|z_k - \hat{z}\|_Z}}_{\to 0}.$$

□

The necessary condition in Theorem 2.3.14 can be expressed as

$$-J'(\hat{z}) \in T(\Sigma, \hat{z})^-,$$

where $T(\Sigma, \hat{z})^-$ denotes the negative dual cone of $T(\Sigma, \hat{z})$, compare Figure 2.2.

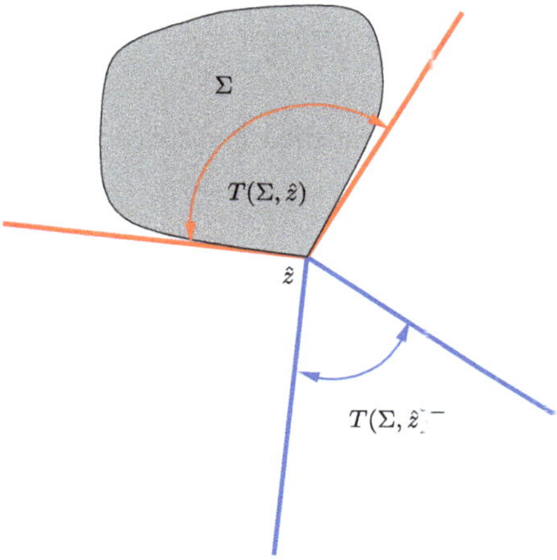

Figure 2.2: Tangent cone to Σ and its negative dual cone in $Z = \mathbb{R}^2$.

In practical applications with a complicated admissible set Σ, the tangent cone $T(\Sigma, \hat{z})$ is not assessable, and the necessary condition in Theorem 2.3.14 is of little practical use. If, however, Σ is sufficiently simple, e. g., a box, then the necessary condition can be exploited.

Example 2.3.15. Consider the following optimization problem:

Minimize

$$J(z) := \int_{-1}^{1} tz(t)dt$$

subject to the constraint

$$z(t) \in \Sigma := \left\{z \in L_\infty([-1,1]) \mid z(t) \in [-1,1] \textit{ for almost every } t \in [-1,1]\right\}.$$

J is Fréchet-differentiable at every $\hat{z} \in L_\infty([-1,1])$ with

$$J'(\hat{z})(d) = \int_{-1}^{1} td(t)dt.$$

We investigate three different candidates $\hat{z}_i$, $i = 1, 2, 3$, and check the necessary condition

$$J'(\hat{z}_i)(d) \geq 0 \quad \text{for all } d \in T(\Sigma, \hat{z}_i).$$

(a) Let $\hat{z}_1 \in \operatorname{int}(\Sigma)$. Then $T(\Sigma, \hat{z}_1) = L_\infty([-1, 1])$. Let $d(t) := -t$ for $t \in [-1, 1]$. Then,

$$J'(\hat{z}_1)(d) = \int_{-1}^{1} td(t)dt = -\int_{-1}^{1} t^2 dt = -\left[\frac{1}{3}t^3\right]_{-1}^{1} = -\frac{2}{3} < 0.$$

Hence, the necessary condition in Theorem 2.3.14 is not satisfied, and $\hat{z}_1$ is not a local minimum.

(b) Let $\hat{z}_2 \equiv 1$. Then,

$$T(\Sigma, \hat{z}_2) = \{d \in L_\infty([-1, 1]) \mid d(t) \leq 0 \text{ for almost every } t \in [-1, 1]\}.$$

Define

$$d(t) := \begin{cases} 1, & \text{if } t \in [-1, 0], \\ 0, & \text{if } t \in (0, 1]. \end{cases}$$

Then,

$$J'(\hat{z}_2)(d) = \int_{-1}^{1} td(t)dt = \int_{-1}^{0} tdt = \left[\frac{1}{2}t^2\right]_{-1}^{0} = -\frac{1}{2} < 0.$$

Hence, the necessary condition in Theorem 2.3.14 is not satisfied, and $\hat{z}_2$ is not a local minimum.

(c) Let

$$\hat{z}_3(t) := \begin{cases} 1, & \text{if } t \in [-1, 0], \\ -1, & \text{if } t \in (0, 1]. \end{cases}$$

Then,

$$T(\Sigma, \hat{z}_3) = \{d \in L_\infty([-1, 1]) \mid d(t)\hat{z}_3(t) \leq 0 \text{ for almost every } t \in [-1, 1]\},$$

and

$$J'(\hat{z}_3)(d) = \int_{-1}^{1} td(t)dt = \int_{-1}^{0} \underbrace{td(t)}_{\geq 0 \text{ a.e.}} dt + \int_{0}^{1} \underbrace{td(t)}_{\geq 0 \text{ a.e.}} dt \geq 0$$

for every $d \in T(\Sigma, \hat{z}_3)$. Hence, the necessary condition in Theorem 2.3.14 is satisfied at $\hat{z}_3$.

2.3.3 Separation theorems

The proof of the necessary Fritz John conditions is based on the separability of two sets by a hyperplane.

Definition 2.3.16 (Hyperplane). (a) Let $h : X \longrightarrow \mathbb{R}$ be a non-zero linear functional and $\gamma \in \mathbb{R}$. The set

$$M := \{x \in X \mid h(x) = \gamma\}$$

is called a *hyperplane*. The set

$$M_+ := \{x \in X \mid h(x) \geq \gamma\}$$

is called *positive halfspace of* M. Similarly,

$$M_- := \{x \in X \mid h(x) \leq \gamma\}$$

is the *negative halfspace of* M.

(b) The hyperplane M *separates* the sets A and B, if

$$h(x) \leq \gamma, \quad \text{for all } x \in A$$
$$h(x) \geq \gamma, \quad \text{for all } x \in B.$$

The hyperplane M *strictly separates* the sets A and B, if

$$h(x) < \gamma, \quad \text{for all } x \in A$$
$$h(x) > \gamma, \quad \text{for all } x \in B.$$

The hyperplane M *properly separates* the sets A and B, if not both sets are contained in M. □

Figure 2.3 illustrates some constellations in $\mathbb{R}^2$.

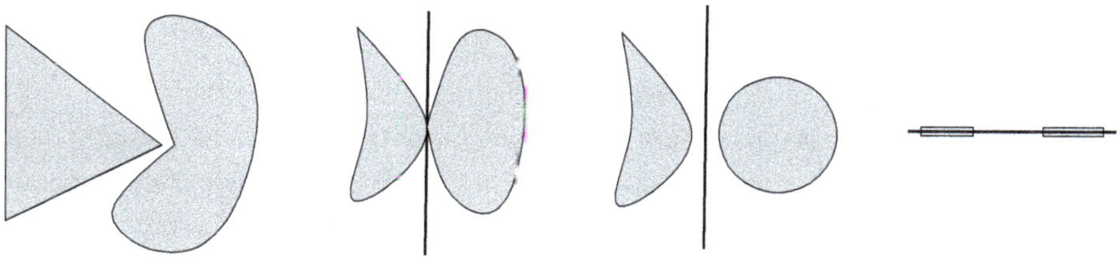

Figure 2.3: Separation of sets by hyperplanes. From left to right, the following cases are illustrated: separation by hyperplanes impossible, strict separation by hyperplanes impossible, strict separation, no proper separation.

i If h is continuous, the hyperplane is closed.

The first result addresses the problem of separating a point from a closed subspace of a Banach space X.

Theorem 2.3.17 (See [326, Cor. III.1.8, p. 98]). *Let X be a Banach space, M a closed subspace of X, and $\hat{x} \in X$, $\hat{x} \notin M$. Then there exists $h \in X^*$ with $h(\hat{x}) \neq 0$ and $h(x) = 0$ for all $x \in M$.* □

Definition 2.3.18 (Affine set, affine hull, relative interior, codimension, defect). Let $K \subseteq X$ be a subset of a vector space X.

(a) K is called *affine set*, if

$$(1-\lambda)x + \lambda y \in K \quad \text{for all } x, y \in K, \lambda \in \mathbb{R}.$$

(b) The set

$$\begin{aligned} \operatorname{aff}(K) &:= \bigcap \{M \mid M \text{ is an affine set and } K \subseteq M\} \\ &= \left\{ \sum_{i=1}^{m} \lambda_i x_i \mid m \in \mathbb{N},\ \sum_{i=1}^{m} \lambda_i = 1,\ x_i \in K,\ i = 1, \ldots, m \right\}. \end{aligned}$$

is called *affine hull* of K.

(c) The set

$$\operatorname{relint}(K) := \{x \in K \mid \text{there exists } \varepsilon > 0 \text{ with } B_\varepsilon(x) \cap \operatorname{aff}(K) \subseteq K\}$$

is called *relative interior of* K. The set $\operatorname{cl}(K) \setminus \operatorname{relint}(K)$ is called *relative boundary of* K.

(d) Let L be a linear subspace of X and $x \in X$. The set $x + L$ is called a *linear manifold*. Let $\{e_i\}_{i \in I}$ be a basis of X and $\{e_j\}_{j \in J}, J \subseteq I$, a basis of L. The *dimension* of the linear manifold $x + L$ is defined as $|J|$, and the *codimension* or *defect* of $x + L$ is defined as $|I \setminus J|$.
The codimension of the set K is defined as the codimension of the linear subspace parallel to the affine hull of K. □

The following very general result holds in vector spaces and can be found in [208, 207]. Notice, however, that relint and int are to be understood in the purely algebraic sense and that the functional defining the hyperplane is not necessarily continuous.

Theorem 2.3.19. *Let A and B be non-empty convex subsets of a vector space X.*

(a) *If $\operatorname{relint}(A) \neq \emptyset$, A has finite defect, and $\operatorname{relint}(A) \cap B = \emptyset$, then there exists a non-zero linear functional h separating A and B.*

(b) *Let* $int(A) \neq \emptyset$. *Then there exists a non-zero linear functional* h *separating* A *and* B, *if and only if* $int(A) \cap B = \emptyset$.

(c) *Let* $relint(A) \neq \emptyset$ *and* $relint(B) \neq \emptyset$. *Then there exists a non-zero linear functional* h *separating* A *and* B, *if and only if either* $A \cup B$ *are contained in one hyperplane, or* $relint(A) \cap relint(B) = \emptyset$. □

The subsequent separation theorem holds in Banach spaces and can be found in [34]. Herein, the functional defining the hyperplane is continuous and thus an element of the dual space. The proof exploits the Hahn–Banach theorem, see [326, Th. III.1.2], [215, p. 111].

Theorem 2.3.20 (See [34, Theorems 2.13, 2.14, 2.17]). *Let* A *and* B *be convex subsets of a Banach space* X.

(a) *Let* $int(A) \neq \emptyset$. *Then there exists a non-zero functional* $h \in X^*$ *separating* A *and* B, *if and only if* $int(A) \cap B = \emptyset$.

(b) *Let* A *and* B *be closed,* A *compact, and* $A \cap B = \emptyset$. *Then there exists a non-zero functional* $h \in X^*$ *separating* A *and* B *strictly.*

(c) *Let* $relint(A) \neq \emptyset$, $x_0 \in X$, *and* $x_0 \notin relint(A)$. *Then there exists a non-zero functional* $h \in X^*$ *separating* A *and* $\{x_0\}$. □

The following results are concerned with separation in $\mathbb{R}^n$ and can be found in [285, Section 11]. It provides a complete characterization of separability of convex sets in $\mathbb{R}^n$.

Theorem 2.3.21. *Let* $A, B \subseteq \mathbb{R}^n$ *be non-empty convex sets.*

(a) *There exists a hyperplane separating* A *and* B *properly, if and only if*

$$relint(A) \cap relint(B) = \emptyset.$$

(b) *There exists a hyperplane separating* A *and* B *strictly, if and only if*

$$0_{\mathbb{R}^n} \notin cl(A - B),$$

i. e. if

$$\inf\{\|a - b\| \mid a \in A,\ b \in B\} > 0.$$

□

2.3.4 First order necessary optimality conditions of Fritz John type

The DAE optimal control problem in Section 2.2 leads to the infinite optimization Problem 2.2.4 respectively Problem 2.3.2. The special structure of the admissible set Σ of Problem 2.3.2 defined by cone constraints and equality constraints, allows deriving the first-order necessary optimality conditions of Fritz John type. These necessary optimality conditions provide the basis for the minimum principle for optimal control problems.

The proof of the Fritz John conditions exploits the following well-known result, see [326, p. 135, Th. IV.3.3].

Theorem 2.3.22 (Open mapping theorem). *Let $T : Z \longrightarrow V$ be a linear, continuous, and surjective operator. Let $S \subseteq Z$ be open. Then $T(S) \subseteq V$ is open.* □

The following result can be found in a slightly more general setting in [209]:

Theorem 2.3.23 (First order necessary optimality conditions, Fritz John conditions). *Let Banach spaces $(Z, \|\cdot\|_Z)$, $(V, \|\cdot\|_V)$, $(W, \|\cdot\|_W)$ be given.*

(a) *Let $S \subseteq Z$ be a closed convex set with $int(S) \neq \emptyset$ and $K \subseteq W$ a closed convex cone with vertex at Θ_W and $int(K) \neq \emptyset$.*
(b) *Let $J : Z \longrightarrow \mathbb{R}$ and $G : Z \longrightarrow W$ be Fréchet-differentiable, and let $H : Z \longrightarrow V$ be continuously Fréchet-differentiable.*
(c) *Let $\hat{z}$ be a local minimum of Problem 2.3.2.*
(d) *Let the image of $H'(\hat{z})$ be not a proper dense subset of V.*

Then there exist nontrivial multipliers $(\ell_0, \mu^*, \lambda^*) \in \mathbb{R}\times W^* \times V^*$, $(\ell_0, \mu^*, \lambda^*) \neq (0, \Theta_{W^*}, \Theta_{V^*})$, such that

$$\ell_0 \geq 0, \tag{2.20}$$

$$\mu^* \in K^+, \tag{2.21}$$

$$\mu^*(G(\hat{z})) = 0, \tag{2.22}$$

$$\ell_0 J'(\hat{z})(d) - \mu^*(G'(\hat{z})(d)) - \lambda^*(H'(\hat{z})(d)) \geq 0, \quad \textit{for all } d \in S - \{\hat{z}\}. \tag{2.23}$$

Proof. Consider the linearized problem around $\hat{z}$:

Minimize

$$J(\hat{z}) + J'(\hat{z})(z - \hat{z})$$

with respect to $z \in S$ subject to the constraints

$$G(\hat{z}) + G'(\hat{z})(z - \hat{z}) \in K,$$
$$H(\hat{z}) + H'(\hat{z})(z - \hat{z}) = \Theta_V.$$

For the first part of the proof, we assume that the mapping $H'(\hat{z})(\cdot) : Z \longrightarrow V$ is surjective and that there exists a feasible $z \in \mathrm{int}(S)$ for the linearized problem satisfying

$$G(\hat{z}) + G'(\hat{z})(z - \hat{z}) \in \mathrm{int}(K), \tag{2.24}$$

$$H(\hat{z}) + H'(\hat{z})(z - \hat{z}) = \Theta_V, \tag{2.25}$$

$$J(\hat{z}) + J'(\hat{z})(z - \hat{z}) < J(\hat{z}). \tag{2.26}$$

(i) Since $H'(\hat{z})$ is surjective and $H'(\hat{z})(z - \hat{z}) = \Theta_V$ holds, we may apply Ljusternik's theorem 2.3.10. Hence, there exist $t_0 > 0$ and a mapping

$$r : [0, t_0] \longrightarrow Z, \quad \lim_{t \downarrow 0} \frac{r(t)}{t} = \Theta_Z, \quad r(0) := \Theta_Z,$$

such that

$$H(\hat{z} + t(z - \hat{z}) + r(t)) = \Theta_V$$

holds for every $t \in [0, t_0]$. This means that the curve

$$z(t) = \hat{z} + t(z - \hat{z}) + r(t)$$

remains feasible for the nonlinear equality constraints for every $t \in [0, t_0]$. Furthermore, it holds

$$z(0) = \hat{z}, \quad z'(0) = z - \hat{z}.$$

The latter holds because

$$z'(0) = \lim_{t \downarrow 0} \frac{z(t) - z(0)}{t} = \lim_{t \downarrow 0} \frac{z(t) - \hat{z}}{t} = z - \hat{z} + \lim_{t \downarrow 0} \frac{r(t)}{t} = z - \hat{z}.$$

(ii) Now we consider the inequality constraints at $\hat{z}$. Since G is Fréchet-differentiable at $\hat{z}$, it holds

$$\begin{aligned} G(z(t)) &= G(\hat{z}) + tG'(\hat{z})(z - \hat{z}) + o(t) \\ &= t\underbrace{\big(G(\hat{z}) + G'(\hat{z})(z - \hat{z})\big)}_{\in \operatorname{int}(K)} + (1 - t)\underbrace{G(\hat{z})}_{\in K} + o(t). \end{aligned}$$

Since $G(\hat{z}) + G'(\hat{z})(z - \hat{z}) \in \operatorname{int}(K)$, there exists $\delta_1 > 0$ such that

$$B_{\delta_1}(G(\hat{z}) - G'(\hat{z})(z - \hat{z})) \subseteq K.$$

Furthermore, the convexity of K yields

$$(1 - t)G(\hat{z}) + tw \in K$$

for all $w \in B_{\delta_1}(G(\hat{z}) + G'(\hat{z})(z - \hat{z}))$ and all $0 \le t \le 1$. Since for $t \downarrow 0$ it holds $o(t)/t \longrightarrow \Theta_W$, there exists $\delta_2 > 0$ with $\|o(t)/t\|_W < \delta_1$ for all $0 < t < \delta_2$. Hence,

$$G(z(t)) = t\underbrace{\big(G(\hat{z}) + G'(\hat{z})(z - \hat{z}) + o(t)/t\big)}_{\in B_{\delta_1}(G(\hat{z}) + G'(\hat{z})(z - \hat{z}))} + (1 - t)G(\hat{z}) \in K.$$

Hence, for sufficiently small $t > 0$ the curve $z(t)$ stays feasible for the nonlinear inequality constraints.

(iii) Now the objective function is investigated. It holds

$$\frac{d}{dt}J(z(t))\big|_{t=0} = J'(\hat{z}) \cdot \frac{dz}{dt}(0) = J'(\hat{z})(z - \hat{z}) \overset{(2.26)}{<} 0.$$

Hence, $z - \hat{z}$ is a direction of descent of J, i. e. it holds $J(z(t)) < J(\hat{z})$ for $t > 0$ sufficiently small. This will contradict the local minimality of $\hat{z}$ for Problem 2.3.2.

(iv) The point z in the linear problem fulfilling (2.24)–(2.26) is assumed to be an interior point of S, see Figure 2.4.

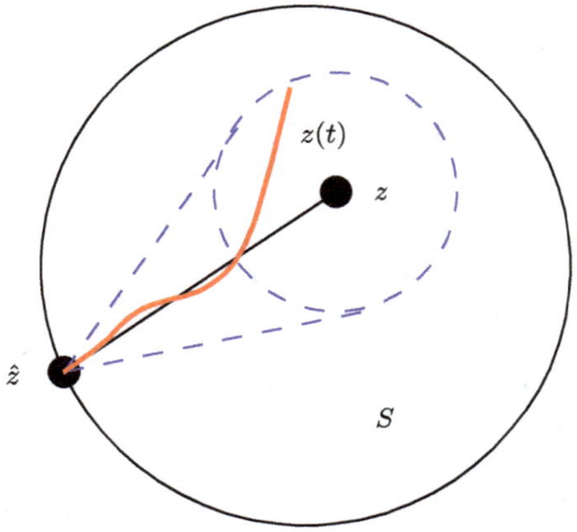

Figure 2.4: Fritz John conditions under set constraints $z \in S$. The assumption $\text{int}(S) \neq \emptyset$ is essential. For sufficiently small $t > 0$, the curve $z(t)$ stays feasible.

Then, since S is assumed to be convex, there exists a neighborhood $B_\delta(z)$ such that $\hat{z} + t(\tilde{z} - \hat{z}) \in S$ for all $0 \le t \le 1$ and all $\tilde{z} \in B_\delta(z)$. Since

$$\lim_{t \downarrow 0} \frac{r(t)}{t} = \Theta_Z,$$

there exists $\varepsilon > 0$ with

$$\left\| \frac{r(t)}{t} \right\|_Z < \delta$$

for all $0 < t < \varepsilon$. Hence,

$$z(t) = \hat{z} + t(z - \hat{z}) + r(t) = \hat{z} + t\Big(\underbrace{z + \frac{r(t)}{t}}_{\in B_\delta(z)} - \hat{z}\Big) \in S$$

for $0 < t < \varepsilon$.

Items (i)–(iv) showed the following: If $\hat{z}$ is a local minimum of Problem 2.3.2 and $H'(\hat{z})$ is surjective, then

$$J'(\hat{z})(d) \geq 0 \quad \text{for all } d \in T(\hat{z}), \tag{2.27}$$

where

$$T(\hat{z}) := \left\{ d \in \operatorname{int}(\operatorname{cone}(S,\hat{z})) \;\middle|\; \begin{array}{l} G'(\hat{z})(d) \in \operatorname{int}(\operatorname{cone}(K, G(\hat{z}))), \\ H'(\hat{z})(d) = \Theta_V \end{array} \right\}.$$

Consider the non-empty convex set

$$A := \left\{ \begin{pmatrix} J'(\hat{z})(d) + r \\ G'(\hat{z})(d) - k \\ H'(\hat{z})(d) \end{pmatrix} \;\middle|\; \begin{array}{l} d \in \operatorname{int}(\operatorname{cone}(S,\hat{z})), \\ k \in \operatorname{int}(\operatorname{cone}(K, G(\hat{z}))), \\ r > 0 \end{array} \right\}.$$

If $H'(\hat{z})$ is not surjective and $\operatorname{im}(H'(\hat{z}))$ is not a proper dense subset of V, the set

$$M := cl\left(\left\{ \begin{pmatrix} r \\ w \\ H'(\hat{z})(z) \end{pmatrix} \;\middle|\; r \in \mathbb{R},\ w \in W,\ z \in Z \right\} \right)$$

is a proper closed subspace of $\mathbb{R} \times W \times V$. According to Theorem 2.3.17, there is a non-zero functional $(\ell_0, \mu^*, \lambda^*) \in (\mathbb{R} \times W \times V)^*$ with $\ell_0 r + \mu^*(w) + \lambda^*(v) = 0$ for all $(r, w, v) \in M$. Hence, the hyperplane

$$\{(r, w, v) \in \mathbb{R} \times W \times V \mid \ell_0 r + \mu^*(w) + \lambda^*(v) = 0\}$$

trivially separates the sets A and the point $(0, \Theta_W, \Theta_V)$ since both are contained in M.

A can be decomposed into $A = A_1 + A_2$ with

$$A_1 := \left\{ \begin{pmatrix} J'(\hat{z})(d) \\ G'(\hat{z})(d) \\ H'(\hat{z})(d) \end{pmatrix} \;\middle|\; d \in \operatorname{int}(\operatorname{cone}(S,\hat{z})) \right\},$$

$$A_2 := \left\{ \begin{pmatrix} r \\ -k \\ \Theta_V \end{pmatrix} \;\middle|\; k \in \operatorname{int}(\operatorname{cone}(K, G(\hat{z}))),\ r > 0 \right\}.$$

If $H'(\hat{z})$ is surjective, the projection of A_1 onto V contains interior points in V according to the open mapping Theorem 2.3.22. Hence, $A_1 + A_2$ contains interior points in $\mathbb{R} \times W \times V$. The considerations in (i)–(iv) showed that $(0, \Theta_W, \Theta_V) \notin \operatorname{int}(A)$. Let us assume that $(0, \Theta_W, \Theta_V) \in \operatorname{int}(A)$. Then there exists $d \in \operatorname{int}(\operatorname{cone}(S,\hat{z}))$ with $J'(\hat{z})(d) < 0$, $G'(\hat{z})(d) \in \operatorname{int}(\operatorname{cone}(K, G(\hat{z})))$, and $H'(\hat{z})(d) = \Theta_V$ contradicting (2.27).

According to the separation Theorem 2.3.20, there exists a hyperplane separating A and $(0, \Theta_W, \Theta_V)$, i. e. there exist multipliers

$$(0, \Theta_{W^*}, \Theta_{V^*}) \neq (\ell_0, \mu^*, \lambda^*) \in \mathbb{R} \times W^* \times V^* = (\mathbb{R} \times W \times V)^*$$

such that

$$0 \leq \ell_0(J'(\hat{z})(d) + r) - \mu^*(G'(\hat{z})(d) - w) - \lambda^*(H'(\hat{z})(d))$$

for all $d \in \operatorname{int}(\operatorname{cone}(S, \hat{z}))$, $w \in \operatorname{int}(\operatorname{cone}(K, G(\hat{z})))$, $r > 0$. Owing to the continuity of the functionals $\ell_0(\cdot)$, $\mu^*(\cdot)$, $\lambda^*(\cdot)$ and the linear operators $J'(\hat{z})(\cdot)$, $G'(\hat{z})(\cdot)$, and $H'(\hat{z})(\cdot)$, this inequality also holds for all $d \in \operatorname{cone}(S, \hat{z})$, $w \in \operatorname{cone}(K, G(\hat{z}))$, $r \geq 0$.

Choosing $d = \Theta_Z \in \operatorname{cone}(S, \hat{z})$ and $w = \Theta_W \in \operatorname{cone}(K, G(\hat{z}))$ yields $\ell_0 r \geq 0$ for all $r \geq 0$ and thus $\ell_0 \geq 0$.

The choices $d = \Theta_Z$ and $r = 0$ imply $\mu^*(w) \geq 0$ for all $w \in \operatorname{cone}(K, G(\hat{z})) = \{k - \alpha G(\hat{z}) \mid k \in K,\ \alpha \geq 0\}$, i. e. $\mu^*(k - \alpha G(\hat{z})) \geq 0$ for all $k \in K$, $\alpha \geq 0$. This in turn implies $\mu^* \in K^+$ (choose $\alpha = 0$) and $\mu^*(G(\hat{z})) = 0$ (choose $k = \Theta_W$, $\alpha = 1$ and observe that $G(\hat{z}) \in K$). □

Definition 2.3.24. Every point $(z, \ell_0, \mu^*, \lambda^*) \in Z \times \mathbb{R} \times W^* \times V^*$, $(\ell_0, \mu^*, \lambda^*) \neq \Theta$, satisfying the Fritz John conditions (2.20)–(2.23) is called *a Fritz John point* of Problem 2.3.2.

Every Fritz John point with $\ell_0 \neq 0$ is called *a Karush–Kuhn–Tucker point (KKT point)* of Problem 2.3.2. The conditions (2.20)–(2.23) with $\ell_0 = 1$ are called *Karush–Kuhn–Tucker conditions (KKT conditions)*.

The multipliers ℓ_0, μ^*, and λ^* are called *Lagrange multipliers* or simply *multipliers*.

The conditions (2.21) and (2.22) are called *complementarity conditions*. □

Notice that $(\ell_0, \mu^*, \lambda^*) = \Theta$ trivially fulfills the Fritz John conditions. The main statement of the theorem is that there exists a *nontrivial* vector $(\ell_0, \mu^*, \lambda^*) \neq \Theta$. Unfortunately, the case $\ell_0 = 0$ may occur. In this case, the objective function J does not enter into in the Fritz John conditions. In case $\ell_0 \neq 0$, without loss of generality, ℓ_0 can be normalized to one as the multipliers enter linearly.

---

i First-order necessary optimality conditions of Fritz John type for general cone constrained problems

Minimize $J(z)$ with respect to $z \in S$ subject to the constraint $G(z) \in K$

can be found in [196]. Essentially, the existence of non-trivial multipliers defining a separating hyperplane can be guaranteed if

$$\operatorname{im}\big(G'(\hat{z})\big) + \operatorname{cone}\big(K, G(\hat{z})\big)$$

is not a proper dense subset of W.

An alternative way to write condition (2.23) is

$$\left(\ell_0 J'(\hat{z}) - G'(\hat{z})^*\left(\mu^*\right) - H'(\hat{z})^*\left(\lambda^*\right)\right)(d) \geq 0 \quad \text{for all } d \in S - \{\hat{z}\},$$

where $G'(\hat{z})^*$ and $H'(\hat{z})^*$ denote the adjoint operators of the continuous linear operators $G'(\hat{z})$ and $H'(\hat{z})$, respectively.

The Fritz John conditions in Theorem 2.3.23 are exploited for the finite optimization problem in Problem 2.3.3 and result in a finite-dimensional version of the Fritz John conditions.

Corollary 2.3.25 (First order necessary optimality conditions for Problem 2.3.3). *Let $\hat{z}$ be a local minimum of Problem 2.3.3 and $S \subseteq \mathbb{R}^{n_z}$ closed and convex with $int(S) \neq \emptyset$. Then there exist multipliers $\ell_0 \geq 0$, $\mu = (\mu_1, \ldots, \mu_{n_G})^\top \in \mathbb{R}^{n_G}$ and $\lambda = (\lambda_1, \ldots, \lambda_{n_H})^\top \in \mathbb{R}^{n_H}$ not all zero such that*

$$L'_z(\hat{z}, \ell_0, \mu, \lambda)(z - \hat{z}) \geq 0 \quad \textit{for all } z \in S, \tag{2.28}$$

$$\mu_i G_i(\hat{z}) = 0, \quad i = 1, \ldots, n_G, \tag{2.29}$$

$$\mu_i \geq 0, \quad i = 1, \ldots, n_G, \tag{2.30}$$

where

$$L(z, \ell_0, \mu, \lambda) := \ell_0 J(z) + \mu^\top G(z) + \lambda^\top H(z).$$

denotes the Lagrange function *for Problem* 2.3.3.

Proof. The assumptions of Theorem 2.3.23 are satisfied because:
(i) The interior of the cone $K = \{k \in \mathbb{R}^{n_G} \mid k \leq 0_{\mathbb{R}^{n_G}}\}$ is non-empty.
(ii) The image of the linear mapping $H'(\hat{z}) : \mathbb{R}^{n_z} \longrightarrow \mathbb{R}^{n_H}$ is always closed. Consequently, $\mathrm{im}(H'(\hat{z}))$ is never a proper dense subset of $V = \mathbb{R}^{n_H}$.

Theorem 2.3.23 yields the existence of non-trivial multipliers ℓ_0, $\tilde{\mu}^* \in W^* = (\mathbb{R}^{n_G})^*$, and $\tilde{\lambda} \in V^* = (\mathbb{R}^{n_H})^*$ with (2.20)–(2.23). As $\mathbb{R}^{n_G}$ and $\mathbb{R}^{n_H}$ are Hilbert spaces, they can be identified with its dual spaces according to Theorem 2.1.10. As a consequence, the multipliers $\tilde{\mu}^*$ and $\tilde{\lambda}^*$ can be expressed as $\tilde{\mu}^* = \tilde{\mu}^\top$ and $\tilde{\lambda}^* = \tilde{\lambda}^\top$ with $\tilde{\mu} \in \mathbb{R}^{n_G}$ and $\tilde{\lambda} \in \mathbb{R}^{n_H}$.

The condition (2.21) reads as $\tilde{\mu}^\top k \geq 0$ for every $k \leq 0_{\mathbb{R}^{n_G}}$. Hence, $\tilde{\mu} \leq 0_{\mathbb{R}^{n_G}}$.

Setting $\mu := -\tilde{\mu}$ and $\lambda := -\tilde{\lambda}$ proves the assertion. □

In infinite spaces, the non-density assumption on $\mathrm{im}(H'(\hat{z}))$ in Theorem 2.3.23 is not satisfied automatically as the image of a linear continuous operator is not closed in general. Next, we investigate a special case, which is particularly important for optimal control problems, and show that the non-density condition holds in this case. To this end, notice that the linearized operator $H'(\hat{z})$ in Section 2.2 is a mapping from the Banach space Z into the Banach space V, which is decomposed into $V := V_1 \times \mathbb{R}^{n_\psi}$ with a Banach space V_1. This special structure of the image space V is exploited in the following result.

Theorem 2.3.26. *Let the linear and continuous operator $H'(\hat{z}) : Z \longrightarrow V$ in Theorem 2.3.23 be given by $H'(\hat{z}) := (T_1, T_2)$, where $T_1 : Z \longrightarrow V_1$ is a linear, continuous, and surjective operator, $T_2 : Z \longrightarrow \mathbb{R}^n$ is linear and continuous, and $V := V_1 \times \mathbb{R}^n$.*

Then $im(H'(\hat{z}))$ is closed in $V = V_1 \times \mathbb{R}^n$.

The proof of Theorem 2.3.26 uses factor spaces.

Definition 2.3.27 (Factor space, quotient space, codimension). Let Z be a vector space and $M \subseteq Z$ a subspace. The *factor space* or *quotient space* Z/M consists of all sets $[z]$, $z \in Z$, where $[z] := z + M$. Vectors $z_1, z_2 \in Z$ are called *M-equivalent*, if $z_1 - z_2 \in M$.

The dimension of Z/M is called *codimension of M* or *defect of M*. □

Properties of factor spaces are summarized in

Corollary 2.3.28. *Let Z be a vector space and $M \subseteq Z$ a subspace. Then:*

(a) *If Z is finite dimensional, then $dim(Z/M) = dim(Z) - dim(M)$.*

(b) *If Z is a Banach space and M a closed subspace of Z, then*

$$\|[z]\|_{Z/M} := \inf\{\|y - z\|_Z \mid y \in M\}$$

defines a norm on Z/M, and $(Z/M, \|\cdot\|_{Z/M})$ is a Banach space.

(c) *The canonical mapping $w : Z \longrightarrow Z/M$, $z \mapsto [z]$, is surjective and $ker(w) = M$.*

(d) *Let $T : Z \longrightarrow V$ be a linear and surjective mapping from the vector space Z onto the vector space V. Then T can be factorized uniquely as $T = \hat{T} \circ w$ with a linear, surjective, and injective mapping $\hat{T} : Z/ker(T) \longrightarrow V$.*

If, in addition, Z and V are Banach spaces and T is continuous, then $\hat{T}$ is continuous, $\hat{T}^{-1}$ exists, and $\hat{T}^{-1}$ is linear and continuous.

Proof. (a) See [182, p. 214].

(b) See [326, Theorem I.3.2].

(c) See [182, p. 213].

(d) See [182, p. 214] for the first part. For the second part, it remains to show that $\hat{T}$ is continuous and that $\hat{T}^{-1}$ exists and is continuous. The continuity follows from $\hat{T}([z]) = T(z)$ and the continuity of T. Since Z and V are Banach spaces, the existence of the inverse operator $\hat{T}^{-1}$ and its continuity follows from Theorem 2.1.15. □

Proof of Theorem 2.3.26. Suppose that $\mathrm{im}(H'(\hat{z}))$ is not closed. Then there exists a sequence $(v_i, r_i) \in \mathrm{im}(H'(\hat{z}))$ with $\lim_{i\to\infty}(v_i, r_i) = (v_0, r_0) \notin \mathrm{im}(H'(\hat{z}))$. Then there exists a (algebraic) hyperplane, which contains $\mathrm{im}(T)$ but not (v_0, r_0).

The existence can be shown as follows: Define the linear subspace

$$M := \{t(v_0, r_0) \mid t \in \mathbb{R}\} \subset V.$$

It holds $M \neq \emptyset$ and $M \neq V$, since $M \not\subset \operatorname{im}(\mathcal{H}'(\hat{z}))$. Hence, V can be written as $V = M \oplus S$ with a subspace $S \subset V$. Define the functional $\ell : V \longrightarrow \mathbb{R}$ by

$$\ell(s + t(v_0, r_0)) := t.$$

Then $\ell(s) = 0$ for $s \in S$ and $\ell(v_0, r_0) = 1$. Furthermore, ℓ is linear because

$$\begin{aligned}
\ell(\lambda(s + t(v_0, r_0))) &= \ell(\underbrace{\lambda s}_{\in S} + \underbrace{\lambda t}_{\in \mathbb{R}}(v_0, r_0)) \\
&= \lambda t \\
&= \lambda \ell(s + t(v_0, r_0)), \\
\ell((s_1 + t_1(v_0, r_0)) + (s_2 + t_2(v_0, r_0))) &= \ell(\underbrace{s_1 + s_2}_{\in S} + \underbrace{(t_1 + t_2)}_{\in \mathbb{R}}(v_0, r_0)) \\
&= t_1 + t_2 \\
&= \ell(s_1 + t_1(v_0, r_0)) + \ell(s_2 + t_2(v_0, r_0)).
\end{aligned}$$

Hence, there exists a linear functional $\ell_v : V_1 \longrightarrow \mathbb{R}$ and a continuous linear functional $\ell_r \in (\mathbb{R}^n)^*$, both not zero, with $\ell(v, r) = \ell_v(v) + \ell_r(r)$ and

$$\begin{aligned}
\ell_v(T_1(z)) + \ell_r(T_2(z)) &= 0 \quad \text{for all } z \in Z, \\
\ell_v(v_0) + \ell_r(r_0) &\neq 0.
\end{aligned}$$

We will show that ℓ_v is actually continuous.

Let $\eta_i \in V_1$ be an arbitrary sequence converging to zero in V_1. Unfortunately, there may be many points in the preimage of η_i under the mapping T_1, i. e. T_1^{-1} is in general no mapping. To circumvent this problem, we consider the factor space $Z/\ker(T_1)$ endowed with the norm $\|[z]\|_{Z/\ker(T_1)}$.

According to (d) in Corollary 2.3.28, T_1 can be uniquely factorized as $T_1 = \hat{T}_1 \circ w$ with $\hat{T}_1 : Z/\ker(T_1) \longrightarrow V_1$ being linear, continuous, surjective, and injective, and $\hat{T}^{-1}$ being linear and continuous.

Hence, to each $\eta_i \in V_1$ there corresponds an element W_i of $Z/\ker(T_1)$. Since the inverse operator is continuous and η_i is a null-sequence, the sequence W_i converges to zero. Furthermore, we can choose a representative $\xi_i \in W_i$ such that

$$\begin{aligned}
&\|\xi_i\|_Z \leq 2\|W_i\|_{Z/\ker(T_1)}, \\
&T_1(\xi_i) = \eta_i
\end{aligned}$$

hold for every $i \in \mathbb{N}$. Since W_i is a null-sequence, the same holds for ξ_i by the first inequality.

This yields

$$\lim_{i\to\infty} \ell_v(\eta_i) = \lim_{i\to\infty} \ell_v(T_1(\xi_i)) = \lim_{i\to\infty} -\ell_r(T_2(\xi_i)) = 0$$

since $\xi_i \longrightarrow \Theta_Z$ and T_2 and ℓ_r are continuous. This shows that ℓ_v is actually continuous in Θ_Z and hence on Z.

In particular, we obtain

$$0 = \lim_{i\to\infty} (\ell_v(v_i) + \ell_r(r_i)) = \ell_v\Big(\lim_{i\to\infty} v_i\Big) + \ell_r\Big(\lim_{i\to\infty} r_i\Big) = \ell_v(v_0) + \ell_r(r_0)$$

in contradiction to $0 \neq \ell_v(v_0) + \ell_r(r_0)$. □

2.3.5 Constraint qualifications and Karush–Kuhn–Tucker conditions

Conditions which ensure that the multiplier ℓ_0 in Theorem 2.3.23 is not zero are called *regularity conditions* or *constraint qualifications*. In this case, without loss of generality, ℓ_0 can be normalized to one due to the linearity in the multipliers.

The following regularity condition was postulated by Robinson [283] in the context of stability analysis for generalized inequalities.

Definition 2.3.29 (Regularity condition of Robinson [283]). The *regularity condition of Robinson* is satisfied at $\hat{z}$, if

$$\begin{pmatrix} \Theta_W \\ \Theta_V \end{pmatrix} \in \operatorname{int} \left\{ \begin{pmatrix} G(\hat{z}) + G'(\hat{z})(z - \hat{z}) - k \\ H'(\hat{z})(z - \hat{z}) \end{pmatrix} \,\middle|\, z \in S,\ k \in K \right\}. \tag{2.31}$$

□

The validity of the regularity condition ensures that the multiplier ℓ_0 is not zero.

Theorem 2.3.30 (Karush–Kuhn–Tucker (KKT) conditions). *Let the assumptions of Theorem* 2.3.23 *be satisfied. In addition, let the regularity condition of Robinson hold at* $\hat{z}$. *Then the assertions of Theorem* 2.3.23 *hold with* $\ell_0 = 1$.

Proof. Let us assume that the necessary conditions are valid with $\ell_0 = 0$:

$$\begin{aligned} \mu^*(k) &\geq 0, && \text{for all } k \in K, \\ \mu^*(G(\hat{z})) &= 0, && \\ \mu^*(G'(\hat{z})(d)) + \lambda^*(H'(\hat{z})(d)) &\leq 0, && \text{for all } d \in S - \{\hat{z}\}. \end{aligned}$$

These conditions imply

$$\mu^*(G(\hat{z}) + G'(\hat{z})(d) - k) + \lambda^*(H'(\hat{z})(d)) \leq 0 \quad \text{for all } d \in S - \{\hat{z}\},\ k \in K.$$

Hence, the functional $\eta^*(w, v) := \mu^*(w) + \lambda^*(v)$ separates (Θ_W, Θ_V) from the set

$$\left\{ \begin{pmatrix} G(\hat{z}) + G'(\hat{z})(z - \hat{z}) - k \\ H'(\hat{z})(z - \hat{z}) \end{pmatrix} \,\middle|\, z \in S,\ k \in K \right\},$$

since $\eta^*(\Theta_W, \Theta_V) = 0$. However, (Θ_W, Θ_V) is assumed to be an interior point of this set. Hence, a separation is impossible, and the assumption $\ell_0 = 0$ was wrong. □

Zowe and Kurcyusz [332] show that the regularity condition of Robinson is stable under perturbations of the constraints and ensures the boundedness of the multipliers.

A sufficient condition for the regularity condition of Robinson is the *surjectivity constraint qualification* or *linear independence constraint qualification (LICQ)*.

Corollary 2.3.31 (Surjectivity constraint qualification, LICQ). *Let $\hat{z} \in int(S)$, and let the operator*

$$T : Z \longrightarrow W \times V, \quad T := (G'(\hat{z}), H'(\hat{z}))$$

be surjective. Then the regularity condition of Robinson (2.31) holds.

Proof. Since $\hat{z} \in \mathrm{int}(S)$, there exists an open ball $B_\varepsilon(\Theta_Z) \subseteq S - \{\hat{z}\}$. The operator T is linear, continuous, surjective, and $T(\Theta_Z) = (\Theta_W, \Theta_V)$. According to the open mapping Theorem 2.3.22, the set $T(B_\varepsilon(\Theta_Z))$ is open in $W \times V$. Hence,

$$\begin{pmatrix} \Theta_W \\ \Theta_V \end{pmatrix} \in \mathrm{int} \left\{ \begin{pmatrix} G'(\hat{z})(z - \hat{z}) \\ H'(\hat{z})(z - \hat{z}) \end{pmatrix} \,\middle|\, z \in S \right\}.$$

Since $G(\hat{z}) \in K$ and thus $\Theta_W \in K - G(\hat{z})$, it follows the regularity condition of Robinson (2.31). □

The subsequent Mangasarian–Fromowitz condition is sufficient for the regularity condition of Robinson.

Corollary 2.3.32 (Mangasarian–Fromowitz constraint qualification (MFCQ)). *Let $G : Z \longrightarrow W$ and $H : Z \longrightarrow V$ be Fréchet-differentiable at $\hat{z} \in S$, $S \subseteq Z$ a closed convex set with $int(S) \neq \emptyset$, $K \subseteq W$ a closed convex cone with vertex at zero and* $\mathrm{int}(K) \neq \emptyset$, *$G(\hat{z}) \in K$, $H(\hat{z}) = \Theta_V$. Furthermore, let the following* Mangasarian–Fromowitz constraint qualification *hold at $\hat{z}$:*

(a) *Let $H'(\hat{z})$ be surjective.*

(b) *Let there exist some $\hat{d} \in int(S - \{\hat{z}\})$ with*

$$H'(\hat{z})(\hat{d}) = \Theta_V, \tag{2.32}$$

$$G'(\hat{z})(\hat{d}) \in int(K - \{G(\hat{z})\}). \tag{2.33}$$

Then the regularity condition of Robinson (2.31) holds.

Proof. Since $G(\hat{z}) + G'(\hat{z})(\hat{d}) \in \text{int}(K)$, there exists a ball $B_{\varepsilon_1}(G(\hat{z}) + G'(\hat{z})(\hat{d}))$ with radius $\varepsilon_1 > 0$ and center $G(\hat{z}) + G'(\hat{z})(\hat{d})$, which lies in $\text{int}(K)$. Since subtraction, viewed as a mapping from $W \times W$ into W, is continuous, and observing that

$$G(\hat{z}) + G'(\hat{z})(\hat{d}) = G(\hat{z}) + G'(\hat{z})(\hat{d}) - \Theta_W,$$

there exist balls $B_{\varepsilon_2}(G(\hat{z}) + G'(\hat{z})(\hat{d}))$ and $B_{\varepsilon_3}(\Theta_W)$ with

$$B_{\varepsilon_2}(G(\hat{z}) + G'(\hat{z})(\hat{d})) - B_{\varepsilon_3}(\Theta_W) \subseteq B_{\varepsilon_1}(G(\hat{z}) + G'(\hat{z})(\hat{d})).$$

Since $G'(\hat{z})$ is a continuous linear mapping, there exists a ball $B_{\delta_1}(\hat{d})$ in Z with

$$G(\hat{z}) + G'(\hat{z})(B_{\delta_1}(\hat{d})) \subseteq B_{\varepsilon_2}(G(\hat{z}) + G'(\hat{z})(\hat{d})).$$

Since $\hat{d} \in \text{int}(S - \{\hat{z}\})$, eventually after diminishing δ_1 to $\delta_2 > 0$, we find

$$G(\hat{z}) + G'(\hat{z})(B_{\delta_2}(\hat{d})) \subseteq B_{\varepsilon_2}(G(\hat{z}) + G'(\hat{z})(\hat{d}))$$

and

$$B_{\delta_2}(\hat{d}) \subseteq \text{int}(S - \{\hat{z}\}).$$

By the open mapping theorem, there exists a ball $B_{\varepsilon_4}(\Theta_V)$ in V with

$$B_{\varepsilon_4}(\Theta_V) \subseteq H'(\hat{z})(B_{\delta_2}(\hat{d})),$$

since $H'(\hat{z})$ is continuous and surjective.

Summarizing, we found the following: For every $\tilde{w} \in B_{\varepsilon_3}(\Theta_W)$, $\tilde{v} \in B_{\varepsilon_4}(\Theta_V)$ there exists $d \in B_{\delta_2}(\hat{d})$, in particular, $d \in \text{int}(S - \{\hat{z}\})$, with

$$H'(\hat{z})(d) = \tilde{v}, \quad G(\hat{z}) + G'(\hat{z})(d) \in B_{\varepsilon_2}(G(\hat{z}) + G'(\hat{z})(\hat{d})),$$

hence

$$G(\hat{z}) + G'(\hat{z})(d) - \tilde{w} \in B_{\varepsilon_1}(G(\hat{z}) + G'(\hat{z})(\hat{d})) \subseteq \text{int}(K),$$

i. e. there exists $k \in \text{int}(K)$ with

$$G(\hat{z}) + G'(\hat{z})(d) - \tilde{w} = k.$$

Thus, we proved

$$\tilde{w} = G(\hat{z}) + G'(\hat{z})(d) - k, \quad \tilde{v} = H'(\hat{z})(d)$$

with some $k \in \text{int}(K)$, $d \in \text{int}(S - \{\hat{z}\})$. □

Condition (b) in Corollary 2.3.32 can be replaced by the following assumption: Let there exist some $\hat{d} \in \text{int}(\text{cone}(S, \hat{z}))$ with

$$H'(\hat{z})(\hat{d}) = \Theta_V,$$
$$G'(\hat{z})(\hat{d}) \in \text{int}\big(\text{cone}\big(K, G(\hat{z})\big)\big).$$

It is possible to show that the above Mangasarian–Fromowitz condition is equivalent to Robinson's condition for problems of type 2.3.2.

Application of the Mangasarian–Fromowitz constraint qualification in Corollary 2.3.32 to the finite optimization Problem 2.3.3 translates as follows:

Definition 2.3.33 (MFCQ for Problem 2.3.3). The constraint qualification of Mangasarian–Fromowitz is satisfied at $\hat{z}$ in Problem 2.3.3, if the following conditions are fulfilled:

(a) The derivatives $H_i'(\hat{z})$, $i = 1, \ldots, n_H$, are linearly independent.

(b) There exists a vector $\hat{d} \in \text{int}(S - \{\hat{z}\})$ with

$$G_i'(\hat{z})(\hat{d}) < 0 \quad \text{for } i \in A(\hat{z}) \quad \text{and} \quad H_i'(\hat{z})(\hat{d}) = 0 \quad \text{for } i = 1, \ldots, n_H.$$ □

Likewise, application of the linear independence constraint qualification in Corollary 2.3.31 to the finite optimization problem 2.3.3 translates as follows:

Definition 2.3.34 (LICQ for Problem 2.3.3). The *linear independence constraint qualification* is satisfied at $\hat{z}$ in Problem 2.3.3, if the following conditions are fulfilled:

(a) $\hat{z} \in \text{int}(S)$.

(b) The derivatives $G_i'(\hat{z})$, $i \in A(\hat{z})$, and $H_i'(\hat{z})$, $i = 1, \ldots, n_H$, are linearly independent. □

The multipliers turn out to be unique if the linear independence constraint qualification holds:

Corollary 2.3.35 (Uniqueness of multipliers). *Let the assumptions of Theorem* 2.3.25 *be satisfied, and let the linear independence constraint qualification hold at* $\hat{z}$.

Then the assertions of Theorem 2.3.25 *hold with* $\ell_0 = 1$ *and, in particular,*

$$\nabla_z L(\hat{z}, \ell_0, \mu, \lambda) = 0_{\mathbb{R}^{n_z}}.$$

Furthermore, the multipliers μ *and* λ *are unique.*

Proof. The first assertion follows directly from Corollary 2.3.31 and Theorem 2.3.30.

The uniqueness of the Lagrange multipliers follows from the following considerations. Assume that there are Lagrange multipliers μ_i, $i = 1, \ldots, n_G$, λ_i, $i = 1, \ldots, n_H$ and $\tilde{\mu}_i$, $i = 1, \ldots, n_G$, $\tilde{\lambda}_i$, $i = 1, \ldots, n_H$, satisfying the KKT conditions. $\hat{z} \in \text{int}(S)$ implies

$$0_{\mathbb{R}^{n_z}}^\top = J'(\hat{z}) + \sum_{i=1}^{n_G} \mu_i G_i'(\hat{z}) + \sum_{i=1}^{n_H} \lambda_i H_i'(\hat{z}),$$
$$0_{\mathbb{R}^{n_z}}^\top = J'(\hat{z}) + \sum_{i=1}^{n_G} \tilde{\mu}_i G_i'(\hat{z}) + \sum_{i=1}^{n_H} \tilde{\lambda}_i H_i'(\hat{z}).$$

Subtracting these equations leads to

$$0_{\mathbb{R}^{n_z}}^\top = \sum_{i=1}^{n_G} (\mu_i - \tilde{\mu}_i) G_i'(\hat{z}) + \sum_{i=1}^{n_H} (\lambda_i - \tilde{\lambda}_i) H_i'(\hat{z}).$$

For inactive inequality constraints, we have $\mu_i = \tilde{\mu}_i = 0$, $i \notin A(\hat{z})$. Since the gradients of the active constraints are assumed to be linearly independent, it follows $0 = \mu_i - \tilde{\mu}_i$, $i \in A(\hat{z})$, and $0 = \lambda_i - \tilde{\lambda}_i$, $i = 1, \ldots, n_H$. Hence, the Lagrange multipliers are unique. □

---

i There exist several other constraint qualifications. One of the weakest conditions is the *constraint qualification of Abadie* postulating

$$T_{\text{lin}}(K, S, \hat{z}) \subseteq T(\Sigma, \hat{z}),$$

where $T_{\text{lin}}(K, S, \hat{z})$ denotes the linearizing cone at $\hat{z}$. The idea behind this condition is to force the linearizing cone and tangent cone to coincide, see Corollary 2.3.13.

The condition of Abadie is weaker as the previously discussed constraint qualifications since those imply the condition of Abadie but not vice versa.

It is important to mention that the tangent cone $T(\Sigma, \hat{z})$ is independent of the representation of the set Σ by inequality and equality constraints, whereas the linearizing cone $T_{\text{lin}}(K, S, \hat{z})$ depends on the functions G and H describing Σ.

The KKT conditions are not only necessary, but also sufficient, if the objective function is convex and the constraints are affine-linear.

Theorem 2.3.36 (Sufficient condition). *Let J be Fréchet-differentiable and convex, and let $G : Z \longrightarrow W$ and $H : Z \longrightarrow V$ be affine-linear functions. Moreover, let $(\hat{z}, \hat{\mu}^*, \lambda^*) \in Z \times W^* \times V^*$ be a KKT point of Problem* 2.3.2*, that is, $\hat{z}$ is feasible and $(\hat{z}, \hat{\mu}^*, \lambda^*)$ satisfies* (2.20)–(2.23) *in Theorem* 2.3.23 *with $\ell_0 = 1$.*

Then $\hat{z}$ is a global minimum of Problem 2.3.2.

Proof. Let z be feasible for Problem 2.3.2. The convexity of J, the affine-linearity of G and H, and the conditions (2.21)–(2.23) yield

$$\begin{aligned} J(z) &\geq J(\hat{z}) + J'(\hat{z})(z - \hat{z}) \\ &\geq J(\hat{z}) + \mu^*\big(\underbrace{G(\hat{z}) + G'(\hat{z})(z - \hat{z})}_{=G(z)\in K}\big) + \lambda^*\big(\underbrace{H'(\hat{z})(z - \hat{z})}_{=\Theta_V}\big) \\ &\geq J(\hat{z}). \end{aligned}$$

As z was an arbitrary feasible point, $\hat{z}$ is a global minimum. □

Second-order necessary and sufficient conditions for infinite optimization problems are discussed in [237, 240]. In [240], the so-called two norm discrepancy is also addressed. This problem occurs if the sufficient conditions do not hold for the norm $\|\cdot\|_Z$, but for an alternate norm $\|\cdot\|_p$ on the space Z. Unfortunately, the appearing functions usually are not differentiable anymore in this alternate norm, and the proof techniques are more intricate.

2.4 Exercises

Prove that $F : \mathbb{R}^{n\times n} \to \mathbb{R}^{n\times n}$ with $F(X) := X^{-1}$ has the Fréchet-derivative $F'(X)(H) = -X^{-1}HX^{-1}$.

Hint: The Neumann series $(I_n - B)^{-1} = \sum_{k=0}^{\infty} B^k$ (for $\|B\| < 1$) or the identity $X^{-1}X = I_n$ can be useful.

Compute the tangent cone for the following sets:

(a) $\{(x,y)^\top \in \mathbb{R}^2 \mid y \geq x^3\}$,

(b) $\{(x,y)^\top \in \mathbb{R}^2 \mid x \geq 0 \text{ or } y \geq 0\}$,

(c) $\{(x,y)^\top \in \mathbb{R}^2 \mid x = 0 \text{ or } y = 0\}$,

(d) $\{(r\cos\varphi, r\sin\varphi)^\top \in \mathbb{R}^2 \mid 0 \leq r \leq 1,\ \pi/4 \leq \varphi \leq 7\pi/4\}$.

Determine for the feasible set

$$\Sigma = \{z \in \mathbb{R}^2 \mid G(z) \leq 0\}$$

the tangent cone and the linearizing cone at $\hat{z}$ with

(a) $G(z) = (z_2 - z_1^5, -z_2)^\top$, $\hat{z} = (0,0)^\top$.

(b) $G(z) = (1 - z_1, 1 - z_1^2 - z_2^2)^\top$, $\hat{z} = (1,0)^\top$.

Are the linear independence constraint qualification and the Mangasarian–Fromowitz constraint qualification satisfied at $\hat{z}$?

Consider the nonlinear optimization problem

$$\begin{aligned} &\textit{Minimize} && z_1 \\ &\textit{subject to} && (z_1 - 4)^2 + z_2^2 \leq 16, \\ &&& (z_1 - 3)^2 + (z_2 - 2)^2 = 13. \end{aligned}$$

Sketch the feasible region and find all KKT points. Which point is optimal?

Solve the nonlinear optimization problem

$$\text{Minimize} \quad z_1^2 + z_2^2 \quad \text{subject to} \quad 1 - z_1 \le 0,\ 1 - z_1^2 - z_2^2 \le 0$$

(a) Graphically.
(b) Using KKT conditions.

Which of the discussed regularity conditions are satisfied in the optimal solution?

(Lipschitz Continuity of Projection)
Let X be a Hilbert space and $M \subseteq X$ closed and convex. Let $P_M : X \longrightarrow M$ be the projection on M satisfying

$$\|x - P_M(x)\|_X \le \|x - m\|_X \quad \text{for all } m \in M.$$

Prove: P_M is Lipschitz continuous on X, that is,

$$\|P_M(x) - P_M(y)\|_X \le \|x - y\|_X \quad \text{for all } x, y \in X.$$

Hint: Use the *projection theorem* [215, Section 3.12, Theorem 1]: Let X be a Hilbert space and $\emptyset \ne M \subseteq X$ a closed and convex set. Then for every vector $y \in X$, there exists a unique vector $x_0 \in M$ with $\|y - x_0\|_X \le \|y - x\|_X$ for all $x \in M$. The condition $\langle y - x_0, x - x_0\rangle_X \le 0$ for all $x \in M$ is necessary and sufficient for the optimality of x_0.

(Equivalence of Variational Inequality and Non-smooth Equation)
Let X be a Hilbert space, $M \subseteq X$ a closed convex set, $\hat{x} \in M$, and $G : X \longrightarrow X$ a mapping. Prove:

$$\left\langle G(\hat{x}), m - \hat{x}\right\rangle_X \ge 0 \quad \text{for all } m \in M$$

holds, if and only if

$$\hat{x} = P_M\big(\hat{x} - \alpha G(\hat{x})\big), \quad (\alpha > 0)$$

where P_M denotes the projection onto M.

Consider the minimum energy problem:

Minimize

$$\int_0^1 u(t)^2 dt$$

with respect to $x, y \in W_{1,2}([0,1])$ and $u \in L_2([0,1])$ subject to the constraints

$$\dot{x}(t) = y(t), \quad x(0) = x(1) = 0,$$
$$\dot{y}(t) = u(t), \quad y(0) = -y(1) = 1.$$

(a) Show that $x(\cdot)$ and $y(\cdot)$ satisfy the constraints, if and only if

$$-1 = \int_0^1 (1-s)u(s)ds,$$

$$-2 = \int_0^1 u(s)ds.$$

(b) Solve the following problem by the Fritz John conditions (check all assumptions first):

Minimize

$$\int_0^1 u(t)^2 dt$$

with respect to $u \in L_2([0,1])$ subject to the constraints in (a).

(c) Solve the problem in (b) by the projection theorem, see [215, Section 3.12, Theorem 1].

Hint: The inhomogeneous linear initial value problem

$$\dot{z}(t) = Az(t) + b(t), \quad z(a) = z_a$$

with $z(t), z_a \in \mathbb{R}^n, A \in \mathbb{R}^{n\times n}$, and $b(t) \in \mathbb{R}^n$ has the solution

$$z(t) = \Phi(t)z_a + \int_a^t \Phi(t)\Phi^{-1}(s)b(s)ds,$$

where Φ solves $\dot{\Phi}(t) = A\Phi(t)$, $\Phi(a) = I_n$.

Prove the following separation theorem:

(a) Let $C \subseteq \mathbb{R}^n$ be a non-empty convex set and $y \in \mathbb{R}^n$ a point that is not in the closure $\bar{C}$ of C. Then there exists a hyperplane $H = \{x \in \mathbb{R}^n \mid a^\top x = y\}$ with $y \in H$ and $a^\top y < \inf_{x\in C} a^\top x$.

(b) Let $C \subseteq \mathbb{R}^n$ be a convex set and $y \in \mathbb{R}^n$ a boundary point of C. Then there exists a hyperplane that contains y and that contains C in one of its closed halfspaces.

Let X be a normed vector space, $D \subseteq X$ open, and $J : D \longrightarrow \mathbb{R}$ a mapping.

(a) *Necessary Optimality Condition:* Let $\hat{x}$ be a local minimum of J and let J be Gâteaux-differentiable at $\hat{x}$. Show that $\delta J(\hat{x})(h) = 0$ holds for all $h \in X$.

(b) Apply (a) to the variational problem

Minimize

$$J(x) := \int_a^b f\big(t, x(t), x'(t)\big)dt$$

with respect to $x \in C_1([a,b])$ subject to

$$x(a) = x_a, \quad x(b) = x_b.$$

Show that the *Euler–Lagrange equation*

$$0 = f'_x\big(t, \hat{x}(t), \hat{x}'(t)\big) - \frac{d}{dt} f'_{x'}\big(t, \hat{x}(t), \hat{x}'(t)\big)$$

(under suitable smoothness assumptions) is necessary for a local minimum $\hat{x}$.

3 Local minimum principles

Optimal control problems subject to ordinary differential equations have a wide range of applications in different disciplines like engineering sciences, chemical engineering, and economics. Necessary conditions known as *maximum principles* or *minimum principles* have been investigated intensively since the 1950s. Early proofs of the maximum principle are given by Pontryagin et al. [274] and Hestenes [162]. Necessary conditions for problems with pure state constraints are discussed in [171, 139, 181, 238, 239, 169, 185]. Optimal control problems with mixed control-state constraints are discussed in [256, 330]. A survey on maximum principles for ODE optimal control problems with state constraints, including an extensive list of references, can be found in [159]. Necessary conditions for variational problems, i. e., smooth optimal control problems, are developed in [41]. Second-order necessary conditions and sufficient conditions are stated in [330]. Sufficient conditions are also presented in [240, 217, 236, 219]. Necessary conditions for optimal control problems subject to index-one DAEs without state constraints and without mixed control-state constraints can be found in [66]. Implicit control systems are discussed in [75]. Necessary conditions for optimal control problems subject to nonlinear quasi-linear DAEs without control and state constraints are presented in [17]. General DAE optimal control problems are discussed in [194]. Necessary and sufficient conditions for linear-quadratic DAE optimal control problems can be found in [247, 248, 192, 197, 17]. Optimal control problems subject to Hessenberg DAEs up to index-three are treated in [287] with results closely related to our results. Their technique for proving the results was based on an index reduction, a technique mentioned already in [162, p. 352]. Semi-explicit index-one systems often occur in process system engineering, see [164], but also in vehicle simulation, see [114], and many other fields of applications. A very important subclass of index-two DAEs is the stabilized descriptor form describing the motion of mechanical multi-body systems.

Necessary conditions are not only interesting from a theoretical point of view but also provide the basis of the indirect approach for solving optimal control problems numerically. In this approach, the minimum principle is exploited and typically leads to a multi-point boundary value problem, which is solved numerically, e. g., by the multiple shooting method. Even for the direct approach, based on a suitable discretization of the optimal control problem, the minimum principle is very important for the post-optimal approximation of adjoints. In this context, the multipliers resulting from the formulation of the necessary Fritz John conditions for the finite-dimensional discretized optimal control problem have to be related to the multipliers of the original infinite-dimensional optimal control problem appropriately. It is evident that this requires the knowledge of necessary conditions for the optimal control problem.

In this chapter, we favor the statement of necessary optimality conditions in terms of *local* minimum principles for optimal control problems subject to index-one and index-two DAEs, pure state constraints, and mixed control-state constraints. The local minimum principles are based on necessary optimality conditions for infinite optimization

https://doi.org/10.1515/9783110797893-003

problems derived in Theorem 2.3.23. The special structure of the optimal control problems under consideration is exploited and allows to obtain suitable representations for the multipliers involved. An additional Mangasarian–Fromowitz constraint qualification for the optimal control problem ensures the regularity of a local minimum.

The term *local* minimum principle is due to the fact, that the necessary optimality conditions can be interpreted as necessary optimality conditions for a *local* minimum of the so-called Hamilton function and the augmented Hamilton function, respectively, compare (3.26), (3.44), and (3.68) below. *Global* minimum principles even state that the optimal control *globally* minimizes the Hamilton function for arbitrary control sets. This is a much stronger statement, and a global minimum principle is discussed in Section 7.1. However, there is a nice relationship between local minimum principles for infinite-dimensional optimal control problems and their finite-dimensional discretizations. The relationship is outlined in Section 5.5. Similar relations for global minimum principles only hold under additional (restrictive) assumptions, see [251] and [169, p. 278].

The main steps in the derivation of local minimum principles can be summarized as follows:

(a) Rewrite the DAE optimal control problem as an infinite optimization problem, compare Section 2.2.
(b) Show the closedness of the image of the linearized equality operator by exploitation of solution formulas for linear DAEs.
(c) Apply the first-order necessary optimality conditions of Fritz John.
(d) Derive explicit representations of the Lagrange multipliers involved.
(e) Use variation lemmas to derive a local minimum principle for the DAE optimal control problem.
(f) Investigate regularity conditions and constraint qualifications.

Although index-one DAEs are easier to handle, we demonstrate the steps (a) to (f) and suitable techniques for semi-explicit index-two DAEs. This problem class already exhibits the main difficulties when dealing with DAE optimal control problems. The proof for the index-one case works similarly with minor adaptions only, and, for this reason, it is omitted here. Details of the proof can be found in [118]. More general DAEs can be handled similarly following essentially the same approach, but with the more technical effort involved, see [17, 194]. All of these approaches impose structural assumptions on the linearized DAE and rely on explicit solution formulas for these linear DAEs.

3.1 Problems without pure state and mixed control-state constraints

Consider the following optimal control problem without pure state constraints and without mixed control-state constraints.

Problem 3.1.1 (Index-two DAE optimal control problem). Let $\mathcal{I} := [t_0, t_f] \subset \mathbb{R}$ be a non-empty compact time interval with $t_0 < t_f$ fixed. Let

$$\begin{aligned}
\varphi &: \mathbb{R}^{n_x} \times \mathbb{R}^{n_x} \longrightarrow \mathbb{R},\\
f_0 &: \mathcal{I} \times \mathbb{R}^{n_x} \times \mathbb{R}^{n_y} \times \mathbb{R}^{n_u} \longrightarrow \mathbb{R},\\
f &: \mathcal{I} \times \mathbb{R}^{n_x} \times \mathbb{R}^{n_y} \times \mathbb{R}^{n_u} \longrightarrow \mathbb{R}^{n_x},\\
g &: \mathcal{I} \times \mathbb{R}^{n_x} \longrightarrow \mathbb{R}^{n_y},\\
\psi &: \mathbb{R}^{n_x} \times \mathbb{R}^{n_x} \longrightarrow \mathbb{R}^{n_\psi}
\end{aligned}$$

be sufficiently smooth functions and $\mathcal{U} \subseteq \mathbb{R}^{n_u}$ a closed convex set with non-empty interior.

Minimize

$$\varphi\big(x(t_0), x(t_f)\big) + \int_{t_0}^{t_f} f_0\big(t, x(t), y(t), u(t)\big)dt$$

with respect to $x \in W^{n_x}_{1,\infty}(\mathcal{I})$, $y \in L^{n_y}_{\infty}(\mathcal{I})$, $u \in L^{n_u}_{\infty}(\mathcal{I})$ *subject to the constraints*

$$\begin{aligned}
\dot{x}(t) &= f\big(t, x(t), y(t), u(t)\big) && \text{a. e. in } \mathcal{I},\\
0_{\mathbb{R}^{n_y}} &= g\big(t, x(t)\big) && \text{in } \mathcal{I},\\
0_{\mathbb{R}^{n_\psi}} &= \psi\big(x(t_0), x(t_f)\big),\\
u(t) &\in \mathcal{U} && \text{a. e. in } \mathcal{I}.
\end{aligned}$$

Using the technique in Example 1.1.17, we may assume that the control u does not appear in the algebraic constraint g in Problem 3.1.1. This assumption is essential for the subsequent analysis. Without this simplification, additional smoothness assumptions for the control have to be imposed.

In applying the first-order necessary optimality conditions in Theorem 2.3.23 to Problem 3.1.1, we make use of the interpretation of Problem 3.1.1 as the infinite optimization problem

Minimize $J(z)$ *subject to* $z \in S$ *and* $H(z) = \Theta_V$

given in Section 2.2 with

$$Z = W^{n_x}_{1,\infty}(\mathcal{I}) \times L^{n_y}_{\infty}(\mathcal{I}) \times L^{n_u}_{\infty}(\mathcal{I}),$$

$$V = L^{n_x}_{\infty}(\mathcal{I}) \times W^{n_y}_{1,\infty}(\mathcal{I}) \times \mathbb{R}^{n_\psi},$$

$$J(x, y, u) = \varphi(x(t_0), x(t_f)) + \int_{t_0}^{t_f} f_0(t, x(t), y(t), u(t))dt,$$

$$H(x, y, u) = \begin{pmatrix} H_1(x, y, u) \\ H_2(x, y, u) \\ H_3(x, y, u) \end{pmatrix} = \begin{pmatrix} f(\cdot, x(\cdot), y(\cdot), u(\cdot)) - \dot{x}(\cdot), \\ g(\cdot, x(\cdot)), \\ -\psi(x(t_0), x(t_f)), \end{pmatrix}, \tag{3.1}$$

$$S = W^{n_x}_{1,\infty}(\mathcal{I}) \times L^{n_y}_{\infty}(\mathcal{I}) \times U_{ad},$$
$$U_{ad} = \{u \in L^{n_u}_{\infty}(\mathcal{I}) \mid u(t) \in \mathcal{U} \text{ almost everywhere in } \mathcal{I}\}.$$

The assumptions in Theorem 2.3.23 need to be checked for the above problem. To this end, let $\hat{z} = (\hat{x}, \hat{y}, \hat{u})$ be a local minimum of Problem 3.1.1.

S is closed and convex with non-empty interior since $\mathcal{U}$ is closed and convex with non-empty interior. Continuous Fréchet-differentiability of J and H at $\hat{z}$ is guaranteed under Assumption 2.2.5 in Section 2.2.

It remains to verify that the image of the linear operator $H'(\hat{z})$ is not a proper dense subset of V. According to Theorem 2.3.26, it suffices to show that the linear and continuous operator T_1 defined by

$$\begin{aligned} T_1(\hat{x}, \hat{y}, \hat{u})(x, y, u) &:= \begin{pmatrix} H_1'(\hat{x}, \hat{y}, \hat{u})(x, y, u) \\ H_2'(\hat{x}, \hat{y}, \hat{u})(x, y, u) \end{pmatrix} \\ &= \begin{pmatrix} f_x'[\cdot]x(\cdot) + f_y'[\cdot]y(\cdot) + f_u'[\cdot]u(\cdot) - \dot{x}(\cdot) \\ g_x'[\cdot]x(\cdot) \end{pmatrix} \end{aligned} \tag{3.2}$$

is surjective.

---

i It is important to point out at this stage that the choice of the image space of H_2 (respectively $g[\cdot]$) with $W^{n_y}_{1,\infty}(\mathcal{I})$ is crucial. If H_2 was considered as a mapping into the continuous functions, then the linearized operator (H_1', H_2') would have a proper dense image in the space of continuous functions and the assumptions of Theorem 2.3.23 are not satisfied. If H_2 instead is considered as a mapping into $W^{n_y}_{1,\infty}(\mathcal{I})$, which is a dense subset of the space of continuous functions, then the following Lemma 3.1.2 allows to show surjectivity of (H_1', H_2'), and Theorem 2.3.23 can be applied.

We exploit

Lemma 3.1.2. *Consider the linear DAE*

$$\dot{x}(t) = A(t)x(t) + B(t)y(t) + h_1(t), \tag{3.3}$$
$$0_{\mathbb{R}^{n_y}} = C(t)x(t) + h_2(t), \tag{3.4}$$

on the compact interval $\mathcal{I} = [t_0, t_f]$ with time-dependent matrix functions

$$A(\cdot) \in L^{n_x \times n_x}_{\infty}(\mathcal{I}), \quad B(\cdot) \in L^{n_x \times n_y}_{\infty}(\mathcal{I}), \quad C(\cdot) \in W^{n_y \times n_x}_{1,\infty}(\mathcal{I}),$$

and time-dependent vector functions

$$h_1(\cdot) \in L^{n_x}_{\infty}(\mathcal{I}), \quad h_2(\cdot) \in W^{n_y}_{1,\infty}(\mathcal{I}).$$

Let $M(t) := C(t) \cdot B(t)$ be non-singular almost everywhere in $\mathcal{I}$, and let $M(\cdot)^{-1}$ be essentially bounded on $\mathcal{I}$. Let $C(t_0)$ have full rank. Then:

(a) *There exists a consistent initial value $x(t_0) = x_0$ satisfying (3.4), and every consistent x_0 possesses the representation*

$$x_0 = \Pi h_2(t_0) + \Gamma w \quad \text{with } w \in \mathbb{R}^{n_x - n_y},$$

where $\Pi \in \mathbb{R}^{n_x \times n_y}$ satisfies $(I + C(t_0)\Pi)h_2(t_0) = 0_{\mathbb{R}^{n_y}}$, and the columns of $\Gamma \in \mathbb{R}^{n_x \times (n_x - n_y)}$ define an orthonormal basis of $\ker(C(t_0))$.

(b) *The initial value problem given by (3.3)–(3.4) together with the consistent initial value $x(t_0) = \Pi h_2(t_0) + \Gamma w$ has a unique solution $x(\cdot) \in W_{1,\infty}^{n_x}(\mathcal{I})$ for every $w \in \mathbb{R}^{n_x - n_y}$, every $h_1(\cdot) \in L_\infty^{n_x}(\mathcal{I})$, and every $h_2(\cdot) \in W_{1,\infty}^{n_y}(\mathcal{I})$. The solution on $\mathcal{I}$ is given by*

$$x(t) = \Phi(t)\left(\Pi h_2(t_0) + \Gamma w + \int_{t_0}^{t} \Phi^{-1}(\tau)h(\tau)d\tau \right), \tag{3.5}$$

$$y(t) = -M(t)^{-1}(q(t) + Q(t)x(t)), \tag{3.6}$$

where the fundamental system $\Phi(t) \in \mathbb{R}^{n_x \times n_x}$ is the unique solution of

$$\dot{\Phi}(t) = \tilde{A}(t)\Phi(t), \quad \Phi(t_0) = I_{n_x}, \tag{3.7}$$

with

$$\begin{aligned} \tilde{A}(t) &:= A(t) - B(t)M(t)^{-1}Q(t), \\ h(t) &:= h_1(t) - B(t)M(t)^{-1}q(t), \\ Q(t) &:= \dot{C}(t) + C(t)A(t), \\ q(t) &:= \dot{h}_2(t) + C(t)h_1(t). \end{aligned}$$

(c) *Let a vector $b \in \mathbb{R}^r$ and matrices $E_0, E_f \in \mathbb{R}^{r \times n_x}$ be given such that*

$$\operatorname{rank}((E_0\Phi(t_0) + E_f\Phi(t_f))\Gamma) = r$$

holds for the fundamental solution Φ from (b) and Γ from (a). Then the boundary value problem given by (3.3)–(3.4) together with the boundary condition

$$E_0 x(t_0) + E_f x(t_f) = b \tag{3.8}$$

has a solution for every $b \in \mathbb{R}^r$.

Proof. (a) Consider the linear equation $0_{\mathbb{R}^{n_y}} = C(t_0)x_0 + h_2(t_0)$ for x_0. Since $C(t_0)$ has full row rank, there exists an orthonormal decomposition

$$C(t_0)^\top = P\begin{pmatrix} R \\ \Theta \end{pmatrix}, \quad P = (\Pi_1, \Gamma) \in \mathbb{R}^{n_x \times n_x},$$

which can be computed using Householder reflections. Herein, $R \in \mathbb{R}^{n_y \times n_y}$ is non-singular, P is orthogonal, $\Pi_1 \in \mathbb{R}^{n_x \times n_y}$ is an orthonormal basis of the image of $C(t_0)^\top$, and $\Gamma \in \mathbb{R}^{n_x \times (n_x - n_y)}$ is an orthonormal basis of the null-space of $C(t_0)$. Every $x_0 \in \mathbb{R}^{n_x}$ can be expressed uniquely as $x_0 = \Pi_1 v + \Gamma w$ with $v \in \mathbb{R}^{n_y}$ and $w \in \mathbb{R}^{n_x - n_y}$. Introducing this expression into (3.4) yields

$$0_{\mathbb{R}^{n_y}} = C(t_0)(\Pi_1 v + \Gamma w) + h_2(t_0) = R^\top v + h_2(t_0),$$

which is satisfied for $v = -R^{-\top} h_2(t_0)$. Hence, consistent values are characterized by

$$x_0 = \Pi h_2(t_0) + \Gamma w, \qquad \Pi := -\Pi_1 R^{-\top}.$$

(b) Differentiation of the algebraic equation (3.4) yields

$$\begin{aligned} 0_{\mathbb{R}^{n_y}} &= \dot{C}(t)x(t) + C(t)\dot{x}(t) + \dot{h}_2(t) \\ &= (\dot{C}(t) + C(t)A(t))x(t) + C(t)B(t)y(t) + \dot{h}_2(t) + C(t)h_1(t) \\ &= Q(t)x(t) + M(t)y(t) + q(t). \end{aligned}$$

The non-singularity of $M(t)$ yields (3.6). Introducing this expression into (3.3) yields the linear ODE

$$\dot{x}(t) = \tilde{A}(t)x(t) + h(t).$$

With (a) the assertion follows as in [161, p. 36].

(c) Part (c) exploits the solution formulas in (a) and (b). The boundary conditions (3.8) are satisfied, if

$$\begin{aligned} b &= E_0 x(t_0) + E_f x(t_f) \\ &= E_0(\Pi h_2(t_0) + \Gamma w) + E_f \Phi(t_f)\left(\Pi h_2(t_0) + \Gamma w + \int_{t_0}^{t_f} \Phi^{-1}(\tau)h(\tau)d\tau \right) \\ &= (E_0 + E_f\Phi(t_f))\Gamma w + (E_0 + E_f\Phi(t_f))\Pi h_2(t_0) + E_f\Phi(t_f) \int_{t_0}^{t_f} \Phi^{-1}(\tau)h(\tau)d\tau. \end{aligned}$$

Rearranging terms and exploiting $\Phi(t_0) = I_{n_x}$ yields

$$(E_0\Phi(t_0) + E_f\Phi(t_f))\Gamma w = b - (E_0 + E_f\Phi(t_f))\Pi h_2(t_0) - E_f\Phi(t_f) \int_{t_0}^{t_f} \Phi^{-1}(\tau)h(\tau)d\tau.$$

This equation is solvable for every $b \in \mathbb{R}^r$, if the matrix $(E_0\Phi(t_0) + E_f\Phi(t_f))\Gamma$ is of rank r. Then for every $b \in \mathbb{R}^r$, there exists w such that (3.8) is satisfied. Application of part (b) completes the proof. □

By Lemma 3.1.2 the linear operator T_1 in (3.2) is surjective, if the following assumption holds:

Assumption 3.1.3. Let the matrix

$$M(t) := g_x'(t, \hat{x}(t)) f_y'(t, \hat{x}(t), \hat{y}(t), \hat{u}(t))$$

be *non-singular* almost everywhere in $\mathcal{I}$. Let $M(\cdot)^{-1}$ be essentially bounded in $\mathcal{I}$. □

Lemma 3.1.2, Assumptions 2.2.5, 3.1.3, and Theorem 2.3.26 guarantee that the assumptions of Theorem 2.3.23 are satisfied, and we prove:

Theorem 3.1.4 (Necessary conditions for Problem 3.1.1). *Let the following assumptions be fulfilled for Problem* 3.1.1*:*

(a) *Assumptions* 2.2.5 *and* 3.1.3 *hold.*
(b) $\mathcal{U} \subseteq \mathbb{R}^{n_u}$ *is closed and convex with non-empty interior.*
(c) $(\hat{x}, \hat{y}, \hat{u})$ *is a local minimum of Problem* 3.1.1.

Then there exist non-trivial multipliers $\ell_0 \geq 0$ and $\lambda^* \in V^*$ with

$$\ell_0 J'(\hat{x}, \hat{y}, \hat{u})(x, y, u) - \lambda^*(H'(\hat{x}, \hat{y}, \hat{u})(x, y, u)) \geq 0 \tag{3.9}$$

for all $(x, y, u) \in S - \{(\hat{x}, \hat{y}, \hat{u})\}$. □

At a first glance, the necessary condition (3.9) seems to be of little practical use since the multiplier $\lambda^* := (\lambda_f^*, \lambda_g^*, \sigma^*)$ is an element of the dual space

$$V^* = L_\infty^{n_x}(\mathcal{I})^* \times W_{1,\infty}^{n_y}(\mathcal{I})^* \times (\mathbb{R}^{n_\psi})^*$$

and as such the components λ_f^* and λ_g^* do not have nice representations. However, as we shall see, exploitation of the variational inequality (3.9) using the derivatives in (2.11)–(2.14) allows us to derive a suitable representation of λ_f^* and λ_g^*. To this end, three separate variational equalities and inequalities are deduced from (3.9) by setting two of the three variations x, y, or u to zero, respectively, and by varying the remaining variation within S. Then (3.9) implies the following relations that hold for *every* $x \in W_{1,\infty}^{n_x}(\mathcal{I})$, *every* $y \in L_\infty^{n_y}(\mathcal{I})$, and *every* $u \in U_{ad} - \{\hat{u}\}$:

$$0 = \kappa_{x_0}' x(t_0) + \kappa_{x_f}' x(t_f) + \int_{t_0}^{t_f} \ell_0 f_{0,x}'[t] x(t) dt + \lambda_f^*(\dot{x}(\cdot) - f_x'[\cdot]x(\cdot)) - \lambda_g^*(g_x'[\cdot]x(\cdot)), \tag{3.10}$$

$$0 = \int_{t_0}^{t_f} \ell_0 f_{0,y}'[t] y(t) dt - \lambda_f^*(f_y'[\cdot]y(\cdot)), \tag{3.11}$$

$$0 \leq \int_{t_0}^{t_f} \ell_0 f_{0,u}'[t] u(t) dt - \lambda_f^*(f_u'[\cdot]u(\cdot)), \tag{3.12}$$

where

$$\kappa(x_0, x_f, \ell_0, \sigma) := \ell_0\varphi(x_0, x_f) + \sigma^\top \psi(x_0, x_f).$$

3.1.1 Representation of multipliers

The two equations (3.10) and (3.11) are exploited to represent the functionals λ_f^* and λ_g^*. To this end, let $h_1 \in L_\infty^{n_x}(\mathcal{I})$ and $h_2 \in W_{1,\infty}^{n_y}(\mathcal{I})$ be arbitrary.

Then by Lemma 3.1.2 (with $w = 0_{\mathbb{R}^{n_x-n_y}}$), the initial value problem

$$\dot{x}(t) = f_x'[t]x(t) + f_y'[t]y(t) + h_1(t), \quad x(t_0) = \Pi h_2(t_0), \tag{3.13}$$

$$0_{\mathbb{R}^{n_y}} = g_x'[t]x(t) + h_2(t) \tag{3.14}$$

possesses the solution

$$x(t) = \Phi(t)\left(\Pi h_2(t_0) + \int_{t_0}^{t} \Phi^{-1}(\tau)h(\tau)d\tau\right), \tag{3.15}$$

$$\begin{aligned} y(t) &= -M(t)^{-1}(q(t) + Q(t)x(t)) \\ &= -M(t)^{-1}\left(q(t) + Q(t)\Phi(t)\left(\Pi h_2(t_0) + \int_{t_0}^{t} \Phi^{-1}(\tau)h(\tau)d\tau\right)\right), \end{aligned} \tag{3.16}$$

where Φ solves (3.7) with

$$\begin{aligned} \tilde{A}(t) &= f_x'[t] - f_y'[t]M(t)^{-1}Q(t), \\ h(t) &= h_1(t) - f_y'[t]M(t)^{-1}q(t), \\ Q(t) &= \frac{d}{dt}g_x'[t] + g_x'[t]f_x'[t], \\ q(t) &= \dot{h}_2(t) + g_x'[t]h_1(t). \end{aligned}$$

Adding equations (3.10) and (3.11) and exploiting the linearity of the functionals leads to

$$0 = \kappa_{x_0}' x(t_0) + \kappa_{x_f}' x(t_f) + \ell_0 \int_{t_0}^{t_f} f_{0,x}'[t]x(t) + f_{0,y}'[t]y(t)dt + \lambda_f^*(h_1(\cdot)) + \lambda_g^*(h_2(\cdot)). \tag{3.17}$$

Introducing the solution formulas (3.15)–(3.16) into equation (3.17) and combining terms leads to the expression

$$0 = \left(\kappa_{x_0}' + \kappa_{x_f}'\Phi(t_f) + \int_{t_0}^{t_f} \ell_0 \hat{f}_0[t]\Phi(t)dt\right)\Pi h_2(t_0)$$

$$
\begin{aligned}
&+ \kappa'_{x_f}\Phi(t_f)\int_{t_0}^{t_f}\Phi^{-1}(t)h(t)dt \\
&+ \int_{t_0}^{t_f}\ell_0\hat{f}_0[t]\Phi(t)\left(\int_{t_0}^{t}\Phi^{-1}(\tau)h(\tau)d\tau\right)dt \\
&- \int_{t_0}^{t_f}\ell_0 f'_{0,y}[t]M(t)^{-1}q(t)dt \\
&+ \lambda_f^*(h_1(\cdot)) + \lambda_g^*(h_2(\cdot)),
\end{aligned}
\tag{3.18}
$$

where

$$
\hat{f}_0[t] := f'_{0,x}[t] - f'_{0,y}[t]M(t)^{-1}Q(t).
$$

Integration by parts yields

$$
\int_{t_0}^{t_f}\hat{f}_0[t]\Phi(t)\left(\int_{t_0}^{t}\Phi^{-1}(\tau)h(\tau)d\tau\right)dt = \int_{t_0}^{t_f}\left(\int_{t}^{t_f}\hat{f}_0[\tau]\Phi(\tau)d\tau\right)\Phi^{-1}(t)h(t)dt
$$

and (3.18) becomes

$$
\lambda_f^*(h_1(\cdot)) + \lambda_g^*(h_2(\cdot)) = -\zeta^\top h_2(t_0) - \int_{t_0}^{t_f}\lambda_f(t)^\top h(t)dt - \int_{t_0}^{t_f}\tilde{\lambda}_g(t)^\top q(t)dt \tag{3.19}
$$

with

$$
\begin{aligned}
\zeta^\top &:= \left(\kappa'_{x_0} + \kappa'_{x_f}\Phi(t_f) + \int_{t_0}^{t_f}\ell_0\hat{f}_0[t]\Phi(t)dt\right)\Pi, \\
\lambda_f(t)^\top &:= \left(\kappa'_{x_f}\Phi(t_f) + \int_{t}^{t_f}\ell_0\hat{f}_0[\tau]\Phi(\tau)d\tau\right)\Phi^{-1}(t), \\
\tilde{\lambda}_g(t)^\top &:= -\ell_0 f'_{0,y}[t]M(t)^{-1}.
\end{aligned}
$$

Recall the definitions of h and q:

$$
\begin{aligned}
h(t) &= h_1(t) - f'_y[t]M(t)^{-1}q(t), \\
q(t) &= \dot{h}_2(t) + g'_x[t]h_1(t).
\end{aligned}
$$

Introducing these expressions into (3.19) and collecting terms yields

$$
\begin{aligned}
&\lambda_f^*(h_1(\cdot)) + \lambda_g^*(h_2(\cdot)) \\
&= -\zeta^\top h_2(t_0) - \int_{t_0}^{t_f} (\lambda_f(t)^\top + \lambda_g(t)^\top g_x'[t])h_1(t)dt - \int_{t_0}^{t_f} \lambda_g(t)^\top \dot{h}_2(t)dt
\end{aligned}
\tag{3.20}
$$

with

$$
\lambda_g(t)^\top := \tilde{\lambda}_g(t)^\top - \lambda_f(t)^\top f_y'[t]M(t)^{-1}.
$$

Setting $h_1 \equiv \Theta$ or $h_2 \equiv \Theta$, respectively, proves the following explicit representations of the functionals λ_f^* and λ_g^*:

Corollary 3.1.5. *Let the assumptions of Theorem* 3.1.4 *be valid. Then there exist* $\zeta \in \mathbb{R}^{n_y}$ *and functions* $\lambda_f \in W_{1,\infty}^{n_x}(\mathcal{I})$, $\lambda_g \in L_\infty^{n_y}(\mathcal{I})$ *with*

$$
\lambda_f^*(h_1(\cdot)) = -\int_{t_0}^{t_f} (\lambda_f(t)^\top + \lambda_g(t)^\top g_x'[t])h_1(t)dt,
$$

$$
\lambda_g^*(h_2(\cdot)) = -\zeta^\top h_2(t_0) - \int_{t_0}^{t_f} \lambda_g(t)^\top \dot{h}_2(t)dt
$$

for every $h_1 \in L_\infty^{n_x}(\mathcal{I})$ *and every* $h_2 \in W_{1,\infty}^{n_y}(\mathcal{I})$. □

Corollary 3.1.5 provides useful representations of the functionals λ_f^* and λ_g^* and shows that these multipliers are more regular than only being elements of the dual spaces of $L_\infty^{n_x}(\mathcal{I})$ and $W_{1,\infty}^{n_y}(\mathcal{I})$.

3.1.2 Local minimum principle

First-order necessary optimality conditions in terms of a local minimum principle are derived for Problem 3.1.1.

Definition 3.1.6 (Hamilton function for Problem 3.1.1). Consider Problem 3.1.1. The *Hamilton function* $\mathcal{H} : \mathcal{I} \times \mathbb{R}^{n_x} \times \mathbb{R}^{n_y} \times \mathbb{R}^{n_u} \times \mathbb{R}^{n_x} \times \mathbb{R}^{n_y} \times \mathbb{R} \longrightarrow \mathbb{R}$ is defined by

$$
\mathcal{H}(t,x,y,u,\lambda_f,\lambda_g,\ell_0) := \ell_0 f_0(t,x,y,u) + \lambda_f^\top f(t,x,y,u) + \lambda_g^\top (g_t'(t,x) + g_x'(t,x)f(t,x,y,u)).
$$

□

We investigate the variational equalities and inequalities (3.10)–(3.12) and use the representations of the functionals λ_f^* and λ_g^* provided by Corollary 3.1.5 with $\zeta \in \mathbb{R}^{n_y}$, $\lambda_f \in W_{1,\infty}^{n_x}(\mathcal{I})$, and $\lambda_g \in L_\infty^{n_y}(\mathcal{I})$.

Investigation of (3.10)

The variational equation (3.10)

$$0 = \kappa'_{x_0}x(t_0) + \kappa'_{x_f}x(t_f) + \int_{t_0}^{t_f} \ell_0 f'_{0,x}[t]x(t)dt + \lambda_f^*(\dot{x}(\cdot) - f'_x[\cdot]x(\cdot)) - \lambda_g^*(g'_x[\cdot]x(\cdot))$$

is valid for all $x \in W_{1,\infty}^{n_x}(\mathcal{I})$ and Corollary 3.1.5 yields

$$\begin{aligned}0 = (\kappa'_{x_0} + \zeta^\top g'_x[t_0])x(t_0) + \kappa'_{x_f}x(t_f) + \int_{t_0}^{t_f} \ell_0 f'_{0,x}[t]x(t)dt \\ + \int_{t_0}^{t_f} (\lambda_f(t)^\top + \lambda_g(t)^\top g'_x[t])(f'_x[t]x(t) - \dot{x}(t))dt \\ + \int_{t_0}^{t_f} \lambda_g(t)^\top \frac{d}{dt}(g'_x[t]x(t))dt.\end{aligned}$$

Moreover, using the Hamilton function, we obtain

$$\begin{aligned}0 &= (\kappa'_{x_0} + \zeta^\top g'_x[t_0])x(t_0) + \kappa'_{x_f}x(t_f) + \int_{t_0}^{t_f} \mathcal{H}'_x[t]x(t)dt - \int_{t_0}^{t_f} \lambda_f(t)^\top \dot{x}(t)dt \\ &= (\kappa'_{x_0} + \zeta^\top g'_x[t_0] + \lambda_f(t_0)^\top)x(t_0) + (\kappa'_{x_f} - \lambda_f(t_f)^\top)x(t_f) + \int_{t_0}^{t_f} (\dot{\lambda}_f(t)^\top + \mathcal{H}'_x[t])x(t)dt,\end{aligned} \tag{3.21}$$

where partial integration was used in the last step. To draw further conclusions from this variational equation, we need a variation lemma. The lemma is provided for a more general setting than actually necessary at this stage because we shall use it again in Section 3.2 in the presence of pure state constraints. For such problems, functions of bounded variation enter the scene.

Lemma 3.1.7 (Variational lemma). *Let $f, g \in L_\infty(\mathcal{I})$, $s \in \mathcal{C}^{n_s}(\mathcal{I})$, and $\mu \in BV^{n_s}(\mathcal{I})$. If*

$$\int_{t_0}^{t_f} f(t)h(t) + g(t)\dot{h}(t)dt + \int_{t_0}^{t_f} (s(t)h(t))^\top d\mu(t) = 0$$

holds for every $h \in W_{1,\infty}(\mathcal{I})$ with $h(t_0) = h(t_f) = 0$, then there exists a constant $c \in \mathbb{R}$ and a function $\hat{g} \in BV(\mathcal{I})$ such that $\hat{g}(t) = g(t)$ almost everywhere in $\mathcal{I}$ and

$$\hat{g}(t) = -\int_t^{t_f} f(\tau)d\tau - \int_t^{t_f} s(\tau)^\top d\mu(\tau) - c$$

Moreover, if the Riemann–Stieltjes integral is not present, then

$$\frac{d}{dt}\hat{g}(t) = f(t) \quad \text{a. e. in } \mathcal{I}.$$

Proof. Define

$$F(t) := \int_t^{t_f} f(\tau)d\tau \quad \text{and} \quad G(t) := \int_t^{t_f} s(\tau)^\top d\mu(\tau).$$

With $h(t_0) = h(t_f) = 0$ it holds

$$\int_{t_0}^{t_f} f(t)h(t)dt = -\int_{t_0}^{t_f} h(t)dF(t) = -[F(t)h(t)]_{t_0}^{t_f} + \int_{t_0}^{t_f} F(t)dh(t) = \int_{t_0}^{t_f} F(t)\dot{h}(t)dt,$$

$$\int_{t_0}^{t_f} (s(t)h(t))^\top d\mu(t) = -\int_{t_0}^{t_f} h(t)dG(t) = -[G(t)h(t)]_{t_0}^{t_f} + \int_{t_0}^{t_f} G(t)dh(t) = \int_{t_0}^{t_f} G(t)\dot{h}(t)dt.$$

Moreover, for any constant c and every function $h \in W_{1,\infty}(\mathcal{I})$ with $h(t_0) = h(t_f) = 0$, it holds

$$\int_{t_0}^{t_f} c\dot{h}(t)dt = 0$$

and hence

$$\begin{aligned} 0 &= \int_{t_0}^{t_f} (f(t)h(t) + (g(t) + c)\dot{h}(t))dt + \int_{t_0}^{t_f} (s(t)h(t))^\top d\mu(t) \\ &= \int_{t_0}^{t_f} (F(t) + G(t) + g(t) + c)\dot{h}(t)dt. \end{aligned}$$

The choices

$$c = \frac{1}{t_f - t_0}\int_{t_0}^{t_f} (-F(t) - G(t) - g(t))dt,$$

$$h(t) = \int_t^{t_f} (F(\tau) + G(\tau) + g(\tau) + c)d\tau$$

satisfy $h(t_0) = h(t_f) = 0$ and yield

$$0 = \int_{t_0}^{t_f} (F(t) + G(t) + g(t) + c)^2 dt.$$

Hence, the integrand vanishes almost everywhere in $\mathcal{I}$, and almost everywhere it holds

$$g(t) = -F(t) - G(t) - c = -\int_t^{t_f} f(\tau)d\tau - \int_t^{t_f} s(\tau)^\top d\mu(\tau) - c = \hat{g}(t).$$

□

Application of Lemma 3.1.7 to (3.21) using variations x with $x(t_0) = x(t_f) = 0_{\mathbb{R}^{n_x}}$ yields the condition

$$\dot{\lambda}_f(t)^\top = -\mathcal{H}'_x[t] \quad \text{a. e. in } \mathcal{I}.$$

Then using variations $x(t_0) \neq 0_{\mathbb{R}^{n_x}}$ and $x(t_f) \neq 0_{\mathbb{R}^{n_x}}$, respectively, yields

$$\lambda_f(t_0)^\top = -(\kappa'_{x_0} + \zeta^\top g'_x[t_0]), \quad \lambda_f(t_f)^\top = \kappa'_{x_f}.$$

Investigation of (3.11)

The variational equation (3.11) becomes

$$\begin{aligned}0 &= \int_{t_0}^{t_f} \ell_0 f'_{0,y}[t]y(t)dt + \int_{t_0}^{t_f} (\lambda_f(t)^\top + \lambda_g(t)^\top g'_x[t])f'_y[t]y(t)dt \\ &= \int_{t_0}^{t_f} \mathcal{H}'_y[t]y(t)dt\end{aligned}$$

and it holds for every $y \in L_\infty^{n_y}(\mathcal{I})$. By Lemma 3.1.7 (with $\mu, g, s \equiv \Theta$ and $W_{1,\infty}(\mathcal{I}) \subset L_\infty(\mathcal{I})$), this immediately implies

$$0_{\mathbb{R}^{n_y}} = \mathcal{H}'_y[t]^\top \quad \text{a. e. in } \mathcal{I}.$$

Investigation of (3.12)

The variational inequality (3.12) becomes

$$\begin{aligned}0 &\leq \int_{t_0}^{t_f} \ell_0 f'_{0,u}[t]u(t)dt + \int_{t_0}^{t_f} (\lambda_f(t)^\top + \lambda_g(t)^\top g'_x[t])f'_u[t]u(t)dt \\ &= \int_{t_0}^{t_f} \mathcal{H}'_u[t]u(t)dt\end{aligned}$$

and it holds for every $u \in U_{ad} - \{\hat{u}\}$. It holds

Lemma 3.1.8 (See [179, p. 102]). *Let*

$$0 \le \int_{t_0}^{t_f} \mathcal{H}'_u[t](u(t) - \hat{u}(t))dt$$

hold for every $u \in U_{ad}$. *Then for almost every* $t \in \mathcal{I}$, *it holds*

$$0 \le \mathcal{H}'_u[t](u - \hat{u}(t))$$

for every $u \in \mathcal{U}$.

Proof. Define the absolutely continuous functions

$$h_1(t) := \int_{t_0}^{t} \mathcal{H}'_u[\tau]d\tau \quad \text{and} \quad h_2(t) := \int_{t_0}^{t} \mathcal{H}'_u[\tau]\hat{u}(\tau)d\tau.$$

Let $\mathcal{J}$ be the set of points $t \in (t_0, t_f)$ with

$$\lim_{\varepsilon \downarrow 0} \frac{h_1(t+\varepsilon) - h_1(t)}{\varepsilon} = \lim_{\varepsilon \downarrow 0} \frac{1}{\varepsilon} \int_{t}^{t+\varepsilon} \mathcal{H}'_u[\tau]d\tau = \mathcal{H}'_u[t],$$

$$\lim_{\varepsilon \downarrow 0} \frac{h_2(t+\varepsilon) - h_2(t)}{\varepsilon} = \lim_{\varepsilon \downarrow 0} \frac{1}{\varepsilon} \int_{t}^{t+\varepsilon} \mathcal{H}'_u[\tau]\hat{u}(\tau)d\tau = \mathcal{H}'_u[t]\hat{u}(t).$$

Since h_1 and h_2 are absolutely continuous and differentiable almost everywhere, the sets $\mathcal{J}$ and $\mathcal{I}$ differ only in a set of measure zero. Now let $t \in \mathcal{J}$ and $u \in \mathcal{U}$ be arbitrary and $\varepsilon > 0$ sufficiently small. Define

$$u_\varepsilon(\tau) := \begin{cases} u, & \text{for } \tau \in [t, t+\varepsilon], \\ \hat{u}(\tau), & \text{for } \tau \notin [t, t+\varepsilon]. \end{cases}$$

Then $u_\varepsilon \in U_{ad}$ and the assertion yields

$$0 \le \frac{1}{\varepsilon} \int_{t_0}^{t_f} \mathcal{H}'_u[\tau](u_\varepsilon(\tau) - \hat{u}(\tau))d\tau = \frac{1}{\varepsilon} \int_{t}^{t+\varepsilon} \mathcal{H}'_u[\tau](u - \hat{u}(\tau))d\tau.$$

Taking the limit $\varepsilon \downarrow 0$ yields the assertion. □

We summarize our findings:

Theorem 3.1.9 (Local minimum principle for Problem 3.1.1). *Let the following assumptions hold for Problem* 3.1.1*:*

(i) *Assumptions 2.2.5 and 3.1.3 hold. g is twice continuously differentiable with respect to all arguments.*
(ii) $\mathcal{U} \subseteq \mathbb{R}^{n_u}$ *is closed and convex with non-empty interior.*
(iii) $(\hat{x}, \hat{y}, \hat{u})$ *is a local minimum of Problem 3.1.1.*

Then there exist multipliers

$$\ell_0 \in \mathbb{R}, \quad \lambda_f \in W_{1,\infty}^{n_x}(\mathcal{I}), \quad \lambda_g \in L_\infty^{n_y}(\mathcal{I}), \quad \zeta \in \mathbb{R}^{n_y}, \quad \sigma \in \mathbb{R}^{n_\psi}$$

such that the following conditions are satisfied:
(a) $\ell_0 \geq 0$, $(\ell_0, \zeta, \sigma, \lambda_f, \lambda_g) \neq \Theta$.
(b) Adjoint equations*: Almost everywhere in $\mathcal{I}$, it holds*

$$\dot{\lambda}_f(t) = -\mathcal{H}_x'(t, \hat{x}(t), \hat{y}(t), \hat{u}(t), \lambda_f(t), \lambda_g(t), \ell_0)^\top, \tag{3.22}$$
$$0_{\mathbb{R}^{n_y}} = \mathcal{H}_y'(t, \hat{x}(t), \hat{y}(t), \hat{u}(t), \lambda_f(t), \lambda_g(t), \ell_0)^\top. \tag{3.23}$$

(c) Transversality conditions*:*

$$\lambda_f(t_0)^\top = -(\ell_0 \varphi_{x_0}'(\hat{x}(t_0), \hat{x}(t_f)) + \sigma^\top \psi_{x_0}'(\hat{x}(t_0), \hat{x}(t_f)) + \zeta^\top g_x'(t_0, \hat{x}(t_0))), \tag{3.24}$$
$$\lambda_f(t_f)^\top = \ell_0 \varphi_{x_f}'(\hat{x}(t_0), \hat{x}(t_f)) + \sigma^\top \psi_{x_f}'(\hat{x}(t_0), \hat{x}(t_f)). \tag{3.25}$$

(d) Stationarity of Hamilton function*: Almost everywhere in $\mathcal{I}$, it holds*

$$\mathcal{H}_u'(t, \hat{x}(t), \hat{y}(t), \hat{u}(t), \lambda_f(t), \lambda_g(t), \ell_0)(u - \hat{u}(t)) \geq 0, \tag{3.26}$$

for all $u \in \mathcal{U}$. □

The adjoint equations (3.22) and (3.23) form a DAE of index one for λ_f and λ_g, where λ_f is the differential variable and λ_g denotes the algebraic variable. This follows from (3.23), which is given by

$$0_{\mathbb{R}^{n_y}} = \ell_0 (f_{0,y}'[t])^\top + (f_y'[t])^\top \lambda_f(t) + (g_x'[t] \cdot f_y'[t])^\top \lambda_g(t).$$

Since $g_x'[t] \cdot f_y'[t]$ is non-singular, by Assumption 3.1.3, we obtain

$$\lambda_g(t) = -((g_x'[t] \cdot f_y'[t])^{-1})^\top (\ell_0 (f_{0,y}'[t])^\top + (f_y'[t])^\top \lambda_f(t)).$$

Notice that the original DAE was index-two according to Assumption 3.1.3.

3.1.3 Constraint qualifications and regularity

The linear independence constraint qualification in Corollary 2.3.31 is translated into the context of Problem 3.1.1. Please note that the linear independence constraint qualification and the Mangasarian–Fromowitz condition in Corollary 2.3.32 coincide for Problem 3.1.1 since no cone constraints are present. The constraint qualifications ensure that the multiplier ℓ_0 in Theorem 3.1.9 is not zero and, without loss of generality, can be normalized to one. Surjectivity of the linear operator $H'(\hat{x}, \hat{y}, \hat{u})$ is guaranteed by

Lemma 3.1.10. *Let Assumption 3.1.3 be valid, and let*

$$rank((\psi'_{x_0}\Phi(t_0) + \psi'_{x_f}\Phi(t_f))\Gamma) = n_\psi,$$

where Φ is the fundamental solution of the homogeneous linear differential equation

$$\dot{\Phi}(t) = \tilde{A}(t)\Phi(t), \quad \Phi(t_0) = I_{n_x}, \quad t \in \mathcal{I}$$

and the columns of Γ constitute an orthonormal basis of the null-space of $g'_x[t_0]$ and

$$\begin{aligned} M(t) &= g'_x[t]f'_y[t], \\ \tilde{A}(t) &= f'_x[t] - f'_y[t]M(t)^{-1}Q(t), \\ Q(t) &= \frac{d}{dt}g'_x[t] + g'_x[t]f'_x[t]. \end{aligned}$$

Then $H'(\hat{x}, \hat{y}, \hat{u})$ with H in (3.1) is surjective.

Proof. Let $h_1 \in L_\infty^{n_x}(\mathcal{I})$, $h_2 \in W_{1,\infty}^{n_y}(\mathcal{I})$, and $h_3 \in \mathbb{R}^{n_\psi}$ be arbitrary. Consider the boundary value problem

$$\begin{aligned} H'_1(\hat{x}, \hat{y}, \hat{u})(x, y, u)(t) &= h_1(t) \quad \text{a. e. in } \mathcal{I}, \\ H'_2(\hat{x}, \hat{y}, \hat{u})(x, y, u)(t) &= h_2(t) \quad \text{in } \mathcal{I}, \\ H'_3(\hat{x}, \hat{y}, \hat{u})(x, y, u) &= h_3. \end{aligned}$$

Evaluation of the derivatives leads to the boundary value problem

$$\begin{aligned} -\dot{x}(t) + f'_x[t]x(t) + f'_y[t]y(t) + f'_u[t]u(t) &= h_1(t) \quad \text{a. e. in } \mathcal{I}, \\ g'_x[t]x(t) &= h_2(t) \quad \text{a. e. in } \mathcal{I}, \\ \psi'_{x_0}x(t_0) + \psi'_{x_f}x(t_f) &= -h_3. \end{aligned}$$

By the rank assumption and Assumption 3.1.3, all assumptions of Lemma 3.1.2 are satisfied, and the boundary value problem is solvable. This shows the surjectivity of the mapping $H'(\hat{x}, \hat{y}, \hat{u})$. □

In control theory, the assumptions in Lemma 3.1.10 guarantee *complete controllability* of the linearized dynamics.

For a specific problem, rather than checking the conditions in Lemma 3.1.10, it is often easier either to find a solution of the necessary optimality conditions in Theorem 3.1.9 with $\ell_0 = 1$ or to assume $\ell_0 = 0$ and lead this assumption to a contradiction.

We investigate some examples.

Example 3.1.11. Consider the following index-two DAE optimal control problem (it is equivalent with the minimum energy problem in Example 1.0.1):

Minimize

$$\frac{1}{2}\int_0^1 u(t)^2 dt$$

subject to the constraints

$$\begin{aligned} \dot{x}_1(t) &= u(t) - y(t), & x_1(0) &= x_1(1) = 0, \\ \dot{x}_2(t) &= u(t), & x_2(0) &= -x_2(1) = 1, \\ \dot{x}_3(t) &= -x_2(t), & x_3(0) &= 0, \\ 0 &= x_1(t) + x_3(t). \end{aligned}$$

Differentiation of the algebraic constraint yields

$$0 = u(t) - y(t) - x_2(t) \quad \Longleftrightarrow \quad y(t) = u(t) - x_2(t).$$

Let $(\hat{x}, \hat{y}, \hat{u})$ be a local minimum. With $x = (x_1, x_2, x_3)^\top$, $\lambda_f = (\lambda_{f,1}, \lambda_{f,2}, \lambda_{f,3})^\top$, and the Hamilton function

$$\mathcal{H}(x, y, u, \lambda_f, \lambda_g, \ell_0) = \frac{\ell_0}{2}u^2 + \lambda_{f,1}(u - y) + \lambda_{f,2}u - \lambda_{f,3}x_2 + \lambda_g(u - y - x_2)$$

the necessary conditions in Theorem 3.1.9 read as follows:

$$\begin{aligned} \dot{\lambda}_{f,1}(t) &= 0, & \lambda_{f,1}(0) &= -\sigma_1 - \zeta,\ \lambda_{f,1}(1) = \sigma_4, \\ \dot{\lambda}_{f,2}(t) &= \lambda_{f,3}(t) + \lambda_g(t), & \lambda_{f,2}(0) &= -\sigma_2,\ \lambda_{f,2}(1) = \sigma_5, \\ \dot{\lambda}_{f,3}(t) &= 0, & \lambda_{f,3}(0) &= -\sigma_3 - \zeta,\ \lambda_{f,3}(1) = 0, \\ 0 &= -\lambda_{f,1}(t) - \lambda_g(t), \\ 0 &= \ell_0\hat{u}(t) + \lambda_{f,1}(t) + \lambda_{f,2}(t) + \lambda_g(t). \end{aligned}$$

The adjoint system has the solution

$$\lambda_{f,3}(t) = 0, \quad \lambda_{f,1}(t) = -\sigma_1 - \zeta, \quad \lambda_g(t) = \sigma_1 + \zeta, \quad \lambda_{f,2}(t) = (\sigma_1 + \zeta)t - \sigma_2$$

for all $t \in [0, 1]$. We investigate the case $\ell_0 = 1$. Then:

$$\hat{u}(t) = -\lambda_{f,1}(t) - \lambda_{f,2}(t) - \lambda_g(t) = -(\sigma_1 + \zeta)t + \sigma_2$$

and hence

$$\hat{x}_2(t) = -\frac{1}{2}(\sigma_1 + \zeta)t^2 + \sigma_2 t + c_3 = \hat{u}(t) - \hat{y}(t),$$
$$\hat{x}_1(t) = -\frac{1}{6}(\sigma_1 + \zeta)t^3 + \frac{1}{2}\sigma_2 t^2 + c_3 t + c_4 = -\hat{x}_3(t).$$

With $\zeta := 0$ and the boundary conditions for $\hat{x}_1$, $\hat{x}_2$, and $\hat{x}_3$, we obtain $c_3 = 1$, $c_4 = 0$, $\sigma_1 = \sigma_3 = \sigma_4 = 0$, $\sigma_2 = -\sigma_5 = -2$. Summarizing, the following functions satisfy the necessary optimality conditions in Theorem 3.1.9:

$$\begin{aligned}
\hat{u}(t) &= -\lambda_{f,2}(t) = -2,\\
\hat{x}_1(t) &= -\hat{x}_3(t) = -t^2 + t,\\
\hat{x}_2(t) &= -2t + 1,\\
\hat{y}(t) &= -3 + 2t,\\
\lambda_{f,1}(t) &= \lambda_{f,3}(t) = \lambda_g(t) = 0,\\
\sigma &= (0, -2, 0, 0, 2)^\top, \quad \zeta = 0.
\end{aligned}$$

According to Theorem 2.3.36, this solution is optimal. □

Example 3.1.12 (Discretized 2D Stokes equation). Consider Example 1.1.11 without control constraints:

Minimize

$$\frac{1}{2}\int_0^T \left\|v_h(t) - v_{d,h}(t)\right\|^2 dt + \frac{\alpha}{2}\int_0^T \left\|u_h(t)\right\|^2 dt$$

subject to

$$\dot{v}_h(t) = A_h v_h(t) + B_h p_h(t) + u_h(t), \quad v_h(0) = \Theta,$$
$$\Theta = B_h^\top v_h(t).$$

The Hamilton function for the problem is given by

$$\begin{aligned}
&H(v_h, p_h, u_h, \lambda_f, \lambda_g, \ell_0)\\
&\quad := \frac{1}{2}\ell_0\|v_h - v_{d,h}\|^2 + \frac{\alpha}{2}\ell_0\|u_h\|^2 + (\lambda_f + B_h\lambda_g)^\top(A_h v_h + B_h p_h + u_h).
\end{aligned}$$

Stationarity of the Hamilton function with respect to u_h yields

$$\Theta = \alpha\ell_0 u_h + (\lambda_f + B_h\lambda_g). \tag{3.27}$$

The adjoint DAE computes to

$$\dot{\lambda}_f = -\ell_0(v_h - v_{d,h}) - A_h^\top(\lambda_f + B_h\lambda_g),$$
$$\Theta = B_h^\top\lambda_f + B_h^\top B_h\lambda_g$$

with the transversality condition

$$\lambda_f(T) = \Theta.$$

Let us assume that $\ell_0 = 0$. Stationarity of the Hamilton function then yields

$$\Theta = \lambda_f + B_h \lambda_g$$

and the adjoint DAE and transversality condition reduce to

$$\dot{\lambda}_f = \Theta, \quad \lambda_f(T) = \Theta.$$

Consequently, $\lambda_f \equiv \Theta$, and since $B_h^\top B_h$ is non-singular, also $\lambda_g \equiv \Theta$. Hence, all multipliers are zero, which contradicts the Fritz John conditions. Hence, ℓ_0 can be set to one. The relation in (3.27) then yields

$$u_h = -\frac{1}{\alpha}(\lambda_f + B_h \lambda_g).$$

Introducing this into the Stokes equation yields a linear two-point DAE boundary value problem that needs to be solved:

$$\begin{aligned}
\dot{v}_h &= A_h v_h + B_h p_h - \frac{1}{\alpha}(\lambda_f + B_h \lambda_g), \\
\dot{\lambda}_f &= -(v_h - v_{d,h}) - A_h^\top(\lambda_f + B_h \lambda_g), \\
\Theta &= B_h^\top v_h, \\
\Theta &= B_h^\top \lambda_f + B_h^\top B_h \lambda_g
\end{aligned}$$

with boundary conditions

$$v_h(0) = \Theta, \quad \lambda_f(T) = \Theta.$$

Figures 3.1–3.2 show the numerical solution for $T = 2$, $\alpha = 10^{-5}$, and the desired flow

$$v_d(t,x,y) = \begin{pmatrix} -q(t,x)q_y'(t,y) \\ q(t,y)q_x'(t,x) \end{pmatrix}, \quad q(t,z) = (1-z)^2(1-\cos(2\pi z t)).$$

□

The following example shows that the case $\ell_0 = 0$ actually may occur:

Example 3.1.13 (Degenerated problem). Consider the following optimal control problem:

Minimize

$$\int_0^1 u(t)dt$$

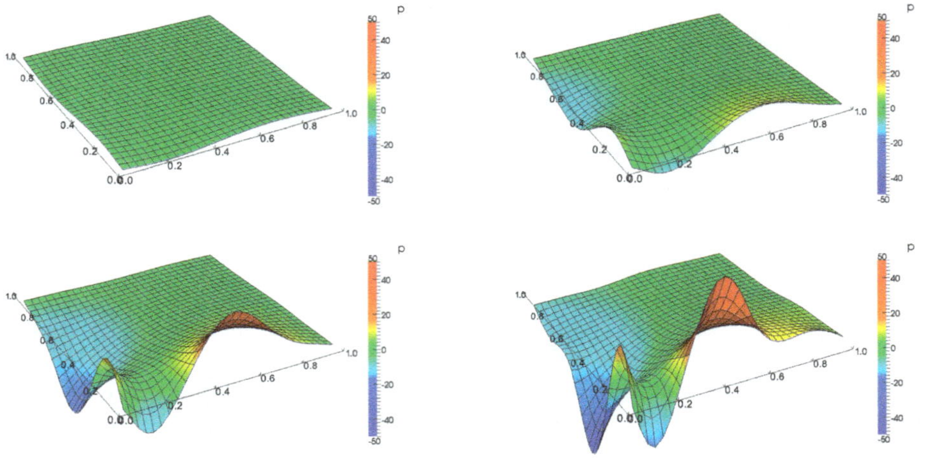

Figure 3.1: Optimal control of Stokes equations. Desired flow (left), controlled flow (middle), and control (right) at $t = 0.6$, $t = 1.0$, $t = 1.4$, and $t = 1.967$.

subject to the constraints

$$\dot{x}(t) = u(t)^2, \quad x(0) = x(1) = 0.$$

Obviously, only $u \equiv 0$ satisfies the constraints, and it is optimal.

With the Hamilton function $\mathcal{H}(x, u, \lambda, \ell_0) = \ell_0 u + \lambda u^2$, it follows

$$\begin{aligned} 0 &= H_u' = \ell_0 + 2\lambda u, \\ \dot{\lambda} &= 0 \quad \Longrightarrow \quad \lambda = \text{const.} \end{aligned}$$

Assume that $\ell_0 \neq 0$. Then $u = -\ell_0/(2\lambda) = \text{const} \neq 0$. However, with this control, we obtain $x(1) > 0$, and u is not feasible. This contradiction shows $\ell_0 = 0$. □

Intuitively, one would expect that the necessary conditions hold with the function $\tilde{\mathcal{H}} := \ell_0 f_0 + \lambda_f^\top f + \lambda_g^\top g$ instead of the Hamilton function $\mathcal{H}$. The following example in [17, Example 3.16] shows that this is not true:

Minimize

$$\frac{1}{2}x_1^2(t_f) + \frac{1}{2}\int_0^{t_f} y(t)^2 + u(t)^2 dt$$

subject to

$$\begin{aligned} \dot{x}_1(t) &= u(t), & x_1(0) &= a, \\ \dot{x}_2(t) &= -y(t) + u(t), & x_2(0) &= 0, \\ 0 &= x_2(t), \end{aligned}$$

where $a \neq 0$.

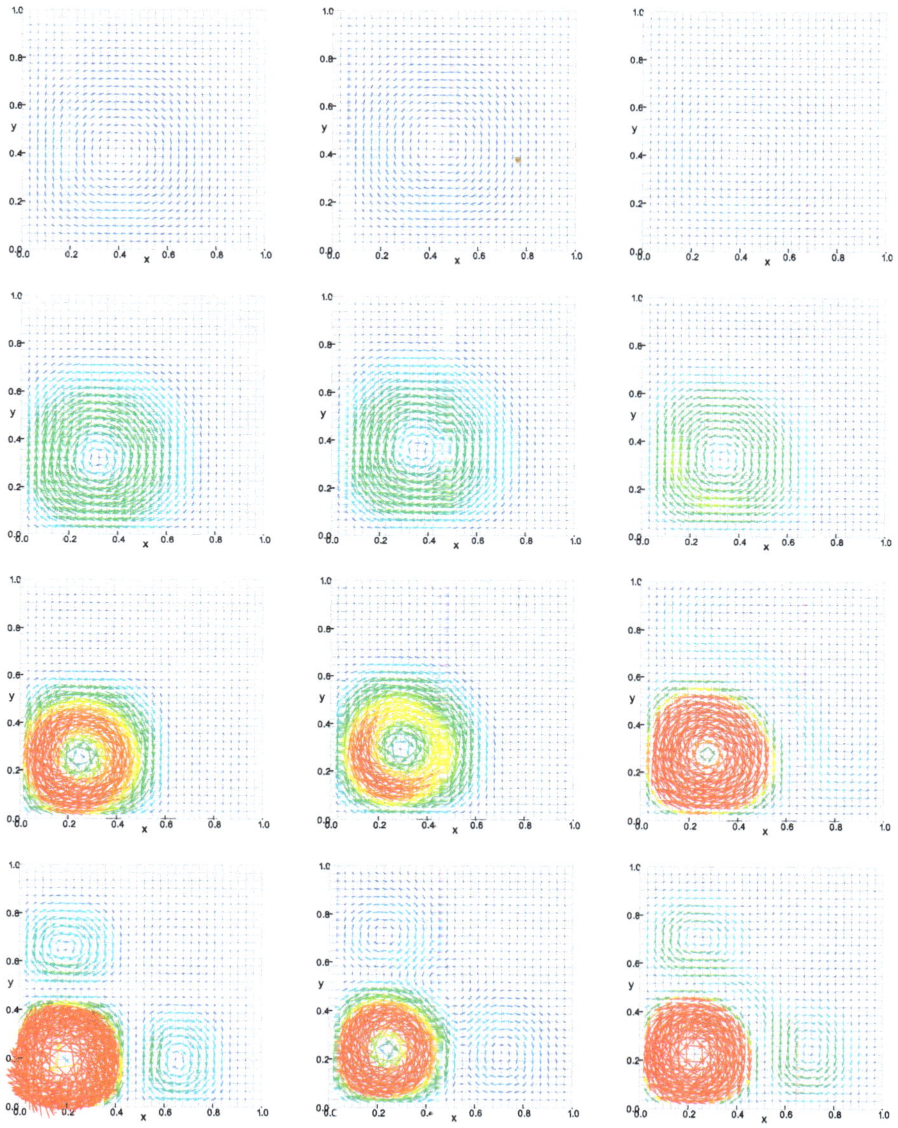

Figure 3.2: Optimal control of Stokes equations. Pressure at $t = 0.6$ (top left), $t = 1.0$ (top right), $t = 1.4$ (bottom left), and $t = 1.967$ (bottom right).

This problem has the solution

$$x_1(t) = a\left(1 - \frac{t}{2+t_f}\right), \quad x_2(t) = 0, \quad y(t) = u(t) = -\frac{a}{2+t_f},$$

but the optimality system with $\tilde{\mathcal{H}}$ instead of the true Hamilton function turns out to be not solvable.

3.2 Problems with pure state constraints

Many applications involve pure state constraints, which we now add to Problem 3.1.1:

Problem 3.2.1 (Index-two DAE optimal control problem with pure state constraints). Let $\mathcal{I} := [t_0, t_f] \subset \mathbb{R}$ be a non-empty compact time interval with $t_0 < t_f$ fixed. Let

$$
\begin{aligned}
\varphi &: \mathbb{R}^{n_x} \times \mathbb{R}^{n_x} \longrightarrow \mathbb{R},\\
f_0 &: \mathcal{I} \times \mathbb{R}^{n_x} \times \mathbb{R}^{n_y} \times \mathbb{R}^{n_u} \longrightarrow \mathbb{R},\\
f &: \mathcal{I} \times \mathbb{R}^{n_x} \times \mathbb{R}^{n_y} \times \mathbb{R}^{n_u} \longrightarrow \mathbb{R}^{n_x},\\
g &: \mathcal{I} \times \mathbb{R}^{n_x} \longrightarrow \mathbb{R}^{n_y},\\
s &: \mathcal{I} \times \mathbb{R}^{n_x} \longrightarrow \mathbb{R}^{n_s},\\
\psi &: \mathbb{R}^{n_x} \times \mathbb{R}^{n_x} \longrightarrow \mathbb{R}^{n_\psi}
\end{aligned}
$$

be sufficiently smooth functions and $\mathcal{U} \subseteq \mathbb{R}^{n_u}$ a closed convex set with non-empty interior.

Minimize

$$
\varphi\big(x(t_0), x(t_f)\big) + \int_{t_0}^{t_f} f_0\big(t, x(t), y(t), u(t)\big)dt
$$

with respect to $x \in W^{n_x}_{1,\infty}(\mathcal{I})$, $y \in L^{n_y}_{\infty}(\mathcal{I})$, $u \in L^{n_u}_{\infty}(\mathcal{I})$ subject to the constraints

$$
\begin{aligned}
\dot{x}(t) &= f\big(t, x(t), y(t), u(t)\big) && \text{a. e. in } \mathcal{I},\\
0_{\mathbb{R}^{n_y}} &= g\big(t, x(t)\big) && \text{in } \mathcal{I},\\
0_{\mathbb{R}^{n_\psi}} &= \psi\big(x(t_0), x(t_f)\big),\\
s\big(t, x(t)\big) &\le 0_{\mathbb{R}^{n_s}} && \text{in } \mathcal{I},\\
u(t) &\in \mathcal{U} && \text{a. e. in } \mathcal{I}.
\end{aligned}
$$

Similarly, as in Section 3.1, we interpret Problem 3.2.1 as the infinite optimization problem.

Minimize $J(z)$ subject to $z \in S$, $G(z) \in K$, and $H(z) = \Theta_V$

given in Section 2.2 with

$$
\begin{aligned}
Z &= W^{n_x}_{1,\infty}(\mathcal{I}) \times L^{n_y}_{\infty}(\mathcal{I}) \times L^{n_u}_{\infty}(\mathcal{I}),\\
V &= L^{n_x}_{\infty}(\mathcal{I}) \times W^{n_y}_{1,\infty}(\mathcal{I}) \times \mathbb{R}^{n_\psi},\\
W &= \mathcal{C}^{n_s}(\mathcal{I}),\\
J(x, y, u) &= \varphi(x(t_0), x(t_f)) + \int_{t_0}^{t_f} f_0(t, x(t), y(t), u(t))dt,
\end{aligned}
$$

$$
\begin{aligned}
H(x,y,u) &= \begin{pmatrix} H_1(x,y,u) \\ H_2(x,y,u) \\ H_3(x,y,u) \end{pmatrix} = \begin{pmatrix} f(\cdot,x(\cdot),y(\cdot),u(\cdot)) - \dot{x}(\cdot), \\ g(\cdot,x(\cdot)), \\ -\psi(x(t_0),x(t_f)), \end{pmatrix}, \\
G(x,y,u) &= -s(\cdot,x(\cdot)), \\
K &= \{k \in C^{n_s}(\mathcal{I}) \mid k(t) \geq 0_{\mathbb{R}^{n_s}} \text{ in } \mathcal{I}\}, \\
S &= W_{1,\infty}^{n_x}(\mathcal{I}) \times L_{\infty}^{n_y}(\mathcal{I}) \times U_{ad}, \\
U_{ad} &= \{u \in L_{\infty}^{n_u}(\mathcal{I}) \mid u(t) \in \mathcal{U} \text{ almost everywhere in } \mathcal{I}\}.
\end{aligned}
$$

Please note that K is a closed convex cone with non-empty interior. With the same reasoning as in Section 3.1, the assumptions in Theorem 2.3.23 are satisfied at a local minimum $\hat{z} = (\hat{x},\hat{y},\hat{u})$ of Problem 3.2.1, and we obtain the following result:

Theorem 3.2.2 (Necessary conditions for Problem 3.2.1). *Let the following assumptions be fulfilled for Problem 3.2.1:*
(i) *Assumptions 2.2.5 and 3.1.3 hold.*
(ii) $\mathcal{U} \subseteq \mathbb{R}^{n_u}$ *is closed and convex with non-empty interior.*
(iii) $(\hat{x},\hat{y},\hat{u})$ *is a local minimum of Problem 3.2.1.*

Then there exist non-trivial multipliers $\ell_0 \geq 0$, $\lambda^* \in V^*$, *and* $\mu^* \in W^*$ *with*

$$\mu^* \in K^+ \quad \text{and} \quad \mu^*(G(\hat{x},\hat{y},\hat{u})) = 0, \tag{3.28}$$

and

$$
\begin{aligned}
0 \leq \ell_0 J'(\hat{x},\hat{y},\hat{u})(x,y,u) &- \mu^*(G'(\hat{x},\hat{y},\hat{u})(x,y,u)) \\
&- \lambda^*(H'(\hat{x},\hat{y},\hat{u})(x,y,u))
\end{aligned} \tag{3.29}
$$

for all $(x,y,u) \in S - \{(\hat{x},\hat{y},\hat{u})\}$. □

Again, the multiplier $\lambda^* := (\lambda_f^*, \lambda_g^*, \sigma^*)$ is an element of the dual space

$$V^* = L_{\infty}^{n_x}(\mathcal{I})^* \times W_{1,\infty}^{n_y}(\mathcal{I})^* \times (\mathbb{R}^{n_\psi})^*$$

and explicit representations are required. The multiplier μ^* is in the dual space $W^* = C^{n_s}(\mathcal{I})^*$. According to Riesz' Representation Theorem 2.1.22, the functional μ^* possesses the explicit representation

$$\mu^*(h) = \int_{t_0}^{t_f} h(t)^\top d\mu(t) := \sum_{i=1}^{n_s} \int_{t_0}^{t_f} h_i(t) d\mu_i(t) \tag{3.30}$$

for every continuous vector function $h \in C^{n_s}(\mathcal{I})$. Herein, μ_i, $i = 1,\dots,n_s$, are functions of bounded variation. To make the representations unique, we choose $\mu \in NBV^{n_s}(\mathcal{I})$, compare Definition 2.1.23. With the explicit representation (3.30), the variational inequality (3.29) yields the following three separate variational equalities and inequalities

$$
\begin{aligned}
0 &= \kappa'_{x_0} x(t_0) + \kappa'_{x_f} x(t_f) + \int_{t_0}^{t_f} \ell_0 f'_{0,x}[t]x(t)dt + \int_{t_0}^{t_f} (s'_x[t]x(t))^\top d\mu(t) \\
&\quad + \lambda_f^*(\dot{x}(\cdot) - f'_x[\cdot]x(\cdot)) - \lambda_g^*(g'_x[\cdot]x(\cdot)), && (3.31)\\
0 &= \int_{t_0}^{t_f} \ell_0 f'_{0,y}[t]y(t)dt - \lambda_f^*(f'_y[\cdot]y(\cdot)), && (3.32)\\
0 &\le \int_{t_0}^{t_f} \ell_0 f'_{0,u}[t]u(t)dt - \lambda_f^*(f'_u[\cdot]u(\cdot)), && (3.33)
\end{aligned}
$$

which hold for *every* $x \in W^{n_x}_{1,\infty}(\mathcal{I})$, *every* $y \in L^{n_y}_{\infty}(\mathcal{I})$, and *every* $u \in U_{ad} - \{\hat{u}\}$.

3.2.1 Representation of multipliers

Explicit representations of the functionals λ_f^* and λ_g^* are obtained similarly as in Subsection 3.1.1. Let $h_1 \in L^{n_x}_{\infty}(\mathcal{I})$ and $h_2 \in W^{n_y}_{1,\infty}(\mathcal{I})$ be arbitrary.

Consider the initial value problem (3.13)–(3.14) and its solution (3.15)–(3.16). Adding equations (3.31) and (3.32), exploiting the linearity of the functionals, introducing the solution formulas (3.15)–(3.16), and using integration by parts as in (3.18) yields

$$
\begin{aligned}
0 &= \left(\kappa'_{x_0} + \kappa'_{x_f}\Phi(t_f) + \int_{t_0}^{t_f} \ell_0 \hat{f}_0[t]\Phi(t)dt + \left(\left(\int_{t_0}^{t_f} (s'_x[t]\Phi(t))^\top d\mu(t) \right)^\top \right) \right) \Pi h_2(t_0) \\
&\quad + \int_{t_0}^{t_f} \left(\kappa'_{x_f}\Phi(t_f) + \int_{t}^{t_f} \ell_0 \hat{f}_0[\tau]\Phi(\tau)d\tau \right) \Phi^{-1}(t)h(t)dt \\
&\quad - \int_{t_0}^{t_f} \ell_0 f'_{0,y}[t]M(t)^{-1}q(t)dt + \int_{t_0}^{t_f} \left(s'_x[t]\Phi(t) \int_{t_0}^{t} \Phi^{-1}(\tau)h(\tau)d\tau \right)^\top d\mu(t) \\
&\quad + \lambda_f^*(h_1(\cdot)) + \lambda_g^*(h_2(\cdot)), && (3.34)
\end{aligned}
$$

where

$$\hat{f}_0[t] := f'_{0,x}[t] - f'_{0,y}[t]M(t)^{-1}Q(t).$$

The Riemann–Stieltjes integral is to be transformed using integration by parts. We exploit

Lemma 3.2.3. *Let $a : \mathcal{I} \longrightarrow \mathbb{R}^n$ be continuous, $b : \mathcal{I} \longrightarrow \mathbb{R}^n$ absolutely continuous, and $\mu : \mathcal{I} \to \mathbb{R}$ of bounded variation. Then:*

$$\int_{t_0}^{t_f} a(t)^\top b(t) d\mu(t) = \left(\int_{t_0}^{t_f} a(\tau) d\mu(\tau) \right)^\top b(t_f) - \int_{t_0}^{t_f} \left(\int_{t_0}^{t} a(\tau) d\mu(\tau) \right)^\top \dot{b}(t) dt.$$

Proof. Using integration by parts for Riemann–Stieltjes integrals leads to

$$\begin{aligned}
&\int_{t_0}^{t_f} a(t)^\top b(t) d\mu(t) \\
&= \sum_{i=1}^{n} \int_{t_0}^{t_f} a_i(t) b_i(t) d\mu(t) \\
&= \sum_{i=1}^{n} \int_{t_0}^{t_f} b_i(t) d\left(\int_{t_0}^{t} a_i(\tau) d\mu(\tau) \right) \\
&= \sum_{i=1}^{n} \left(\left[\left(\int_{t_0}^{t} a_i(\tau) d\mu(\tau) \right) \cdot b_i(t) \right]_{t_0}^{t_f} - \int_{t_0}^{t_f} \left(\int_{t_0}^{t} a_i(\tau) d\mu(\tau) \right) db_i(t) \right) \\
&= \left[\left(\int_{t_0}^{t} a(\tau) d\mu(\tau) \right)^\top b(t) \right]_{t_0}^{t_f} - \int_{t_0}^{t_f} \left(\int_{t_0}^{t} a(\tau) d\mu(\tau) \right)^\top \dot{b}(t) dt \\
&= \left(\int_{t_0}^{t_f} a(\tau) d\mu(\tau) \right)^\top b(t_f) - \int_{t_0}^{t_f} \left(\int_{t_0}^{t} a(\tau) d\mu(\tau) \right)^\top \dot{b}(t) dt.
\end{aligned}$$

□

Component wise application of Lemma 3.2.3 with

$$a(t)^\top = a_i(t)^\top = s'_{i,x}[t]\Phi(t), \quad i = 1, \dots, n_s, \quad \text{and} \quad b(t) = \int_{t_0}^{t} \Phi^{-1}(\tau) h(\tau) d\tau$$

yields

$$\int_{t_0}^{t_f} \left(s'_x[t]\Phi(t) \int_{t_0}^{t} \Phi^{-1}(\tau) h(\tau) d\tau \right)^\top d\mu(t) = \int_{t_0}^{t_f} \left(\int_{t}^{t_f} (s'_x[\tau]\Phi(\tau))^\top d\mu(\tau) \right)^\top \Phi^{-1}(t) h(t) dt.$$

Introducing this relation into (3.34), using the definitions of h and q, and collecting terms yields

$$\lambda_f^*(h_1(\cdot)) + \lambda_g^*(h_2(\cdot)) = -\zeta^\top h_2(t_0) - \int_{t_0}^{t_f} (\lambda_f(t)^\top + \lambda_g(t)^\top g_x'[t])h_1(t)dt - \int_{t_0}^{t_f} \lambda_g(t)^\top \dot{h}_2(t)dt \tag{3.35}$$

with

$$\zeta^\top := \left(\kappa_{x_0}' + \kappa_{x_f}'\Phi(t_f) + \int_{t_0}^{t_f} \ell_0 \hat{f}_0[t]\Phi(t)dt + \left(\int_{t_0}^{t_f} (s_x'[t]\Phi(t))^\top d\mu(t)\right)^\top\right)\Pi,$$

$$\lambda_f(t)^\top := \left(\kappa_{x_f}'\Phi(t_f) + \int_t^{t_f} \ell_0 \hat{f}_0[\tau]\Phi(\tau)d\tau + \left(\int_t^{t_f} (s_x'[\tau]\Phi(\tau))^\top d\mu(\tau)\right)^\top\right)\Phi^{-1}(t),$$

$$\lambda_g(t)^\top := -(\ell_0 f_{0,y}'[t] + \lambda_f(t)^\top f_y'[t])M(t)^{-1}.$$

Setting $h_1 \equiv \Theta$ or $h_2 \equiv \Theta$, respectively, proves the following explicit representations of the functionals λ_f^* and λ_g^*:

Corollary 3.2.4. *Let the assumptions of Theorem 3.2.2 be valid. Then there exist $\zeta \in \mathbb{R}^{n_y}$ and functions $\lambda_f \in BV^{n_x}(\mathcal{I})$, $\lambda_g \in L_\infty^{n_y}(\mathcal{I})$ with*

$$\lambda_f^*(h_1(\cdot)) = -\int_{t_0}^{t_f} (\lambda_f(t)^\top + \lambda_g(t)^\top g_x'[t])h_1(t)dt,$$

$$\lambda_g^*(h_2(\cdot)) = -\zeta^\top h_2(t_0) - \int_{t_0}^{t_f} \lambda_g(t)^\top \dot{h}_2(t)dt$$

for every $h_1 \in L_\infty^{n_x}(\mathcal{I})$ and every $h_2 \in W_{1,\infty}^{n_y}(\mathcal{I})$. □

3.2.2 Local minimum principle

We investigate the variational equalities and inequalities (3.31)–(3.33) and use the representations of the functionals λ_f^* and λ_g^* provided by Corollary 3.2.4 with $\zeta \in \mathbb{R}^{n_y}$, $\lambda_f \in BV^{n_x}(\mathcal{I})$, and $\lambda_g \in L_\infty^{n_y}(\mathcal{I})$.

Investigation of (3.31)

The variational equation (3.31) holds for all $x \in W_{1,\infty}^{n_x}(\mathcal{I})$ and becomes

$$0 = \kappa_{x_0}'x(t_0) + \kappa_{x_f}'x(t_f) + \int_{t_0}^{t_f} \ell_0 f_{0,x}'[t]x(t)dt + \int_{t_0}^{t_f} (s_x'[t]x(t))^\top d\mu(t)$$

$$
\begin{aligned}
&+ \lambda_f^*(\dot{x}(\cdot) - f_x'[\cdot]x(\cdot)) - \lambda_g^*(g_x'[\cdot]x(\cdot)) \\
&= (\kappa_{x_0}' + \zeta^\top g_x'[t_0])x(t_0) + \kappa_{x_f}' x(t_f) + \int_{t_0}^{t_f} \ell_0 f_{0,x}'[t]x(t)dt \\
&+ \int_{t_0}^{t_f} (\lambda_f(t)^\top + \lambda_g(t)^\top g_x'[t])(f_x'[t]x(t) - \dot{x}(t))dt \\
&+ \int_{t_0}^{t_f} \lambda_g(t)^\top \frac{d}{dt}(g_x'[t]x(t))dt + \int_{t_0}^{t_f} (s_x'[t]x(t))^\top d\mu(t) \\
&= (\kappa_{x_0}' + \zeta^\top g_x'[t_0])x(t_0) + \kappa_{x_f}' x(t_f) \\
&+ \int_{t_0}^{t_f} \mathcal{H}_x'[t]x(t)dt + \int_{t_0}^{t_f} (s_x'[t]x(t))^\top d\mu(t) - \int_{t_0}^{t_f} \lambda_f(t)^\top \dot{x}(t)dt. \qquad (3.36)
\end{aligned}
$$

Let $x \in W_{1,\infty}^{n_x}(\mathcal{I})$ be an arbitrary variation with $x(t_0) = x(t_f) = 0_{\mathbb{R}^{n_x}}$. Then Lemma 3.1.7 applied to (3.36) yields

$$
C = \lambda_f(t) - \int_t^{t_f} \mathcal{H}_x'[\tau]^\top d\tau - \int_t^{t_f} s_x'[\tau]^\top d\mu(\tau)
$$

with some constant vector $C \in \mathbb{R}^{n_x}$. Evaluation at $t = t_f$ yields $C = \lambda_f(t_f)$ and

$$
\lambda_f(t) = \lambda_f(t_f) + \int_t^{t_f} \mathcal{H}_x'[\tau]^\top d\tau + \int_t^{t_f} s_x'[\tau]^\top d\mu(\tau) \qquad (3.37)
$$

for $t \in \mathcal{I}$.

Application of a computation rule for Riemann–Stieltjes integrals to Equation (3.36) yields

$$
\begin{aligned}
0 &= (\kappa_{x_0}' + \zeta^\top g_x'[t_0])x(t_0) - \kappa_{x_f}' x(t_f) \\
&+ \int_{t_0}^{t_f} \mathcal{H}_x'[t]x(t)dt + \int_{t_0}^{t_f} (s_x'[t]x(t))^\top d\mu(t) - \int_{t_0}^{t_f} \lambda_f(t)^\top dx(t)
\end{aligned}
$$

and integration by parts of the last term yields

$$
\begin{aligned}
0 &= (\kappa_{x_0}' + \zeta^\top g_x'[t_0] + \lambda_f(t_0)^\top)x(t_0) + (\kappa_{x_f}' - \lambda_f(t_f)^\top)x(t_f) \\
&+ \int_{t_0}^{t_f} \mathcal{H}_x'[t]x(t)dt + \int_{t_0}^{t_f} (s_x'[t]x(t))^\top d\mu(t) + \int_{t_0}^{t_f} x(t)^\top d\lambda_f(t).
\end{aligned}
$$

Introducing (3.37) into the last integral leads to

$$0 = (\kappa'_{x_0} + \zeta^\top g'_x[t_0] + \lambda_f(t_0)^\top)x(t_0) + (\kappa'_{x_f} - \lambda_f(t_f)^\top)x(t_f),$$

which holds for every variation x. It follows

$$\lambda_f(t_0)^\top = -(\kappa'_{x_0} + \zeta^\top g'_x[t_0]), \quad \lambda_f(t_f)^\top = \kappa'_{x_f}.$$

Investigation of (3.32) and (3.33)

The variational equation (3.32) and the variational inequality (3.33) can be analyzed literally as in Subsection 3.1.2.

Investigation of the complementarity condition (3.28)

The complementarity condition (3.28) reads as

$$\mu^* \in K^+ \Longleftrightarrow \int_{t_0}^{t_f} k(t)^\top d\mu(t) \geq 0 \quad \text{for } k \in \mathcal{C}^{n_s}(\mathcal{I}),\ k(t) \geq 0_{\mathbb{R}^{n_s}} \text{ in } \mathcal{I},$$

$$\mu^*(G(\hat{x}, \hat{y}, \hat{u})) = 0 \Longleftrightarrow \int_{t_0}^{t_f} s(t, \hat{x}(t))^\top d\mu(t) = 0.$$

Lemma 3.2.5. *Let $\mu \in BV(\mathcal{I})$. If*

$$\int_{t_0}^{t_f} f(t) d\mu(t) \geq 0$$

holds for every non-negative $f \in \mathcal{C}(\mathcal{I})$, then μ is non-decreasing in $\mathcal{I}$.

Proof. According to the assumption of the lemma, for every non-negative continuous function f, it holds

$$S_f := \int_{t_0}^{t_f} f(t) d\mu(t) \geq 0.$$

Hence, for every non-negative continuous function and every $\varepsilon > 0$, there exists $\delta > 0$ such that for every partition $\mathbb{G}_N = \{t_0 < t_1 < \cdots < t_N = t_f\}$ with

$$\max_{i=1,\ldots,N} \{t_i - t_{i-1}\} < \delta$$

it holds

$$-\varepsilon \leq -\varepsilon + S_f < \sum_{i=1}^{N} f(\xi_i)(\mu(t_i) - \mu(t_{i-1})) < \varepsilon + S_f, \tag{3.38}$$

where ξ_i is an arbitrary point in $[t_{i-1}, t_i]$, $i = 1, \dots, N$.

Assume that μ is not non-decreasing. Then there are points $\underline{t} < \overline{t}$ with $\mu(\underline{t}) = \mu(\overline{t}) + \gamma$ and $\gamma > 0$. Let $\varepsilon := \gamma/2$ and $\delta > 0$ be arbitrary and

$$\mathbb{G}_N := \{t_0 < t_1 < \cdots < t_p := \underline{t} < \cdots < t_q := \overline{t} < \cdots < t_N = t_f\}$$

with

$$\max_{i=1,\dots,N} \{t_i - t_{i-1}\} < \delta.$$

Then there exists a continuous non-negative function f with $f(t) \equiv 1$ in $[\underline{t}, \overline{t}]$, $f(t_i) = 0$ for $i \notin \{p, \dots, q\}$, and f linear in $[t_{p-1}, t_p]$ and $[t_q, t_{q+1}]$. Then

$$\sum_{i=1}^{N} f(t_{i-1})(\mu(t_i) - \mu(t_{i-1})) = \mu(\overline{t}) - \mu(\underline{t}) = -\gamma < -\varepsilon.$$

This contradicts (3.38). □

Lemma 3.2.6. *Let $\mu \in BV(\mathcal{I})$ be non-decreasing, and let $f \in C(\mathcal{I})$ be non-positive. If*

$$\int_{t_0}^{t_f} f(t)\,d\mu(t) = 0$$

holds, then μ is constant on every interval $[a, b] \subseteq \mathcal{I}$ with $a < b$ and $f(t) < 0$ in $[a, b]$.

Proof. Since $S_f = 0$ by assumption, we find that for every $\varepsilon > 0$, there exists $\delta > 0$ such that for every partition $\mathbb{G}_N = \{t_0 < t_1 < \cdots < t_N = t_f\}$ with

$$\max_{i=1,\dots,N} \{t_i - t_{i-1}\} < \delta$$

it holds

$$-\varepsilon < \sum_{i=1}^{N} f(\xi_i)(\mu(t_i) - \mu(t_{i-1})) < \varepsilon, \tag{3.39}$$

where ξ_i is an arbitrary point in $[t_{i-1}, t_i]$, $i = 1, \dots, N$.

Assume that μ is not constant in intervals $[a, b]$ of the above type.

Then there exist points $\underline{t} < \overline{t}$ with $f(t) \le -\alpha < 0$ for all $t \in [\underline{t}, \overline{t}]$ and $\mu(\underline{t}) = \mu(\overline{t}) - \gamma$, $\gamma > 0$ (because μ is non-decreasing). Choose $\varepsilon := \alpha\gamma/2$. Let $\delta > 0$ be arbitrary and

$$\mathbb{G}_N = \{t_0 < t_1 < \cdots < t_p = \underline{t} < \cdots < t_q = \overline{t} < \cdots < t_N = t_f\}$$

with $\max_{i=1,\dots,N}\{t_i - t_{i-1}\} < \delta$. Then,

$$\begin{aligned}\sum_{i=1}^{N} f(\xi_{i-1})(\mu(t_i)-\mu(t_{i-1})) &= \sum_{i=1}^{p} \underbrace{f(\xi_{i-1})}_{\leq 0}\underbrace{(\mu(t_i)-\mu(t_{i-1}))}_{\geq 0} \\ &\quad + \sum_{i=p+1}^{q} f(\xi_{i-1})(\mu(t_i)-\mu(t_{i-1})) \\ &\quad + \sum_{i=q+1}^{N} \underbrace{f(\xi_{i-1})}_{\leq 0}\underbrace{(\mu(t_i)-\mu(t_{i-1}))}_{\geq 0} \\ &\leq \sum_{i=p+1}^{q} f(\xi_{i-1})(\mu(t_i)-\mu(t_{i-1})) \\ &\leq -\alpha \sum_{i=p+1}^{q} (\mu(t_i)-\mu(t_{i-1})) \\ &= -\alpha(\mu(\bar{t})-\mu(\underline{t})) \\ &= -\alpha\gamma < -\varepsilon.\end{aligned}$$

This contradicts (3.39). □

Application of Lemmas 3.2.5 and 3.2.6 to the complementarity system together with our previous findings yields

Theorem 3.2.7 (Local minimum principle for Problem 3.2.1). *Let the following assumptions hold for Problem* 3.2.1*:*

(i) *Assumptions* 2.2.5 *and* 3.1.3 *hold. g is twice continuously differentiable with respect to all arguments.*
(ii) $\mathcal{U} \subseteq \mathbb{R}^{n_u}$ *is closed and convex with non-empty interior.*
(iii) $(\hat{x}, \hat{y}, \hat{u})$ *is a local minimum of Problem* 3.2.1.

Then there exist multipliers

$$\ell_0 \in \mathbb{R}, \quad \lambda_f \in BV^{n_x}(\mathcal{I}), \quad \lambda_g \in L_\infty^{n_y}(\mathcal{I}), \quad \mu \in NBV^{n_s}(\mathcal{I}), \quad \zeta \in \mathbb{R}^{n_y}, \quad \sigma \in \mathbb{R}^{n_\psi}$$

such that the following conditions are satisfied:

(a) $\ell_0 \geq 0$, $(\ell_0, \zeta, \sigma, \lambda_f, \lambda_g, \mu) \neq \Theta$,
(b) Adjoint equations*: Almost everywhere in* $\mathcal{I}$ *it holds*

$$\lambda_f(t) = \lambda_f(t_f) + \int_t^{t_f} \mathcal{H}'_x(\tau, \hat{x}(\tau), \hat{y}(\tau), \hat{u}(\tau), \lambda_f(\tau), \lambda_g(\tau), \ell_0)^\top d\tau, + \int_t^{t_f} s'_x(\tau, \hat{x}(\tau))^\top d\mu(\tau) \tag{3.40}$$

$$0_{\mathbb{R}^{n_y}} = \mathcal{H}'_y(t, \hat{x}(t), \hat{y}(t), \hat{u}(t), \lambda_f(t), \lambda_g(t), \ell_0)^\top. \tag{3.41}$$

(c) Transversality conditions:

$$\lambda_f(t_0)^\top = -(\ell_0\varphi'_{x_0}(\hat{x}(t_0),\hat{x}(t_f)) + \sigma^\top\psi'_{x_0}(\hat{x}(t_0),\hat{x}(t_f)) + \zeta^\top g'_x(t_0,\hat{x}(t_0))), \quad (3.42)$$

$$\lambda_f(t_f)^\top = \quad \ell_0\varphi'_{x_f}(\hat{x}(t_0),\hat{x}(t_f)) + \sigma^\top\psi'_{x_f}(\hat{x}(t_0),\hat{x}(t_f)). \quad (3.43)$$

(d) Stationarity of Hamilton function: *Almost everywhere in $\mathcal{I}$ it holds*

$$\mathcal{H}'_u(t,\hat{x}(t),\hat{y}(t),\hat{u}(t),\lambda_f(t),\lambda_g(t),\ell_0)(u-\hat{u}(t)) \geq 0, \quad (3.44)$$

for all $u \in \mathcal{U}$.

(e) Complementarity condition: *μ_i, $i \in \{1,\dots,n_s\}$, is non-decreasing on $\mathcal{I}$ and constant on every interval (t_1,t_2) with $t_1 < t_2$ and $s_i(t,\hat{x}(t)) < 0$ for all $t \in (t_1,t_2)$.* □

The constraint qualification of Mangasarian–Fromowitz in Corollary 2.3.32 together with Lemma 3.1.10 translates as follows for Problem 3.2.1 since the interior of

$$K = \{k \in \mathcal{C}^{n_s}(\mathcal{I}) \mid k(t) \geq 0_{\mathbb{R}^{n_s}} \text{ in } \mathcal{I}\}$$

is given by

$$\operatorname{int}(K) = \{k \in \mathcal{C}^{n_s}(\mathcal{I}) \mid k(t) > 0_{\mathbb{R}^{n_s}} \text{ in } \mathcal{I}\}.$$

Lemma 3.2.8. *Let the assumptions of Theorem 3.2.7 and Lemma 3.1.10 hold. Let there exist $x \in W^{n_x}_{1,\infty}(\mathcal{I})$, $y \in L^{n_y}_\infty(\mathcal{I})$, and $u \in \operatorname{int}(\mathcal{U}_{ad} - \{\hat{u}\})$ with*

$$\begin{aligned} s[t] + s'_x[t]x(t) &< 0_{\mathbb{R}^{n_s}} && \text{in } \mathcal{I},\\ f'_x[t]x(t) + f'_y[t]y(t) + f'_u[t]u(t) - \dot{x}(t) &= 0_{\mathbb{R}^{n_x}} && \text{a. e. in } \mathcal{I},\\ g'_x[t]x(t) &= 0_{\mathbb{R}^{n_y}} && \text{in } \mathcal{I},\\ \psi'_{x_0}x(t_0) + \psi'_{x_f}x(t_f) &= 0_{\mathbb{R}^{n_\psi}}. \end{aligned}$$

Then the Mangasarian–Fromowitz constraint qualification holds for Problem 3.2.1, and $\ell_0 = 1$ can be chosen in Theorem 3.2.7. □

3.2.3 Finding controls on active state constraint arcs

We outline how the control can be computed on an active state constraint arc and need some terminology, which is illustrated in Figure 3.3.

Definition 3.2.9 (Active/inactive/boundary arc, junction/contact/touch point). Let $(x(t), y(t), u(t))$ be feasible for Problem 3.2.1.

(a) The constraint s_j, $j \in \{1,\dots,n_s\}$, is called *active for $t \in [t_1,t_2] \subseteq \mathcal{I}$*, if $s_j(t,x(t)) = 0$ holds for $t \in [t_1,t_2]$. If $s_j(t,x(t)) < 0$ at $t \in \mathcal{I}$, then s_j is called *inactive at t*.

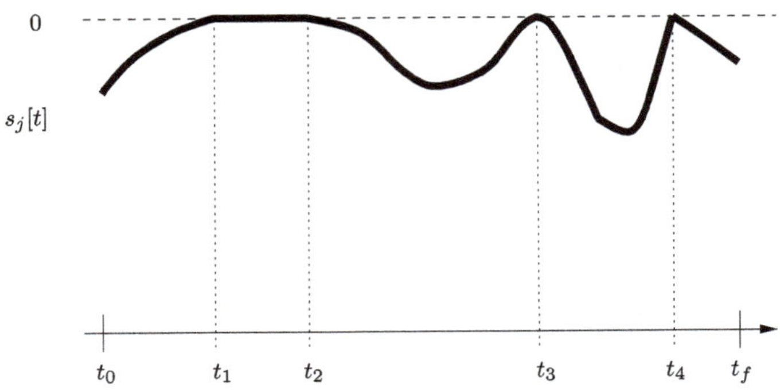

Figure 3.3: Boundary arc $[t_1, t_2]$, touch point t_3, and contact point t_4.

(b) If s_j is active on $[t_1, t_2] \subseteq \mathcal{I}$ with $t_1 < t_2$, then $[t_1, t_2]$ is called *boundary arc of* s_j.
If s_j is inactive on $[t_1, t_2] \subseteq \mathcal{I}$ with $t_1 < t_2$, then $[t_1, t_2]$ is called *free arc of* s_j.
The point t_1 is called *junction point* of the boundary arc $[t_1, t_2]$ of s_j, if there is $\delta > 0$ such that s_j is inactive for all $t \in [t_1 - \delta, t_1)$.
Likewise, t_2 is called *junction point* of the boundary arc $[t_1, t_2]$ of s_j, if there is $\delta > 0$ such that s_j is inactive for all $t \in (t_2, t_2 + \delta]$.

(c) A point $t_1 \in \mathcal{I}$ is called *contact point of* s_j, if s_j is active at t_1 and there is $\delta > 0$ such that s_j is inactive for all $t \in [t_1 - \delta, t_1 + \delta] \setminus \{t_1\}$.

(d) A contact point t_1 of s_j is called *touch point*, if $\frac{d}{dt} s_j$ is continuous at t_1. □

Applying the idea of the differentiation index for DAEs to boundary arcs of state constraints yields the *order of a state constraint*. The order counts how often the active constraint needs to be differentiated with respect to time until the respective derivative can be solved for the control. In this respect, a boundary arc of a state constraint can be interpreted as a DAE itself, with the control playing the role of an algebraic variable.

To simplify notation, we restrict the discussion to scalar controls ($n_u = 1$) and a single state constraint ($n_s = 1$). The multi-dimensional case essentially inherits the same structural difficulties as for general DAEs.

Let $[t_1, t_2] \subseteq \mathcal{I}$ be a boundary arc of the state constraint $s(t, \hat{x}(t)) \leq 0$, that is

$$s(t, \hat{x}(t)) = 0 \quad \text{for all } t \in [t_1, t_2].$$

We exploit that the algebraic variable $\hat{y} = Y(\hat{x}, \hat{u})$ under Assumption 3.1.3 can be expressed by the implicit function theorem as a function of $\hat{x}$ and $\hat{u}$, which is determined by the first derivative of the algebraic constraint

$$0_{\mathbb{R}^{n_y}} = g_t'(t, \hat{x}(t)) + g_x'(t, \hat{x}(t)) f(t, \hat{x}(t), \hat{y}(t), \hat{u}(t)).$$

Define the functions $S^{(k)}(t, x, u)$, $0 \leq k \leq \infty$, recursively by

$$S^{(0)}(t, x, u) := s(t, x)$$

and

$$S^{(k+1)}(t,x,u) := \nabla_t S^{(k)}(t,x,u) - \nabla_x S^{(k)}(t,x,u)^\top f(t,x,Y(x,u),u)$$

for $k = 0, 1, \dots$.

Until u appears explicitly for the first time, $S^{(k)}$ evaluated at $\hat{x}(t)$ and $\hat{u}(t)$ is just the kth time derivative of s on the boundary arc, and it holds

$$S^{(k)}(t, \hat{x}(t), \hat{u}(t)) = 0 \quad \text{for all } t \in [t_1, t_2],\ k = 0, 1, 2, \dots.$$

Two cases may occur:

(a) There exists $k < \infty$ with

$$\frac{\partial}{\partial u} S^{(j)}(t,x,u) = 0 \quad \text{for all } 0 \le j \le k-1, \tag{3.45}$$

and

$$\frac{\partial}{\partial u} S^{(k)}(t,x,u) \neq 0. \tag{3.46}$$

In this case, the equation

$$S^{(k)}(t,x,u) = 0 \quad \text{for } t \in [t_1, t_2] \tag{3.47}$$

can be solved for u by the implicit function theorem. The boundary control is thus given implicitly by $\hat{u} = u_{\text{boundary}}(t,x)$ as a function of t and x.

(b) For every $0 \le k \le \infty$, it holds

$$\frac{\partial}{\partial u} S^{(k)}(t,x,u) = 0.$$

In this case, $\hat{u}$ cannot be determined on the boundary arc.

Definition 3.2.10 (Order of a state constraint). Let there exist $k < \infty$ with (3.45) and (3.46). Then k is called *order of the state constraint*. The special case $k = 0$ corresponds to a mixed control-state constraint. □

This proves

Corollary 3.2.11. *Let there exist $k < \infty$ with (3.45) and (3.46). Then the boundary control in $[t_1, t_2]$ is implicitly defined by (3.47) with*

$$\hat{u}(t) = u_{\text{boundary}}(t, \hat{x}(t)), \quad t \in [t_1, t_2].$$ □

3.2.4 Jump conditions for the adjoint

The integral equation (3.40) involves a Riemann–Stieltjes integral, making the evaluation of the necessary optimality conditions in Theorem 3.2.7 difficult. Since λ_f is a function of bounded variation, it is differentiable almost everywhere, and this allows us to derive a piecewise-defined differential equation for λ_f with additional jump conditions.

To this end, recall that the multiplier μ is of bounded variation. Hence, it has at most countably many jump points and μ can be expressed as $\mu = \mu_a + \mu_d + \mu_s$, where μ_a is absolutely continuous, μ_d is a jump function, and μ_s is singular (continuous, non-constant, $\dot{\mu}_s(t) = 0$ almost everywhere in $\mathcal{I}$). Hence, the adjoint equation (3.40) can be written as

$$\lambda_f(t) = \lambda_f(t_f) + \int_t^{t_f} \mathcal{H}'_x[\tau]^\top d\tau + \int_t^{t_f} s'_x[\tau]^\top d\mu_a(\tau) + \int_t^{t_f} s'_x[\tau]^\top d\mu_d(\tau) + \int_t^{t_f} s'_x[\tau]^\top d\mu_s(\tau)$$

for all $t \in \mathcal{I}$. Notice that λ_f is continuous from the right in the open interval (t_0, t_f) since μ is normalized. Let $\{t_j\}, j \in \mathcal{J}$, be the set of jump points of μ. Then at every jump point t_j, it holds

$$\lim_{\varepsilon \downarrow 0} \left(\int_{t_j}^{t_f} s'_x[\tau]^\top d\mu_d(\tau) - \int_{t_j-\varepsilon}^{t_f} s'_x[\tau]^\top d\mu_d(\tau) \right) = -s'_x[t_j]^\top (\mu_d(t_j) - \mu_d(t_j-)),$$

according to (2.3), where $\mu_d(t_j-)$ denotes the left-sided limit of μ_d at t_j. Since μ_a is absolutely continuous and μ_s is continuous, we obtain the *jump-condition*

$$\lambda_f(t_j) - \lambda_f(t_j-) = -s'_x[t_j]^\top (\mu(t_j) - \mu(t_j-)), \quad j \in \mathcal{J}.$$

Since every function of bounded variation is differentiable almost everywhere, μ and λ_f are differentiable almost everywhere. Exploitation of Lemma 2.1.25 proves

Corollary 3.2.12. *Let the assumptions of Theorem* 3.2.7 *hold. Then λ_f is differentiable almost everywhere in $\mathcal{I}$ with*

$$\dot{\lambda}_f(t) = -\mathcal{H}'_x(t, \hat{x}(t), \hat{y}(t), \hat{u}(t), \lambda_f(t), \lambda_g(t), \ell_0)^\top - s'_x(t, \hat{x}(t))^\top \dot{\mu}(t). \tag{3.48}$$

Furthermore, the jump conditions

$$\lambda_f(t_j) - \lambda_f(t_j-) = -s'_x(t_j, \hat{x}(t_j))^\top (\mu(t_j) - \mu(t_j-)) \tag{3.49}$$

hold at every point $t_j \in (t_0, t_f)$ of discontinuity of the multiplier μ. □

Notice that μ in (3.48) can be replaced by the absolutely continuous component μ_a, since the derivatives of the jump component μ_d and the singular component μ_s are zero almost everywhere. It should be noted that (3.48) and (3.49) are implications from the integral equation (3.40), but singular components get lost in the former, and both systems are not necessarily equivalent. Notice further that the jump points of μ are located on boundary arcs of active state constraints.

Example 3.2.13. We add a pure state constraint to Example 3.1.11 and obtain:

Minimize

$$\frac{1}{2}\int_0^1 u(t)^2 dt$$

subject to the constraints

$$\begin{aligned} \dot{x}_1(t) &= u(t) - y(t), & x_1(0) &= x_1(1) = 0, \\ \dot{x}_2(t) &= u(t), & x_2(0) &= -x_2(1) = 1, \\ \dot{x}_3(t) &= -x_2(t), & x_3(0) &= 0, \\ 0 &= x_1(t) + x_3(t), \\ x_1(t) - \ell &\leq 0. \end{aligned}$$

For $\ell > 1/4$ the state constraint does not get active, and the optimal solution is provided in Example 3.1.11. Hence, let $0 < \ell \leq 1/4$.

Differentiation of the algebraic constraint yields

$$0 = u - y - x_2 \quad \Longrightarrow \quad y = u - x_2 =: Y(u, x_2).$$

Suppose the optimal solution possesses exactly one boundary arc $[t_1, t_2] \subseteq [0, 1]$ with $\hat{x}_1(t) = \ell$ for $t \in [t_1, t_2]$. We obtain

$$\begin{aligned} S^{(0)}(x, u) &= x_1 - \ell, \\ S^{(1)}(x, u) &= u - y = u - (u - x_2) = x_2, \\ S^{(2)}(x, u) &= u. \end{aligned}$$

Hence, the order of the state constraint is $k = 2$, and the boundary control is given by $u = u_{\text{boundary}} \equiv 0$. In particular, it holds $\hat{x}_1(t) = \ell$ and $\hat{x}_2(t) = 0$ on the boundary arc $[t_1, t_2]$.

Owing to the boundary conditions $x_1(0) = x_1(1) = 0$, the state constraint is not active in neighborhoods of $t = 0$ and $t = 1$. Hence, $\mu(t)$ is constant in these neighborhoods, $\dot{\mu}(t) \equiv 0$, and the integral adjoint equation reduces to an adjoint differential equation, which has already been analyzed in Example 3.1.11. Together with the initial conditions, we obtain

$$\hat{u}(t) = \begin{cases} c_1 t + c_2, & \text{if } t \in [0, t_1), \\ 0, & \text{if } t \in (t_1, t_2), \\ \bar{c}_1 t + \bar{c}_2, & \text{if } t \in (t_1, 1], \end{cases}$$

$$\hat{x}_1(t) = -\hat{x}_3(t) = \begin{cases} \frac{1}{6} c_1 t^3 + \frac{1}{2} c_2 t^2 + t, & \text{if } t \in [0, t_1), \\ \ell, & \text{if } t \in [t_1, t_2], \\ \frac{1}{6} \bar{c}_1 t^3 + \frac{1}{2} \bar{c}_2 t^2 + \bar{c}_3 t + \bar{c}_4, & \text{if } t \in (t_1, 1], \end{cases}$$

$$\hat{x}_2(t) = \begin{cases} \frac{1}{2} c_1 t^2 + c_2 t + 1, & \text{if } t \in [0, t_1), \\ 0, & \text{if } t \in [t_1, t_2], \\ \frac{1}{2} \bar{c}_1 t^2 + \bar{c}_2 t + \bar{c}_3, & \text{if } t \in (t_1, 1] \end{cases}$$

with suitable constants $c_1, c_2, \bar{c}_1, \bar{c}_2, \bar{c}_3, \bar{c}_4 \in \mathbb{R}$.

The boundary conditions yield

$$\begin{aligned} 0 = \hat{x}_1(1) &= \frac{1}{6} \bar{c}_1 + \frac{1}{2} \bar{c}_2 + \bar{c}_3 + \bar{c}_4, \\ -1 = \hat{x}_2(1) &= \frac{1}{2} \bar{c}_1 + \bar{c}_2 + \bar{c}_3. \end{aligned}$$

Continuity conditions at t_1 and t_2 yield

$$\begin{aligned} \ell = \hat{x}_1(t_1) &= \frac{1}{6} c_1 t_1^3 + \frac{1}{2} c_2 t_1^2 + t_1, \\ \ell = \hat{x}_1(t_2) &= \frac{1}{6} \bar{c}_1 t_2^3 + \frac{1}{2} \bar{c}_2 t_2^2 + \bar{c}_3 t_2 + \bar{c}_4, \\ 0 = \hat{x}_2(t_1) &= \frac{1}{2} c_1 t_1^2 + c_2 t_1 + 1, \\ 0 = \hat{x}_2(t_2) &= \frac{1}{2} \bar{c}_1 t_2^2 + \bar{c}_2 t_2 + \bar{c}_3. \end{aligned}$$

The stationarity condition for the Hamilton function (3.44) with $\ell_0 = 1$ and the algebraic adjoint equation (3.41) yield

$$\lambda_{f,2}(t) = -\hat{u}(t).$$

As $\lambda_{f,2}$ is continuous from the right, it follows

$$\lambda_{f,2}(t_1) = \lambda_{f,2}(t_2-) = 0, \quad \lambda_{f,2}(t_1-) = -(c_1 t_1 + c_2), \quad \lambda_{f,2}(t_2) = -(\bar{c}_1 t_2 + \bar{c}_2).$$

Moreover, the jump condition (3.49) yields

$$\begin{aligned} 0 &= \lambda_{f,2}(t_1) - \lambda_{f,2}(t_1-) = c_1 t_1 + c_2, \\ 0 &= \lambda_{f,2}(t_2) - \lambda_{f,2}(t_2-) = -(\bar{c}_1 t_2 + \bar{c}_2), \end{aligned}$$

and hence $\hat{u}$ is continuous at t_1 and t_2. The boundary conditions, continuity conditions, and jump conditions provide eight conditions for the eight unknowns $c_1, c_2, \bar{c}_1, \bar{c}_2, \bar{c}_3, \bar{c}_4, t_1, t_2$.

A solution is given by $t_1 = 3\ell$, $t_2 = 1 - 3\ell$, and

$$(c_1, c_2, \bar{c}_1, \bar{c}_2, \bar{c}_3, \bar{c}_4) = \left(\frac{2}{9\ell^2}, -\frac{2}{3\ell}, -\frac{2}{9\ell^2}, -\frac{2(3\ell-1)}{9\ell^2}, -\frac{1-6\ell+9\ell^2}{9\ell^2}, \frac{-9\ell+27\ell^2+1}{27\ell^2}\right).$$

This solution holds for $0 < t_1 \le t_2 < 1$, which implies $0 < \ell \le 1/6$.

For $1/6 < \ell \le 1/4$, the solution possesses a contact point at $t = 1/2$, and it can be computed similarly. We leave the details to the reader. □

A further example with pure state constraints working directly with the multiplier μ is discussed in full detail in Subsection 5.5.1.

3.3 Problems with mixed control-state constraints

Many applications involve mixed control-state constraints, which we now add to Problem 3.1.1:

Problem 3.3.1 (Index-two DAE OCP with mixed control-state constraints). Let $\mathcal{I} := [t_0, t_f] \subset \mathbb{R}$ be a non-empty compact time interval with $t_0 < t_f$ fixed. Let

$$\begin{aligned}
\varphi &: \mathbb{R}^{n_x} \times \mathbb{R}^{n_x} \longrightarrow \mathbb{R},\\
f_0 &: \mathcal{I} \times \mathbb{R}^{n_x} \times \mathbb{R}^{n_y} \times \mathbb{R}^{n_u} \longrightarrow \mathbb{R},\\
f &: \mathcal{I} \times \mathbb{R}^{n_x} \times \mathbb{R}^{n_y} \times \mathbb{R}^{n_u} \longrightarrow \mathbb{R}^{n_x},\\
g &: \mathcal{I} \times \mathbb{R}^{n_x} \longrightarrow \mathbb{R}^{n_y},\\
c &: \mathcal{I} \times \mathbb{R}^{n_x} \times \mathbb{R}^{n_y} \times \mathbb{R}^{n_u} \longrightarrow \mathbb{R}^{n_c},\\
\psi &: \mathbb{R}^{n_x} \times \mathbb{R}^{n_x} \longrightarrow \mathbb{R}^{n_\psi}
\end{aligned}$$

be sufficiently smooth functions.

Minimize

$$\varphi\big(x(t_0), x(t_f)\big) + \int_{t_0}^{t_f} f_0\big(t, x(t), y(t), u(t)\big)dt$$

with respect to $x \in W^{n_x}_{1,\infty}(\mathcal{I})$, $y \in L^{n_y}_{\infty}(\mathcal{I})$, $u \in L^{n_u}_{\infty}(\mathcal{I})$ subject to the constraints

$$\begin{aligned}
\dot{x}(t) &= f\big(t, x(t), y(t), u(t)\big) && \text{a. e. in } \mathcal{I},\\
0_{\mathbb{R}^{n_y}} &= g\big(t, x(t)\big) && \text{in } \mathcal{I},\\
0_{\mathbb{R}^{n_\psi}} &= \psi\big(x(t_0), x(t_f)\big),\\
c\big(t, x(t), y(t), u(t)\big) &\le 0_{\mathbb{R}^{n_c}} && \text{a. e. in } \mathcal{I}.
\end{aligned}$$

Similarly as in Section 3.1, we interpret Problem 3.3.1 as the infinite optimization problem

Minimize $J(z)$ subject to $z \in S$, $G(z) \in K$, and $H(z) = \Theta_V$

given in Section 2.2 with $S = Z$ and

$$\begin{aligned}
Z &= W^{n_x}_{1,\infty}(\mathcal{I}) \times L^{n_y}_{\infty}(\mathcal{I}) \times L^{n_u}_{\infty}(\mathcal{I}),\\
V &= L^{n_x}_{\infty}(\mathcal{I}) \times W^{n_y}_{1,\infty}(\mathcal{I}) \times \mathbb{R}^{n_\psi},\\
W &= L^{n_c}_{\infty}(\mathcal{I}),\\
J(x,y,u) &= \varphi(x(t_0), x(t_f)) + \int_{t_0}^{t_f} f_0(t, x(t), y(t), u(t))dt,\\
H(x,y,u) &= \begin{pmatrix} H_1(x,y,u) \\ H_2(x,y,u) \\ H_3(x,y,u) \end{pmatrix} = \begin{pmatrix} f(\cdot, x(\cdot), y(\cdot), u(\cdot)) - \dot{x}(\cdot), \\ g(\cdot, x(\cdot)), \\ -\psi(x(t_0), x(t_f)), \end{pmatrix},\\
G(x,y,u) &= -c(\cdot, x(\cdot), y(\cdot), u(\cdot)),\\
K &= \{k \in L^{n_c}_{\infty}(\mathcal{I}) \mid k(t) \geq 0_{\mathbb{R}^{n_c}} \text{ a. e. in } \mathcal{I}\}.
\end{aligned}$$

Please note that K is a closed convex cone with non-empty interior. With the same reasoning as in Section 3.1, the assumptions in Theorem 2.3.23 are satisfied at a local minimum $\hat{z} = (\hat{x}, \hat{y}, \hat{u})$ of Problem 3.3.1. Taking into account $S = Z$, the variational inequality in Theorem 2.3.23 becomes an equation, and we obtain the following result.

Theorem 3.3.2 (Necessary conditions for Problem 3.3.1). *Let the following assumptions be fulfilled for Problem* 3.3.1*:*
(i) *Assumptions* 2.2.5 *and* 3.1.3 *hold.*
(ii) $(\hat{x}, \hat{y}, \hat{u})$ *is a local minimum of Problem* 3.3.1.

Then there exist non-trivial multipliers $\ell_0 \geq 0$, $\lambda^* \in V^*$, and $\eta^* \in W^*$ with

$$\eta^* \in K^+ \quad \textit{and} \quad \eta^*(G(\hat{x}, \hat{y}, \hat{u})) = 0, \tag{3.50}$$

and

$$\begin{aligned}
0 = \ell_0 J'(\hat{x}, \hat{y}, \hat{u})(x,y,u) &- \eta^*(G'(\hat{x}, \hat{y}, \hat{u})(x,y,u))\\
&- \lambda^*(H'(\hat{x}, \hat{y}, \hat{u})(x,y,u))
\end{aligned} \tag{3.51}$$

for all $(x,y,u) \in Z$. □

The multipliers $\lambda^* := (\lambda_f^*, \lambda_g^*, \sigma^*)$ and η^* are elements of the dual spaces

$$V^* = L_\infty^{n_x}(\mathcal{I})^* \times W_{1,\infty}^{n_y}(\mathcal{I})^* \times (\mathbb{R}^{n_\psi})^* \quad \text{and} \quad W^* = L_\infty^{n_c}(\mathcal{I})^*$$

and explicit representations are required. The variational equation (3.51) yields

$$0 = \kappa'_{x_0} x(t_0) + \kappa'_{x_f} x(t_f) + \int_{t_0}^{t_f} \ell_0 f'_{0,x}[t]x(t)dt$$
$$+ \lambda_f^*(\dot{x}(\cdot) - f'_x[\cdot]x(\cdot)) - \lambda_g^*(g'_x[\cdot]x(\cdot)) + \eta^*(c'_x[\cdot]x(\cdot)), \tag{3.52}$$

$$0 = \int_{t_0}^{t_f} \ell_0 f'_{0,y}[t]y(t)dt - \lambda_f^*(f'_y[\cdot]y(\cdot)) + \eta^*(c'_y[\cdot]y(\cdot)), \tag{3.53}$$

$$0 = \int_{t_0}^{t_f} \ell_0 f'_{0,u}[t]u(t)dt - \lambda_f^*(f'_u[\cdot]u(\cdot)) + \eta^*(c'_u[\cdot]u(\cdot)), \tag{3.54}$$

which hold for *every* $x \in W_{1,\infty}^{n_x}(\mathcal{I})$, *every* $y \in L_\infty^{n_y}(\mathcal{I})$, and *every* $u \in L_\infty^{n_u}(\mathcal{I})$.

3.3.1 Representation of multipliers

Explicit representations of the functionals λ_f^*, λ_g^*, and η^* are obtained similarly as in Subsection 3.1.1. Let $h_1 \in L_\infty^{n_x}(\mathcal{I})$, $h_2 \in W_{1,\infty}^{n_y}(\mathcal{I})$, and $h_3 \in L_\infty^{n_c}(\mathcal{I})$ be arbitrary.

Consider the initial value problem

$$\dot{x}(t) = f'_x[t]x(t) + f'_y[t]y(t) + f'_u[t]u(t) + h_1(t), \quad x(t_0) = \Pi h_2(t_0), \tag{3.55}$$

$$0_{\mathbb{R}^{n_y}} = g'_x[t]x(t) + h_2(t), \tag{3.56}$$

and the equation

$$h_3(t) = c'_x[t]x(t) + c'_y[t]y(t) + c'_u[t]u(t). \tag{3.57}$$

We aim at solving (3.55)–(3.57) for x, y, and u and need regularity assumptions.

Assumption 3.3.3 (Pseudo-inverse). It holds $\operatorname{rank}(c'_u[t]) = n_c$ almost everywhere in $\mathcal{I}$ and the *pseudo-inverse of* $c'_u[t]$,

$$(c'_u[t])^+ := c'_u[t]^\top (c'_u[t]c'_u[t]^\top)^{-1},$$

is essentially bounded in $\mathcal{I}$. □

Assumption 3.3.3 allows solving (3.57) for u, and almost everywhere, it holds

$$u(t) = (c_u'[t])^+(h_3(t) - c_x'[t]x(t) - c_y'[t]y(t)).$$

Introducing this into (3.55) yields

$$\dot{x}(t) = \hat{f}_x[t]x(t) + \hat{f}_y[t]y(t) + \hat{h}_1(t), \quad x(t_0) = \Pi h_2(t_0), \tag{3.58}$$

$$0_{\mathbb{R}^{n_y}} = g_x'[t]x(t) + h_2(t) \tag{3.59}$$

with

$$\hat{f}_x[t] := f_x'[t] - f_u'[t](c_u'[t])^+c_x'[t],$$
$$\hat{f}_y[t] := f_y'[t] - f_u'[t](c_u'[t])^+c_y'[t],$$
$$\hat{h}_1(t) := f_u'[t](c_u'[t])^+h_3(t) + h_1(t).$$

Differentiation of (3.59) yields

$$0_{\mathbb{R}^{n_y}} = \hat{Q}(t)x(t) + g_x'[t](\hat{f}_y[t]y(t) + \hat{h}_1(t)) + \dot{h}_2(t)$$

with

$$\hat{Q}(t) := \left(\frac{d}{dt}g_x'[t]\right) + g_x'[t]\hat{f}_x[t]$$

and the DAE (3.58)–(3.59) has index two under

Assumption 3.3.4 (Compatibility). The matrix

$$\hat{M}(t) := g_x'[t] \cdot \hat{f}_y[t] = g_x'[t] \cdot (f_y'[t] - f_u'[t](c_u'[t])^+c_y'[t]) \tag{3.60}$$

is non-singular with essentially bounded inverse $\hat{M}^{-1}$ almost everywhere in $\mathcal{I}$. □

Now we are in the same setting as in Subsection 3.1.1. Using the solution formulas in Lemma 3.1.2 for (3.58)–(3.59), adding equations (3.52)–(3.54), exploiting the linearity of the functionals, introducing the solution formulas, and using integration by parts as in (3.18) finally yields

$$\begin{aligned}
&\lambda_f^*(h_1(\cdot)) + \lambda_g^*(h_2(\cdot)) + \eta^*(h_3(\cdot)) \\
&\quad = -\zeta^\top h_2(t_0) - \int_{t_0}^{t_f} (\lambda_f(t)^\top + \lambda_g(t)^\top g_x'[t])h_1(t)dt - \int_{t_0}^{t_f} \lambda_g(t)^\top \dot{h}_2(t)dt + \int_{t_0}^{t_f} \eta(t)^\top h_3(t)dt
\end{aligned}$$

with

$$\zeta^\top := \left(\kappa'_{x_0} + \kappa'_{x_f}\Phi(t_f) + \int_{t_0}^{t_f} \ell_0 \hat{f}_0[t]\Phi(t)dt \right)\Pi,$$

$$\lambda_f(t)^\top := \left(\kappa'_{x_f}\Phi(t_f) + \int_{t}^{t_f} \ell_0 \hat{f}_0[\tau]\Phi(\tau)d\tau \right)\Phi^{-1}(t),$$

$$\lambda_g(t)^\top := -\ell_0 v_0(t)\hat{M}(t)^{-1} - \lambda_f(t)^\top \hat{f}_y[t]\hat{M}(t)^{-1},$$

$$\eta(t)^\top = -\ell_0 f'_{0,u}[t](c'_u[t])^+ - (\lambda_f(t)^\top + \lambda_g(t)^\top g'_x[t])f'_u[t](c'_u[t])^+,$$

$$v_0(t) := f'_{0,y}[t] - f'_{0,u}[t](c'_u[t])^- c'_y[t],$$

$$\hat{f}_0[t] := f'_{0,x}[t] - v_0(t)\hat{M}(t)^{-1}\hat{Q}(t) - f'_{0,u}[t](c'_u[t])^+ c'_x[t].$$

Setting $h_1 \equiv \Theta$, $h_2 \equiv \Theta$, or $h_3 \equiv \Theta$, respectively, proves

Corollary 3.3.5. *Let Assumptions 3.3.3, 3.3.4, and the assumptions of Theorem 3.3.2 hold. Then there exist $\zeta \in \mathbb{R}^{n_y}$ and functions $\lambda_f \in W^{n_x}_{1,\infty}(\mathcal{I})$, $\lambda_g \in L^{n_y}_\infty(\mathcal{I})$, and $\eta \in L^{n_c}_\infty(\mathcal{I})$ with*

$$\lambda_f^*(h_1(\cdot)) = -\int_{t_0}^{t_f} (\lambda_f(t)^\top + \lambda_g(t)^\top g'_x[t])h_1(t)dt, \tag{3.61}$$

$$\lambda_g^*(h_2(\cdot)) = -\zeta^\top h_2(t_0) - \int_{t_0}^{t_f} \lambda_g(t)^\top \dot{h}_2(t)dt, \tag{3.62}$$

$$\eta^*(h_3(\cdot)) = \int_{t_0}^{t_f} \eta(t)^\top h_3(t)dt \tag{3.63}$$

for every $h_1 \in L^{n_x}_\infty(\mathcal{I})$, every $h_2 \in W^{n_y}_{1,\infty}(\mathcal{I})$, and every $h_3 \in L^{n_c}_\infty(\mathcal{I})$. □

3.3.2 Local minimum principle

The variational equations (3.52)–(3.54) can be analyzed as in Subsection 3.1.2 using the representations of the functionals λ_f^*, λ_g^*, and η^* provided by Corollary 3.3.5, if the augmented Hamilton function is used instead of the Hamilton function.

Definition 3.3.6 (Augmented Hamilton function for Problem 3.3.1). Consider the Problem 3.3.1. The *augmented Hamilton function* $\hat{\mathcal{H}} : \mathcal{I} \times \mathbb{R}^{n_x} \times \mathbb{R}^{n_y} \times \mathbb{R}^{n_u} \times \mathbb{R}^{n_x} \times \mathbb{R}^{n_y} \times \mathbb{R}^{n_c} \times \mathbb{R} \longrightarrow \mathbb{R}$ is defined by

$$\hat{\mathcal{H}}(t, x, y, u, \lambda_f, \lambda_g, \eta, \ell_0) := \mathcal{H}(t, x, y, u, \lambda_f, \lambda_g, \ell_0) + \eta^\top c(t, x, y, u).$$ □

Investigation of the complementarity condition (3.50)

The complementarity condition (3.50) reads as

$$\eta^* \in K^+ \iff \int_{t_0}^{t_f} \eta(t)^\top k(t)dt \geq 0 \quad \text{for } k \in L_\infty^{n_c}(\mathcal{I}),\ k(t) \geq 0_{\mathbb{R}^{n_s}} \text{ a. e. in } \mathcal{I},$$

$$\eta^*(G(\hat{x},\hat{y},\hat{u})) = 0 \iff \int_{t_0}^{t_f} \eta(t)^\top c(t,\hat{x}(t),\hat{y}(t),\hat{u}(t))dt = 0.$$

Lemma 3.3.7. *Let $f \in L_1(\mathcal{I})$. If*

$$\int_{t_0}^{t_f} f(t)h(t)dt \geq 0$$

holds for every $h \in L_\infty(\mathcal{I})$ with $h(t) \geq 0$ almost everywhere in $\mathcal{I}$, then $f(t) \geq 0$ almost everywhere in $\mathcal{I}$.

Proof. Assume the contrary, i. e. there is a set $N \subseteq \mathcal{I}$ with positive measure and $f(t) < 0$ for $t \in N$. Choose h to be one on N and zero otherwise. Then

$$\int_{t_0}^{t_f} f(t)h(t)dt = \int_N f(t)dt < 0$$

in contradiction to the assumption. □

Application of Lemma 3.3.7 to the complementarity system together with our previous findings yields

Theorem 3.3.8 (Local minimum principle for Problem 3.3.1). *Let the following assumptions hold for Problem* 3.3.1*:*

(i) *Assumptions* 2.2.5, 3.1.3, 3.3.3, *and* 3.3.4 *hold. g is twice continuously differentiable with respect to all arguments.*
(ii) *$(\hat{x},\hat{y},\hat{u})$ is a local minimum of Problem* 3.3.1.

Then there exist multipliers

$$\ell_0 \in \mathbb{R}, \quad \lambda_f \in W_{1,\infty}^{n_x}(\mathcal{I}), \quad \lambda_g \in L_\infty^{n_y}(\mathcal{I}), \quad \eta \in L_\infty^{n_c}(\mathcal{I}), \quad \zeta \in \mathbb{R}^{n_y}, \quad \sigma \in \mathbb{R}^{n_\psi}$$

such that the following conditions are satisfied:

(a) $\ell_0 \geq 0$, $(\ell_0,\zeta,\sigma,\lambda_f,\lambda_g,\eta) \neq \Theta$.
(b) Adjoint equations*: Almost everywhere in $\mathcal{I}$, it holds*

$$\dot{\lambda}_f(t) = -\hat{\mathcal{H}}_x'(t,\hat{x}(t),\hat{y}(t),\hat{u}(t),\lambda_f(t),\lambda_g(t),\eta(t),\ell_0)^\top \tag{3.64}$$

$$0_{\mathbb{R}^{n_y}} = \hat{\mathcal{H}}'_y(t, \hat{x}(t), \hat{y}(t), \hat{u}(t), \lambda_f(t), \lambda_g(t), \eta(t), \ell_0)^\top. \tag{3.65}$$

(c) Transversality conditions:

$$\lambda_f(t_0)^\top = -(\ell_0 \varphi'_{x_0}(\hat{x}(t_0), \hat{x}(t_f)) + \sigma^\top \psi'_{x_0}(\hat{x}(t_0), \hat{x}(t_f)) + \zeta^\top g'_x(t_0, \hat{x}(t_0))), \tag{3.66}$$

$$\lambda_f(t_f)^\top = \ell_0 \varphi'_{x_f}(\hat{x}(t_0), \hat{x}(t_f)) + \sigma^\top \psi'_{x_f}(\hat{x}(t_0), \hat{x}(t_f)). \tag{3.67}$$

(d) Stationarity of Hamilton function: *Almost everywhere in $\mathcal{I}$, it holds*

$$\hat{\mathcal{H}}'_u(t, \hat{x}(t), \hat{y}(t), \hat{u}(t), \lambda_f(t), \lambda_g(t), \eta(t), \ell_0) = 0^\top_{\mathbb{R}^{n_u}}. \tag{3.68}$$

(e) Complementarity condition: *Almost everywhere in $\mathcal{I}$, it holds*

$$\eta(t) \geq 0_{\mathbb{R}^{n_c}} \quad \text{and} \quad \eta(t)^\top c(t, \hat{x}(t), \hat{y}(t), \hat{u}(t)) = 0. \qquad \square$$

The necessary optimality conditions of Theorems 3.3.8 and 3.2.7 can be combined, if pure state constraints are added to Problem 3.3.1, see [121].

The rank assumption on c'_u in Assumption 3.3.3 is often too strong – it even does not hold for simple box constraints – but it can be weakened. It is sufficient to assume the existence of $\alpha > 0$ and $\beta > 0$ with

$$\|A_\alpha(t)^\top d\| \geq \beta \|d\| \quad \text{for all } d,$$

where

$$I_\alpha(t) := \{i \in \{1, \ldots, n_c\} \mid c_i(t, \hat{x}(t), \hat{y}(t), \hat{u}(t)) \geq -\alpha\},$$

$$A_\alpha(t) := \left(c'_{i,u}(t, \hat{x}(t), \hat{y}(t), \hat{u}(t))\right)_{i \in I_\alpha(t)},$$

compare [218]. A detailed proof for optimal control problems subject to higher index Hessenberg DAEs and mixed control-state constraints is given in [225, Chapter 3].

The constraint qualification of Mangasarian–Fromowitz in Corollary 2.3.32 together with Lemma 3.1.10 translates as follows for Problem 3.3.1 since the interior of

$$K = \{k \in L^{n_c}_\infty(\mathcal{I}) \mid k(t) \geq 0_{\mathbb{R}^{n_c}} \text{ a. e. in } \mathcal{I}\}$$

is given by

$$\begin{aligned} \operatorname{int}(K) &= \{k \in L^{n_c}_\infty(\mathcal{I}) \mid \text{there exists } \varepsilon > 0 \text{ with } B_\varepsilon(k) \subseteq K\} \\ &= \{k \in L^{n_c}_\infty(\mathcal{I}) \mid \text{there exists } \varepsilon > 0 \text{ with } k_i(t) \geq \varepsilon \text{ a. e. in } \mathcal{I}, i = 1, \ldots, n_c\}. \end{aligned}$$

Lemma 3.3.9. *Let the assumptions of Theorem 3.3.8 and Lemma 3.1.10 hold. Let there exist $x \in W^{n_x}_{1,\infty}(\mathcal{I})$, $y \in L^{n_y}_\infty(\mathcal{I})$, and $u \in L^{n_u}_\infty(\mathcal{I})$ with*

$$
\begin{aligned}
c[t] + c_x'[t]x(t) + c_y'[t]y(t) + c_u'[t]u(t) &\leq -\varepsilon \cdot e && \text{a. e. in } \mathcal{I}, \\
f_x'[t]x(t) + f_y'[t]y(t) + f_u'[t]u(t) - \dot{x}(t) &= 0_{\mathbb{R}^{n_x}} && \text{a. e. in } \mathcal{I}, \\
g_x'[t]x(t) &= 0_{\mathbb{R}^{n_y}} && \text{in } \mathcal{I}, \\
\psi_{x_0}' x(t_0) + \psi_{x_f}' x(t_f) &= 0_{\mathbb{R}^{n_\psi}},
\end{aligned}
$$

where $e = (1, \dots, 1)^\top \in \mathbb{R}^{n_c}$.

Then the Mangasarian–Fromowitz constraint qualification is satisfied for Problem 3.3.1*, and* $\ell_0 = 1$ *can be chosen in Theorem* 3.3.8. □

3.4 Summary of local minimum principles for index-one problems

We summarize local minimum principles for optimal control problems subject to index-one DAEs.

Problem 3.4.1 (Index-one DAE optimal control problem). Let $\mathcal{I} := [t_0, t_f] \subset \mathbb{R}$ be a compact interval with $t_0 < t_f$ fixed.

Minimize

$$
\varphi(x(t_0), x(t_f)) + \int_{t_0}^{t_f} f_0(t, x(t), y(t), u(t)) dt
$$

with respect to $x \in W_{1,\infty}^{n_x}(\mathcal{I})$, $y \in L_\infty^{n_y}(\mathcal{I})$, $u \in L_\infty^{n_u}(\mathcal{I})$ *subject to the constraints*

$$
\begin{aligned}
\dot{x}(t) &= f(t, x(t), y(t), u(t)) && \text{a. e. in } \mathcal{I}, \\
0_{\mathbb{R}^{n_y}} &= g(t, x(t), y(t), u(t)) && \text{a. e. in } \mathcal{I}, \\
\psi(x(t_0), x(t_f)) &= 0_{\mathbb{R}^{n_\psi}}, && && (3.69) \\
c(t, x(t), y(t), u(t)) &\leq 0_{\mathbb{R}^{n_c}} && \text{a. e. in } \mathcal{I}, && (3.70) \\
s(t, x(t)) &\leq 0_{\mathbb{R}^{n_s}} && \text{in } \mathcal{I}, \\
u(t) &\in \mathcal{U} && \text{a. e. in } \mathcal{I}.
\end{aligned}
$$

Let $(\hat{x}, \hat{y}, \hat{u})$ be a local minimum of Problem 3.4.1. We assume:

Assumption 3.4.2 (Index-one assumption). Almost everywhere in $\mathcal{I}$, the matrix

$$
M(t) := g_y'(t, \hat{x}(t), \hat{y}(t), \hat{u}(t))
$$

is non-singular, and M^{-1} is essentially bounded. □

Since the proof of the local minimum principles for Problem 3.4.1 works basically the same way as in the index-two case, we leave the details to the reader.

The Hamilton function and the augmented Hamilton function for Problem 3.4.1 read as

$$\mathcal{H}(t,x,y,u,\lambda_f,\lambda_g,\ell_0) := \ell_0 f_0(t,x,y,u) + \lambda_f^\top f(t,x,y,u) + \lambda_g^\top g(t,x,y,u),$$
$$\hat{\mathcal{H}}(t,x,y,u,\lambda_f,\lambda_g,\eta,\ell_0) := \mathcal{H}(t,x,y,u,\lambda_f,\lambda_g,\ell_0) + \eta^\top c(t,x,y,u).$$

Theorem 3.4.3 (Local minimum principle for index-one optimal control problems without mixed control-state constraints). *Let the following assumptions hold for Problem* 3.4.1*:*

(i) *Assumptions* 2.2.5 *and* 3.4.2 *hold.*

(ii) $\mathcal{U} \subseteq \mathbb{R}^{n_u}$ *is a closed and convex set with non-empty interior.*

(iii) $(\hat{x},\hat{y},\hat{u})$ *is a local minimum of Problem* 3.4.1.

(iv) *There are no mixed control-state constraints* (3.70) *in Problem* 3.4.1.

Then there exist multipliers

$$\ell_0 \in \mathbb{R}, \quad \sigma \in \mathbb{R}^{n_\psi}, \quad \lambda_f \in BV^{n_x}(\mathcal{I}), \quad \lambda_g \in L_\infty^{n_y}(\mathcal{I}), \quad \mu \in NBV^{n_s}(\mathcal{I}),$$

such that the following conditions are satisfied:

(a) $\ell_0 \geq 0$, $(\ell_0,\sigma,\lambda_f,\lambda_g,\mu) \neq \Theta$.

(b) Adjoint equations*: Almost everywhere in $\mathcal{I}$, it holds*

$$\lambda_f(t) = \lambda_f(t_f) + \int_t^{t_f} \mathcal{H}_x'(\tau,\hat{x}(\tau),\hat{y}(\tau),\hat{u}(\tau),\lambda_f(\tau),\lambda_g(\tau),\ell_0)^\top d\tau + \int_t^{t_f} s_x'(\tau,\hat{x}(\tau))^\top d\mu(\tau), \tag{3.71}$$

$$0_{\mathbb{R}^{n_y}} = \mathcal{H}_y'(t,\hat{x}(t),\hat{y}(t),\hat{u}(t),\lambda_f(t),\lambda_g(t),\ell_0)^\top. \tag{3.72}$$

(c) Transversality conditions*:*

$$\lambda_f(t_0)^\top = -(\ell_0 \varphi_{x_0}'(\hat{x}(t_0),\hat{x}(t_f)) + \sigma^\top \psi_{x_0}'(\hat{x}(t_0),\hat{x}(t_f))), \tag{3.73}$$
$$\lambda_f(t_f)^\top = \ell_0 \varphi_{x_f}'(\hat{x}(t_0),\hat{x}(t_f)) + \sigma^\top \psi_{x_f}'(\hat{x}(t_0),\hat{x}(t_f)). \tag{3.74}$$

(d) Stationarity of Hamilton function*: Almost everywhere in $\mathcal{I}$, it holds*

$$\mathcal{H}_u'(t,\hat{x}(t),\hat{y}(t),\hat{u}(t),\lambda_f(t),\lambda_g(t),\ell_0)(u-\hat{u}(t)) \geq 0 \tag{3.75}$$

for all $u \in \mathcal{U}$.

(e) Complementarity condition*: μ_i, $i \in \{1,\dots,n_s\}$, is monotonically increasing on $\mathcal{I}$ and constant on every interval (t_1,t_2) with $t_1 < t_2$ and $s_i(t,\hat{x}(t)) < 0$ for all $t \in (t_1,t_2)$.* □

In the presence of mixed control-state constraints, the following result holds:

Theorem 3.4.4 (Local minimum principle for index-one optimal control problems with mixed control-state constraints). *Let the following assumptions hold for Problem* 3.4.1*:*

(i) *Assumptions 2.2.5 and 3.4.2 hold.*
(ii) $\mathcal{U} = \mathbb{R}^{n_u}$.
(iii) $(\hat{x}, \hat{y}, \hat{u})$ *is a local minimum of Problem 3.4.1.*
(iv) *Almost everywhere in* $\mathcal{I}$, *it holds*

$$rank(c'_u(t, \hat{x}(t), \hat{y}(t), \hat{u}(t))) = n_c.$$

(v) *The pseudo-inverse of* $c'_u[t]$,

$$(c'_u[t])^+ = c'_u[t]^\top (c'_u[t] c'_u[t]^\top)^{-1},$$

is essentially bounded, and the matrix

$$g'_y[t] - g'_u[t](c'_u[t])^+ c'_y[t]$$

is non-singular almost everywhere with essentially bounded inverse in $\mathcal{I}$.

Then there exist multipliers

$$\ell_0 \in \mathbb{R}, \quad \sigma \in \mathbb{R}^{n_\psi}, \quad \lambda_f \in BV^{n_x}(\mathcal{I}), \quad \lambda_g \in L_\infty^{n_y}(\mathcal{I}), \quad \eta \in L_\infty^{n_c}(\mathcal{I}), \quad \mu \in NBV^{n_s}(\mathcal{I}),$$

such that the following conditions are satisfied:
(a) $\ell_0 \geq 0$, $(\ell_0, \sigma, \lambda_f, \lambda_g, \eta, \mu) \neq \Theta$.
(b) Adjoint equations: *Almost everywhere in* $\mathcal{I}$, *it holds*

$$\lambda_f(t) = \lambda_f(t_f) + \int_t^{t_f} \hat{\mathcal{H}}'_x(\tau, \hat{x}(\tau), \hat{y}(\tau), \hat{u}(\tau), \lambda_f(\tau), \lambda_g(\tau), \eta(\tau), \ell_0)^\top d\tau + \int_t^{t_f} s'_x(\tau, \hat{x}(\tau))^\top d\mu(\tau),$$
$$0_{\mathbb{R}^{n_y}} = \hat{\mathcal{H}}'_y(t, \hat{x}(t), \hat{y}(t), \hat{u}(t), \lambda_f(t), \lambda_g(t), \eta(t), \ell_0)^\top.$$

(c) Transversality conditions:

$$\lambda_f(t_0)^\top = -(\ell_0 \varphi'_{x_0}(\hat{x}(t_0), \hat{x}(t_f)) + \sigma^\top \psi'_{x_0}(\hat{x}(t_0), \hat{x}(t_f))),$$
$$\lambda_f(t_f)^\top = \quad \ell_0 \varphi'_{x_f}(\hat{x}(t_0), \hat{x}(t_f)) + \sigma^\top \psi'_{x_f}(\hat{x}(t_0), \hat{x}(t_f)).$$

(d) Stationarity of augmented Hamilton function: *Almost everywhere in* $\mathcal{I}$, *it holds*

$$\hat{\mathcal{H}}'_u(t, \hat{x}(t), \hat{y}(t), \hat{u}(t), \lambda_f(t), \lambda_g(t), \eta(t), \ell_0) = 0_{\mathbb{R}^{n_u}}^\top.$$

(e) Complementarity conditions: *Almost everywhere in* $\mathcal{I}$, *it holds*

$$\eta(t)^\top c(t, \hat{x}(t), \hat{y}(t), \hat{u}(t)) = 0 \quad \text{and} \quad \eta(t) \geq 0_{\mathbb{R}^{n_c}}.$$

μ_i, $i \in \{1, \dots, n_s\}$, *is monotonically increasing on* $\mathcal{I}$ *and constant on every interval* (t_1, t_2) *with* $t_1 < t_2$ *and* $s_i(t, \hat{x}(t)) < 0$ *for all* $t \in (t_1, t_2)$. □

λ_f in Theorem 3.4.3 is differentiable almost everywhere in $\mathcal{I}$ with

$$\dot{\lambda}_f(t) = -\mathcal{H}'_x(t,\hat{x}(t),\hat{y}(t),\hat{u}(t),\lambda_f(t),\lambda_g(t),\ell_0)^\top - s'_x(t,\hat{x}(t))^\top \dot{\mu}(t).$$

Furthermore, the jump conditions

$$\lambda_f(t_j) - \lambda_f(t_j-) = -s'_x(t_j,\hat{x}(t_j))^\top (\mu(t_j) - \mu(t_j-)) \tag{3.76}$$

hold at every point $t_j \in (t_0,t_f)$ of discontinuity of the multiplier μ. An analog result holds for Theorem 3.4.4.

Finally, we state a regularity condition that allows to set $\ell_0 = 1$.

Theorem 3.4.5. *Let the assumptions of Theorems* 3.4.3 *or* 3.4.4 *hold, and let*

$$\operatorname{rank}(\psi'_{x_0}\Phi(t_0) + \psi'_{x_f}\Phi(t_f)) = n_\psi,$$

where Φ is the fundamental solution of the homogeneous linear differential equation

$$\dot{\Phi}(t) = \tilde{A}(t)\Phi(t), \quad \Phi(t_0) = I_{n_x}, \quad t \in \mathcal{I},$$

with

$$\tilde{A}(t) := f'_x[t] - f'_y[t](g'_y[t])^{-1} g'_x[t].$$

Furthermore, let there exist $\varepsilon > 0$, $x \in W^{n_x}_{1,\infty}(\mathcal{I})$, $y \in L^{n_y}_\infty(\mathcal{I})$, and $u \in int(U_{ad} - \{\hat{u}\})$ satisfying

$$\begin{aligned}
c[t] + c'_x[t]x(t) + c'_y[t]y(t) + c'_u[t]u(t) &\le -\varepsilon \cdot e && \text{a. e. in } \mathcal{I},\\
s[t] + s'_x[t]x(t) &< 0_{\mathbb{R}^{n_s}} && \text{in } \mathcal{I},\\
f'_x[t]x(t) + f'_y[t]y(t) + f'_u[t]u(t) - \dot{x}(t) &= 0_{\mathbb{R}^{n_x}} && \text{a. e. in } \mathcal{I},\\
g'_x[t]x(t) + g'_y[t]y(t) + g'_u[t]u(t) &= 0_{\mathbb{R}^{n_y}} && \text{a. e. in } \mathcal{I},\\
\psi'_{x_0}x(t_0) + \psi'_{x_f}x(t_f) &= 0_{\mathbb{R}^{n_\psi}},
\end{aligned}$$

where $e = (1,\dots,1)^\top \in \mathbb{R}^{n_c}$. Then it holds $\ell_0 = 1$ in Theorems 3.4.3 *or* 3.4.4, *respectively.* □

The necessary conditions in Theorem 3.1.9 are identical with the necessary conditions in Theorem 3.4.3, if the latter are applied in the index-two DAE optimal control problem 3.1.1 to the mathematically equivalent index reduced index-one DAE

$$\begin{aligned}
\dot{x}(t) &= f(t,x(t),y(t),u(t)),\\
0_{\mathbb{R}^{n_y}} &= g'_t(t,x(t)) + g'_x(t,x(t))f(t,x(t),y(t),u(t))
\end{aligned}$$

together with the additional boundary condition $g(t_0,x(t_0)) = 0_{\mathbb{R}^{n_y}}$.

3.5 Exercises

Solve the following optimal control problem:

Minimize

$$\frac{1}{2}\int_0^2 u(t)^2 dt$$

subject to

$$\dot{x}_1(t) = x_2(t), \qquad x_1(0) = 0,\ x_1(2) = 1,$$
$$\dot{x}_2(t) = -4 + 0.8 \cdot x_2(t) + u(t), \qquad x_2(0) = 0,\ x_2(2) = 0.$$

Solve the following optimal control problem with free final time t_f and control β:

Minimize

$$t_f = \int_0^{t_f} 1\, dt$$

subject to

$$\dot{x}(t) = u(t), \qquad x(0) = 1,\ x(t_f) = 0,$$
$$\dot{u}(t) = \cos\big(\beta(t)\big), \qquad u(0) = 1,$$
$$\dot{y}(t) = v(t), \qquad y(0) = -1,\ y(t_f) = 0,$$
$$\dot{v}(t) = \sin\big(\beta(t)\big), \qquad v(0) = 1.$$

Solve the following optimal control problem:

Minimize

$$t_f = \int_0^{t_f} 1\, dt$$

subject to

$$\dot{x}(t) = v(t), \quad x(0) = 1,\ x(t_f) = 0,$$
$$\dot{v}(t) = u(t), \quad v(0) = 0,\ v(t_f) = 0,$$

and

$$-1 \le u(t) \le 1.$$

Solve the following optimal control problem:

Minimize

$$t_f = \int_0^{t_f} 1\,dt$$

subject to

$$\begin{aligned} \dot{h}(t) &= v(t), & h(0) &= 0,\ h(t_f) = h_f, \\ \dot{v}(t) &= -g + u(t), & v(0) &= 0, \\ \dot{x}(t) &= u(t), & x(0) &= 0,\ x(t_f) = x_f, \\ u(t) &\in [0, u_{\max}], \end{aligned}$$

where g denotes acceleration due to gravity.

(Rolling Disc by H. Gruschinski)
Write down and evaluate necessary optimality conditions for the following optimal control problem:

Minimize

$$\frac{1}{2}\int_0^1 u(t)^2 dt$$

subject to the constraints

$$\begin{aligned} \dot{x}_1(t) &= x_3(t) - y_1(t)r, & x_1(0) &= 0,\ x_1(1) = 4, \\ \dot{x}_2(t) &= x_4(t) + y_1(t), & x_2(0) &= 0,\ x_2(1) = 1, \\ J\dot{x}_3(t) &= u(t) - ry_2(t), & x_3(0) &= 0,\ x_3(1) = 0, \\ m\dot{x}_4(t) &= -c_D x_4(t) - c_S x_2(t) + y_2(t), & x_4(0) &= 0,\ x_4(1) = 0, \\ 0 &= rx_1(t) - x_2(t), \\ 0 &= rx_3(t) - x_4(t). \end{aligned}$$

Parameters: $r = 0.25, J = 0.05, m = 0.05, c_D = 0.05, c_S = 2.5$.

Solve Example 3.2.13 with $1/6 < \ell < 1/4$.

Write down and evaluate necessary optimality conditions for the following optimal control problem:

Minimize

$$\frac{1}{2}\int_0^1 u(t)^2 dt$$

subject to the constraints

$$\begin{aligned} \dot{x}_1(t) &= x_2(t), & x_1(0) &= x_1(1) = 0, \\ \dot{x}_2(t) &= y(t), & x_2(0) &= -x_2(1) = 1, \\ 0 &= y(t) - u(t), \\ u(t) &\le 10(t-0.5)^2 - 2.5. \end{aligned}$$

Let $\mathcal{I} := [t_0, t_f]$ be a compact interval with $t_0 < t_f$. Let $\hat{x} \in W^{n_x}_{1,\infty}(\mathcal{I})$, $\hat{u} \in L^{n_u}_{\infty}(\mathcal{I})$, and $s \in L^{n_c}_{\infty}(\mathcal{I})$, s non-negative, be functions with

$$\begin{aligned} c_i(\hat{x}(t), \hat{u}(t)) &\le 0, \quad i = 1, \dots, n_c, \\ c_i(\hat{x}(t), \hat{u}(t)) + \frac{1}{2} s_i(t)^2 &= 0, \quad i = 1, \dots, n_c. \end{aligned}$$

For $\alpha > 0$ and $t \in \mathcal{I}$, define

$$\begin{aligned} I_\alpha(t) &:= \{i \in \{1, \dots, n_c\} \mid c_i(\hat{x}(t), \hat{u}(t)) \ge -\alpha\}, \\ A(t) &:= c_u'(\hat{x}(t), \hat{u}(t)), \\ A_\alpha(t) &:= \left(c_{i,u}'(\hat{x}(t), \hat{u}(t))\right)_{i \in I_\alpha(t)}, \\ S(t) &:= \operatorname{diag}(s_i(t),\ i = 1, \dots, n_c), \\ S_\alpha(t) &:= \operatorname{diag}(s_i(t),\ i \in I_\alpha(t)). \end{aligned}$$

Prove that the following statements are equivalent:

(a) There exists $\beta > 0$ with

$$\left\| \begin{pmatrix} A(t) & S(t) \end{pmatrix}^\top d \right\| \ge \beta \|d\|$$

for every $d \in \mathbb{R}^{n_c}$ for almost every $t \in \mathcal{I}$.

(b) There exists $\hat{\alpha} > 0$ and $\hat{\beta} > 0$ with

$$\left\| A_{\hat{\alpha}}(t)^\top d \right\| \ge \hat{\beta} \|d\|$$

for every d of appropriate dimension for almost every $t \in \mathcal{I}$.

The mixed control-state inequality constraints

$$c_i(x(t), u(t)) \le 0, \quad i = 1, \dots, n_c,$$

can be equivalently replaced by the equality constraints

$$0 = \phi_i(x(t), u(t), s(t)) := c_i(x(t), u(t)) + \frac{1}{2} s_i(t)^2, \quad i = 1, \dots, n_c,$$

where s_i, $i = 1, \dots, n_c$, denote slack variables.

Apply and evaluate the Fritz John conditions in Theorem 2.3.23 under the usual Assumption 2.2.5 to the following equality constrained optimal control problem on $\mathcal{I} := [0,1]$ with $x_0 \in \mathbb{R}^{n_x}$ given. To this end, use the linear independence constraint qualification.

Minimize

$$\varphi\big(x(t_0), x(t_f)\big) + \int_0^1 f_0\big(x(t), u(t)\big)dt$$

with respect to $x \in W_{1,\infty}^{n_x}(\mathcal{I})$, $u \in L_\infty^{n_u}(\mathcal{I})$, $s \in L_\infty^{n_c}(\mathcal{I})$ *subject to the constraints*

$$\begin{aligned} \dot{x}(t) &= f\big(x(t), u(t)\big), & x(0) &= x_0, \\ 0 &= \phi_i\big(x(t), u(t), s(t)\big), & i &= 1, \ldots, n_c. \end{aligned}$$

Consider the variational problem

$$\int_a^b f\big(t, x(t), x'(t)\big)dt \quad \longrightarrow \quad \min.$$

(a) Restate the variational problem as an optimal control problem.
(b) Evaluate the local minimum principle for the optimal control problem in (a).
(c) Solve the following problems:

$$\begin{aligned} &\text{(a)} \quad f\big(t, x, x'\big) = 1 - \big(x'\big)^2, \\ &\text{(b)} \quad f\big(t, x, x'\big) = \big(1 - \big(x'\big)^2\big)^2, \\ &\text{(c)} \quad f\big(t, x, x'\big) = \big(x'\big)^2 - x^2, \\ &\text{(d)} \quad f\big(t, x, x'\big) = \sin\big(tx'\big). \end{aligned}$$

Consider the following optimal control problem with interior point conditions:

Minimize

$$\varphi\big(t_s, x(t_0), x(t_s), x(t_f)\big)$$

with respect to $t_s \in [t_0, t_f]$, $x \in W_{1,\infty}^{n_x}([t_0, t_f])$, $u \in L_\infty^{n_u}([t_0, t_f])$ *subject to*

$$\begin{aligned} \dot{x}(t) &= \begin{cases} f_1(x(t), u(t)), & \text{if } t \in [t_0, t_s), \\ f_2(x(t), u(t)), & \text{if } t \in [t_s, t_f], \end{cases} \\ 0_{\mathbb{R}^{n_\psi}} &= \psi\big(t_s, x(t_0), x(t_s), x(t_f)\big). \end{aligned}$$

Herein, $t_s \in [t_0, t_f]$ is free, and φ, ψ, f_i, $i = 1, 2$, are sufficiently smooth functions.

(a) Use the transformation techniques of Section 1.2 to transform the problem into an equivalent optimal control problem on the time interval $[0, 1]$.

(b) Apply the minimum principle to the transformed problem in (a) and formulate the resulting two-point boundary value problem.
(c) Transform the conditions in (b) back to the original formulation.

Prove Theorems 3.4.3 and 3.4.4 by repeating the steps in Sections 3.1, 3.2, and 3.3. To this end, formulate and prove a similar result as in Lemma 3.1.2 for the index-one case.

Consider the linear operator $T : W_{1,2}^{n_x}([t_0, t_f]) \longrightarrow L_2^{n_x}([t_0, t_f]) \times \mathbb{R}^{n_x}$ defined by

$$T(x)(\cdot) := \begin{pmatrix} \dot{x}(\cdot) - A(\cdot)x(\cdot) \\ x(0) \end{pmatrix},$$

where $A \in L_\infty^{n_x \times n_x}([t_0, t_f])$ is a given matrix function.

Compute the adjoint operator $T^* : L_2^{n_x}([t_0, t_f])^* \times \mathbb{R}^{n_x} \longrightarrow W_{1,2}^{n_x}([t_0, t_f])^*$, which is defined in part (g) of Definition 2.1.9.

4 Discretization methods for ODEs and DAEs

Discretization methods for ODEs and DAEs are a key tool for the numerical simulation of dynamic systems, but they are also the basis of most numerical algorithms in optimal control. For instance, the indirect method requires to solve boundary value problems by, e. g., a multiple shooting technique, the direct discretization method for optimal control problems in Chapter 5, and the function space approaches in Chapter 8 require to solve the dynamic equations (and adjoint equations) repeatedly for given control inputs. Moreover, gradient information is often provided by a sensitivity analysis of the dynamic system with respect to input parameters.

Many discretization methods originally designed for ODEs, e. g., Runge–Kutta, multi-step, or extrapolation methods, can be adapted in a fairly straightforward way to solve DAEs. Naturally, the resulting methods are implicit due to the inherently implicit character of DAEs, and solving nonlinear systems of equations in each integration step is necessary. The nonlinear systems are typically solved by Newton's method or well-known variants, e. g., globalized or simplified Newton or quasi-Newton methods. To reduce the computational effort of Newton's method for nonlinear equations, methods based on a suitable linearization are considered, e. g., linearized implicit Runge–Kutta methods.

For notational convenience, we will write down the methods for general DAE initial value problems of the following type:

Problem 4.0.1 (DAE initial value problem). Let $\mathcal{I} := [t_0, t_f] \subset \mathbb{R}$ be a compact time interval with $t_0 < t_f$, $F : \mathcal{I} \times \mathbb{R}^{n_z} \times \mathbb{R}^{n_z} \longrightarrow \mathbb{R}^{n_z}$ a sufficiently smooth function, and z_0 a consistent initial value.

Find a solution z of the *DAE initial value problem*

$$F\big(t, z(t), \dot{z}(t)\big) = 0_{\mathbb{R}^{n_z}}, \quad z(t_0) = z_0. \tag{4.1}$$

Although most methods can be conveniently motivated for (4.1), it is important to note that this does not imply automatically that the resulting discretizations actually converge to a solution. In contrast to the ODE case, where a quite general convergence theory exists, for DAEs, convergence depends on the structure of the system in general. Convergence results assume a certain index or a special structure of (4.1), for instance, the Hessenberg structure.

The discretization methods can be classified as one-step and multi-step. Next, we will discuss Runge–Kutta (one-step) and BDF methods (multi-step), both being well investigated in the literature, see [104, 105, 270, 109, 108, 214, 37, 38, 272, 155, 14, 172, 173, 306, 151, 12, 324, 57, 191, 234, 233, 303] and the monographs [35, 154, 193]. As we will use integration methods merely as a tool within optimal control algorithms, we do not intend to provide a detailed convergence analysis. We rather motivate the construction of methods and provide some general concepts for a possible convergence analysis. A general discretization theory can be found in [311].

https://doi.org/10.1515/9783110797893-004

All subsequent discretization methods for the approximate solution of Problem 4.0.1 have in common a discretization of the time interval $\mathcal{I}$ by a *grid*

$$\mathbb{G}_N := \{a = t_0 < t_1 < t_2 < \cdots < t_{N-1} < t_N = b\} \tag{4.2}$$

with *grid points* t_i, $i = 0,1,\ldots,N$, $N \in \mathbb{N}$, and *step-sizes* $h_i := t_{i+1} - t_i$, $i = 0,1,\ldots,N-1$. Often equidistant grids with step-sizes $h_i := h := (b-a)/N$ for $i = 0,\ldots,N-1$ are considered for optimal control problems. Equidistant grids, however, are not efficient in many practical applications, and, instead, adaptive grids are chosen, which are designed automatically depending on the solution by suitable step-size selection algorithms. The working principle of such step-size selection algorithms is illustrated in Section 4.4.

A discretization method constructs a *grid function* $z_N : \mathbb{G}_N \longrightarrow \mathbb{R}^{n_z}$ with $t \mapsto z_N(t)$ for $t \in \mathbb{G}_N$, which approximates the solution $\hat{z}$ of the initial value problem (4.1) at least on the grid $\mathbb{G}_N$, that is

$$z_N(t_i) \approx \hat{z}(t_i), \quad i = 0,\ldots,N.$$

As the grid function is only defined on the grid $\mathbb{G}_N$ by its $N+1$ values, the abbreviation

$$z_i := z_N(t_i), \quad i = 0,\ldots,N,$$

is used.

The interpretation of the values z_i, $i = 0,\ldots,N$, as a grid function is nevertheless useful, since for the approximation error a grid function z_N has to be related to a solution $\hat{z}$ of the initial value problem defined on $\mathcal{I}$. This relation can be done in two ways: Either a grid function z_N is extended to the whole interval $\mathcal{I}$ by suitable interpolation schemes, or the solution of the initial value problem is restricted to the grid $\mathbb{G}_N$. We follow the latter approach. From a mathematical point of view, it is particularly interesting whether the sequence $\{z_N\}_{N\in\mathbb{N}}$ of grid functions for $N \longrightarrow \infty$ converges to a solution $\hat{z}$ of the initial value problem 4.1. This requires that the *grid size*

$$h := \max_{i=0,\ldots,N-1} h_i$$

tends to zero such that the whole of the interval $\mathcal{I}$ is actually covered by grid points. However, before we analyze the convergence of discretization methods, we have a look at common one-step methods.

4.1 Discretization by one-step methods

We discuss so-called one-step methods, particularly Runge–Kutta methods for the numerical solution of ODEs and DAEs.

4.1.1 The Euler method

The working principle of a one-step method is illustrated using the simplest one-step method – the *explicit Euler method* – for the ODE initial value problem

$$\dot{z}(t) = f(t, z(t)), \quad z(t_0) = z_0, \tag{4.3}$$

which is a special case of (4.1).

The explicit Euler method works step-by-step through the grid points in $\mathbb{G}_N$ starting at t_0 with the given initial value $\hat{z}(t_0) = z_0$, which defines the value of the grid function $z_N(t_0) := z_0$. In t_0 the derivative of the solution is known, it is $\frac{d}{dt}\hat{z}(t_0) = f(t_0, z_0)$. At $t_1 = t_0 + h_0$, the solution can be approximated by Taylor expansion (sufficient smoothness is assumed) by

$$\hat{z}(t_1) \approx \hat{z}(t_0) + \frac{d}{dt}\hat{z}(t_0)(t_1 - t_0) = z_0 + h_0 f(t_0, z_0).$$

This value is used as an approximation at t_1:

$$z_1 = z_N(t_1) := z_0 + h_0 f(t_0, z_0).$$

In the approximation z_1, one can evaluate f again to obtain an approximation z_2 at t_2. This procedure is repeated until the end point $t_N = t_f$ is reached, compare Figure 4.1.

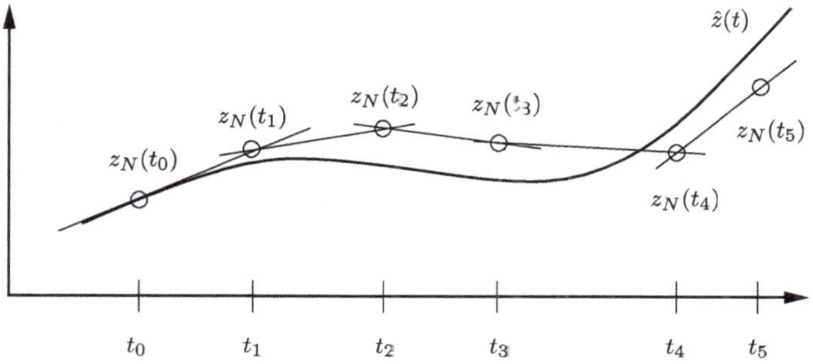

Figure 4.1: Idea of the explicit Euler method: Approximation by local linearization.

Summarizing, the explicit Euler method works as follows:

Algorithm 4.1.1 (Explicit Euler method).

(0) Let the initial value problem (4.3) be given. Choose a grid $\mathbb{G}_N$ according to (4.2).

(1) Set $z_N(t_0) = z_0$.

(2) Compute for $i = 0, 1, \ldots, N-1$:

$$z_N(t_{i+1}) := z_N(t_i) + h_i f(t_i, z_N(t_i)).$$

□

The computation of $z_N(t_{i+1})$ in Algorithm 4.1.1 only depends on the value $z_N(t_i)$ at the previous grid point. Hence, the method is a *one-step method*. In contrast, *multi-step methods* not only use $z_N(t_i)$ but also $z_N(t_{i-1}), z_N(t_{i-2}), \ldots$.

Alternative motivations:

(a) The Euler method can be motivated alternatively by approximating the derivative $\dot{z}(t_i)$ by the finite difference scheme

$$f(t_i, z(t_i)) = \dot{z}(t_i) \approx \frac{z(t_{i+1}) - z(t_i)}{h_i}.$$

(b) A further motivation is based on the expression

$$z(t_{i+1}) - z(t_i) = \int_{t_i}^{t_{i+1}} f(t, z(t))dt$$

for the solution z of (4.3). The explicit Euler method appears, if the integral is approximated by $(t_{i+1} - t_i)f(t_i, z(t_i))$.

Instead of using Taylor expansion at t_{i+1} around t_i, we could have used Taylor expansion in a backward sense as follows:

$$z_0 = \hat{z}(t_0) \approx \hat{z}(t_1) + \frac{d}{dt}\hat{z}(t_1)(t_0 - t_1) = z_1 - h_0 f(t_1, z_1),$$

and

$$z_1 = z_0 + h_0 f(t_1, z_1),$$

respectively. This rule yields the *implicit Euler method*:

Algorithm 4.1.2 (Implicit Euler method).

(0) Let the initial value problem (4.3) be given. Choose a grid $\mathbb{G}_N$ according to (4.2).

(1) Set $z_N(t_0) = z_0$.

(2) Compute for $i = 0, 1, \ldots, N-1$:

$$z_N(t_{i+1}) := z_N(t_i) + h_i f(t_{i+1}, z_N(t_{i+1})). \tag{4.4}$$

□

The difficulty for the implicit Euler method is that (4.4) is a *nonlinear* equation for the unknown value $z_N(t_{i+1})$. This nonlinear equation can be solved by fixed-point iteration or Newton's method. Since this has to be done in every step, the computational effort for the implicit Euler method is much higher than for the explicit Euler method.

The implicit Euler method, however, has better stability properties (A-stability) as the explicit Euler method. Moreover, the explicit Euler method is useful for ODEs, but its application to DAEs is crucial. For instance, consider the semi-explicit DAE

$$\dot{x}(t) = f(t, x(t), y(t)),$$
$$0_{\mathbb{R}^{n_y}} = g(t, x(t))$$

with a consistent initial value $(x_0, y_0)^\top$. Application of the explicit Euler method to the differential equation yields

$$x_{i+1} = x_i + h_i f(t_i, x_i, y_i), \quad i = 0, 1, \dots, N-1.$$

However, no indication is given on how y_i should be chosen, and for an insufficient choice of y_i, the approximations x_{i+1}, $i = 0, 1, \dots, N-1$, in general, do not satisfy the algebraic constraints $g(t_{i+1}, x_{i+1}) = 0_{\mathbb{R}^{n_y}}$, $i = 0, 1, \dots, N-1$, anymore.

Application of the implicit Euler method yields

$$x_{i+1} = x_i + h_i f(t_{i+1}, x_{i+1}, y_{i+1}),$$
$$0_{\mathbb{R}^{n_y}} = g(t_{i+1}, x_{i+1}),$$

for $i = 0, 1, \dots, N-1$. This is a nonlinear equation for x_{i+1} and y_{i+1}, and its solution satisfies the algebraic constraint. Hence, the implicit Euler method can be extended to the general DAE in Problem 4.0.1 in a straightforward way by exploiting the relation

$$\dot{z}(t_i) \approx \frac{z(t_{i+1}) - z(t_i)}{h_i}.$$

Algorithm 4.1.3 (Implicit Euler method for Problem 4.0.1).

(0) Let the initial value problem 4.0.1 be given. Choose a grid $\mathbb{G}_N$ according to (4.2).

(1) Set $z_N(t_0) = z_0$.

(2) Solve the nonlinear equation

$$F\left(t_{i+1}, z_N(t_{i+1}), \frac{z_N(t_{i+1}) - z_N(t_i)}{h_i}\right) = 0_{\mathbb{R}^{n_z}} \tag{4.5}$$

with respect to $z_N(t_{i+1})$ for $i = 0, 1, \dots, N-1$. □

The implicit Euler method is one-step since the computation of $z_N(t_{i+1})$ only depends on the previous value $z_N(t_i)$, but the dependence is given implicitly.

4.1.2 Runge–Kutta methods

The first approximation z_1 in the Euler method is already subject to an approximation error with respect to the true solution $\hat{z}(t_1)$. This approximation error is further propagated in every step. Runge–Kutta methods aim at reducing the approximation error using higher-order approximations in each step. We first motivate the construction principle of Runge–Kutta methods for the ODE (4.3) and extend it later to the DAE in Problem 4.0.1.

To this end, consider the integral representation

$$z(t_{i+1}) - z(t_i) = \int_{t_i}^{t_{i+1}} f(t, z(t))dt$$

for a solution of the ODE (4.3).

Approximation of the integral by the trapezoidal rule yields

$$z(t_{i+1}) - z(t_i) = \int_{t_i}^{t_{i+1}} f(t, z(t))dt \approx \frac{h_i}{2}(f(t_i, z(t_i)) + f(t_{i+1}, z(t_{i+1}))).$$

As $z(t_{i+1})$ is determined implicitly by this equation, $z(t_{i+1})$ in the latter term is approximated by an explicit Euler step, and we obtain

$$z(t_{i+1}) - z(t_i) \approx \frac{h_i}{2}(f(t_i, z(t_i)) + f(t_{i+1}, z(t_i) + h_i f(t_i, z(t_i)))).$$

This is an explicit method – *Heun's method.*

Algorithm 4.1.4 (Method of Heun).
(0) Let the initial value problem 4.3 be given. Choose a grid $\mathbb{G}_N$ according to (4.2).
(1) Set $z_N(t_0) = z_0$.
(2) Compute for $i = 0, 1, \ldots, N-1$:

$$\begin{aligned} k_1 &:= f(t_i, z_N(t_i)), \\ k_2 &:= f(t_i + h_i, z_N(t_i) + h_i k_1), \\ z_N(t_{i+1}) &:= z_N(t_i) + \frac{h_i}{2}(k_1 + k_2). \end{aligned}$$

□

As the trapezoidal rule is a better approximation than the Riemann sum approximation used in Euler's method, one can expect that Heun's method possesses better approximation properties as the explicit Euler method. The method of Heun and both of the Euler methods are particular examples of *Runge–Kutta methods* defined in

Definition 4.1.5 (Runge–Kutta method for ODEs). Consider (4.3). Let $\mathbb{G}_N$ be a grid as in (4.2). For $s \in \mathbb{N}$ and coefficients b_j, c_j, a_{ij}, $i, j = 1, \ldots, s$, the *s-stage Runge–Kutta method* is defined by

$$z_N(t_{i+1}) := z_N(t_i) + h_i \sum_{j=1}^{s} b_j k_j(t_i, z_N(t_i); h_i) \tag{4.6}$$

with the *stage derivatives*

$$k_j(t_i, z_N(t_i); h_i) := f\left(t_i + c_j h_i, z_N(t_i) + h_i \sum_{\ell=1}^{s} a_{j\ell} k_\ell(t_i, z_N(t_i); h_i)\right)$$

for $j = 1, \ldots, s$.

The s-stage Runge–Kutta method is called *explicit*, if $a_{ij} = 0$ for $j \geq i$. Otherwise, it is called *implicit*. □

The stage derivative k_j in an explicit Runge–Kutta method depends only on $k_1, \ldots, k_{j-1}$, and thus the stage derivatives can be computed one after the other starting with k_1.

Implicit Runge–Kutta methods require the solution of a system of $n_z \cdot s$ nonlinear equations in each integration step.

Runge–Kutta methods are defined by the so-called *Butcher array*:

$$\begin{array}{c|cccc} c_1 & a_{11} & a_{12} & \cdots & a_{1s} \\ c_2 & a_{21} & a_{22} & \cdots & a_{2s} \\ \vdots & \vdots & \vdots & \ddots & \vdots \\ c_s & a_{s1} & a_{s2} & \cdots & a_{ss} \\ \hline & b_1 & b_2 & \cdots & b_s \end{array} \quad \Leftrightarrow \quad \begin{array}{c|c} c & A \\ \hline & b^\top \end{array}$$

For explicit methods, only the left lower triangular part of the matrix A without the diagonal is non-zero, and the Butcher array reduces to

$$\begin{array}{c|ccccc} c_1 & & & & & \\ c_2 & a_{21} & & & & \\ c_3 & a_{31} & a_{32} & & & \\ \vdots & \vdots & \vdots & \ddots & & \\ c_s & a_{s1} & a_{s2} & \cdots & a_{s,s-1} & \\ \hline & b_1 & b_2 & \cdots & b_{s-1} & b_s \end{array}$$

Example 4.1.6 (Classic 4-stage Runge–Kutta method). The classic explicit 4-stage Runge–Kutta method has the Butcher array

$$\begin{array}{c|cccc} 0 & & & & \\ 1/2 & 1/2 & & & \\ 1/2 & 0 & 1/2 & & \\ 1 & 0 & 0 & 1 & \\ \hline & 1/6 & 1/3 & 1/3 & 1/6 \end{array}$$

and the method is given by

$$
\begin{aligned}
z_N(t_{i+1}) = z_N(t_i) + h_i\left(\frac{1}{6}k_1 + \frac{1}{3}k_2 + \frac{1}{3}k_3 + \frac{1}{6}k_4\right) \\
k_1 &= f(t_i, z_N(t_i)), \\
k_2 &= f\left(t_i + \frac{h_i}{2}, z_N(t_i) + \frac{h_i}{2}k_1\right), \\
k_3 &= f\left(t_i + \frac{h_i}{2}, z_N(t_i) + \frac{h_i}{2}k_2\right), \\
k_4 &= f(t_i + h_i, z_N(t_i) + h_i k_3).
\end{aligned}
$$

□

Example 4.1.7 (RADAUIIA method). A commonly used implicit 2-stage Runge–Kutta method is the *RADAUIIA method* with the Butcher array

$$
\begin{array}{c|cc}
1/3 & 5/12 & -1/12 \\
1 & 3/4 & 1/4 \\
\hline
 & 3/4 & 1/4
\end{array}
$$

and

$$
\begin{aligned}
z_N(t_{i+1}) = z_N(t_i) + h_i\left(\frac{3}{4}k_1 + \frac{1}{4}k_2\right) \\
k_1 &= f\left(t_i + \frac{h_i}{3}, z_N(t_i) + h_i\left(\frac{5}{12}k_1 - \frac{1}{12}k_2\right)\right), \\
k_2 &= f\left(t_i + h_i, z_N(t_i) + h_i\left(\frac{3}{4}k_1 + \frac{1}{4}k_2\right)\right).
\end{aligned}
$$

□

Example 4.1.8. The initial value problem

$$
\begin{aligned}
\dot{z}_1(t) &= z_2(t), & z_1(0) &= 0, \\
\dot{z}_2(t) &= -4z_1(t) + 3\cos(2t), & z_2(0) &= 0,
\end{aligned}
$$

possesses the solution

$$
\begin{aligned}
\hat{z}_1(t) &= \frac{3}{4}t\sin(2t), \\
\hat{z}_2(t) &= \frac{3}{4}\sin(2t) + \frac{3}{2}t\cos(2t).
\end{aligned}
$$

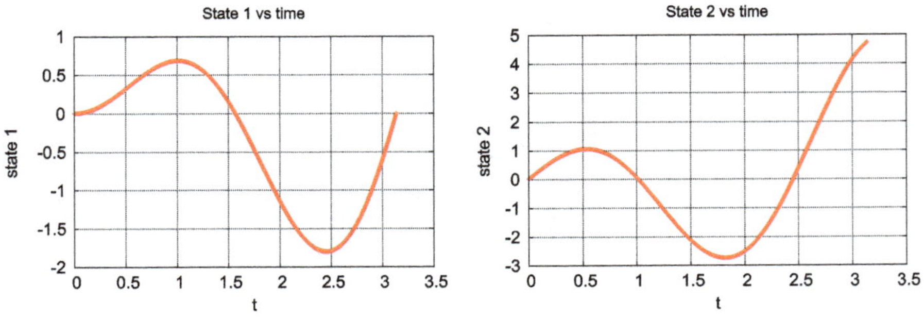

Approximations z_N are computed for the explicit Euler method, the method of Heun, and the classic Runge–Kutta method for $\mathcal{I} = [0, \pi]$ on grids $\mathbb{G}_N$ with $N = 5, 10, 20, 40, 80, 160, 320, 640, 1280$.

We compare the errors

$$\|z_N - \hat{z}\|_\infty = \max_{t_i \in \mathbb{G}_N} \|z_N(t_i) - \hat{z}(t_i)\|_2,$$

estimate the approximation order p with the ansatz $\|z_N - \hat{z}\|_\infty = C(1/N)^p$ by

$$\frac{\|z_N - \hat{z}\|_\infty}{\|z_{2N} - \hat{z}\|_\infty} = \frac{1/N^p}{1/(2N)^p} = 2^p \quad \Longrightarrow \quad p = \log_2\left(\frac{\|z_N - \hat{z}\|_\infty}{\|z_{2N} - \hat{z}\|_\infty}\right),$$

and obtain the following values:

N	Euler error	order *p*	Heun error	order *p*	RK error	order *p*
5	0.189e+02	0.000e+00	0.612e+01	0.000e+00	0.330e+00	0.000e+00
10	0.646e+01	0.155e+01	0.102e+01	0.258e+01	0.218e-01	0.392e+01
20	0.281e+01	0.120e+01	0.245e+00	0.206e+01	0.133e-02	0.404e+01
40	0.137e+01	0.103e+01	0.606e-01	0.202e+01	0.815e-04	0.403e+01
80	0.660e+00	0.106e+01	0.151e-01	0.201e+01	0.504e-05	0.401e+01
160	0.322e+00	0.104e+01	0.375e-02	0.201e+01	0.314e-06	0.401e+01
320	0.159e+00	0.102e+01	0.936e-03	0.200e+01	0.196e-07	0.400e+01
640	0.788e-01	0.101e+01	0.234e-03	0.200e+01	0.122e-08	0.400e+01
1280	0.393e-01	0.101e+01	0.585e-04	0.200e+01	0.762e-10	0.400e+01

It can be observed nicely that Euler's method has approximation order one, Heun's method has order two, and the classic Runge–Kutta method has order four. Euler's method reaches an accuracy of approximately $4 \cdot 10^{-2}$ for $N = 1280$ and requires 1280 function evaluations of the right-hand-side f. Heun's method reaches this accuracy already with $N = 80$ and needs only 160 function evaluations. The classic Runge–Kutta method already reaches the same accuracy with $N = 10$ with only 40 function evaluations; hence, it is the most efficient method of the three methods. □

Implicit Runge–Kutta methods are suitable for the general DAE in Problem 4.0.1, if the stage derivatives k_j, $j = 1, \ldots, s$, in Definition 4.1.5 are interpreted as derivatives of the solution at intermediate points $t_i + c_j h_i$:

Definition 4.1.9 (Runge–Kutta method for DAEs). Consider Problem 4.0.1. Let $\mathbb{G}_N$ be a grid as in (4.2). For $s \in \mathbb{N}$ and coefficients $b_j, c_j, a_{ij}, i, j = 1, \ldots, s$, the *s-stage Runge–Kutta method* is defined by

$$z_N(t_{i+1}) := z_N(t_i) + h_i \sum_{j=1}^{s} b_j k_j(t_i, z_N(t_i); h_i), \tag{4.7}$$

where the stage derivatives $k_j = k_j(t_i, z_N(t_i), h_i)$, $j = 1, \dots, s$, are implicitly defined by the nonlinear system of equations

$$G(k, t_i, z_N(t_i), h_i) := \begin{pmatrix} F(t_i + c_1 h_i, z_{i+1}^{(1)}, k_1) \\ \vdots \\ F(t_i + c_s h_i, z_{i+1}^{(s)}, k_s) \end{pmatrix} = 0_{\mathbb{R}^{s \cdot n_z}} \tag{4.8}$$

for $k = (k_1, \dots, k_s)^\top \in \mathbb{R}^{s \cdot n_z}$ and the *stage approximations*

$$z_{i+1}^{(\ell)} = z_N(t_i) + h_i \sum_{j=1}^{s} a_{\ell j}\, k_j(t_i, z_N(t_i), h_i). \quad \ell = 1, \dots, s. \tag{4.9}$$

Equation (4.8) is solved numerically by the (globalized) Newton method defined by

$$\begin{aligned} G_k'(k^{(\ell)}, t_i, z_N(t_i), h_i)\, \Delta^{(\ell)} &= -G(k^{(\ell)}, t_i, z_N(t_i), h_i), \\ k^{(\ell+1)} &= k^{(\ell)} + \lambda_\ell \Delta^{(\ell)}, \quad \ell = 0, 1, \dots, \end{aligned} \tag{4.10}$$

with a suitable step-size $0 < \lambda_\ell \leq 1$, provided that the Jacobian matrix G_k' is non-singular. The Jacobian matrix G_k' at $(k^{(\ell)}, t_i, z_N(t_i), h_i)$ is given by

$$G_k' = \begin{pmatrix} F_{\dot z}'(\xi_1^{(\ell)}) & & \\ & \ddots & \\ & & F_{\dot z}'(\xi_s^{(\ell)}) \end{pmatrix} + h_i \begin{pmatrix} a_{11} F_z'(\xi_1^{(\ell)}) & \cdots & a_{1s} F_z'(\xi_1^{(\ell)}) \\ \vdots & \ddots & \vdots \\ a_{s1} F_z'(\xi_s^{(\ell)}) & \cdots & a_{ss} F_z'(\xi_s^{(\ell)}) \end{pmatrix}, \tag{4.11}$$

where

$$\xi_\nu^{(\ell)} = \left(t_i + c_\nu h_i, z_N(t_i) + h_i \sum_{j=1}^{s} a_{\nu j} k_j^{(\ell)}, k_\nu^{(\ell)} \right), \quad \nu = 1, \dots, s.$$

A Runge–Kutta method is called *stiffly accurate*, if it satisfies

$$c_s = 1 \quad \text{and} \quad a_{sj} = b_j \quad \text{for } j = 1, \dots, s,$$

see [315, 155]. For instance, the RADAUIIA method is stiffly accurate. These methods are particularly well-suited for DAEs, since the stage $z_{i+1}^{(s)}$ and the new approximation $z_N(t_{i+1})$ coincide. Since $z_{i+1}^{(s)}$ satisfies the DAE at $t_i + h$, so does $z_N(t_{i+1})$.

Given a Runge–Kutta method that is not stiffly accurate, the approximation (4.7) usually will not satisfy the DAE (4.1), although the stages $z_{i+1}^{(\ell)}$, $\ell = 1, \dots, s$, satisfy it. A projected Runge–Kutta methods for semi-explicit DAEs is used in [14] to circumvent this problem. Herein, $z_N(t_{i+1})$ is projected onto the algebraic equations, and the projected value is used as an approximation at t_{i+1}.

The dimension of the equation (4.10) can be reduced for half-explicit Runge–Kutta methods, see [11, 12]. Implicit Runge–Kutta method in connection with implicit index-one DAEs of type $F(\dot{x}, x, y) = 0$ is considered in [276]. Convergence of implicit Runge–Kutta methods for index-one DAEs (4.1) was shown in [35], for semi-explicit index-two DAEs in [38], and for semi-explicit Hessenberg index-three DAEs in [172, 173]. Consistent initial values and certain regularity and stability conditions of the Runge–Kutta methods are assumed in the proofs, compare [155].

4.1.3 General one-step method

The Euler methods and Runge–Kutta methods are one-step:

Definition 4.1.10 (One-step method). Let $\Phi : \mathbb{R} \times \mathbb{R}^{n_z} \times \mathbb{R} \longrightarrow \mathbb{R}^{n_z}$ be a given continuous function and $\mathbb{G}_N$ a grid according to (4.2). The method

$$z_N(t_0) = z_0, \tag{4.12}$$

$$z_N(t_{i+1}) = z_N(t_i) + h_i \Phi(t_i, z_N(t_i), h_i), \quad i = 0, \ldots, N-1, \tag{4.13}$$

for the approximate solution of Problem 4.0.1 is called *one-step method*, and Φ is called *increment function*. □

The notion 'one-step method' refers to the fact that Φ only depends on the previous approximation $z_N(t_i)$ and not on $z_N(t_{i-1}), z_N(t_{i-2}), \ldots$.

Example 4.1.11 (Increment functions). The increment function of the explicit Euler method applied to (4.3) is

$$\Phi(t, z, h) := f(t, z),$$

that of Heun's method is

$$\Phi(t, z, h) = \frac{1}{2}(f(t, z) + f(t + h, z + hf(t, z))).$$

The increment function of the implicit Euler method for (4.3) is formally defined implicitly by

$$\Phi(t, z, h) = f(t + h, z + h\Phi(t, z, h)).$$

The increment function of a Runge–Kutta method is given by

$$\Phi(t, z, h) = \sum_{j=1}^{s} b_j k_j(t, z, h),$$

where the stage derivatives for implicit methods are defined implicitly. □

4.1.4 Consistency, stability, and convergence of one-step methods

Next, the discussion is restricted to equidistant grids $\mathbb{G}_N$ with step-sizes $h = (t_f - t_0)/N$, $N \in \mathbb{N}$, for the one-step method (4.12)–(4.13).

Let $\hat{z}$ denote the exact solution of Problem 4.0.1. Define the *restriction operator* onto the grid $\mathbb{G}_N$ by

$$\Delta_N : \{z : \mathcal{I} \longrightarrow \mathbb{R}^{n_z}\} \longrightarrow \{z_N : \mathbb{G}_N \longrightarrow \mathbb{R}^{n_z}\}, \quad \Delta_N(z)(t) = z(t) \quad \text{for } t \in \mathbb{G}_N.$$

On the space of all grid functions, a norm is defined by

$$\|z_N\|_\infty = \max_{t_i \in \mathbb{G}_N} \|z_N(t_i)\|_2.$$

With this norm, the global error and the term convergence can be defined:

Definition 4.1.12 (Global error, convergence). The *global error* $e_N : \mathbb{G}_N \longrightarrow \mathbb{R}^{n_z}$ is defined by

$$e_N := z_N - \Delta_N(\hat{z}).$$

The one-step method (4.12)–(4.13) is called *convergent*, if

$$\lim_{N \to \infty} \|e_N\|_\infty = 0.$$

The one-step method (4.12)–(4.13) has *order of convergence p*, if

$$\|e_N\|_\infty = \mathcal{O}\left(\frac{1}{N^p}\right) \quad \text{as } N \longrightarrow \infty.$$

□

Next to the global error, the local discretization error is important.

Definition 4.1.13 (Local discretization error, consistency). Let $\hat{z} \in \mathbb{R}^{n_z}$ be a consistent initial value for the DAE in (4.1) and $\hat{t} \in \mathcal{I}$. Consider the one-step method (4.12)–(4.13). Let y denote the solution of the initial value problem

$$F(t, y(t), \dot{y}(t)) = 0_{\mathbb{R}^{n_z}}, \quad y(\hat{t}) = \hat{z}.$$

The *local discretization error at* $(\hat{t}, \hat{z})$ is defined by

$$\ell_h(\hat{t}, \hat{z}) := \frac{y(\hat{t} + h) - \hat{z}}{h} - \Phi(\hat{t}, \hat{z}, h).$$

The one-step method is called *consistent in a solution z of Problem* 4.0.1, if

$$\lim_{h \to 0} \left(\max_{t \in [t_0, t_f - h]} \|\ell_h(t, z(t))\| \right) = 0.$$

The one-step method has *order of consistency p in a solution z of Problem* 4.0.1, if there exist constants $C > 0$ and $h_0 > 0$ independent of h with

$$\max_{t\in[t_0,t_f-h]} \|\ell_h(t, z(t))\| \le Ch^p \quad \text{for all } 0 < h \le h_0.$$

The local discretization error indicates how good the true solution y of the initial value problem with consistent initial value $(\hat{t}, \hat{z})$ satisfies the one-step method, compare Figure 4.2. Notice that it is not enough for consistency, if merely the local error $\|y(\hat{t} + h) - z_N(\hat{t} + h)\|$ tends to zero as $h \longrightarrow 0$.

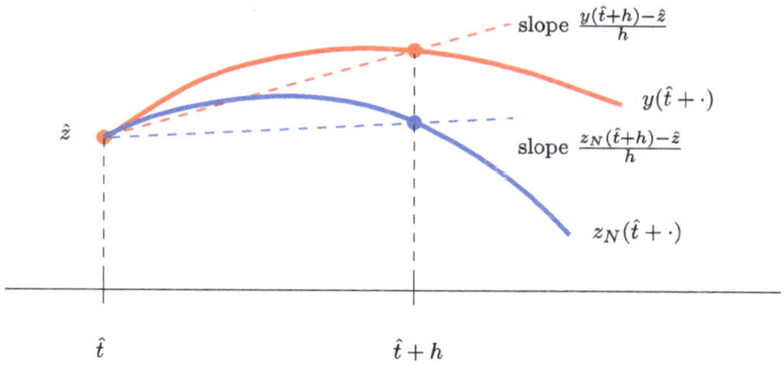

Figure 4.2: Local discretization error and consistency: $z_N(\hat{t} + \cdot)$ denotes the solution of the one-step method as a function of the step-size h. The slopes need to converge towards each other as h tends to zero in order to have a consistent one-step method.

Since

$$\ell_h(\hat{t}, \hat{z}) = \frac{y(\hat{t} + h) - (\hat{z} + h\Phi(\hat{t}, \hat{z}, h))}{h} = \frac{y(\hat{t} + h) - z_N(\hat{t} + h)}{h},$$

the local discretization error can be regarded as local error per unit step. Note that the local error in the numerator is of order $p + 1$, if the method has order of consistency p.

Consistency can be defined independently of the solution z of Problem 4.0.1. To this end, it needs to hold a condition like

$$\lim_{h\to 0} \left(\sup_{(\hat{t},\hat{z})\in\mathcal{I}\times D} \|\ell_h(\hat{t}, \hat{z})\| \right) = 0$$

for a suitable set $D \subseteq \mathbb{R}^{n_z}$.

In the ODE case, (4.3) consistency is equivalent with

$$\lim_{h\to 0} \Phi\big(t, z(t), h\big) = f\big(t, z(t)\big)$$

that has to hold uniformly in t, since

$$\lim_{h\to 0} \frac{z(t+h) - z(t)}{h} = \dot{z}(t) = f\big(t, z(t)\big).$$

Example 4.1.14. Let z denote the solution of the initial value problem (4.3). It satisfies

$$\lim_{h\to 0} \frac{z(t+h) - z(t)}{h} = \dot{z}(t) = f(t, z(t)).$$

Consider the explicit Euler method

$$z_N(t_{i+1}) = z_N(t_i) + hf(t_i, z_N(t_i))$$

with f being continuous. It holds

$$\lim_{h\to 0} \|\ell_h(t, z(t))\| = \lim_{h\to 0} \left\| \frac{z(t+h) - z(t)}{h} - f(t, z(t)) \right\| = 0$$

uniformly in t, since $\dot{z}$ is continuous in the compact set $\mathcal{I}$. Hence, the explicit Euler method is consistent in the ODE case.

If f is continuously differentiable (and z twice continuously differentiable), then Taylor expansion of z at $t \in [t_0, t_f - h]$ with $h > 0$ yields

$$\begin{aligned}
\|\ell_h(t, z(t))\| &= \left\| \frac{z(t+h) - z(t)}{h} - f(t, z(t)) \right\| \\
&= \left\| \frac{\dot{z}(t)h + \frac{1}{2}\int_0^1 \ddot{z}(t+sh)h^2 ds}{h} - f(t, z(t)) \right\| \\
&= \left\| \frac{f(t, z(t))h + \frac{1}{2}\int_0^1 \ddot{z}(t+sh)h^2 ds}{h} - f(t, z(t)) \right\| \\
&\leq Ch
\end{aligned}$$

with $C = \max_{t\in\mathcal{I}} \|\ddot{z}(t)\|$. Hence, it holds

$$\max_{t\in[t_0, t_f - h]} \|\ell_h(t, z(t))\| \leq Ch$$

and the explicit Euler method is consistent of order one in the ODE case. A higher order cannot be achieved by the explicit Euler method in general. □

Taylor expansion is used to prove the order of consistency for Runge–Kutta methods.

Example 4.1.15. Consider again (4.3). Let f be twice continuously differentiable (and z three times continuously differentiable). Taylor expansion of z at $t \in [t_0, t_f - h]$ with $h > 0$ yields

$$\begin{aligned} z(t+h) &= z(t) + h\dot{z}(t) + \frac{h^2}{2}\ddot{z}(t) + \mathcal{O}(h^3) \\ &= z(t) + hf + \frac{h^2}{2}(f_t + f_z \cdot f) + \mathcal{O}(h^3), \end{aligned}$$

where f and its derivatives are evaluated at $(t, z(t))$. The increment function of Heun's method is given by

$$\Phi(t, z, h) = \frac{1}{2}(f(t, z) + f(t + h, z + hf(t, z))).$$

Taylor expansion of the right-hand side yields

$$f(t + h, z(t) + hf(t, z(t))) = f + h(f_t + f_z \cdot f) + \mathcal{O}(h^2),$$

where f and its derivatives are evaluated at $(t, z(t))$. The local discretization error computes to

$$\begin{aligned} \|\ell_h(t, z(t))\| &= \left\| \frac{z(t+h) - z(t)}{h} - \Phi(t, z(t), h) \right\| \\ &= \left\| f + \frac{h}{2}(f_t + f_z \cdot f) - \frac{1}{2}(f + f + hf_t + hf_z \cdot f) + \mathcal{O}(h^2) \right\| \\ &= \mathcal{O}(h^2). \end{aligned}$$

Hence, Heun's method has consistency order two, if f is twice continuously differentiable. □

Example 4.1.16. Given a sufficient smoothness of f in (4.3) by Taylor expansion the following order conditions for Runge–Kutta methods can be obtained, compare [315, p. 50]:

$$\begin{aligned} p = 1 &: \sum_{i=1}^{s} b_i = 1, \\ p = 2 &: \sum_{i=1}^{s} b_i c_i = \frac{1}{2}, \\ p = 3 &: \sum_{i=1}^{s} b_i c_i^2 = \frac{1}{3}, \quad \sum_{i,j=1}^{s} b_i a_{ij} c_j = \frac{1}{6}, \end{aligned}$$

$$p = 4: \sum_{i=1}^{s} b_i c_i^3 = \frac{1}{4}, \quad \sum_{i,j=1}^{s} b_i c_i a_{ij} c_j = \frac{1}{8}, \quad \sum_{i,j=1}^{s} b_i a_{ij} c_j^2 = \frac{1}{12}, \quad \sum_{i,j,k=1}^{s} b_i a_{ij} a_{jk} c_k = \frac{1}{24}.$$

Herein, the node conditions

$$c_i = \sum_{j=1}^{s} a_{ij}, \quad i = 1, \ldots, s$$

are supposed to hold, and for order p, all conditions for order $1, \ldots, p-1$ are supposed to be satisfied as well. □

Consistency alone is not sufficient for convergence of a one-step method. Stability is required on top of it.

Definition 4.1.17 (Stability). Let $\{z_N\}_{N\in\mathbb{N}}$ be a sequence of grid functions with (4.12)–(4.13) and $h = (t_f - t_0)/N, N \in \mathbb{N}$. Moreover, let $\{y_N\}_{N\in\mathbb{N}}$ be grid functions $y_N : \mathbb{G}_N \longrightarrow \mathbb{R}^n$ with

$$\delta_N(t_0) := y_N(t_0) - z_0,$$
$$\delta_N(t_i) := \frac{y_N(t_i) - y_N(t_{i-1})}{h} - \Phi(t_{i-1}, y_N(t_{i-1}), h), \quad i = 1, \ldots, N.$$

The function $\delta_N : \mathbb{G}_N \longrightarrow \mathbb{R}^{n_z}$ is called *defect of* y_N.

The one-step method is called *stable at* $\{z_N\}_{N\in\mathbb{N}}$, if there exist constants $S, R \geq 0$ independent of N such that for almost all $h = (t_f - t_0)/N, N \in \mathbb{N}$, it holds:

Given

$$\|\delta_N\|_\infty < R,$$

it holds

$$\|y_N - z_N\|_\infty \leq S\|\delta_N\|_\infty.$$

The constant R is called *stability threshold*, and S is called *stability bound*. □

Consistency and stability ensure convergence of the one-step method.

Theorem 4.1.18 (Convergence). *Let the one-step method be consistent in a solution $\hat{z}$ of Problem* 4.0.1 *and stable in the sequence $\{z_N\}_{N\in\mathbb{N}}$ generated by the one-step method.*

Then the one-step method is convergent.

If the one-step method is consistent of order p in $\hat{z}$, then it is convergent of order p.

Proof. A solution $\hat{z}$ of the initial value problem does not satisfy the computation rule of the one-step method with step-size h exactly and yields a defect

$$\delta_N(t_0) = \hat{z}(t_0) - z_0 = 0,$$
$$\delta_N(t_i) = \frac{\hat{z}(t_i) - \hat{z}(t_{i-1})}{h} - \Phi(t_{i-1}, \hat{z}(t_{i-1}), h), \quad i = 1, \ldots, N.$$

For sufficiently small step-sizes $h = (t_f - t_0)/N$, it is always possible to guarantee $\|\delta_N\|_\infty < R$ as the method is consistent, and with $\delta_N(t_{i+1}) = \ell_h(t_i, \hat{z}(t_i))$, it follows

$$0 = \lim_{h\to 0}\Big(\max_{i=0,\dots,N-1}\|\ell_h(t_i,\hat{z}(t_i))\|\Big) = \lim_{N\to\infty}\Big(\max_{i=0,\dots,N-1}\|\delta_N(t_{i+1})\|\Big).$$

Since the method is stable, it follows (with $y_N = \Delta_N(\hat{z})$ in Definition 4.1.17)

$$\|e_N\|_\infty = \|z_N - \Delta_N(\hat{z})\|_\infty \le S\|\delta_N\|_\infty.$$

Since $\delta_N(t_0) = 0$ and $\delta_N(t_i) = \ell_N(t_{i-1}, \hat{z}(t_{i-1}))$, $i = 1,\dots,N$, consistency implies

$$\lim_{N\to\infty}\|\delta_N\|_\infty = 0$$

and consistency of order p yields

$$\|\delta_N\|_\infty = \mathcal{O}\Big(\frac{1}{N^p}\Big) \quad \text{as } N \longrightarrow \infty.$$

This proves convergence and convergence of order p, respectively. □

The stability definition is somewhat abstract. In order to find a sufficient condition for stability, we exploit the following auxiliary result.

Lemma 4.1.19 (Discrete Gronwall lemma). *For figures $h > 0$, $L > 0$, $a_k \ge 0$, $d_k \ge 0$, $k = 1,\dots,N$, related by*

$$a_k \le (1 + hL)a_{k-1} + hd_k, \quad k = 1,\dots,N,$$

it holds

$$a_k \le \exp(khL)\Big(a_0 + kh\max_{j=1,\dots,k} d_j\Big), \quad k = 0,\dots,N.$$

Proof. We use the induction principle to proof the assertion.

For $k = 0$, the assertion is true. Let the assertion hold for $k \in \mathbb{N}_0$. Then

$$\begin{aligned}
a_{k+1} &\le (1 + hL)a_k + hd_{k+1}\\
&\overset{1+hL>0}{\le} (1 + hL)\exp(khL)\Big(a_0 + kh\max_{j=1,\dots,k} d_j\Big) + hd_{k+1}\\
&\overset{1+hL\le\exp(hL)}{\le} \exp((k+1)hL)\Big(a_0 + kh\max_{j=1,\dots,k} d_j\Big) + hd_{k+1}\\
&\overset{\exp((k+1)hL)\ge 1}{\le} \exp((k+1)hL)\Big(a_0 + kh\max_{j=1,\dots,k} d_j + hd_{k+1}\Big)\\
&\le \exp((k+1)hL)\Big(a_0 + (k+1)h\max_{j=1,\dots,k+1} d_j\Big).
\end{aligned}$$

□

The following result provides a sufficient condition for stability.

Lemma 4.1.20. *Let constants $h_0 > 0$ and $L > 0$ be given such that the increment function Φ of the one-step method satisfies the Lipschitz condition*

$$\|\Phi(t,y,h) - \Phi(t,z,h)\| \le L\|y - z\|$$

for all $y, z \in \mathbb{R}^{n_z}$, all $0 < h \le h_0$, and all $t \in \mathcal{I}$.

Then the one-step method is stable.

Proof. Let $\mathbb{G}_N$ be a grid with $0 < h = \frac{t_f - t_0}{N} \le h_0$. Moreover, let y_N be a grid function with defect δ_N and

$$\|\delta_N\|_\infty < R.$$

Then for all $j = 1, \ldots, N$, it holds

$$\begin{aligned}
\|y_N(t_0) - z_N(t_0)\| &= \|z_0 + \delta_N(t_0) - z_0\| = \|\delta_N(t_0)\|, \\
\|y_N(t_j) - z_N(t_j)\| &= \|y_N(t_{j-1}) + h\Phi(t_{j-1}, y_N(t_{j-1}), h) + h\delta_N(t_j) \\
&\quad - z_N(t_{j-1}) - h\Phi(t_{j-1}, z_N(t_{j-1}), h)\| \\
&\le \|y_N(t_{j-1}) - z_N(t_{j-1})\| + h\|\Phi(t_{j-1}, y_N(t_{j-1}), h) - \Phi(t_{j-1}, z_N(t_{j-1}), h)\| \\
&\quad + h\|\delta_N(t_j)\| \\
&\le (1 + hL)\|y_N(t_{j-1}) - z_N(t_{j-1})\| + h\|\delta_N(t_j)\|.
\end{aligned}$$

Application of the discrete Gronwall Lemma 4.1.19 with $a_k := \|y_N(t_k) - z_N(t_k)\|$, $d_k := \|\delta_N(t_k)\|$ yields

$$\begin{aligned}
\|y_N(t_j) - z_N(t_j)\| &\le \exp(jhL)\Big(\|y_N(t_0) - z_N(t_0)\| + jh \max_{k=1,\ldots,j} \|\delta_N(t_k)\|\Big) \\
&= \exp(jhL)\Big(\|\delta_N(t_0)\| + jh \max_{k=1,\ldots,j} \|\delta_N(t_k)\|\Big) \\
&\le C_j \exp(jhL) \max_{k=0,\ldots,j} \|\delta_N(t_k)\|
\end{aligned}$$

with $C_j = 1 + jh$ for $j = 0, \ldots, N$. With $t_j - t_0 = jh \le Nh = t_f - t_0$ it follows

$$\|y_N - z_N\|_\infty \le S\|\delta_N\|_\infty$$

with $S := C_N \exp((t_f - t_0)L)$. □

For (explicit) Runge–Kutta methods, Lipschitz continuity of Φ follows from Lipschitz continuity of f in (4.3).

---

i The assumptions in Lemma 4.1.20 can be weakened. It is sufficient that Φ satisfies the Lipschitz condition only locally around the true solution $\hat{z}$ of the initial value problem.

The definition of stability implicitly assumes that z_N satisfies (4.12)–(4.13) exactly, but, in practice, rounding errors occur. The above convergence concept can be extended such that perturbations are taken into account, see [311, 67]. A very general discretization theory has been developed in [311].

In addition to the stability definition used here, further (and different) stability definitions play an important role for the numerical solution. For stiff ODEs and DAEs, A-stable or A(α)-stable methods are required.

4.2 Backward Differentiation Formulas (BDF)

The Backward Differentiation Formulas (BDF) are introduced in [65] and [105] and belong to the class of implicit linear multi-step methods.

We consider the integration step from t_{i+k-1} to t_{i+k}. Given approximations

$$z_N(t_i), \quad z_N(t_{i+1}), \quad \dots, \quad z_N(t_{i+k-1})$$

for

$$\hat{z}(t_i), \quad \hat{z}(t_{i+1}), \quad \dots, \quad \hat{z}(t_{i+k-1})$$

and the approximation $z_N(t_{i+k})$ for $\hat{z}(t_{i+k})$, which has to be determined, we compute the interpolating polynomial $Q(t)$ of degree k with

$$Q(t_{i+j}) = z_N(t_{i+j}), \quad j = 0, \dots, k.$$

The unknown value $z_N(t_{i+k})$ is determined by the postulation that the interpolating polynomial Q satisfies the DAE (4.1) at the time point t_{i+k}, i. e. $z_N(t_{i+k})$ satisfies

$$F(t_{i+k}, Q(t_{i+k}), \dot{Q}(t_{i+k})) = 0_{\mathbb{R}^{n_z}}. \tag{4.14}$$

The derivative $\dot{Q}(t_{i+k})$ can be expressed as

$$\dot{Q}(t_{i+k}) = \frac{1}{h_{i+k-1}} \sum_{j=0}^{k} \alpha_j z_N(t_{i+j}), \tag{4.15}$$

where the coefficients α_j depend on the step-sizes $h_{i+j-1} = t_{i+j} - t_{i+j-1}, j = 1, \dots, k$. Equation (4.14) together with (4.15) yields

$$F\left(t_{i+k}, z_N(t_{i+k}), \frac{1}{h_{i+k-1}} \sum_{j=0}^{k} \alpha_j z_N(t_{i+j}) \right) = 0_{\mathbb{R}^{n_z}}, \tag{4.16}$$

which is a nonlinear equation for $z = z_N(t_{i+k})$. Equation (4.16) is solved numerically by the (globalized) Newton method given by

$$\begin{aligned}\left(F'_z + \frac{\alpha_k}{h_{i+k-1}} F'_{\dot{z}}\right)\Delta^{(\ell)} &= -F,\\ z^{(\ell+1)} &= z^{(\ell)} + \lambda_\ell \Delta^{(\ell)}, \quad \ell = 0, 1, \ldots,\end{aligned} \tag{4.17}$$

where $0 < \lambda_\ell \leq 1$ is a suitable step-size, and the functions are evaluated at

$$\left(t_{i+k}, z^{(\ell)}, \frac{1}{h_{i+k-1}}\left(\alpha_k z^{(\ell)} + \sum_{j=0}^{k-1} \alpha_j z_N(t_{i+j})\right)\right).$$

BDF methods are appealing, since they only require to solve an n_z-dimensional nonlinear system instead of an $(n_z \cdot s)$-dimensional system that arises for Runge–Kutta methods.

Brenan et al. [35] develop in their code DASSL an efficient algorithm to set up the interpolating polynomial using a modified version of Newton's divided differences. In addition, an automatic step-size and order control algorithm is implemented based on the estimation of the local discretization error. A further improvement of the numerical performance is achieved by using some sort of simplified Newton's method, where the iteration matrix

$$F'_z + \frac{\alpha_k}{h_{i+k-1}} \cdot F'_{\dot{z}}$$

is held constant as long as possible, even for several integration steps. For higher-index systems, the estimation procedure for the local discretization error has to be modified, see [271]. In practice, the higher-index algebraic components are assigned very large error tolerances or are simply neglected in the error estimator. BDF methods are applied to overdetermined DAEs in [102, 103]. The implementation is called ODASSL and is an extension of DASSL.

Convergence of BDF methods up to order 6 for index-one DAEs (4.1) assuming constant step-sizes was proven in [109, 272]. Convergence results up to order 6 assuming constant step-sizes for DAEs arising in electric circuit simulation, fluid mechanics, and mechanics can be found in [214]. Convergence of Hessenberg DAEs (1.39) with index two and three and constant step-sizes is shown in [37]. Convergence results using non-constant step-sizes and non-constant order for index-two DAEs (1.39) and index-one DAEs (1.37)–(1.38) are obtained in [108]. Herein, it is also shown that BDF with non-constant step-sizes applied to index-three Hessenberg DAEs may yield wrong results in certain components of z. All these results assume consistent initial values.

4.3 Linearized implicit Runge–Kutta methods

The nonlinear system (4.8) has to be solved in each integration step. On top of that, the DAE (4.1) itself has to be solved many times during the iterative solution of discretized optimal control problems. Since large step-sizes h may cause Newton's method to fail, the implicit Runge–Kutta method is typically extended by an algorithm for automatic step-size selection. The resulting method may be quite time consuming in practical computations. To reduce the computational cost for integration of (4.1), we introduce linearized implicit Runge–Kutta (LRK) methods, which only require to solve linear equations in each integration step and allow the use of fixed step-sizes during integration. A related idea is used for the construction of Rosenbrock–Wanner (ROW) methods, compare [284, 280, 153].

Numerical experiments show that these methods often lead to a speed-up in the numerical solution (with reasonable accuracy) when compared to nonlinear implicit Runge–Kutta or BDF methods with step-size and order selection discussed in Section 4.2.

Consider the implicit Runge–Kutta method (4.6) with stage derivatives defined implicitly by (4.8). The linearized implicit Runge–Kutta method is based on the idea to perform only one iteration of Newton's method for (4.8). This leads to

$$G_k'(k^{(0)}, t_i, z_N(t_i), h_i) \cdot (k - k^{(0)}) = -G(k^{(0)}, t_i, z_N(t_i), h_i), \tag{4.18}$$

which is a linear equation for the stage derivatives $k = (k_1, \dots, k_s)$. The derivative G_k' is given by (4.11) and the new approximation $z_N(t_{i+1})$ by (4.6).

The vector $k^{(0)} = (k_1^{(0)}, \dots, k_s^{(0)})$ denotes an initial guess for k and will be specified below. It turns out that the choice of $k^{(0)}$ is important in view of the order of convergence.

To gain more insights into the convergence properties of the linearized Runge–Kutta method, the discussion is restricted to the ODE (4.3). Later, an extension to DAEs is suggested yielding promising numerical results although a convergence proof could not be obtained yet.

For the explicit ODE in (4.3), the derivative G_k' becomes

$$G_k'(k^{(0)}, t_i, z_N(t_i), h_i) = \begin{pmatrix} I & & \\ & \ddots & \\ & & I \end{pmatrix} - h_i \begin{pmatrix} a_{11} f_z'(v_1) & \cdots & a_{1s} f_z'(v_1) \\ \vdots & \ddots & \vdots \\ a_{s1} f_z'(v_s) & \cdots & a_{ss} f_z'(v_s) \end{pmatrix}, \tag{4.19}$$

where

$$v_\ell = \left(t_i + c_\ell h_i, z_N(t_i) + h_i \sum_{j=1}^{s} a_{\ell j} k_j^{(0)} \right), \quad \ell = 1, \dots, s.$$

Writing down the linear equation (4.18) in its components, we obtain

$$k_\ell = f(v_\ell) + h_i \sum_{j=1}^{s} a_{\ell j} f_z'(v_\ell)(k_j - k_j^{(0)}), \quad \ell = 1, \dots, s. \tag{4.20}$$

It turns out that depending on the choice of the initial guess $k^{(0)}$, different method with different approximation properties arise.

The simplest idea is to use $k^{(0)} := 0_{\mathbb{R}^{s \cdot n_z}}$, i. e. z_i is used as predictor for $z_{i+1}^{(\ell)}$, $\ell = 1, \ldots, s$, in (4.9). It turns out that the resulting method has a maximum order of consistency equal to two. However, the method is not invariant under autonomization, i. e. it will not produce the same results for (4.3) and the equivalent autonomous system

$$\begin{pmatrix} \dot{z}(t) \\ \dot{T}(t) \end{pmatrix} = \begin{pmatrix} f(T(t), z(t)) \\ 1 \end{pmatrix}, \quad \begin{pmatrix} z(0) \\ T(0) \end{pmatrix} = \begin{pmatrix} z_0 \\ t_0 \end{pmatrix}, \tag{4.21}$$

see [116]. Hence, we do not follow this choice.

Instead, we use the initial guess $k^{(0)} = (k_1^{(0)}, \ldots, k_s^{(0)})$ with $k_\ell^{(0)} = f(t_i, z_i)$ for all $\ell = 1, \ldots, s$, which means, that we use

$$z_i + h_i \sum_{j=1}^{s} a_{\ell j} f(t_i, z_i)$$

as predictor for $z_{i+1}^{(\ell)}$, $\ell = 1, \ldots, s$, in (4.9). Equation (4.20) reduces to

$$k_\ell = f(v_\ell) + h_i \sum_{j=1}^{s} a_{\ell j} f_z'(v_\ell)(k_j - f(t_i, z_i)), \quad \ell = 1, \ldots, s, \tag{4.22}$$

where

$$v_\ell = \left(t_i + c_\ell h_i, z_i + h_i \sum_{j=1}^{s} a_{\ell j} f(t_i, z_i) \right), \quad \ell = 1, \ldots, s.$$

It turns out that the method (4.22) is invariant under autonomization under appropriate assumptions, which are identical to those in [72, Lemma 4.16, p. 138]:

Lemma 4.3.1. *Let the coefficients fulfill the conditions*

$$\sum_{j=1}^{s} b_j = 1 \tag{4.23}$$

and

$$c_\ell = \sum_{j=1}^{s} a_{\ell j}, \quad \ell = 1, \ldots, s. \tag{4.24}$$

Then the LRK method (4.22) is invariant under autonomization.

Proof. Application of the LRK method to (4.3) and (4.21) yields (4.22) and

$$\begin{pmatrix} k_\ell^z \\ k_\ell^T \end{pmatrix} = \begin{pmatrix} f(\tilde{v}_\ell) + h_i \sum_{j=1}^{s} a_{\ell j}(f_z'(\tilde{v}_\ell)(k_j^z - f(T_i, z_i)) + f_t'(\tilde{v}_\ell)(k_j^T - 1)) \\ 1 \end{pmatrix}$$

for $\ell = 1, \ldots, s$, respectively, where

$$\tilde{v}_\ell = \left(T_i + h_i \sum_{j=1}^{s} a_{\ell j} \cdot 1, z_i + h_i \sum_{j=1}^{s} a_{\ell j} f(T_i, z_i) \right), \quad \ell = 1, \ldots, s.$$

k_ℓ^z is equal to k_ℓ in (4.22), if $\tilde{v}_\ell = v_\ell$ and $T_i = t_i$ hold for all ℓ and all i. The latter is fulfilled if (4.23) holds, since then

$$T_{i+1} = T_i + h_i \sum_{j=1}^{s} b_j k_j^T = T_i + h_i \sum_{j=1}^{s} b_j = T_i + h_i = t_{i+1}.$$

The first condition requires that

$$T_i + h_i \sum_{j=1}^{s} a_{\ell j} = t_i + c_\ell h_i,$$

which is satisfied if (4.24) holds. □

Equation (4.22) in matrix notation is given by

$$(I_{s \cdot n_z} - h_i B_2(t_i, z_i, h_i))k = r(t_i, z_i, h_i), \tag{4.25}$$

where

$$\begin{aligned} B_2(t_i, z_i, h_i) &:= \begin{pmatrix} a^1 \otimes f_z'(t_i + c_1 h_i, z_i + h_i c_1 f(t_i, z_i)) \\ \vdots \\ a^s \otimes f_z'(t_i + c_s h_i, z_i + h_i c_s f(t_i, z_i)) \end{pmatrix}, \\ r(t_i, z_i, h_i) &:= -h_i B_2(t_i, z_i, h_i) \cdot (e \otimes f(t_i, z_i)) + \begin{pmatrix} f(t_i + c_1 h_i, z_i + h_i c_1 f(t_i, z_i)) \\ \vdots \\ f(t_i + c_s h_i, z_i + h_i c_s f(t_i, z_i)) \end{pmatrix}, \end{aligned} \tag{4.26}$$

a^ℓ denotes the ℓth row of $A = (a_{\ell j})_{\ell,j=1,\ldots,s}$, $e = (1, \ldots, 1)^\top \in \mathbb{R}^s$, (4.24) is exploited, and $\otimes$ is the Kronecker product defined in

Definition 4.3.2 (Kronecker product). Let $A \in \mathbb{R}^{m \times n}$ and $B \in \mathbb{R}^{p \times q}$. The *Kronecker product* $\otimes$ is defined as follows:

$$A \otimes B := \begin{pmatrix} a_{11}B & a_{12}B & \cdots & a_{1m}B \\ a_{21}B & a_{22}B & \cdots & a_{2m}B \\ \vdots & \vdots & \ddots & \vdots \\ a_{n1}B & a_{n2}B & \cdots & a_{nm}B \end{pmatrix} \in \mathbb{R}^{(np)\times(mq)}.$$

□

A detailed investigation of the local discretization error reveals that the maximal attainable order of consistency is 4:

Theorem 4.3.3 (See [116]). *Let f possess continuous and bounded partial derivatives up to order four, and let (4.24) be valid. The LRK method (4.22) is consistent of order 1, if*

$$\sum_{j=1}^{s} b_j = 1$$

holds. It is consistent of order 2, if, in addition,

$$\sum_{j=1}^{s} b_j c_j = \frac{1}{2}$$

holds. It is consistent of order 3, if, in addition,

$$\sum_{i,j=1}^{s} b_i a_{ij} c_j = \frac{1}{6}, \quad \sum_{i=1}^{s} b_i c_i^2 = \frac{1}{3}$$

hold. It is consistent of order 4, if, in addition,

$$\sum_{i,j,l=1}^{s} b_i a_{ij} a_{jl} c_l = \frac{1}{24}, \quad \sum_{i,j=1}^{s} b_i a_{ij} c_j^2 = \frac{1}{12}, \quad \sum_{i,j=1}^{s} b_i c_i a_{ij} c_j = \frac{1}{8}, \quad \sum_{i=1}^{s} b_i c_i^3 = \frac{1}{4},$$

hold. $p = 4$ is the maximal attainable order of consistency. □

The above consistency conditions are the usual conditions known for general implicit Runge–Kutta (IRK) methods. It has to be mentioned that the LRK method is A-stable, if the nonlinear IRK method (4.6) is A-stable, since A-stability is defined for linear differential equations. For linear differential equations, the LRK method coincides with the IRK method. Notice that the LRK method (4.22) is a one-step method. Moreover, it can be shown that the increment function is locally Lipschitz continuous, and hence, the method is stable and convergent.

The method can be extended to the DAE in (4.1). To this end, consider the integration step from t_i to t_{i+1}. The choice $k_\ell^{(0)} = f(t_i, z_N(t_i))$, $\ell = 1, \ldots, s$, in the ODE case cannot be applied directly to DAEs as the right-hand-side f defining the derivative of z is not explicitly available. Numerical experiments suggest that for stiffly accurate Runge–Kutta methods the choice

$$k_\ell^{(0)} := k_s(t_{i-1}, z_N(t_{i-1}), h_{i-1}), \quad \ell = 1, \dots, s, \tag{4.27}$$

seems to be reasonable. Recall that stiffly accurate methods, for instance, the RADAUIIA method, satisfy

$$c_s = 1 \quad \text{and} \quad a_{sj} = b_j, \quad j = 1, \dots, s.$$

The quantity $k_s(t_{i-1}, z_N(t_{i-1}), h_{i-1})$ is calculated in the previous integration step from t_{i-1} to t_i and can be interpreted as a derivative at the time point $t_{i-1} + c_s h_{i-1} = t_i$ and hence plays at least approximately the role of f in the ODE case. For Runge–Kutta methods, which are not stiffly accurate, an initial guess of type

$$k_\ell^{(0)} := \frac{z_N(t_i) - z_N(t_{i-1})}{h_{i-1}}, \quad \ell = 1, \dots, s,$$

seems to be reasonable since the right side is an approximation of the derivative at t_i.

A formal proof of convergence has not been established by now for these DAE extensions. In numerical computations, however, the method has performed well as the following example indicates.

Example 4.3.4 (Pendulum problem). Computational results for the mathematical pendulum problem with $m = 1$ [kg], $\ell = 1$ [m] are to be presented, compare [116]. The equations of motion of the pendulum are given by the index-three Hessenberg DAE

$$\begin{aligned}
0 &= \dot{x}_1(t) - x_3(t), \\
0 &= \dot{x}_2(t) - x_4(t), \\
0 &= m \cdot \dot{x}_3(t) - (\qquad\quad - 2 \cdot x_1(t) \cdot y(t)), \\
0 &= m \cdot \dot{x}_4(t) - (-m \cdot g - 2 \cdot x_2(t) \cdot y(t)), \\
0 &= x_1(t)^2 + x_2(t)^2 - \ell^2.
\end{aligned}$$

The index reduced index-two DAE arises, if the last equation is replaced by its time derivative

$$0 = 2(x_1(t)x_3(t) + x_2(t)x_4(t)).$$

For the following numerical tests, we used the initial value

$$(x_1(0), x_2(0), x_3(0), x_4(0), y(0)) = (1, 0, 0, 0, 0).$$

Table 4.1 shows computational results for the linearized 2-stage RADAUIIA method with Butcher array

1/3	5/12	−1/12
1	3/4	1/4
	3/4	1/4

applied to the index reduced index-two pendulum example with initial guess given by (4.27) and fixed step-sizes $h = 1/N$ on the time interval $[0, 1]$. The estimated order of convergence agrees with the order for the nonlinear RADAUIIA method (4.6) and (4.8) derived in [155] for Hessenberg systems, i. e., order three for the differential variables x_1, x_2, x_3, x_4 and order two for the algebraic variable y.

Table 4.1: Order of convergence for the linearized 2-stage RADAUIIA method applied to the index reduced index-two pendulum test example.

N	max. ERR $x_1, \ldots, x_4$	max. ERR y	Order $x_1, \ldots, x_4$	Order y
10	0.48083e-01	0.72142e+00	0.29990e+01	0.19993e+01
20	0.60145e-02	0.18044e+00	0.29999e+01	0.20000e+01
40	0.75184e-03	0.45111e-01	0.30000e+01	0.20000e+01
80	0.93981e-04	0.11278e-01	0.30000e+01	0.20000e+01
160	0.11748e-04	0.28194e-02	0.30000e+01	0.20000e+01
320	0.14684e-05	0.70485e-03	0.30000e+01	0.20000e+01
640	0.18356e-06	0.17621e-03	0.30000e+01	0.20000e+01
1280	0.22944e-07	0.44053e-04	0.30000e+01	0.20000e+01
2560	0.28681e-08	0.11013e-04	0.30000e+01	0.20000e+01

Table 4.2 shows computational results for the linearized 2-stage RADAUIIA method applied to the index-three pendulum example with initial guess given by (4.27) and fixed step-sizes $h = 1/N$ on the time interval $[0, 1]$. The computed order of convergence agrees with the order for the nonlinear RADAUIIA method (4.6) and (4.8) in [155] for Hessenberg systems, i. e., order three for the positions x_1, x_2, order two for the velocities x_3, x_4, and order one for the algebraic variable y.

Table 4.2: Order of convergence for the linearized 2-stage RADAUIIA method applied to the index-three pendulum test example.

N	max. eRR x_1, x_2	max. ERR x_3, x_4	max ERR y	Order x_1, x_2	Order x_3, x_4	Order y
10	0.633e-01	0.352e+00	0.180e+02	0.307e+01	0.332e+01	0.292e+01
20	0.754e-02	0.352e-01	0.238e+01	0.319e+01	0.255e+01	0.144e+01
40	0.825e-03	0.602e-02	0.875e+00	0.308e+01	0.223e+01	0.119e+01
80	0.972e-04	0.128e-02	0.385e+00	0.303e+01	0.204e+01	0.109e+01
160	0.119e-04	0.311e-03	0.180e+00	0.302e+01	0.202e+01	0.105e+01
320	0.147e-05	0.767e-04	0.872e-01	0.301e+01	0.201e+01	0.102e+01
640	0.183e-06	0.190e-04	0.429e-01	0.300e+01	0.200e+01	0.101e+01
1280	0.228e-07	0.475e-05	0.212e-01	0.299e+01	0.200e+01	0.101e+01
2560	0.286e-08	0.118e-05	0.106e-01	0.289e+01	0.200e+01	0.100e+01

Similar computations for the linearized implicit Euler method always yield order one for all components. Similar computations for the linearized 3-stage RADAUIIA method for the index reduced index-two pendulum example yield only order three for the differential components x_1, x_2 and x_3, x_4, and order two for y. According to [155], the nonlinear method has orders 5 and 3, respectively.

In each case, the reference solution was obtained by RADAU5 with absolute and relative integration tolerance $atol = rtol = 10^{-12}$ and GGL-stabilization. □

The use of the linearized implicit Runge–Kutta method for the discretization of optimal control problems often allows to speed up the solution process significantly. An illustration of this statement can be found in [119], where an optimal control problem resulting from automobile test-driving is solved numerically. Table 4.3 summarizes the CPU times for the numerical solution of the latter optimal control problem obtained by the third-order linearized 2-stage RADAUIIA method with constant step-size h and a standard BDF method (DASSL) with automatic step-size and order selection. The relative error in Table 4.3 denotes the relative error in the respective optimal objective function values of the discretized optimal control problem. The number N denotes the number of discretization points used to discretize the optimal control problem. The speedup is the ratio of columns 4 and 2. Table 4.3 shows that the LRK method, on average, is ten times faster than the BDF method. A comparison of the accuracies of the respective solutions reveals that the quality of the LRK solution is as good as the BDF solution in this example.

Table 4.3: CPU times for the numerical solution of a discretized optimal control problem by the linearized 2-stage RADAUIIA method and the BDF method, respectively, for different numbers N of control grid points.

N	CPU LRK (in [s])	OBJ LRK	CPU BDF (in [s])	OBJ BDF	RELERR OBJ	SPEEDUP FACTOR
26	2.50	7.718303	26.63	7.718305	0.00000026	10.7
51	8.15	7.787998	120.99	7.787981	0.00000218	14.8
101	18.02	7.806801	208.40	7.806798	0.00000038	11.6
201	21.24	7.819053	171.31	7.819052	0.00000013	8.1
251	198.34	7.817618	1691.15	7.817618	0.00000000	8.5
401	615.31	7.828956	4800.09	7.828956	0.00000000	7.8

4.4 Automatic step-size selection

Owing to efficiency reasons, it is often not advisable to use a fixed step-size h. Instead, algorithms for automatic and adapted step-size selection are required. We demonstrate the benefit of such a strategy with the following example.

Example 4.4.1. The following initial value problem is a model for the motion of a satellite in the Earth–Moon system and yields a periodic orbit, the so-called Ahrenstorf-Orbit:

$$\ddot{x}(t) = x(t) + 2\dot{y}(t) - \bar{\mu}\frac{x(t)+\mu}{D_1} - \mu\frac{x(t)-\bar{\mu}}{D_2},$$
$$\ddot{y}(t) = y(t) - 2\dot{x}(t) - \bar{\mu}\frac{y(t)}{D_1} - \mu\frac{y(t)}{D_2},$$

with $\mu = 0.012277471$, $\bar{\mu} = 1 - \mu$,

$$D_1 = \sqrt{((x(t)+\mu)^2 + y(t)^2)^3}, \quad D_2 = \sqrt{((x(t)-\bar{\mu})^2 + y(t)^2)^3},$$

and

$$x(0) = 0.994, \quad y(0) = 0,$$
$$\dot{x}(0) = 0, \quad \dot{y}(0) = -2.001585106379.$$

The second-order ODE can be transformed into an equivalent first-order ODE letting $x_1 := x$, $x_2 := y$, $x_3 := \dot{x} = \dot{x}_1$, and $x_4 := \dot{y} = \dot{x}_2$:

$$\dot{x}_1(t) = x_3(t),$$
$$\dot{x}_2(t) = x_4(t),$$
$$\dot{x}_3(t) = x(t) + 2x_4(t) - \bar{\mu}\frac{x_1(t)+\mu}{D_1} - \mu\frac{x_1(t)-\bar{\mu}}{D_2},$$
$$\dot{x}_4(t) = x_2(t) - 2x_3(t) - \bar{\mu}\frac{x_2(t)}{D_1} - \mu\frac{x_2(t)}{D_2},$$

where

$$D_1 = \sqrt{((x_1(t)+\mu)^2 + x_2(t)^2)^3}, \quad D_2 = \sqrt{((x_1(t)-\bar{\mu})^2 + x_2(t)^2)^3},$$

and

$$x_1(0) = 0.994, \qquad x_2(0) = 0,$$
$$x_3(0) = 0, \qquad x_4(0) = -2.001585106379.$$

Figure 4.3 shows the components $(x_1(t), x_2(t))$ of the numerical solution, which was obtained by the classic Runge–Kutta method on the interval $[0, t_f]$ with t_f = 17.065216560158 and fixed step-size $h = t_f/N$. Only for $N = 10000$ (or larger values of N), the numerical solution coincides with the reference solution below. For $N = 10000$, about 40000 evaluations of the right-hand side of the ODE are necessary. The reference solution in Figure 4.4 has been obtained by a Runge–Kutta method of order two that uses an automatic step-size selection strategy and only requires 6368 function evaluations.

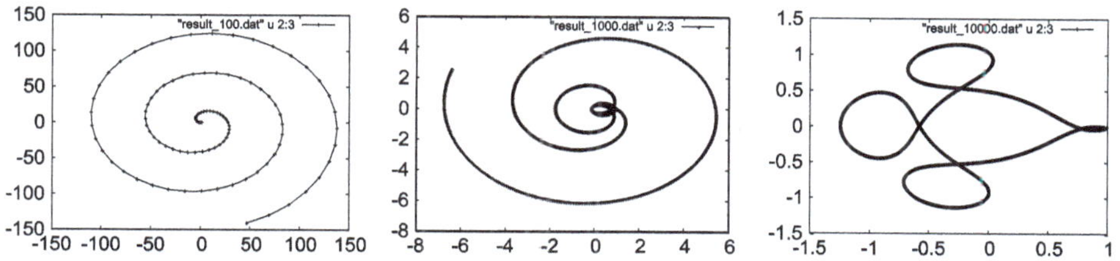

Figure 4.3: Numerical solutions for $N = 100, 1000, 10000$.

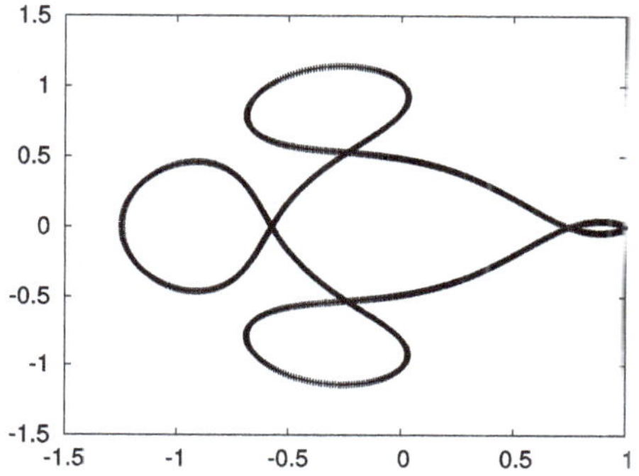

Figure 4.4: Reference solution with step-size selection.

The example shows that a fairly small step-size is needed to obtain a sufficiently good approximation, if a fixed step-size is used. The algorithm with automatic step-size selection is much more efficient as it chooses appropriate step-sizes whose size can be estimated by the distance of the points in the figure. □

Common step-size selection strategies are based on numerical estimates of the local discretization error with the aim of keeping this error below a given tolerance. A comparatively large step-size can be used in those regions where the solution does not change dramatically. In those regions, where the solution changes substantially, a comparatively small step-size has to be chosen in order to achieve a certain accuracy.

What ought an automatic step-size selection strategy be able to achieve? **?**

To answer this, assume that a single-step method has reached grid point t_i. Furthermore, let h be a suggested step-size for the next integration step. Then an automatic step-size selection strategy should be able to achieve the following items:

(1) With regard to the step from t_i to $t_{i+1} = t_i + h$ with step-size h, it has to be decided whether h is acceptable. If so, then the integration step will be performed with the underlying single-step method and step-size h.

(2) If h is accepted, then a new step-size h_{new} for the next step from t_{i+1} to $t_{i+1} + h_{\text{new}}$ has to be suggested.
(3) If h is not accepted, then a new step-size h_{new} has to be suggested, and the step from t_i to $t_i + h_{\text{new}}$ has to be repeated.

So-called *embedded Runge–Kutta methods* are used to achieve the above requirements. Two Runge–Kutta methods are called embedded, if they possess *neighboring order* p and $q = p+1$ or p and $q = p-1$, and if they differ only in the weight vector b, but, otherwise, use the same vector c and matrix A in the Butcher array.

An example is given by the following embedded method of order $p(q) = 2(3)$:

0				
1/4	1/4			
27/40	−189/800	729/800		
1	214/891	1/33	650/891	
*RK*1	214/891	1/33	650/891	0
*RK*2	533/2106	0	800/1053	−1/78

The computation of approximations η_{i+1} with $RK1$ and $\hat{\eta}_{i+1}$ with $RK2$ for the same step-size h is not expensive as the node vector c and the matrix A coincide for both methods, and, thus, the stage approximations have to be computed only once.

Now let two Runge–Kutta methods of neighboring orders p and $p+1$ and increment functions Φ and $\bar{\Phi}$, respectively, be given.

Assume that the one-step method has reached t_i with approximation z_i.

Applying both methods in (t_i, z_i) with step-size h yields approximations

$$\begin{aligned} \eta_h &= z_i + h\Phi(t_i, z_i, h), \\ \bar{\eta}_h &= z_i + h\bar{\Phi}(t_i, z_i, h). \end{aligned}$$

The local errors satisfy

$$\begin{aligned} y(t_i + h) - \eta_h &= C(t_i)h^{p+1} + \mathcal{O}(h^{p+2}), \\ y(t_i + h) - \bar{\eta}_h &= \bar{C}(t_i)h^{p+2} + \mathcal{O}(h^{p+3}), \end{aligned}$$

where y denotes the true solution of the IVP

$$F(t, y(t), \dot{y}(t)) = 0_{\mathbb{R}^{n_z}}, \quad y(t_i) = z_i.$$

Subtracting the second equation from the first yields

$$C(t_i) = \frac{1}{h^{p+1}}(\bar{\eta}_h - \eta_h) + \mathcal{O}(h).$$

This yields an estimate of the local error for the first method with lower order p:

$$y(t_i + h) - \eta_h = (\bar{\eta}_h - \eta_h) + \mathcal{O}(h^{p+2}).$$

Is h acceptable? ?

If

$$err := \|\bar{\eta}_h - \eta_h\| \le tol$$

is satisfied for a user defined tolerance $tol > 0$, then the step-size h is considered to be acceptable, and the step is accepted.

If not, then h needs to be adapted. With the above considerations, we obtain

$$\begin{aligned} y(t_i + h_{\text{new}}) - \eta_{h_{\text{new}}} &= C(t_i)h_{\text{new}}^{p+1} + \mathcal{O}(h_{\text{new}}^{p+2}) \\ &= (\bar{\eta}_h - \eta_h)\left(\frac{h_{\text{new}}}{h}\right)^{p+1} + \mathcal{O}(hh_{\text{new}}^{p+1}) + \mathcal{O}(h_{\text{new}}^{p+2}) \end{aligned}$$

for the step from t_i to $t_i + h_{\text{new}}$ using step-size h_{new}. Neglecting higher-order terms and forcing the local error to be less than or equal to tol yields

$$\|\bar{\eta}_h - \eta_h\|\left(\frac{h_{\text{new}}}{h}\right)^{p+1} \le tol$$

and finally, the new step-size has to satisfy

$$h_{\text{new}} \le \left(\frac{tol}{err}\right)^{\frac{1}{p+1}} \cdot h. \tag{4.28}$$

How to choose h_{new} after a successful step? ?

For the step from t_{i+1} with approximation z_{i+1} to $t_{i+1} + h_{\text{new}}$, it holds

$$y(t_{i+1} + h_{\text{new}}) - \eta_{h_{\text{new}}} = C(t_{i+1})h_{\text{new}}^{p+1} + \mathcal{O}(h_{\text{new}}^{p+2}).$$

If C is differentiable, i. e. if Φ is sufficiently smooth, then $C(t_{i+1}) = C(t_i + h) = C(t_i) + \mathcal{O}(h)$ and thus

$$y(t_{i+1} + h_{\text{new}}) - \eta_{h_{\text{new}}} = C(t_i)h_{\text{new}}^{p+1} + \mathcal{O}(hh_{\text{new}}^{p+1}) + \mathcal{O}(h_{\text{new}}^{p+2}).$$

Using a similar reasoning as above, we obtain again the step-size suggestion h_{new} according to (4.28).

Together with some additional technical details regarding scaling and security factors to prevent frequent step-size changes, we obtain the following step-size selection algorithm that is specifically designed for the ODE initial value problem (4.3), compare [315, p. 62].

Algorithm 4.4.2 (Automatic step-size selection).

(0) Initialization: $t = t_0$, $z = z_0$. Choose initial step-size h.

(1) If $t + h > t_f$, set $h = t_f - t$.

(2) Starting with z, compute with RK_1 and RK_2, respectively, approximations η and $\bar{\eta}$ at $t + h$.

(3) Compute err and h_{neu} according to

$$err = \max_{i=1,\dots,n_z} \left(\frac{|\eta_i - \bar{\eta}_i|}{sk_i} \right)$$

with scaling factors $sk_i = atol + \max(|\eta_i|, |z_i|) \cdot rtol$, absolute error tolerance $atol = 10^{-7}$, and relative error tolerance $rtol = 10^{-7}$ (η_i, $\bar{\eta}_i$, and z_i denote the components of η, $\bar{\eta}$, and z, respectively), as well as

$$h_{\text{new}} = \min(\alpha_{\max}, \max(\alpha_{\min}, \alpha \cdot (1/err)^{1/(1+p)})) \cdot h$$

with $\alpha_{\max} = 1.5$, $\alpha_{\min} = 0.2$, and $\alpha = 0.8$.

(4) If $h_{\text{new}} < h_{\min} := 10^{-8}$, STOP with error message.

(5) If $err \leq 1$ (step is accepted):
  (i) Set $z = \eta$, $t = t + h$.
  (ii) If $|t - t_f| < 10^{-8}$, STOP with success.
  (iii) Set $h = h_{\text{new}}$ and go to (1).
  If $err > 1$ (step is repeated): Set $h = h_{\text{new}}$ and go to (1). □

Specific automatic step-size selection algorithms for DAEs often require additional safeguards considering different approximation orders of the differential and algebraic state components, see [155, 154, 270, 35]. Owing to the different approximation orders of differential and algebraic states, the error estimator needs to be scaled by suitable powers of h for the different state components.

4.5 Computation of consistent initial values

The computation of consistent initial values for a given DAE is a pre-requisite for solving boundary value problems and optimal control problems.

Several attempts have been made for consistent initialization of general DAEs, see [262, 205, 157, 56, 142, 9, 40, 89, 112, 125]. The main difficulty in the context of general unstructured DAEs is to identify differential equations and algebraic constraints. Once this has been achieved, most methods typically use projection methods.

We restrict the discussion to Hessenberg DAEs of order $k > 1$ as in (1.39) in Definition 1.1.14. Furthermore, in view of the subsequent Chapter 5 on discretization methods for DAE optimal control problems, we assume that the control u in (1.39) is parameterized (respectively discretized) by a finite-dimensional vector $\bar{w} \in \mathbb{R}^{n_w}$ leading to a functional relation of type $u = u(t;\bar{w})$. Moreover, the control parameterization $u(t;\bar{w})$ is supposed to be $k-1$ times continuously differentiable with respect to time t, and $\bar{w}$ is supposed to be a fixed vector in the sequel:

$$\begin{array}{rcl}
\dot{x}_1(t) & = & f_1(t,\ y(t),\ x_1(t),\ x_2(t),\ \dots,\ x_{k-2}(t),\ x_{k-1}(t),\ u(t;\bar{w})),\\
\dot{x}_2(t) & = & f_2(t,\ x_1(t),\ x_2(t),\ \dots,\ x_{k-2}(t),\ x_{k-1}(t),\ u(t;\bar{w})),\\
 & \vdots & \ddots \quad \vdots \quad \vdots \quad \vdots\\
\dot{x}_{k-1}(t) & = & f_{k-1}(t,\ x_{k-2}(t),\ x_{k-1}(t),\ u(t;\bar{w})),\\
0_{\mathbb{R}^{n_y}} & = & g(t,\ x_{k-1}(t),\ u(t;\bar{w})).
\end{array} \tag{4.29}$$

Herein, the differential variable is $x(t) := (x_1(t),\dots,x_{k-1}(t))^\top \in \mathbb{R}^{n_x}$, and the algebraic variable is $y(t) \in \mathbb{R}^{n_y}$.

According to Definition 1.1.16, a consistent initial value $(\bar{x},\bar{y}) := (x(t_0), y(t_0))$ for the Hessenberg DAE at $t = t_0$ has to satisfy not only the algebraic constraint but also the hidden constraints

$$\begin{aligned}
0_{\mathbb{R}^{n_y}} &= G_j(\bar{x},\bar{w})\\
&:= g^{(j)}(t_0,\bar{x}_{k-1-j},\dots,\bar{x}_{k-1}, u(t_0;\bar{w}),\dot{u}(t_0;\bar{w}),\dots,u^{(j)}(t_0;\bar{w}))
\end{aligned} \tag{4.30}$$

for $j = 0,1,\dots,k-2$, and

$$\begin{aligned}
0_{\mathbb{R}^{n_y}} &= G_{k-1}(\bar{y},\bar{x},\bar{w})\\
&:= g^{(k-1)}(t_0,\bar{y},\bar{x},u(t_0;\bar{w}),\dot{u}(t_0;\bar{w}),\dots,u^{(k-1)}(t_0;\bar{w})),
\end{aligned} \tag{4.31}$$

compare (1.43)–(1.44).

Notice that only G_{k-1} depends on the algebraic variable y, while G_j, $j = 0,\dots,k-2$, do not depend on y. The matrix

$$\frac{\partial}{\partial y} G_{k-1}(y,x,\bar{w}) = g'_{x_{k-1}}(\cdot)\cdot f'_{k-1,x_{k-2}}(\cdot)\cdots f'_{2,x_1}(\cdot)\cdot f'_{1,y}(\cdot), \tag{4.32}$$

compare (1.40), is supposed to be non-singular.

4.5.1 Projection method for consistent initial values

A projection technique is used to compute a consistent initial value for a given (inconsistent) vector, compare [112, 125, 113]. To this end, consider the Hessenberg DAE (4.29), and let $(\bar{x}, \bar{y})$ be an arbitrary (inconsistent) vector.

The computation of a consistent pair $x := x(t_0)$ and $y := y(t_0)$ at $t = t_0$ is performed consecutively. First, x is computed such that the hidden algebraic constraints (4.30) are satisfied. Second, y is computed such that (4.31) is satisfied for a given x:

Algorithm 4.5.1 (Projection algorithm for consistent initial values).

(0) Consider the DAE (4.29). Let $(\bar{x}, \bar{y})$ be given.

(1) Solve the *constrained least-squares problem LSQ($\bar{x}, \bar{w}$)*:

Minimize

$$\frac{1}{2}\|x - \bar{x}\|^2$$

with respect to $x \in \mathbb{R}^{n_x}$ *subject to the constraints*

$$G_j(x, \bar{w}) = 0_{\mathbb{R}^{n_y}}, \quad j = 0, \ldots, k-2.$$

Let $X_0(\bar{x}, \bar{w})$ denote the solution of the least-squares problem LSQ($\bar{x}, \bar{w}$).

(2) Solve the nonlinear equation

$$G_{k-1}(y, X_0(\bar{x}, \bar{w}), \bar{w}) = 0_{\mathbb{R}^{n_y}}$$

with respect to y. Let $Y_0(\bar{x}, \bar{w})$ denote the solution.

(iv) Use $x = X_0(\bar{x}, \bar{w})$ and $y = Y_0(\bar{x}, \bar{w})$ as consistent initial value for (4.29) at t_0. □

The least-squares problem LSQ($\bar{x}, \bar{w}$) in step (1) is a *parametric nonlinear optimization problem* with parameters $\bar{x}$ and $\bar{w}$ entering the problem. Under certain regularity assumptions, see Theorem 6.1.4 in Section 6.1.1, it can be shown that the solution $X_0(\bar{x}, \bar{w})$ depends continuously differentiable on the parameters $\bar{x}$ and $\bar{w}$ and the *sensitivities* $X'_{0,x}(\bar{x}, \bar{w})$ and $X'_{0,w}(\bar{x}, \bar{w})$ can be computed by solving the linear system (6.3) in Theorem 6.1.4.

As the Jacobian matrix in (4.32) is supposed to be non-singular, by the implicit function theorem, there exists a function $Y : \mathbb{R}^{n_x} \times \mathbb{R}^{n_w} \longrightarrow \mathbb{R}^{n_y}$ with

$$G_{k-1}(Y(x, w), x, w) = 0_{\mathbb{R}^{n_y}}$$

for all (x, w) in some neighborhood of $(X_0(\bar{x}, \bar{w}), \bar{w})$. Differentiation with respect to x and w and exploitation of the non-singularity of $G'_{k-1,y}$ yields

$$Y'_x(x, w) = -G'_{k-1,y}(y, x, w)^{-1} G'_{k-1,x}(y, x, w),$$

$$Y'_w(x,w) = -G'_{k-1,y}(y,x,w)^{-1} G'_{k-1,w}(y,x,w)$$

at $y = Y(x,w)$ in this neighborhood.

Differentiation of $Y_0(x,w) := Y(X_0(x,w),w)$ with respect to x and w yields

$$\begin{aligned}
Y'_{0,x}(\bar{x},\bar{w}) &= Y'_x(X_0(\bar{x},\bar{w}),\bar{w}) \cdot X'_{0,x}(\bar{x},\bar{w}),\\
Y'_{0,w}(\bar{x},\bar{w}) &= Y'_x(X_0(\bar{x},\bar{w}),\bar{w}) \cdot X'_{0,w}(\bar{x},\bar{w}) + Y'_w(X_0(\bar{x},\bar{w}),\bar{w}).
\end{aligned}$$

Hence, Algorithm 4.5.1 not only provides the consistent initial value $(X_0(\bar{x},\bar{w}), Y_0(\bar{x},\bar{w}))$ for the Hessenberg DAE (4.29), but, by the sensitivity analysis outlined above, it also provides the derivatives

$$X'_{0,x}, \quad X'_{0,w}, \quad Y'_{0,x}, \quad Y'_{0,w}.$$

The projection algorithm 4.5.1 can be extended to general DAEs (4.1), provided it is possible to identify the algebraic constraints and the hidden constraints. Hence, from now on, it is tacitly assumed that there exists a method to compute a consistent initial value $Z_0(z,w)$ for the general DAE (5.1) for given initial guess z and control parameterization w. The function Z_0 is assumed to be at least continuously differentiable.

4.5.2 Consistent initial values via relaxation

A relaxation approach was suggested in [303] for DAE boundary value problems. Instead of projecting inconsistent initial values onto algebraic constraints, the idea is to modify the algebraic constraints in such a way that the modified constraints become consistent for given initial values. Hence, not the initial value is changed, but the DAE itself is changed.

Theorem 4.5.2. *Let $\bar{x} = (\bar{x}_1,\dots,\bar{x}_{k-1})^\top \in \mathbb{R}^{n_x}$ and $\bar{y} = \bar{x}_0 \in \mathbb{R}^{n_y}$ be arbitrary vectors.*

Then $x(t_0) := \bar{x}$ and $y(t_0) := \bar{y}$ are consistent with the relaxed Hessenberg DAE of order k defined by

$$\begin{array}{rclllllll}
\dot{x}_1(t) &=& f_1(t, & y(t), & x_1(t), & x_2(t), & \dots, & x_{k-2}(t), & x_{k-1}(t), \quad u(t;\bar{w})),\\
\dot{x}_2(t) &=& f_2(t, & & x_1(t), & x_2(t), & \dots, & x_{k-2}(t), & x_{k-1}(t), \quad u(t;\bar{w})),\\
\vdots & & & & & \ddots & & \vdots & \vdots \qquad\qquad \vdots\\
\dot{x}_{k-1}(t) &=& f_{k-1}(t, & & & & & x_{k-2}(t), & x_{k-1}(t), \quad u(t;\bar{w})),\\
0_{\mathbb{R}^{n_y}} &=& g_{\mathrm{rel}}(t, & & & & & & x_{k-1}(t), \quad u(t;\bar{w})),
\end{array} \tag{4.33}$$

with

$$g_{\mathrm{rel}}(t, x_{k-1}(t), u(t;\bar{w})) \\ := g(t, x_{k-1}(t), u(t;\bar{w})) - \sum_{j=0}^{k-2} \frac{(t-t_0)^j}{j!} G_j(\bar{x},\bar{w}) - \frac{(t-t_0)^{k-1}}{(k-1)!} G_{k-1}(\bar{y},\bar{x},\bar{w}). \tag{4.34}$$

Proof. The hidden constraints of the relaxed algebraic constraint (4.34) evaluated at $t = t_0$, $\bar{x}$, and $\bar{y}$ compute to

$$\frac{d^j}{dt^j} g_{\mathrm{rel}}(t, x_{k-1}(t), u(t;\bar{w})) = G_j(\bar{x},\bar{w}) - G_j(\bar{x},\bar{w}) = 0_{\mathbb{R}^{n_y}}, \quad j = 0,\dots,k-2,$$

$$\frac{d^{k-1}}{dt^{k-1}} g_{\mathrm{rel}}(t, x_{k-1}(t), u(t;\bar{w})) = G_{k-1}(\bar{y},\bar{x},\bar{w}) - G_{k-1}(\bar{y},\bar{x},\bar{w}) = 0_{\mathbb{R}^{n_y}}.$$

Hence, $(\bar{x},\bar{y})$ is consistent. □

Typically, the relaxation approach is used within some superordinate iterative algorithm like the multiple shooting method, which requires to solve many initial value problems. The advantage of the relaxation approach is that time-consuming projection onto algebraic constraints to achieve consistency is unnecessary for each iteration of the superordinate algorithm. There is a downside, though.

Apparently solutions of (4.29) and (4.33) do not coincide, if $(\bar{x},\bar{y})$ is not consistent, since different DAEs are solved. Hence, consistency of $(\bar{x},\bar{y})$ has to be enforced by adding the equality constraints

$$G_j(\bar{x},\bar{w}) = 0_{\mathbb{R}^{n_y}}, \quad j = 0,\dots,k-2, \quad G_{k-1}(\bar{y},\bar{x},\bar{w}) = 0_{\mathbb{R}^{n_y}} \tag{4.35}$$

to the superordinate algorithm in order to ensure consistency at least in a solution of the algorithm, but not necessarily at intermediate steps. So, the difficulty of computing consistent initial values is not solved by the relaxation but is shifted to another level. It depends on the superordinate algorithm and how it deals with these constraints. For instance, if the superordinate algorithm is an optimization algorithm, then the constraints (4.35) can be added simply to the problem formulation. If the superordinate algorithm is a multiple shooting algorithm, then the constraint violation can be minimized using a Gauss–Newton algorithm, see [303] for details.

i The term $G_{k-1}(\bar{y},\bar{x},\bar{w})$ in (4.34) can be omitted, if a consistent algebraic variable $\bar{y}$ for a given $\bar{x}$ is computed by solving equation (4.31) for $\bar{y}$.

i Consistent initial values for the differential components of the so-called sensitivity DAE associated with the relaxed DAE (4.33), see Subsections 4.6.1 and 5.4.1, are given by

$$S^x(t_0) := \frac{\partial x(t;\bar{x},\bar{w})}{\partial(\bar{x},\bar{w})} := \begin{pmatrix} I_{n_x} & \Theta_{n_x \times n_w} \end{pmatrix}.$$

4.6 Shooting techniques for boundary value problems

The knowledge of necessary conditions in terms of a minimum principle for optimal control problems subject to state and/or control constraints gives rise to the *indirect approach*. The indirect approach attempts to exploit the minimum principle and usually results in a boundary value problem for the state and adjoint variables.

Example 4.6.1. Evaluation of the local minimum principle in Theorem 3.1.9 with $\mathcal{U} = \mathbb{R}^{n_u}$ yields the two-point DAE boundary value problem

$$
\begin{aligned}
\dot{x}(t) &= f(t, x(t), y(t), u(t)),\\
\dot{\lambda}_f(t) &= -\mathcal{H}'_x(t, x(t), y(t), u(t), \lambda_f(t), \lambda_g(t), \ell_0)^\top,\\
0_{\mathbb{R}^{n_y}} &= g(t, x(t)),\\
0_{\mathbb{R}^{n_y}} &= \mathcal{H}'_y(t, x(t), y(t), u(t), \lambda_f(t), \lambda_g(t), \ell_0)^\top,\\
0_{\mathbb{R}^{n_u}} &= \mathcal{H}'_u(t, x(t), y(t), u(t), \lambda_f(t), \lambda_g(t), \ell_0)^\top,\\
0_{\mathbb{R}^{n_\psi}} &= \psi(x(t_0), x(t_f)),\\
\lambda_f(t_0)^\top &= -(\ell_0 \varphi'_{x_0}(x(t_0), x(t_f)) + \sigma^\top \psi'_{x_0}(x(t_0), x(t_f)) + \zeta^\top g'_x(t_0, \hat{x}(t_0))),\\
\lambda_f(t_f)^\top &= \ell_0 \varphi'_{x_f}(x(t_0), x(t_f)) + \sigma^\top \psi'_{x_f}(x(t_0), x(t_f)).
\end{aligned}
$$

Herein, y, λ_g, and u are algebraic variables, and x, λ_f are differential variables. A particular example for the pendulum is given in Example 4.6.5. □

We consider

Problem 4.6.2 (Two-point DAE boundary value problem). Let $\mathcal{I} := [t_0, t_f] \subset \mathbb{R}$ be a compact interval with $t_0 < t_f$, let $f : \mathcal{I} \times \mathbb{R}^{n_x} \times \mathbb{R}^{n_y} \longrightarrow \mathbb{R}^{n_x}$, $g : \mathcal{I} \times \mathbb{R}^{n_x} \times \mathbb{R}^{n_y} \longrightarrow \mathbb{R}^{n_y}$, and $r : \mathbb{R}^{n_x} \times \mathbb{R}^{n_x} \longrightarrow \mathbb{R}^{n_x}$ be sufficiently smooth functions. Find a solution (x, y) of the boundary value problem

$$
\begin{aligned}
\dot{x}(t) &= f(t, x(t), y(t)),\\
0_{\mathbb{R}^{n_y}} &= g(t, x(t), y(t)),\\
0_{\mathbb{R}^{n_x}} &= r(x(t_0), x(t_f))
\end{aligned}
$$

in the interval $\mathcal{I}$.

If state or control constraints are present in the optimal control problem, then additional interior point conditions at unknown switching points occur, and the necessary optimality conditions lead to a multi-point point boundary value problem. By the same transformation techniques as in Section 1.2, multi-point boundary value problems can be transformed into equivalent two-point boundary value problems.

4.6.1 Single shooting method using projections

Consider the two point boundary value problem 4.6.2. The single shooting method is based on the repeated solution of initial value problems for different initial values. The initial values are iteratively updated until the boundary conditions are satisfied, see Figure 4.5.

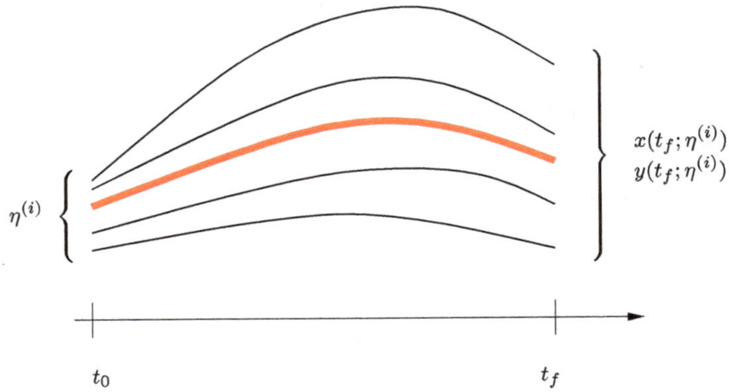

Figure 4.5: Idea of the single shooting method: Repeated solution of initial value problems with different initial values $\eta^{(i)}$. The solution of the boundary value problem is colored red.

Next, we assume that a procedure for the computation of a consistent initial value for the DAE in (4.6.2) is available. More specifically, the method needs to be able to provide a consistent initial value $x(t_0) = X_0(\eta)$ and $y(t_0) = Y_0(\eta)$ for a given estimate $\eta \in \mathbb{R}^{n_x}$ of the differential component $x(t_0)$. The mappings X_0 and Y_0 are supposed to be continuously differentiable. The projection method in Section 4.5.1 can be used to realize X_0 and Y_0 for Hessenberg DAEs.

For a given estimate $\eta \in \mathbb{R}^{n_x}$, let $x(t;X_0(\eta))$ and $y(t;Y_0(\eta))$ denote the solution of the initial value problem

$$\begin{aligned} \dot{x}(t) &= f(t,x(t),y(t)), \quad & x(t_0) &= X_0(\eta), \\ 0_{\mathbb{R}^{n_y}} &= g(t,x(t),y(t)), \quad & y(t_0) &= Y_0(\eta) \end{aligned}$$

on $\mathcal{I}$.

In order to satisfy the boundary condition, η has to be chosen such that

$$\begin{aligned} T(\eta) &:= r(x(t_0;X_0(\eta)),x(t_f;X_0(\eta))) \\ &= r(X_0(\eta),x(t_f;X_0(\eta))) \\ &= 0_{\mathbb{R}^{n_x}}. \end{aligned} \tag{4.36}$$

Equation (4.36) is a *nonlinear equation* for η and application of Newton's method yields

Algorithm 4.6.3 (Single shooting method using projections).

(0) Choose $\eta^{(0)} \in \mathbb{R}^{n_x}$ and set $i := 0$.

(1) Compute the projections $X_0(\eta^{(i)})$ and $Y_0(\eta^{(i)})$ and solve the initial value problem

$$\begin{aligned} \dot{x}(t) &= f(t, x(t), y(t)), \quad & x(t_0) &= X_0(\eta^{(i)}), \\ 0_{\mathbb{R}^{n_y}} &= g(t, x(t), y(t)), \quad & y(t_0) &= Y_0(\eta^{(i)}) \end{aligned}$$

on $\mathcal{I}$.

Compute $T(\eta^{(i)})$ and the Jacobian matrix

$$T'(\eta^{(i)}) = r'_{x_0}[\eta^{(i)}] \cdot X'_0(\eta^{(i)}) + r'_{x_f}[\eta^{(i)}] \cdot S^x(t_f),$$

where

$$\begin{aligned} r'_{x_0}[\eta^{(i)}] &:= r'_{x_0}(X_0(\eta^{(i)}), x(t_f; X_0(\eta^{(i)}))), \\ r'_{x_f}[\eta^{(i)}] &:= r'_{x_f}(X_0(\eta^{(i)}), x(t_f; X_0(\eta^{(i)}))), \\ S^x(t_f) &:= \frac{\partial x}{\partial \eta}(t_f; X_0(\eta^{(i)})). \end{aligned}$$

(2) If $T(\eta^{(i)}) = 0_{\mathbb{R}^{n_x}}$, STOP.

(3) Compute the Newton direction $d^{(i)}$ from the linear equation

$$T'(\eta^{(i)})d = -T(\eta^{(i)}).$$

(4) Set $\eta^{(i+1)} = \eta^{(i)} + d^{(i)}$, $i \leftarrow i + 1$, and go to (1). □

The derivative $T'(\eta^{(i)})$ in step (1) of Algorithm 4.6.3 can be computed as follows:

(i) *Finite difference approximation:*

$$\frac{\partial T}{\partial \eta_j}(\eta^{(i)}) \approx \frac{T(\eta^{(i)} + he_j) - T(\eta^{(i)})}{h}, \quad j = 1, \ldots, n_x,$$

with the jth unity vector $e_j \in \mathbb{R}^{n_x}$.

(ii) *Sensitivity DAE:* Formal differentiation of the DAE in $\mathcal{I}$ with respect to η yields the linear matrix DAE

$$\begin{aligned} \dot{S}^x(t) &= f'_x[t]S^x(t) - f'_y[t]S^y(t), \quad & S^x(t_0) &= X'_0(\eta^{(i)}), \\ 0_{\mathbb{R}^{n_y \times n_x}} &= g'_x[t]S^x(t) + g'_y[t]S^y(t), \quad & S^y(t_0) &= Y'_0(\eta^{(i)}) \end{aligned}$$

for the sensitivities

$$S^x(t) := \frac{\partial x}{\partial \eta}(t; X_0(\eta^{(i)})), \quad S^y(t) := \frac{\partial y}{\partial \eta}(t; Y_0(\eta^{(i)})),$$

where the derivatives of f and g are evaluated at $x(t; X_0(\eta^{(i)}))$ and $y(t; Y_0(\eta^{(i)}))$.

The finite difference approach in (i) requires to solve $n_x + 1$ nonlinear DAE initial value problems, while the sensitivity DAE in (ii) requires to solve one nonlinear DAE initial value problem and n_x linear DAE initial value problems. Hence, the sensitivity DAE approach can be considered to be more efficient.

Since the single shooting method in Algorithm 4.6.3 essentially is Newton's method applied to the nonlinear equation $T(\eta) = 0_{\mathbb{R}^{n_x}}$, the well-known convergence results of Newton's method hold, and one can expect a locally superlinear convergence rate, if $T'(\hat{\eta})$ is non-singular in a zero $\hat{\eta}$ of T, and a locally quadratic convergence rate, if in addition T' is locally Lipschitz continuous in $\hat{\eta}$.

The Jacobian $T'(\eta^{(i)})$ in step (1) is non-singular, if the matrix

$$r'_{x_0}[\eta^{(i)}] \cdot S^x(t_0) + r'_{x_f}[\eta^{(i)}] \cdot S^x(t_f)$$

is non-singular. The non-singularity of this matrix is closely related to the regularity condition of Mangasarian–Fromowitz for optimal control problems in Lemma 3.1.10!

We consider a simple example with an ODE.

Example 4.6.4. Consider the following optimal control problem:

Minimize

$$\frac{5}{2}\left(x(1) - 1\right)^2 + \frac{1}{2}\int_0^1 u(t)^2 + x(t)^3 dt$$

subject to

$$\dot{x}(t) = u(t) - r(t), \quad x(0) = 4$$

with $r(t) = 15\exp(-2t)$.

The minimum principle leads to the boundary value problem

$$\begin{aligned} \dot{x}(t) &= -\lambda(t) - r(t), & x(0) - 4 &= 0, \\ \dot{\lambda}(t) &= -\frac{3}{2}x(t)^2, & \lambda(1) - 5\left(x(1) - 1\right) &= 0. \end{aligned}$$

The single shooting method in Algorithm 4.6.3 is applied with initial guess $\eta^{(0)} := (4, -5)^\top$ and yields the output in Figure 4.6. □

Example 4.6.5. We apply the local minimum principle in Theorem 3.1.9 to the following index-two DAE optimal control problem:

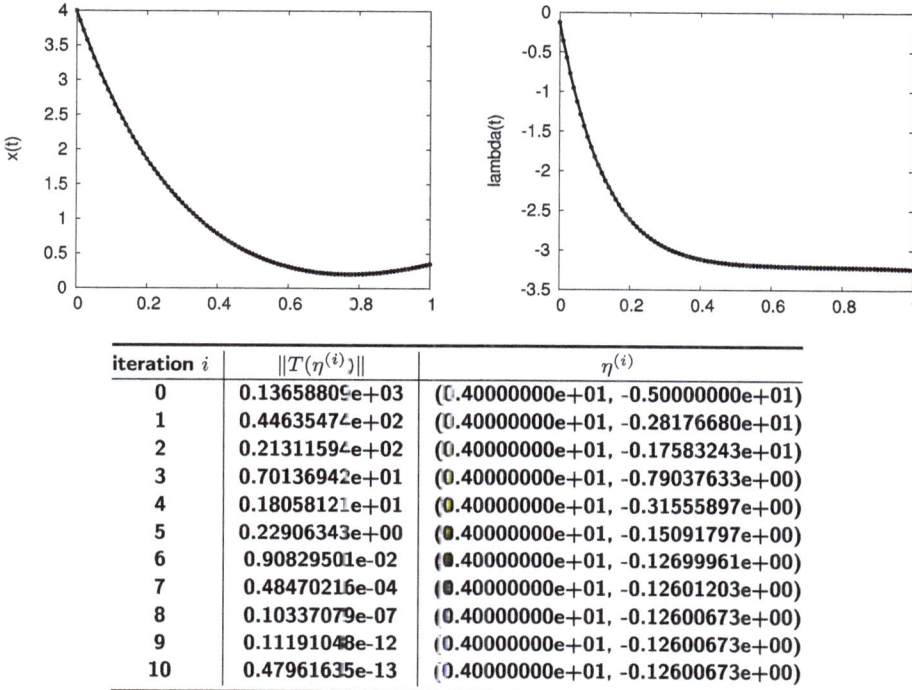

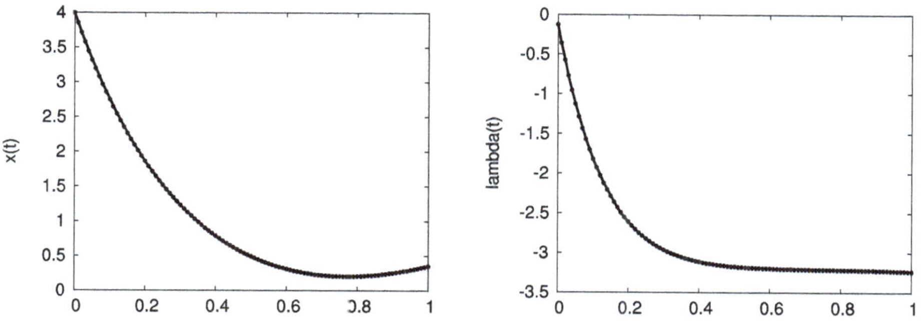

iteration i	$\|T(\eta^{(i)})\|$	$\eta^{(i)}$
0	0.13658809e+03	(0.40000000e+01, -0.50000000e+01)
1	0.44635474e+02	(0.40000000e+01, -0.28176680e+01)
2	0.21311594e+02	(0.40000000e+01, -0.17583243e+01)
3	0.70136942e+01	(0.40000000e+01, -0.79037633e+00)
4	0.18058121e+01	(0.40000000e+01, -0.31555897e+00)
5	0.22906343e+00	(0.40000000e+01, -0.15091797e+00)
6	0.90829501e-02	(0.40000000e+01, -0.12699961e+00)
7	0.48470216e-04	(0.40000000e+01, -0.12601203e+00)
8	0.10337079e-07	(0.40000000e+01, -0.12600673e+00)
9	0.11191048e-12	(0.40000000e+01, -0.12600673e+00)
10	0.47961635e-13	(0.40000000e+01, -0.12600673e+00)

Figure 4.6: Output of the Newton method in single shooting method: State x (top left), adjoint $\lambda = -u$ (top right), and norm $\|T(\eta^{(i)})\|$ for iterations $i = 0, \ldots, 10$ of the method (bottom).

Minimize

$$\int_0^3 u(t)^2 dt$$

subject to the constraints

$$\dot{x}_1(t) = x_3(t) - 2x_1(t)y_2(t), \tag{4.37}$$

$$\dot{x}_2(t) = x_4(t) - 2x_2(t)y_2(t), \tag{4.38}$$

$$\dot{x}_3(t) = \quad -2x_1(t)y_1(t) + u(t)x_2(t), \tag{4.39}$$

$$\dot{x}_4(t) = \quad -g - 2x_2(t)y_1(t) - u(t)x_1(t), \tag{4.40}$$

$$0 = x_1(t)x_3(t) + x_2(t)x_4(t), \tag{4.41}$$

$$0 = x_1(t)^2 + x_2(t)^2 - 1, \tag{4.42}$$

and

$$\psi\big(x(0), x(3)\big) := \big(x_1(0) - 1, x_2(0), x_3(0), x_4(0), x_1(3), x_3(3)\big)^\top = 0_{\mathbb{R}^6}. \tag{4.43}$$

Herein, $g = 9.81$ denotes acceleration due to gravity. The control u is not restricted and $\mathcal{U} = \mathbb{R}$. With $x = (x_1, x_2, x_3, x_4)^\top, y = (y_1, y_2)^\top, f_0(u) = u^2$, and

$$f(x,y,u) = (x_3 - 2x_1y_2, x_4 - 2x_2y_2, -2x_1y_1 + ux_2, -g - 2x_2y_1 - ux_1)^\top,$$
$$g(x) = (x_1x_3 + x_2x_4, x_1^2 + x_2^2 - 1)^\top$$

the problem has the structure of Problem 3.1.1. The matrix

$$\begin{aligned} g_x'(x) \cdot f_y'(x,y,u) &= \begin{pmatrix} x_3 & x_4 & x_1 & x_2 \\ 2x_1 & 2x_2 & 0 & 0 \end{pmatrix} \begin{pmatrix} 0 & -2x_1 \\ 0 & -2x_2 \\ -2x_1 & 0 \\ -2x_2 & 0 \end{pmatrix} \\ &= \begin{pmatrix} -2(x_1^2 + x_2^2) & -2(x_1x_3 + x_2x_4) \\ 0 & -4(x_1^2 + x_2^2) \end{pmatrix} \\ &= \begin{pmatrix} -2 & 0 \\ 0 & -4 \end{pmatrix} \end{aligned}$$

is non-singular in a local minimum; hence, the DAE has index two, and Assumption 3.1.3 is satisfied. The remaining assumptions of Theorem 3.1.9 are also satisfied, and necessarily there exist functions $\lambda_f = (\lambda_{f,1}, \lambda_{f,2}, \lambda_{f,3}, \lambda_{f,4})^\top \in W_{1,\infty}^4([0,3])$, $\lambda_g = (\lambda_{g,1}, \lambda_{g,2})^\top \in L_\infty^2([0,3])$, and vectors $\zeta = (\zeta_1, \zeta_2)^\top$ and $\sigma = (\sigma_1, \dots, \sigma_6)^\top$ such that the adjoint equations (3.22)–(3.23), the transversality conditions (3.24)–(3.25), and the optimality condition (3.26) are satisfied. The Hamilton function is given by

$$\begin{aligned} &\mathcal{H}(x,y,u,\lambda_f,\lambda_g,\ell_0) \\ &= \ell_0 u^2 + \lambda_{f,1}(x_3 - 2x_1y_2) + \lambda_{f,2}(x_4 - 2x_2y_2) \\ &\quad + \lambda_{f,3}(-2x_1y_1 + ux_2) + \lambda_{f,4}(-g - 2x_2y_1 - ux_1) \\ &\quad + \lambda_{g,1}(-gx_2 - 2y_1(x_1^2 + x_2^2) + x_3^2 + x_4^2 - 2y_2(x_1x_3 + x_2x_4)) \\ &\quad + \lambda_{g,2}(2(x_1x_3 + x_2x_4) - 4y_2(x_1^2 + x_2^2)). \end{aligned}$$

Next, we assume that $\ell_0 = 1$. Then the stationarity condition (3.26) yields

$$0 = 2u + \lambda_{f,3}x_2 - \lambda_{f,4}x_1 \quad \Longrightarrow \quad u = \frac{\lambda_{f,4}x_1 - \lambda_{f,3}x_2}{2}. \tag{4.44}$$

The transversality conditions (3.24)–(3.25) are given by

$$\lambda_f(0) = (-\sigma_1 - 2\zeta_2, -\sigma_2, -\sigma_3 - \zeta_1, -\sigma_4)^\top, \quad \lambda_f(3) = (\sigma_5, 0, \sigma_6, 0)^\top.$$

The adjoint equations (3.22)–(3.23) yield

$$\dot{\lambda}_{f,1} = 2(\lambda_{f,1}y_2 + \lambda_{f,3}y_1) + \lambda_{f,4}u - \lambda_{g,1}(-4y_1x_1 - 2y_2x_3) - \lambda_{g,2}(2x_3 - 8y_2x_1), \tag{4.45}$$
$$\dot{\lambda}_{f,2} = 2(\lambda_{f,2}y_2 + \lambda_{f,4}y_1) - \lambda_{f,3}u - \lambda_{g,1}(-g - 4y_1x_2 - 2y_2x_4) - \lambda_{g,2}(2x_4 - 8y_2x_2), \tag{4.46}$$
$$\dot{\lambda}_{f,3} = -\lambda_{f,1} - \lambda_{g,1}(2x_3 - 2x_1y_2) - 2\lambda_{g,2}x_1, \tag{4.47}$$

$$\dot{\lambda}_{f,4} = -\lambda_{f,2} - \lambda_{g,1}(2x_4 - 2x_2y_2) - 2\lambda_{g,2}x_2; \tag{4.48}$$

$$0 = -2(\lambda_{f,3}x_1 + \lambda_{f,4}x_2 + \lambda_{g,1}(x_1^2 + x_2^2)); \tag{4.49}$$

$$0 = -2(\lambda_{f,1}x_1 + \lambda_{f,2}x_2 + \lambda_{g,1}(x_1x_3 + x_2x_4) + 2\lambda_{g,2}(x_1^2 + x_2^2)). \tag{4.50}$$

Notice that consistent initial values for $\lambda_{g,1}(0)$ and $\lambda_{g,2}(0)$ could be calculated from (4.49)–(4.50) and (4.41)–(4.42) by

$$\lambda_{g,1} = -\lambda_{f,3}x_1 - \lambda_{f,4}x_2, \quad \lambda_{g,2} = \frac{-\lambda_{f,1}x_1 - \lambda_{f,2}x_2}{2}.$$

The differential equations (4.37)–(4.42) and (4.45)–(4.50) with u replaced by (4.44) together with the boundary conditions (4.43) and $\lambda_{f,2}(3) = 0$, $\lambda_{f,4}(3) = 0$ form a two point boundary value problem (BVP). Notice that the DAE system has index-one constraints (4.49)–(4.50) for λ_g as well as index-two constraints (4.41)–(4.42) for y.

The BVP is solved numerically by the single shooting method.

The sensitivity DAE is used to compute the Jacobian. Figures 4.7, 4.8, 4.9 show the numerical solution obtained from Newton's method. Notice that the initial conditions in (4.43) and the algebraic equations (4.41) and (4.42) contain redundant information. Hence, the multipliers σ and ζ are not unique, e. g., one may set $\zeta = 0_{\mathbb{R}^2}$. In order to obtain unique multipliers, one could dispense with the first and third initial condition in (4.43), since these are determined by (4.41) and (4.42). Finally, the output of Newton's method is depicted. □

iteration i	$\lVert T(\eta^{(i)})\rVert$
0	0.4328974468004262e+01
1	0.4240642504525387e+01
2	0.4193829901526831e+01
3	0.4144234257066851e+01
4	0.4112295471233999e+01
5	0.4044231539950495e+01
⋮	
29	0.1478490489094870e-06
30	0.4656214333457540e-08
31	0.3079526214273727e-09
32	0.1222130887100730e-09
33	0.1222130887100730e-09
34	0.4512679350279913e-10
35	0.4512679350279913e-10
36	0.3975362397848623e-10

The final approximate solution is

$$\sigma_1 = 15.20606630397864$$
$$\sigma_2 = -13.10630040572912$$
$$\sigma_3 = 19.62000001208976$$
$$\sigma_4 = -0.27699589145618$$

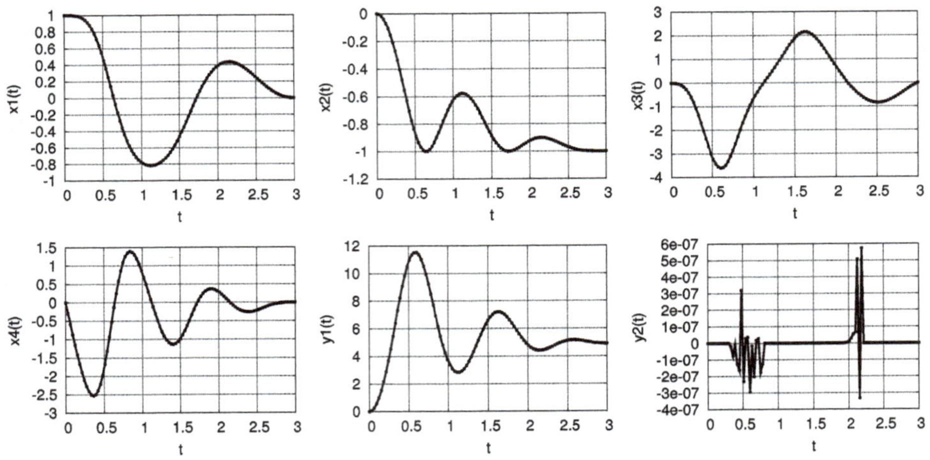

Figure 4.7: Numerical solution of BVP resulting from the minimum principle: Differential state $x(t)$ and algebraic state $y(t)$ for $t \in [0, 3]$.

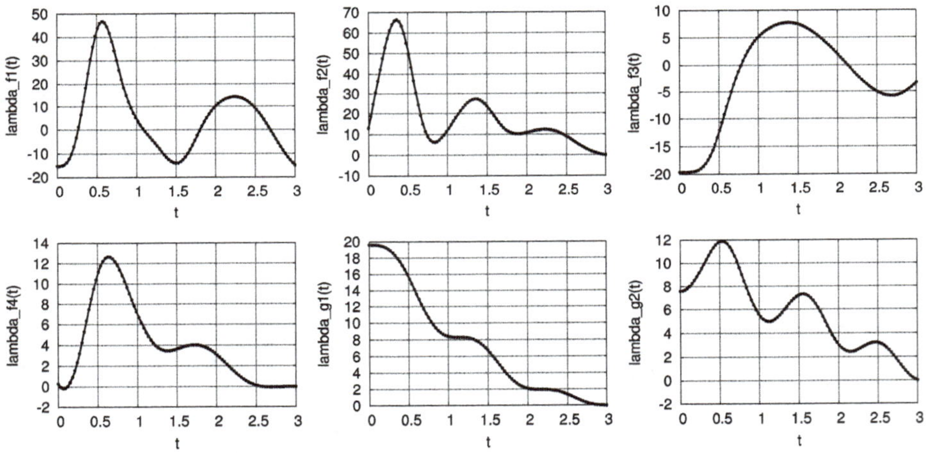

Figure 4.8: Numerical solution of BVP resulting from the minimum principle: Adjoint variables $\lambda_f(t)$ and $\lambda_g(t)$ for $t \in [0, 3]$.

4.6.2 Single shooting method using relaxations

If it is not possible or too expensive to use projections onto consistent initial values, then alternatively the relaxation approach in Section 4.5.2 can be exploited to solve the boundary value problem 4.6.2, compare [303].

To this end, we assume that a relaxation procedure is available that relaxes the algebraic constraint in Problem 4.6.2 in such a way that arbitrary $\mu_x \in \mathbb{R}^{n_x}$ and $\mu_y \in \mathbb{R}^{n_y}$ become consistent for the relaxed DAE

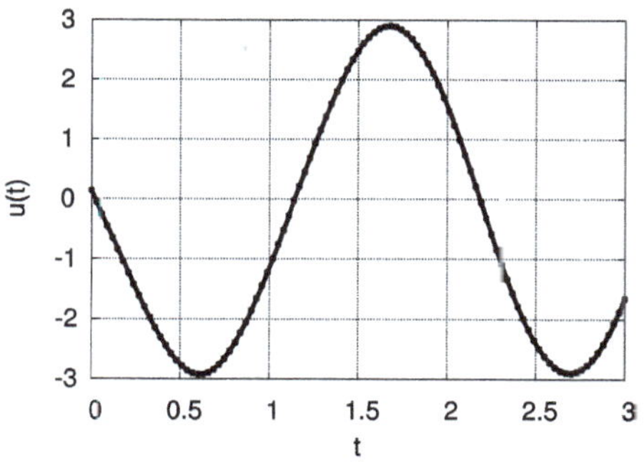

Figure 4.9: Numerical solution of BVP resulting from the minimum principle: Control $u(t)$ for $t \in [0,3]$.

$$0_{\mathbb{R}^{n_y}} = g_{\text{rel}}(t, x(t), y(t)), \quad t \in \mathcal{I}.$$

For Hessenberg DAEs the relaxation in (4.33) can be used.

For a given estimate $\mu := (\mu_x, \mu_y)^\top \in \mathbb{R}^{n_x+n_y}$, let $x(t;\mu)$ and $y(t;\mu)$ denote the solution of the initial value problem

$$\begin{aligned} \dot{x}(t) &= f(t, x(t), y(t)), & x(t_0) &= \mu_x, \\ 0_{\mathbb{R}^{n_y}} &= g_{\text{rel}}(t, x(t), y(t)), & y(t_0) &= \mu_y \end{aligned}$$

on $\mathcal{I}$.

In order to satisfy the boundary condition, μ has to fulfill the nonlinear equation

$$\tilde{T}(\mu) = r(\mu_x, x(t_f;\mu)) = 0_{\mathbb{R}^{n_x}}.$$

However, solving this equation is not sufficient to solve the boundary value problem 4.6.2, because a solution $\hat{\mu}$ with $\tilde{T}(\hat{\mu}) = 0_{\mathbb{R}^{n_x}}$ is in general not consistent with the original DAE. Hence, consistency has to be enforced by additional measures. For the Hessenberg DAE (4.29) with hidden algebraic constraints (4.30) and (4.31) and its relaxation (4.33), enforcing consistency can be approached by solving the following constrained optimization problem:

Minimize

$$J(\mu) := \sum_{j=0}^{k-2} \left\| G_j(\mu_x) \right\|^2 + \left\| G_{k-1}(\mu_y, \mu_x) \right\|^2$$

with respect to $\mu \in \mathbb{R}^{n_x+n_y}$ subject to the constraint

$$\tilde{T}(\mu) = 0_{\mathbb{R}^{n_x}}.$$

This optimization problem can be solved by, e. g., the SQP method in Section 5.2.2. However, recall that a solution $\hat{\mu}$ of the above optimization problem solves the boundary value problem 4.6.2 only if $J(\hat{\mu}) = 0$; otherwise, $\hat{\mu}$ is not consistent. Hence, a global minimum is sought, but SQP methods – depending on the initial guess – are likely to end in a local minimum only.

4.6.3 Multiple shooting method

The single shooting method in Section 4.6.1 occasionally suffers from instability problems. In the worst case, it might even not be possible to solve the initial value problem for a given initial guess $\eta^{(i)}$ on the whole interval $\mathcal{I}$.

The *multiple shooting method* tries to circumvent this difficulty by introducing intermediate *shooting nodes*

$$t_0 < t_1 < \cdots < t_{N-1} < t_N = t_f$$

and solving initial value problems on each subinterval $[t_j, t_{j+1}]$, $j = 0, 1, \ldots, N-1$, with initial values η_j, $j = 0, 1, \ldots, N-1$, given at each time point t_j, $j = 0, 1, \ldots, N-1$.

The number and position of required shooting nodes depend on the problem.

As in Section 4.6.1, we make use of the procedure for the computation of a consistent initial value for the DAE in Problem 4.6.2 by means of the functions $X_0(\cdot)$ and $Y_0(\cdot)$. In the multiple shooting approach, each of the values η_j, $j = 0, 1, \ldots, N-1$, needs to be projected onto a consistent initial value before integration can take place.

As indicated in Figure 4.10, the initial value problem

$$\begin{aligned} \dot{x}(t) &= f(t, x(t), y(t)), \quad & x(t_j) &= X_0(\eta_j), \\ 0_{\mathbb{R}^{n_y}} &= g(t, x(t), y(t)), \quad & y(t_j) &= Y_0(\eta_j) \end{aligned}$$

has to be solved in each subinterval $\mathcal{I}_j := [t_j, t_{j+1})$, $j = 0, \ldots, N-1$.

Let $x_j(t; X_0(\eta_j))$ and $y_j(t; Y_0(\eta_j))$ denote the solution in $\mathcal{I}_j$. The composite functions

$$\begin{aligned} & x(t; X_0(\eta_0), \ldots, X_0(\eta_{N-1})) \\ & \quad := \begin{cases} x_j(t; X_0(\eta_j)), & \text{if } t \in \mathcal{I}_j,\ j = 0, \ldots, N-1, \\ x_{N-1}(t_N; X_0(\eta_{N-1})), & \text{if } t = t_f, \end{cases} \\ & y(t; Y_0(\eta_0), \ldots, Y_0(\eta_{N-1})) \\ & \quad := \begin{cases} y_j(t; Y_0(\eta_j)), & \text{if } t \in \mathcal{I}_j,\ j = 0, \ldots, N-1, \\ y_{N-1}(t_N; Y_0(\eta_{N-1})), & \text{if } t = t_f \end{cases} \end{aligned}$$

have to satisfy the boundary conditions and continuity conditions for the differential states at the shooting nodes:

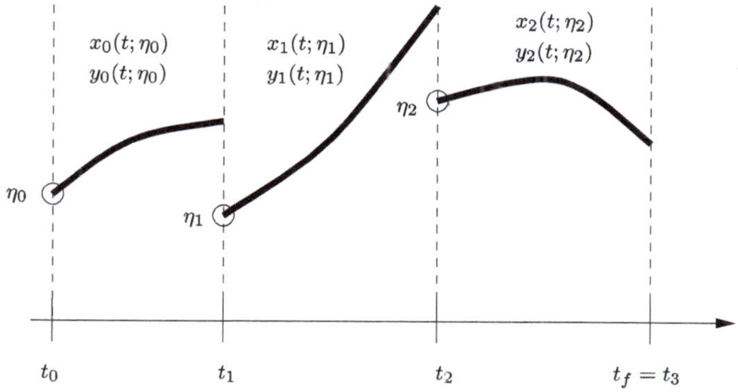

Figure 4.10: Idea of the multiple shooting method: Subsequent solution of initial value problems on subintervals.

$$T(\eta_0,\dots,\eta_{N-1}) := \begin{pmatrix} x_0(t_1;X_0(\eta_0)) - \eta_1 \\ x_1(t_2;X_0(\eta_1)) - \eta_2 \\ \vdots \\ x_{N-2}(t_{N-1};X_0(\eta_{N-2})) - \eta_{N-1} \\ r(X_0(\eta_0), x_{N-1}(t_N;X_0(\eta_{N-1}))) \end{pmatrix} = 0_{\mathbb{R}^{N\cdot n_x}}. \tag{4.51}$$

Equation (4.51) is again a *nonlinear equation* for the function T in the variable $\eta := (\eta_0,\dots,\eta_{N-1})^\top \in \mathbb{R}^{N\cdot n_x}$. The single shooting method arises as a special case for $N = 1$. Note that $\hat{\eta}_j$, $j = 1,\dots,N-1$, are consistent in a zero $\hat{\eta}$ of T because of the continuity conditions.

The dimension of the nonlinear equation (4.51) grows with the number of shooting nodes. The Jacobian $T'(\eta)$, however, exhibits a sparse structure, which should be exploited in numerical computations:

$$T'(\eta) = \begin{pmatrix} S_0^x(t_1) & -I_{n_x} & & & \\ & S_1^x(t_2) & -I_{n_x} & & \\ & & \ddots & \ddots & \\ & & & S_{N-2}^x(t_{N-1}) & -I_{n_x} \\ A\cdot X_0'(\eta_0) & & & & B\cdot S_{N-1}^x(t_N) \end{pmatrix} \tag{4.52}$$

with

$$\begin{aligned} S_j^x(t_{j+1}) &:= \frac{\partial x_j}{\partial \eta_j}(t_{j+1};X_0(\eta_j)), \quad j = 0,1,\dots,N-1, \\ A &:= r'_{x_0}(X_0(\eta_0), x_{N-1}(t_N;X_0(\eta_{N-1}))), \\ B &:= r'_{x_f}(X_0(\eta_0), x_{N-1}(t_N;X_0(\eta_{N-1}))). \end{aligned}$$

Application of Newton's method to the nonlinear equation (4.51) yields the following algorithm.

Algorithm 4.6.6 (Multiple shooting method).

(0) Choose an initial guess $\eta^{(0)} = (\eta_0^{(0)}, \dots, \eta_{N-1}^{(0)})^\top \in \mathbb{R}^{N \cdot n_x}$ and set $i := 0$.

(1) For $j = 0, \dots, N-1$ solve the initial value problems

$$\begin{aligned} \dot{x}_j(t) &= f(t, x_j(t), y_j(t)), & x_j(t_j) &= X_0(\eta_j^{(i)}), \\ 0_{\mathbb{R}^{n_y}} &= g(t, x_j(t), y_j(t)), & y_j(t_j) &= Y_0(\eta_j^{(i)}) \end{aligned}$$

and the sensitivity DAEs

$$\begin{aligned} \dot{S}_j^x(t) &= f_x'[t] S_j^x(t) + f_y'[t] S_j^y(t), & S_j^x(t_j) &= X_0'(\eta_j^{(i)}), \\ 0_{\mathbb{R}^{n_y \times n_x}} &= g_x'[t] S_j^x(t) + g_y'[t] S_j^y(t), & S_j^y(t_j) &= Y_0'(\eta_j^{(i)}) \end{aligned}$$

in $\mathcal{I}_j$, where the derivatives of f and g are evaluated at x_j and y_j.

Compute $T(\eta^{(i)})$ and $T'(\eta^{(i)})$ according to (4.51) and (4.52), respectively.

(2) If $T(\eta^{(i)}) = 0_{\mathbb{R}^{N \cdot n_x}}$, STOP.

(3) Compute the Newton direction $d^{(i)}$ from the linear equation

$$T'(\eta^{(i)}) d = -T(\eta^{(i)}).$$

(4) Set $\eta^{(i+1)} := \eta^{(i)} + d^{(i)}$, $i \leftarrow i + 1$, and go to (1). □

Example 4.6.7. Consider again Example 4.6.4 and the boundary value problem

$$\begin{aligned} \dot{x}(t) &= -\lambda(t) - r(t), & x(0) - 4 &= 0, \\ \dot{\lambda}(t) &= -\frac{3}{2} x(t)^2, & \lambda(1) - 5(x(1) - 1) &= 0. \end{aligned}$$

The multiple shooting method in Algorithm 4.6.6 is applied with $N = 5$ and initial guess $\eta_j^{(0)} := (4, -5)^\top, j = 0, \dots, 4$. It yields the output in Figure 4.11. □

Various other numerical solution approaches for boundary value problems are available, e. g., finite difference methods, collocation methods, and finite element methods. Details can be found in [313, 13]. Detailed descriptions of boundary value solvers for DAEs can be found in [199, 15, 200, 314, 195, 189, 190]. Multiple shooting methods and particularly the implementations MUMUS, see [163], and BOUNDSCO, see [258], have shown their capability in several practical applications from optimal control.

---

i Instead of using the true Jacobian matrix $T'(\eta^{(i)})$, whose computation is costly as the sensitivity DAE has to be solved, one can replace it in iteration i by a matrix J_i, which is updated by a Broyden rank-1 update:

$$J_{i+1} := J_i + \frac{(z - J_i d) d^\top}{d^\top d}, \quad d := \eta^{(i+1)} - \eta^{(i)}, \quad z := T(\eta^{(i+1)}) - T(\eta^{(i)}).$$

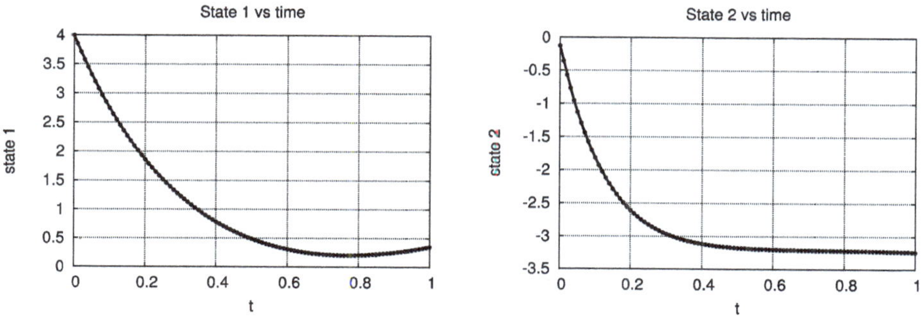

Final approximate solution:

iteration i	$\|T(\eta^{(i)})\|$
0	0.3106e+02
1	0.1099e+02
2	0.2869e+01
3	0.3732e+00
4	0.7544e-02
5	0.3530e-05
6	0.1044e-11

$$\eta^{(6)} = \begin{pmatrix} 4.000000000000000 \\ -0.126006611401308 \\ 1.861143297606159 \\ -2.605573076115534 \\ 0.789168032504686 \\ -3.111928004046596 \\ 0.311481417299404 \\ -3.196300177167728 \\ 0.207939098565028 \\ -3.213118469008610 \end{pmatrix}$$

Figure 4.11: Output of the Newton method in multiple shooting method: State x (top left), adjoint $\lambda = -u$ (top right), and norm $\|T(\eta^{(i)})\|$ for iterations $i = 0, \dots, 6$ of the method (bottom).

The damped Newton method can be used to extend the radius of convergence of the Newton method. To this end, a step-size $\alpha_i > 0$ is introduced in

$$\eta^{(i+1)} := \eta^{(i)} + \alpha_i d^{(i)}, \quad i = 0, 1, 2, \dots .$$

The step-size can be obtained by one-dimensional (approximate) minimization of the function

$$\varphi(\alpha) := \frac{1}{2}\|T(\eta^{(i)} + \alpha d^{(i)})\|_2^2.$$

A substantial problem in solving the boundary value problem resulting from an optimal control problem should not be underestimated: A sufficiently good initial guess of the switching structure (sequence of active, inactive, and singular arcs) and the state and adjoint variables at the shooting nodes is required. There is no general rule on how to obtain such an initial guess. Often, homotopy methods are used, for which a sequence of neighboring problems is solved depending on a homotopy parameter. The direct discretization methods in Chapter 5 can be used alternatively to compute a sufficiently good approximation, which is then refined by indirect methods.

4.7 Exercises

Verify that $z_1(t) = t^2 - t\cos t + \sin t$, $z_2(t) = -t + \cos t$ solve the linear DAE

$$\begin{pmatrix} 1 & t \\ 0 & 0 \end{pmatrix}\begin{pmatrix} \dot{z}_1(t) \\ \dot{z}_2(t) \end{pmatrix} + \begin{pmatrix} 0 & 0 \\ 1 & t \end{pmatrix}\begin{pmatrix} z_1(t) \\ z_2(t) \end{pmatrix} - \begin{pmatrix} t \\ \sin t \end{pmatrix} = 0_{\mathbb{R}^2}.$$

What happens if the implicit Euler method is applied to the DAE?

(compare [35]) Let $c \in \mathbb{R}$, $c \neq -1$, be a constant and $u_1(\cdot)$ and $u_2(\cdot)$ sufficiently smooth functions. Consider the DAE

$$\begin{pmatrix} 1 & ct \\ 0 & 0 \end{pmatrix}\begin{pmatrix} \dot{z}_1(t) \\ \dot{z}_2(t) \end{pmatrix} + \begin{pmatrix} 0 & 1+c \\ 1 & ct \end{pmatrix}\begin{pmatrix} z_1(t) \\ z_2(t) \end{pmatrix} - \begin{pmatrix} u_1(t) \\ u_2(t) \end{pmatrix} = 0_{\mathbb{R}^2}.$$

(a) Compute the solution of the DAE for given c, $u_1(\cdot)$, and $u_2(\cdot)$.
(b) Find the perturbation index and the differentiation index on a compact interval $[t_0, t_f]$ with $t_0 < t_f$.
(c) Let $u_1(t) = \exp(-t)$ and $u_2(t) \equiv 0$. Investigate the limit of the solution $z_1(t)$ and $z_2(t)$ in (a) as $t \to \infty$. Apply the implicit Euler method with step-size $h > 0$ for the grid points $t_i = ih$, $i \in \mathbb{N}_0$, and investigate the limit of the numerical solution $z_{1,i} \approx z_1(t_i)$ and $z_{2,i} \approx z_2(t_i)$ as $i \to \infty$ depending on the constant c.

(Solvability of Implicit Runge–Kutta Methods for Explicit ODEs) Let $f : [t_0, t_f] \times \mathbb{R}^{n_z} \longrightarrow \mathbb{R}^{n_z}$, $(t, z) \mapsto f(t, z)$, be Lipschitz continuous with respect to z with Lipschitz constant L.

(a) Show that there exists $h_0 > 0$ such that the equation

$$z = x + hf(t, z)$$

possesses a unique solution for all $x \in \mathbb{R}^n$, all $t_0 \leq t \leq t_f$, and all $0 < h \leq h_0$.

(b) Let constants $c_i \in \mathbb{R}$ and $a_{ij} \in \mathbb{R}$ for $1 \leq i, j \leq s$ and $s \in \mathbb{N}$ be given. Show that there exists $h_0 > 0$ such that the nonlinear equations

$$k_i = f\left(t + c_i h, x + h\sum_{j=1}^{s} a_{ij}k_j\right), \quad i = 1, \dots, s,$$

possess a unique solution $k = (k_1, \dots, k_s)$ for all $x \in \mathbb{R}^n$, all $t_0 \leq t \leq t_f$, and all $0 < h \leq h_0$.

Consider the initial value problem (4.3), where $f : \mathbb{R} \times \mathbb{R}^{n_z} \longrightarrow \mathbb{R}^{n_z}$ is continuous and has continuous and bounded partial derivatives up to order three.

Use Taylor expansion of the local discretization error to show that the Runge–Kutta method

$$z_{i+1} = z_i + \frac{h}{6}(k_1 + 4k_2 + k_3),$$
$$k_1 = f(t_i, z_i),$$

$$k_2 = f\left(t_i + \frac{h}{2}, z_i + \frac{h}{2}k_1\right),$$
$$k_3 = f(t_i + h, z_i - hk_1 + 2hk_2),$$

has order of consistency three.

(Collocation and Implicit Runge–Kutta Method) The idea of the *collocation method* for the solution of the initial value problem (4.3) is to construct for the integration step $t_i \to t_{i+1}$ a polynomial $p : [t_i, t_{i+1}] \longrightarrow \mathbb{R}^{n_z}$ of maximum degree s, which satisfies the following conditions:

$$p(t_i) = z_i,$$
$$\dot{p}(\tau_k) = f(\tau_k, p(\tau_k)), \quad k = 1, \ldots, s,$$

where $t_i \leq \tau_1 < \tau_2 < \cdots \tau_s \leq t_{i+1}$ are given *collocation points*. As an approximation of $z(t_{i+1})$ at t_{i+1}, one uses the value $z_{i+1} := p(t_{i+1})$.

Show that the collocation method with $s = 2$, $\tau_1 = t_i$, and $\tau_2 = t_{i+1}$ actually is an implicit Runge–Kutta method (the implicit trapezoidal rule).

Let $\mathcal{I} = [t_0, t_f]$, $t_0 < t_f$, $A \in L_\infty^{n_x \times n_x}(\mathcal{I})$, $B \in L_\infty^{n_x \times n_y}(\mathcal{I})$, $C \in L_\infty^{n_y \times n_x}(\mathcal{I})$, and $D \in L_\infty^{n_y \times n_y}(\mathcal{I})$ with D non-singular and $D^{-1} \in L_\infty^{n_y \times n_y}(\mathcal{I})$ be given.

Consider the linear index-one DAE

$$\dot{x}(t) = A(t)x(t) + B(t)y(t),$$
$$0_{\mathbb{R}^{n_y}} = C(t)x(t) + D(t)y(t).$$

(a) Apply the implicit Euler method to the DAE and find the increment function.
(b) Show that the implicit Euler method is consistent of order one.
(c) Show that the implicit Euler method is convergent of order one.

Implement Algorithm 4.4.2 and test it for the initial value problem in Example 4.4.1. Use the embedded Runge–Kutta method of order $p(q) = 2(3)$ given by the following Butcher array:

0				
1/4	1/4			
27/40	−189/800	729/800		
1	214/891	1/33	650/891	
*RK*1	214/891	1/33	650/891	0
*RK*2	533/2106	0	800/1053	−1/78

Consider the parametric initial value problem

$$\dot{z}(t) = f\big(t, z(t), p\big), \quad z(t_0) = Z_0(p).$$

Let $Z_0 : \mathbb{R}^{n_p} \longrightarrow \mathbb{R}^{n_z}$ be continuous, and let $f : [t_0, t_f] \times \mathbb{R}^{n_z} \times \mathbb{R}^{n_p} \longrightarrow \mathbb{R}^{n_z}$ satisfy the Lipschitz condition

$$\big\|f(t, z_1, p_1 - f(t, z_2, p_2)\big\| \le L\big(\|z_1 - z_2\| + \|p_1 - p_2\|\big)$$

for all $t \in [t_0, t_f]$, $z_1, z_2 \in \mathbb{R}^{n_z}$, $p_1, p_2 \in \mathbb{R}^{n_p}$.

Prove that the solution $z(t; p)$ depends continuously on p for all $t \in [t_0, t_f]$, that is

$$\lim_{p \to \hat{p}} z(t; p) = z(t; \hat{p}) \quad \text{for all } t \in [t_0, t_f],\ \hat{p} \in \mathbb{R}^{n_p}.$$

If Z_0 is even Lipschitz continuous, then there is a constant S with

$$\big\|z(t; p_1) - z(t; p_2)\big\| \le S\|p_1 - p_2\| \quad \text{for all } t \in [t_0, t_f],\ p_1, p_2 \in \mathbb{R}^{n_p}.$$

Prove an analog statement for Hessenberg DAEs (1.39) in Definition 1.1.14.
Hint: Use the implicit function theorem and Gronwall's lemma.

Consider the parametric initial value problem

$$\dot{z}(t) = f\big(t, z(t), p\big), \quad z(t_0) = Z_0(p).$$

Let $Z_0 : \mathbb{R}^{n_p} \longrightarrow \mathbb{R}^{n_z}$ and $f : [t_0, t_f] \times \mathbb{R}^{n_z} \times \mathbb{R}^{n_p} \longrightarrow \mathbb{R}^{n_z}$ be continuous with continuous partial derivatives with respect to z and p. Let $z(t; p)$ denote the solution for a given parameter p.

Prove that the sensitivity $S(t) := \frac{\partial z(t;p)}{\partial p}$ exists and satisfies the *sensitivity differential equation*

$$\begin{aligned} S(t_0) &= Z_0'(p), \\ S'(t) &= f_z'\big(t, z(t; p), p\big) \cdot S(t) + f_p'\big(t, z(t; p), p\big). \end{aligned}$$

(Sensitivity Analysis using Sensitivity Equation)
Let $f : [t_0, t_f] \times \mathbb{R}^{n_z} \times \mathbb{R}^{n_p} \longrightarrow \mathbb{R}^{n_z}$ and $Z_0 : \mathbb{R}^{n_p} \longrightarrow \mathbb{R}^{n_z}$ be sufficiently smooth functions and

$$\begin{aligned} z(t_0) &= Z_0(p), \\ \dot{z}(t) &= f\big(t, z(t), p\big), \quad t \in [t_0, t_f], \end{aligned}$$

a parametric initial value problem with solution $z(t; p)$.

(a) Write a program that solves the initial value problem and the sensitivity ODE

$$\begin{aligned} S(t_0) &= Z_0'(p), \\ \dot{S}(t) &= f_z'\big(t, z(t; p), p\big) \cdot S(t) + f_p'\big(t, z(t; p), p\big), \quad t \in [t_0, t_f], \end{aligned}$$

simultaneously for the classic 4-stage Runge–Kutta method with Butcher array

$$\begin{array}{c|cccc}
0 & & & & \\
1/2 & 1/2 & & & \\
1/2 & 0 & 1/2 & & \\
1 & 0 & 0 & 1 & \\
\hline
 & 1/6 & 1/3 & 1/3 & 1/6
\end{array}$$

(b) Use (a) to compute the derivative

$$\frac{d}{dp}\varphi\big(z(t_f;p),p\big),$$

of the function $\varphi : \mathbb{R}^{n_z} \times \mathbb{R}^{n_p} \longrightarrow \mathbb{R}$, $(z,p) \mapsto \varphi(z,p)$.
Test the program for $[t_0, t_f] = [0, 10]$, step-size $h = 0.1$, $\varphi(z_1, z_2) := z_1^2 - z_2$, and

$$\begin{aligned}
z_1(0) &= p_1, \\
z_2(0) &= p_2, \\
\dot{z}_1(t) &= z_2(t), \\
\dot{z}_2(t) &= -\frac{p_3}{p_4}\sin\big(z_1(t)\big)
\end{aligned}$$

with $p_1 = \frac{\pi}{2}$, $p_2 = 0$, $p_3 = 9.81$, $p_4 = 1$.

The class of the so-called s-stage SDIRK methods (*singly diagonally implicit Runge–Kutta method*) for the DAE

$$F\big(t, z(t), \dot{z}(t)\big) = 0_{\mathbb{R}^{n_z}}$$

is defined by the Butcher array

$$\begin{array}{c|ccccc}
\gamma & \gamma & & & & \\
c_2 & a_{21} & \gamma & & & \\
\vdots & \vdots & & \ddots & & \\
c_s & a_{s1} & \cdots & a_{s,s-1} & \gamma \\
\hline
 & b_1 & \cdots & b_{s-1} & b_s
\end{array}$$

Starting from $z_i \approx z(t_i) \in \mathbb{R}^{n_z}$, the integration step $t_i \longrightarrow t_i + h$ requires to solve the $n_z \cdot s$-dimensional nonlinear equation

$$F\big(t_i + c_j h, z_{i+1}^{(\ell)}, k_\ell\big) = 0_{\mathbb{R}^{n_z}}, \quad \ell = 1, \ldots, s,$$

where

$$z_{i+1}^{(\ell)} = z_i + h\left(\gamma k_\ell + \sum_{j=1}^{\ell-1} a_{\ell j} k_j\right), \quad \ell = 1, \ldots, s.$$

Formulate an algorithm using Newton's method, which only requires to solve n_z-dimensional nonlinear equations. Why is the algorithm particularly efficient for the simplified Newton method, which works with a fixed Jacobian matrix?

The motion of a ball subject to a linear spring force, linear damping, and an external force $F(t)$ obeys the second-order differential equation

$$m\ddot{x}(t) + d\dot{x}(t) + cx(t) = F(t),$$

where $c > 0$, $d > 0$, and $m > 0$ are constants, compare figure. Boundary values are given by $x(0) = x_0$ and $x(b) = x_b$.

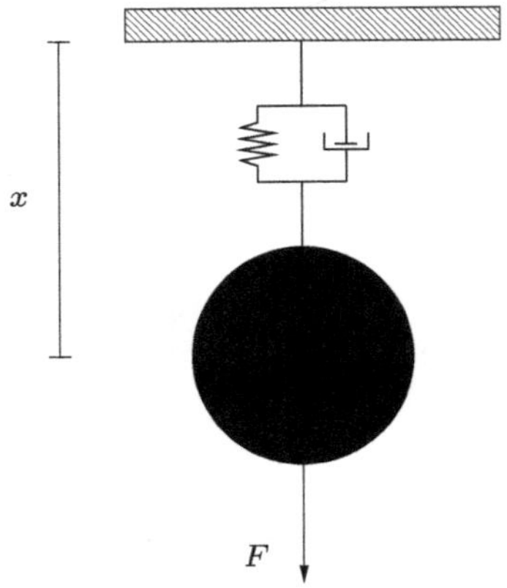

(a) Discuss the solvability of the boundary value problem for $m = d = c = 1$ and $F(t) \equiv 0$ with respect to b, x_0, x_b.

(b) Consider the initial value problem

$$\ddot{x}(t) + \dot{x}(t) + x(t) = 0, \quad x(0) = 1, \quad \dot{x}(0) = \beta.$$

Formulate an equation $r(\beta) = 0$, such that the solution of the initial value problem (as a function of β) solves the boundary value problem for $b = 1$, $x_0 = x_b = 1$.

(c) Apply Newton's method to solve the equation $r(\beta) = 0$ in (b).

(Finite Difference Approximation)
Consider the boundary value problem

$$-\ddot{x}(t) = \left(1 + t^2\right) x(t) + 1 \quad \text{for all } t \in [-1, 1],$$
$$x(-1) = x(1) = 0.$$

(a) Derive a formula for the approximation of $\ddot{x}(t_i)$, $i = 1, \dots, N-1$, using central finite differences on the grid $\mathbb{G}_N = \{t_i := -1 + ih \mid i = 0, \dots, N\}$ with step-size $h := 2/N$. Use the boundary values at $t = -1$ and $t = 1$.

(b) Introduce the formulas into the differential equation and formulate the resulting system of linear equations for the approximations $x_1, \dots, x_{N-1} \in \mathbb{R}$ of $x(t_1), \dots, x(t_{N-1})$.

(c) Solve the linear equation for $N = 5, 10, 20, 40, 80, 160, 320$ and estimate the order of convergence numerically.

Consider Example 3.1.11 in Subsection 3.1.3. Solve the boundary value problem

$$
\begin{aligned}
\dot{x}_1(t) &= u(t) - y(t), & x_1(0) &= x_1(1) = 0,\\
\dot{x}_2(t) &= u(t), & x_2(0) &= -x_2(1) = 1,\\
\dot{x}_3(t) &= -x_2(t), & x_3(0) &= 0,\\
\dot{\lambda}_{f,1}(t) &= 0, & &\\
\dot{\lambda}_{f,2}(t) &= \lambda_{f,3}(t) + \lambda_g(t), & &\\
\dot{\lambda}_{f,3}(t) &= 0, & \lambda_{f,3}(1) &= 0,\\
0 &= x_1(t) + x_3(t), & &\\
0 &= -\lambda_{f,1}(t) - \lambda_g(t), & &\\
0 &= u(t) + \lambda_{f,1}(t) + \lambda_{f,2}(t) + \lambda_g(t) & &
\end{aligned}
$$

by the single shooting method.

5 Discretization of optimal control problems

Based on solving boundary value problems resulting from a minimum principle (see Section 4.6), the indirect approach usually leads to highly accurate numerical solutions, provided that the multiple shooting method (or any other method for boundary value problems) converges. It occasionally suffers from two major drawbacks, though. The first drawback is that a very good initial guess for the approximate solution is needed to achieve convergence. The construction of a good initial guess is complicated, since this requires, among other things, a good estimate of the switching structure of the problem. The second drawback is that high-dimensional systems are often cumbersome and require sophisticated knowledge of necessary or sufficient conditions to set up the optimality system, even if algebra packages like MAPLE are used.

In contrast to indirect methods, the *direct methods*, see [36, 148, 92, 113], are based on a suitable discretization of the infinite-dimensional optimal control problem. The resulting finite-dimensional optimization problem can be solved by suitable methods from nonlinear programming, for instance, sequential quadratic programming (SQP). This approach is particularly advantageous for large-scale problems and users who do not have deep insights into optimal control theory. Moreover, numerical experiences show that direct methods are also robust and sufficiently accurate. Furthermore, by comparing the necessary conditions for the discretized problem with those of the original problem, it is possible to derive approximations for adjoint variables based on the multipliers of the discretized problem. Consequently, direct and indirect approaches can often be combined by using the direct approach for finding an accurate initial guess (including switching structure and adjoints) for the indirect approach.

Next, we will focus on the direct discretization approach for DAE optimal control problems. This section aims to develop a general framework for the numerical solution of such problems. Therefore, the optimal control problem statement is kept as general as possible without assuming a special structure of the DAE in advance. Hence, as in Chapter 4, most of the time, we will deal with general DAEs. This has the notational advantage that we do not have to distinguish between differential and algebraic variables as far as it is not essential. For numerical computations, however, there are certain restrictions. First, from Chapter 3, we can conclude that control and algebraic variables have the same properties (both are L_∞-functions). Hence, it makes no sense – at least theoretically – to allow that φ and ψ in the following Problem 5.0.1 depend on pointwise evaluations of the algebraic variables. Second, the numerical computation of consistent initial values only works efficiently for certain subclasses of the DAE (5.1), like Hessenberg DAEs in Definition 1.1.14. Third, as has been mentioned already in Chapter 4, numerical methods again only work well for certain subclasses of (5.1), for instance, Hessenberg DAEs up to index three. Keeping these restrictions in mind, we consider

https://doi.org/10.1515/9783110797893-005

Problem 5.0.1 (General DAE optimal control problem). Let $\mathcal{I} := [t_0, t_f] \subset \mathbb{R}$ be a compact interval with fixed time points $t_0 < t_f$, and let

$$\begin{aligned}
\varphi &: \mathbb{R}^{n_z} \times \mathbb{R}^{n_z} \longrightarrow \mathbb{R}, \\
F &: \mathcal{I} \times \mathbb{R}^{n_z} \times \mathbb{R}^{n_z} \times \mathbb{R}^{n_u} \longrightarrow \mathbb{R}^{n_z}, \\
c &: \mathcal{I} \times \mathbb{R}^{n_z} \times \mathbb{R}^{n_u} \longrightarrow \mathbb{R}^{n_c}, \\
s &: \mathcal{I} \times \mathbb{R}^{n_z} \longrightarrow \mathbb{R}^{n_s}, \\
\psi &: \mathbb{R}^{n_z} \times \mathbb{R}^{n_z} \longrightarrow \mathbb{R}^{n_\psi}
\end{aligned}$$

be sufficiently smooth functions.

Minimize

$$\varphi\big(z(t_0), z(t_f)\big)$$

with respect to $z : \mathcal{I} \longrightarrow \mathbb{R}^{n_z}$ *and* $u : \mathcal{I} \longrightarrow \mathbb{R}^{n_u}$ *subject to the constraints*

$$F\big(t, z(t), \dot{z}(t), u(t)\big) = 0_{\mathbb{R}^{n_z}} \quad \text{a. e. in } \mathcal{I}, \tag{5.1}$$

$$\psi\big(z(t_0), z(t_f)\big) = 0_{\mathbb{R}^{n_\psi}}, \tag{5.2}$$

$$c\big(t, z(t), u(t)\big) \le 0_{\mathbb{R}^{n_c}} \quad \text{a. e. in } \mathcal{I}, \tag{5.3}$$

$$s\big(t, z(t)\big) \le 0_{\mathbb{R}^{n_s}} \quad \text{in } \mathcal{I}. \tag{5.4}$$

5.1 Direct discretization methods

Direct discretization methods are based on a discretization of the infinite-dimensional optimal control problem. The resulting discretized problem will be a finite-dimensional optimization problem. All subsequently discussed methods work on the (not necessarily equidistant) grid

$$\mathbb{G}_N := \{t_0 < t_1 < \cdots < t_N = t_f\} \tag{5.5}$$

with step-sizes $h_j = t_{j+1} - t_j$, $j = 0, \ldots, N-1$, and mesh-size $h := \max_{j=0,\ldots,N-1} h_j$. Often, $\mathbb{G}_N$ will be an equidistant partition of the interval $[t_0, t_f]$ with constant step-size $h = (t_f - t_0)/N$ and grid points $t_i = t_0 + ih$, $i = 0, \ldots, N$.

A direct discretization method is essentially defined by the following operations:

(a) *Control discretization:*
The control space $L_\infty^{n_u}(\mathcal{I})$ is replaced by some M-dimensional subspace

$$U^M \subset L_\infty^{n_u}(\mathcal{I})$$

where $M \in \mathbb{N}$ is finite. The dimension M usually depends on the number N of intervals in $\mathbb{G}_N$, i. e. $M = \mathcal{O}(N)$.

Let $\mathcal{B} := \{B_1, \ldots, B_M\}$ be a basis of U^M. Then every $u_M \in U^M$ is defined by

$$u_M(\cdot) := \sum_{i=1}^{M} w_i B_i(\cdot) \tag{5.6}$$

with coefficients $w := (w_1, \ldots, w_M)^\top \in \mathbb{R}^M$. The dependence on the vector w is indicated by the notation

$$u_M(t) := u_M(t; w) := u_M(t; w_1, \ldots, w_M). \tag{5.7}$$

Furthermore, we may identify u_M and w.

(b) *State discretization:*
The DAE is discretized by a suitable discretization scheme, for instance, an one-step method in Definition 4.1.10 or a multi-step method (4.16).

(c) *Constraint discretization:*
The control and state constraints are only evaluated on the grid $\mathbb{G}_N$.

(d) *Optimizer:*
The resulting discretized optimal control problem has to be solved numerically, for instance, by the SQP method in Section 5.2.2.

(e) *Calculation of derivatives:*
Gradient-based optimizers need the gradient of the objective function and the Jacobian of the constraints. Derivatives can be obtained in different ways, for instance, through finite difference approximations, by solving sensitivity or adjoint equations, or by algorithmic differentiation.

Next, we illustrate the direct discretization method with the generic one-step method in Definition 4.1.10, which is supposed to be appropriate for the DAE (5.1).

For a given consistent initial value $z_N(t_0)$ and a given control approximation u_M as in (5.6) and (5.7), the one-step method generates values

$$z_N(t_{j+1}) = z_N(t_j) + h_j \Phi(t_j, z_N(t_j), w, h_j), \quad j = 0, 1, \ldots, N-1.$$

Notice that Φ depends on the control parameterization w. As usual, we use the abbreviation $z_j := z_N(t_j)$ for $j = 0, \ldots, N$ and likewise for other grid functions.

We distinguish two approaches for the discretization of Problem 5.0.1: the *full discretization approach* and the *reduced discretization approach*. Both approaches use a procedure for the calculation of a consistent initial value

$$Z_0 : \mathbb{R}^{n_z} \times \mathbb{R}^M \longrightarrow \mathbb{R}^{n_z}, \quad (z_0, w) \mapsto Z_0(z_0, w),$$

that maps a given control parameterization w and some guess z_0 of a consistent initial value into the set of consistent initial values. It is assumed throughout that such a method is available, for instance, the projection method in Section 4.5.1 can be used for Hessenberg DAEs.

5.1.1 Full discretization approach

We obtain a full discretization of the optimal control problem 5.0.1 by replacing the DAE by the one-step method and discretizing the constraints on the grid $\mathbb{G}_N$:

Problem 5.1.1 (Full discretization).
Minimize

$$\varphi(z_0, z_N)$$

with respect to $z_N : \mathbb{G}_N \longrightarrow \mathbb{R}^{n_z}$, $t_j \mapsto z_N(t_j) =: z_j$, *and* $w \in \mathbb{R}^M$ *subject to the constraints*

$$\begin{aligned}
Z_0(z_0, w) - z_0 &= 0_{\mathbb{R}^{n_z}}, \\
z_j + h_j \Phi(t_j, z_j, w, h_j) - z_{j+1} &= 0_{\mathbb{R}^{n_z}}, \quad j = 0, 1, \ldots, N-1, \\
\psi(z_0, z_N) &= 0_{\mathbb{R}^{n_\psi}}, \\
c\big(t_j, z_j, u_M(t_j; w)\big) &\leq 0_{\mathbb{R}^{n_c}}, \quad j = 0, 1, \ldots, N, \\
s(t_j, z_j) &\leq 0_{\mathbb{R}^{n_s}}, \quad j = 0, 1, \ldots, N.
\end{aligned}$$

Alternative full discretization methods, which do not use the consistency map Z_0, arise by imposing additional constraints, which restrict initial values to the set of consistent initial values. To this end, for Hessenberg DAEs the constraint $Z_0(z_0, w) = z_0$ can be replaced by the hidden algebraic constraints (4.30)–(4.31) in order to guarantee consistency.

Problem 5.1.1 is a finite-dimensional optimization problem of type

$$\begin{gathered}
\textit{Minimize } J(\bar{z}) \textit{ with respect to } \bar{z} \in \mathbb{R}^{n_{\bar{z}}} \textit{ subject to} \\
G(\bar{z}) \leq 0_{\mathbb{R}^{n_G}}, \quad H(\bar{z}) = 0_{\mathbb{R}^{n_H}}
\end{gathered} \tag{5.8}$$

with

$$\bar{z} := (z_0, z_1, \ldots, z_N, w)^\top \in \mathbb{R}^{n_z(N+1)+M},$$

$$J(\bar{z}) := \varphi(z_0, z_N),$$

$$G(\bar{z}) := \begin{pmatrix} c(t_0, z_0, u_M(t_0; w)) \\ \vdots \\ c(t_N, z_N, u_M(t_N; w)) \\ s(t_0, z_0) \\ \vdots \\ s(t_N, z_N) \end{pmatrix} \in \mathbb{R}^{(n_c+n_s)(N+1)},$$

$$H(\bar{z}) := \begin{pmatrix} Z_0(z_0, w) - z_0 \\ z_0 + h_0 \Phi(t_0, z_0, w, h_0) - z_1 \\ \vdots \\ z_{N-1} + h_{N-1}\Phi(t_{N-1}, z_{N-1}, w, h_{N-1}) - z_N \\ \psi(z_0, z_N) \end{pmatrix} \in \mathbb{R}^{n_z(N+1)+n_\psi}.$$

The size of Problem (5.8) depends on n_z, n_s, n_c, n_ψ, N, and M and can become very large. In practice, dimensions up to a million of optimization variables and constraints, or even more, are not unrealistic. On the other hand, it is easy to compute derivatives with respect to the optimization variable $\bar{z}$:

$$J'(\bar{z}) = (\varphi'_{z_0} \mid 0^\top_{\mathbb{R}^{n_z}} \mid \cdots \mid 0^\top_{\mathbb{R}^{n_z}} \mid \varphi'_{z_f} \mid 0^\top_{\mathbb{R}^M}), \tag{5.9}$$

$$G'(\bar{z}) = \left(\begin{array}{ccc|c} c'_z[t_0] & & & c'_u[t_0] \cdot u'_{M,w}(t_0; w) \\ & \ddots & & \vdots \\ & & c'_z[t_N] & c'_u[t_N] \cdot u'_{M,w}(t_N; w) \\ \hline s'_z[t_0] & & & \\ & \ddots & & \Theta \\ & & s'_z[t_N] & \end{array}\right), \tag{5.10}$$

$$H'(\bar{z}) = \left(\begin{array}{cccc|c} Z'_{0,z_0} - I_{n_z} & & & & Z'_{0,w} \\ M_0 & -I_{n_z} & & & h_0 \Phi'_w[t_0] \\ & \ddots & & & \vdots \\ & & M_{N-1} & -I_{n_z} & h_{N-1}\Phi'_w[t_{N-1}] \\ \hline \psi'_{z_0} & & & \psi'_{z_f} & \Theta \end{array}\right), \tag{5.11}$$

where $M_j := I_{n_z} + h_j \Phi'_z(t_j, z_j, w, h_j)$, $j = 0, \ldots, N-1$.

The optimization problem in (5.8) is large-scale, but the gradient and the Jacobians in (5.9)–(5.11) exhibit a sparse structure, which has to be exploited in numerical optimization algorithms using appropriate techniques from numerical linear algebra, see [27, 28], the comments in Section 5.2.2, and the details in Section 5.3.

The full discretization approach as outlined above has to be understood as a generic concept. Particularly for implicit integration methods, e. g., implicit Runge–Kutta methods, the equations

$$z_j + h_j \Phi(t_j, z_j, w, h_j) - z_{j+1} = 0_{\mathbb{R}^{n_z}} \quad (j = 0, 1, \ldots, N-1)$$

are replaced or augmented, respectively, by the implicit equations defining the integration scheme. The following examples illustrate some Runge–Kutta discretization schemes.

Example 5.1.2 (Discretization by implicit Runge–Kutta methods). Consider the following DAE optimal control problem:

Minimize

$$\varphi(z(t_0), z(t_f))$$

subject to the constraints

$$\begin{aligned} F(t, z(t), \dot{z}(t), u(t)) &= 0_{\mathbb{R}^{n_z}}, \\ \psi(z(t_0), z(t_f)) &= 0_{\mathbb{R}^{n_\psi}}, \\ c(t, z(t), u(t)) &\leq 0_{\mathbb{R}^{n_c}}, \\ s(t, z(t)) &\leq 0_{\mathbb{R}^{n_s}}. \end{aligned}$$

Application of the implicit Euler method and a piecewise constant control approximation on the grid $\mathbb{G}_N$ yields a fully discretized problem:

Minimize

$$\varphi(z_0, z_N)$$

with respect to

$$z_0, \ldots, z_N, u_0, \ldots, u_N$$

subject to the constraints

$$\begin{aligned} Z_0(z_0, u_0) - z_0 &= 0_{\mathbb{R}^{n_z}}, \\ F\left(t_{j+1}, z_{j+1}, \frac{z_{j+1} - z_j}{h_j}, u_{j-1}\right) &= 0_{\mathbb{R}^{n_z}}, \quad j = 0, \ldots, N-1, \\ \psi(z_0, z_N) &= 0_{\mathbb{R}^{n_\psi}}, \\ c(t_j, z_j, u_j) &\leq 0_{\mathbb{R}^{n_c}}, \quad j = 0, \ldots, N, \\ s(t_j, z_j) &\leq 0_{\mathbb{R}^{n_s}}, \quad j = 0, \ldots, N. \end{aligned}$$

Application of the RADAUIIA method, see Example 4.1.7, and a control approximation $u_M(\cdot; w)$ on the grid $\mathbb{G}_N$ yields the following fully discretized problem:

Minimize

$$\varphi(z_0, z_N)$$

with respect to

$$z_0, \ldots, z_N, k_{1,0}, \ldots, k_{1,N-1}, k_{2,0}, \ldots, k_{2,N-1}, w$$

subject to the constraints

$$
\begin{aligned}
Z_0(z_0, w) - z_0 &= 0_{\mathbb{R}^{n_z}}, \\
z_j + h_j\left(\frac{3}{4}k_{1,j} + \frac{1}{4}k_{2,j}\right) - z_{j+1} &= 0_{\mathbb{R}^{n_z}}, \quad j = 0, \dots, N-1, \\
F\left(t_j + \frac{h_j}{3}, z_{j+1}^{(1)}, k_{1,j}, u_M(t_j + h_j/3; w)\right) &= 0_{\mathbb{R}^{n_z}}, \quad j = 0, \dots, N-1, \\
F\big(t_j + h_j, z_{j+1}^{(2)}, k_{2,j}, u_M(t_j + h_j; w)\big) &= 0_{\mathbb{R}^{n_z}}, \quad j = 0, \dots, N-1, \\
\psi(z_0, z_N) &= 0_{\mathbb{R}^{n_\psi}}, \\
c\big(t_j, z_j, u_M(t_j; w)\big) &\le 0_{\mathbb{R}^{n_c}}, \quad j = 0, \dots, N, \\
s(t_j, z_j) &\le 0_{\mathbb{R}^{n_s}}, \quad j = 0, \dots, N.
\end{aligned}
$$

Herein, the stage approximations are given by

$$
\begin{aligned}
z_{j+1}^{(1)} &:= z_j + h_j\left(\frac{5}{12}k_{1,j} - \frac{1}{12}k_{2,j}\right), \\
z_{j+1}^{(2)} &:= z_j + h_j\left(\frac{3}{4}k_{1,j} + \frac{1}{4}k_{2,j}\right), \quad j = 0, \dots, N-1.
\end{aligned}
$$

Application of a general s-stage Runge–Kutta method, see Definition 4.1.9, and a control approximation $u_M(\cdot; w)$ on the grid $\mathbb{G}_N$ yields the following fully discretized problem:

Minimize

$$\varphi(z_0, z_N)$$

with respect to

$$z_j, j = 0, \dots, N, \quad k_{\ell,j}, j = 0, \dots, N-1, \ell = 1, \dots, s, \quad w$$

subject to the constraints

$$
\begin{aligned}
Z_0(z_0, w) - z_0 &= 0_{\mathbb{R}^{n_z}}, \\
z_j + h_j \sum_{i=1}^{s} b_k k_{i,j} - z_{j+1} &= 0_{\mathbb{R}^{n_z}}, \quad j = 0, \dots, N-1, \\
F\big(t_j + c_\ell h_j, z_{j+1}^{(\ell)}, k_{\ell,j}, u_M(t_j + c_\ell h_j; w)\big) &= 0_{\mathbb{R}^{n_z}}, \quad j = 0, \dots, N-1, \ \ell = 1, \dots, s, \\
\psi(z_0, z_N) &= 0_{\mathbb{R}^{n_\psi}}, \\
c\big(t_j, z_j, u_M(t_j; w)\big) &\le 0_{\mathbb{R}^{n_c}}, \quad j = 0, \dots, N, \\
s(t_j, z_j) &\le 0_{\mathbb{R}^{n_s}}, \quad j = 0, \dots, N.
\end{aligned}
$$

Herein, the stage approximations are given by

$$z_{j+1}^{(\ell)} := z_j + h_j \sum_{i=1}^{s} a_{\ell i} k_{i,j}, \quad \ell = 1, \dots, s, j = 0, \dots, N-1. \qquad \square$$

Example 5.1.3 (Collocation schemes for ODEs). Consider the following ODE optimal control problem:

Minimize

$$\varphi(x(t_0), x(t_f))$$

subject to the constraints

$$\begin{aligned} f(t, x(t), u(t)) - \dot{x}(t) &= 0_{\mathbb{R}^{n_x}}, \\ \psi(x(t_0), x(t_f)) &= 0_{\mathbb{R}^{n_\psi}}, \\ c(t, x(t), u(t)) &\leq 0_{\mathbb{R}^{n_c}}, \\ s(t, x(t)) &\leq 0_{\mathbb{R}^{n_s}}. \end{aligned}$$

Application of the trapezoidal rule and a continuous, piecewise linear control approximation on the grid $\mathbb{G}_N$ yields a fully discretized problem:

Minimize

$$\varphi(x_0, x_N)$$

with respect to

$$x_0, \dots, x_N, u_0, \dots, u_N$$

subject to the constraints

$$\begin{aligned} x_j + \frac{h_j}{2}(f_j + f_{j+1}) - x_{j+1} &= 0_{\mathbb{R}^{n_x}}, \quad j = 0, \dots, N-1, \\ \psi(x_0, x_N) &= 0_{\mathbb{R}^{n_\psi}}, \\ c(t_j, x_j, u_j) &\leq 0_{\mathbb{R}^{n_c}}, \quad j = 0, \dots, N, \\ s(t_j, x_j) &\leq 0_{\mathbb{R}^{n_s}}, \quad j = 0, \dots, N. \end{aligned}$$

The *classic collocation discretization* occurs for the Simpson rule and a continuous, piecewise linear control approximation on $\mathbb{G}_N$, compare [26]:

Minimize

$$\varphi(x_0, x_N)$$

with respect to

$$x_0, \dots, x_N, u_0, \dots, u_N$$

subject to the constraints

$$\begin{aligned} x_j + \frac{h_j}{6}(f_j + f_{j+\frac{1}{2}} + f_{j+1}) - x_{j+1} &= 0_{\mathbb{R}^{n_x}}, \quad j = 0, \dots, N-1, \\ \psi(x_0, x_N) &= 0_{\mathbb{R}^{n_\psi}}, \\ c(t_j, x_j, u_j) &\leq 0_{\mathbb{R}^{n_c}}, \quad j = 0, \dots, N, \\ s(t_j, x_j) &\leq 0_{\mathbb{R}^{n_s}}, \quad j = 0, \dots, N. \end{aligned}$$

Herein,

$$
\begin{aligned}
f_j &:= f(t_j, x_j, u_j), \\
f_{j+1} &:= f(t_{j+1}, x_{j+1}, u_{j+1}), \\
f_{j+\frac{1}{2}} &:= f\left(t_j + \frac{h_j}{2}, x_{j+\frac{1}{2}}, u_{j+\frac{1}{2}}\right), \\
x_{j+\frac{1}{2}} &:= \frac{1}{2}(x_j + x_{j+1}) + \frac{h_j}{8}(f_j - f_{j+1}), \\
u_{j+\frac{1}{2}} &:= \frac{h_j}{2}(u_j + u_{j+1}).
\end{aligned}
$$
□

---

i In fact, there are different possibilities to implement the constraints for Runge–Kutta methods. For instance, the trapezoidal rule in Example 5.1.3 could be implemented in condensed form as

$$x_j + \frac{h_j}{2}\big(f(t_j, x_j, u_j) + f(t_{j+1}, x_{j+1}, u_{j+1})\big) - x_{j+1} = 0_{\mathbb{R}^{n_x}}, \quad j = 0, \dots, N-1,$$

or alternatively as

$$
\begin{aligned}
x_j + \frac{h_j}{2}(k_{1,j} + k_{2,j}) - x_{j+1} &= 0_{\mathbb{R}^{n_x}}, \quad j = 0, \dots, N-1, \\
f(t_j, x_j, u_j) - k_{1,j} &= 0_{\mathbb{R}^{n_x}}, \quad j = 0, \dots, N-1, \\
f\left(t_{j+1}, x_j + \frac{h_j}{2}(k_{1,j} + k_{2,j}), u_{j+1}\right) - k_{2,j} &= 0_{\mathbb{R}^{n_x}}, \quad j = 0, \dots, N-1.
\end{aligned}
$$

The resulting fully discretized problem for the latter option is larger in terms of variables and constraints, but with some additional sparsity.

5.1.2 Reduced discretization approach

The reduced discretization approach is based on the full discretization 5.1.1, but the equations resulting from the one-step method are not explicitly imposed as equality constraints in the discretized optimization problem. Instead, it is exploited that z_{i+1} is completely defined by (t_i, z_i, w), and the one-step equations can be solved recursively:

$$
\begin{aligned}
\tilde{z}_0 &= Z_0(z_0, w), \\
z_1 &= \tilde{z}_0 + h_0\Phi(t_0, \tilde{z}_0, w, h_0) \\
&= Z_0(z_0, w) + h_0\Phi(t_0, Z_0(z_0, w), w, h_0) \\
&=: Z_1(z_0, w), \\
z_2 &= z_1 + h_1\Phi(t_1, z_1, w, h_1) \\
&= Z_1(z_0, w) + h_1\Phi(t_1, Z_1(z_0, w), w, h_1) \\
&=: Z_2(z_0, w), \qquad (5.12)
\end{aligned}
$$

$$
\begin{aligned}
&\vdots\\
z_N &= z_{N-1} + h_{N-1}\Phi(t_{N-1}, z_{N-1}, w, h_{N-1})\\
&= Z_{N-1}(z_0, w) + h_{N-1}\Phi(t_{N-1}, Z_{N-1}(z_0, w), w, h_{N-1})\\
&=: Z_N(z_0, w).
\end{aligned}
$$

Of course, (5.12) is just the formal procedure of solving the DAE for given consistent initial value and control parameterization w. Herein, the procedure for computing a consistent initial value $Z_0(z_0, w)$ for z_0 and w is used again. We obtain:

Problem 5.1.4 (Reduced discretization).
Minimize

$$\varphi\big(Z_0(z_0, w), Z_N(z_0, w)\big)$$

with respect to $z_0 \in \mathbb{R}^{n_z}$ *and* $w \in \mathbb{R}^M$ *subject to the constraints*

$$
\begin{aligned}
\psi\big(Z_0(z_0, w), Z_N(z_0, w)\big) &= 0_{\mathbb{R}^{n_\psi}},\\
c\big(t_j, Z_j(z_0, w), u_M(t_j; w)\big) &\le 0_{\mathbb{R}^{n_c}}, \quad j = 0, 1, \dots, N,\\
s\big(t_j, Z_j(z_0, w)\big) &\le 0_{\mathbb{R}^{n_s}}, \quad j = 0, 1, \dots, N.
\end{aligned}
$$

An alternative reduced discretization method, which does not use the consistency map Z_0, is obtained using the relaxation method in Section 4.5.2. Herein, consistency is ensured by modifying the DAE and imposing additional constraints as described in Section 4.5.2.

Problem 5.1.4 is again a finite-dimensional nonlinear optimization problem of type (5.8) with

$$
\begin{aligned}
\bar{z} &:= (z_0, w)^\top \in \mathbb{R}^{n_z+M},\\
J(\bar{z}) &:= \varphi(Z_0(z_0, w), Z_N(z_0, w)), \qquad (5.13)\\
G(\bar{z}) &:= \begin{pmatrix} c(t_0, Z_0(z_0, w), u_M(t_0; w))\\ \vdots \\ c(t_N, Z_N(z_0, w), u_M(t_N; w))\\ s(t_0, Z_0(z_0, w))\\ \vdots \\ s(t_N, Z_N(z_0, w)) \end{pmatrix} \in \mathbb{R}^{(n_c+n_s)(N+1)}, \qquad (5.14)\\
H(\bar{z}) &:= \psi(Z_0(z_0, w), Z_N(z_0, w)) \in \mathbb{R}^{n_\psi}. \qquad (5.15)
\end{aligned}
$$

The size of Problem 5.1.4 is small compared to Problem 5.1.1, since most of the equality constraints as well as the variables $z_1, \dots, z_N$ are eliminated. Just the initial value z_0

remains an optimization variable. Unfortunately, the derivatives are not sparse anymore, and it is more involved to compute them by the chain rule:

$$\begin{aligned} J'(\bar{z}) &= (\varphi'_{z_0} \cdot Z'_{0,z_0} + \varphi'_{z_f} \cdot Z'_{N,z_0} \mid \varphi'_{z_0} \cdot Z'_{0,w} + \varphi'_{z_f} \cdot Z'_{N,w}), \\ H'(\bar{z}) &= (\psi'_{z_0} \cdot Z'_{0,z_0} + \psi'_{z_f} \cdot Z'_{N,z_0} \mid \psi'_{z_0} \cdot Z'_{0,w} + \psi'_{z_f} \cdot Z'_{N,w}), \end{aligned} \tag{5.16}$$

$$G'(\bar{z}) = \left(\begin{array}{c|c} c'_z[t_0] \cdot Z'_{0,z_0} & c'_z[t_0] \cdot Z'_{0,w} + c'_u[t_0] \cdot u'_{M,w}(t_0; w) \\ c'_z[t_1] \cdot Z'_{1,z_0} & c'_z[t_1] \cdot Z'_{1,w} + c'_u[t_1] \cdot u'_{M,w}(t_1; w) \\ \vdots & \vdots \\ c'_z[t_N] \cdot Z'_{N,z_0} & c'_z[t_N] \cdot Z'_{N,w} + c'_u[t_N] \cdot u'_{M,w}(t_N; w) \\ \hline s'_z[t_0] \cdot Z'_{0,z_0} & s'_z[t_0] \cdot Z'_{0,w} \\ s'_z[t_1] \cdot Z'_{1,z_0} & s'_z[t_1] \cdot Z'_{1,w} \\ \vdots & \vdots \\ s'_z[t_N] \cdot Z'_{N,z_0} & s'_z[t_N] \cdot Z'_{N,w} \end{array} \right). \tag{5.17}$$

The sensitivities

$$Z'_{i,z_0}(z_0, w), \quad Z'_{i,w}(z_0, w), \quad i = 0, \ldots, N, \tag{5.18}$$

will be computed in Section 5.4.

---

i Based on the same ideas, it is possible and often necessary to extend the reduced discretization method to a reduced multiple shooting method by introducing multiple shooting nodes. Details can be found in [112, 113]. Related methods based on the multiple shooting idea are discussed in [70, 29, 313, 206, 303, 164, 258].

5.1.3 Control discretization

Control parameterizations can be obtained in various ways. Some authors use interpolating cubic splines [184], Hermite polynomials [19], or polynomials [286, 175]. These parameterizations have in common that the components of the control parameter w influence the parameterized control function on the whole time interval, that is, changing one component w_i of w usually affects $u_M(t; w)$ for all $t \in \mathcal{I}$. This may lead to numerical instabilities, especially if the optimal control is not smooth, and in addition, sparsity of the derivative $u'_{M,w}(\cdot; w)$ is destroyed.

Owing to these potential drawbacks, a control parameterization with basis functions having *local support* is preferred in the sequel. To this end, *B-spline representations* are found to be particularly useful.

Definition 5.1.5 (B-spline). Let $k \in \mathbb{N}$ and $\mathbb{G}_N$ as in (5.5). Define the auxiliary grid

$$\mathbb{G}_N^k := \{\tau_i \mid i = 1, \dots, N + 2k - 1\} \tag{5.19}$$

with auxiliary grid points

$$\tau_i := \begin{cases} t_0, & \text{if } 1 \le i \le k, \\ t_{i-k}, & \text{if } k+1 \le i \le N+k-1, \\ t_N, & \text{if } N+k \le i \le N+2k-1. \end{cases}$$

The *elementary B-splines* $B_i^k(\cdot)$ *of order* k, $i = 1, \dots, N+k-1$, are defined by the recursion

$$\begin{aligned} B_i^1(t) &:= \begin{cases} 1, & \text{if } \tau_i \le t < \tau_{i+1}, \\ 0, & \text{otherwise}, \end{cases} \\ B_i^k(t) &:= \frac{t - \tau_i}{\tau_{i+k-1} - \tau_i} B_i^{k-1}(t) + \frac{\tau_{i+k} - t}{\tau_{i+k} - \tau_{i+1}} B_{i+1}^{k-1}(t). \end{aligned} \tag{5.20}$$

Herein, the convention $0/0 = 0$ is used whenever auxiliary grid points coincide in the recursion (5.20). □

The evaluation of the recursion (5.20) is well-conditioned, see [73]. The elementary B-splines $B_i^k(\cdot)$, $i = 1, \dots, N+k-1$, restricted to the intervals $[t_j, t_{j+1}]$, $j = 0, \dots, N-1$, are polynomials of degree at most $k-1$, i. e.

$$B_i^k(\cdot)|_{[t_j,t_{j+1}]} \in \mathcal{P}_{k-1}([t_j, t_{j+1}]),$$

and they form a basis of the space of splines

$$\{s(\cdot) \in \mathcal{C}_{k-2}(\mathcal{I}) \mid s(\cdot)|_{[t_j,t_{j+1}]} \in \mathcal{P}_{k-1}([t_j, t_{j+1}]) \text{ for } j = 0, \dots, N-1\}.$$

Obviously, the elementary B-splines are essentially bounded, i. e. $B_i^k(\cdot) \in L_\infty(\mathcal{I})$.

For $k \ge 2$, it holds $B_i^k(\cdot) \in \mathcal{C}_{k-2}(\mathcal{I})$, and for $k \ge 3$, the derivative obeys the recursion

$$\frac{d}{dt} B_i^k(t) = \frac{k-1}{\tau_{i+k-1} - \tau_i} B_i^{k-1}(t) - \frac{k-1}{\tau_{i+k} - \tau_{i+1}} B_{i+1}^{k-1}(t).$$

Furthermore, the elementary B-splines possess local support $\operatorname{supp}(B_i^k) \subset [\tau_i, \tau_{i+k}]$ with

$$B_i^k(t) \begin{cases} > 0, & \text{if } t \in (\tau_i, \tau_{i+k}), \\ = 0, & \text{otherwise}, \end{cases} \quad \text{for } k > 1.$$

The cases $k = 1$ or $k = 2$ frequently occur in numerical computations. For $k = 1$, the elementary B-splines are piecewise constant functions, while for $k = 2$, we obtain the continuous and piecewise linear functions

$$B_i^2(t) = \begin{cases} \frac{t-\tau_i}{\tau_{i+1}-\tau_i}, & \text{if } \tau_i \leq t < \tau_{i+1}, \\ \frac{\tau_{i+2}-t}{\tau_{i+2}-\tau_{i+1}}, & \text{if } \tau_{i+1} \leq t < \tau_{i+2}, \\ 0, & \text{otherwise.} \end{cases}$$

In some situations, it might be necessary to have a continuously differentiable function, e. g.,

$$B_i^3(t) = \begin{cases} \frac{(t-\tau_i)^2}{(\tau_{i+2}-\tau_i)(\tau_{i+1}-\tau_i)}, & \text{if } t \in [\tau_i, \tau_{i+1}), \\ \frac{(t-\tau_i)(\tau_{i+2}-t)}{(\tau_{i+2}-\tau_i)(\tau_{i+2}-\tau_{i+1})} + \frac{(\tau_{i+3}-t)(t-\tau_{i+1})}{(\tau_{i+3}-\tau_{i+1})(\tau_{i+2}-\tau_{i+1})}, & \text{if } t \in [\tau_{i+1}, \tau_{i+2}), \\ \frac{(\tau_{i+3}-t)^2}{(\tau_{i+3}-\tau_{i+1})(\tau_{i+3}-\tau_{i+2})}, & \text{if } t \in [\tau_{i+2}, \tau_{i+3}), \\ 0, & \text{otherwise.} \end{cases}$$

Figure 5.1 visualizes B-splines of orders $k = 2, 3, 4$.

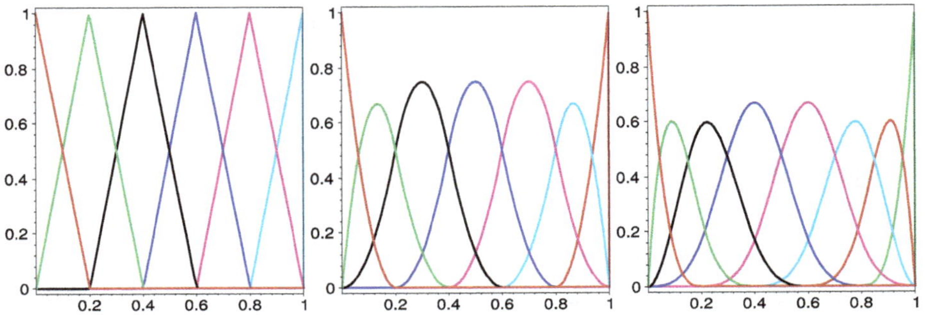

Figure 5.1: Elementary B-splines of order $k = 2$ (left), $k = 3$ (middle), and $k = 4$ (right) with $[t_0, t_f] = [0, 1]$ and $N = 5$ on an equidistant grid.

A control parameterization is obtained by choosing $u_M(\cdot)$ for some fixed $k \in \mathbb{N}$ and a given grid $\mathbb{G}_N$ with $N \in \mathbb{N}$ from the set

$$U^M := \left\{ \sum_{i=1}^{N+k-1} w_i B_i^k(\cdot) \mid w_i \in \mathbb{R}^{n_u}, \ i = 1, \ldots, N + k - 1 \right\}. \tag{5.21}$$

The number $k \in \mathbb{N}$ controls the smoothness of the control parameterization. Each $u_M(\cdot) \in U^M$ is determined by $M := n_u(N+k-1)$ control parameters $w := (w_1, \ldots, w_{N+k-1}) \in \mathbb{R}^{n_u(N+k-1)}$, and hence, U^M is a finite-dimensional subspace of $L_\infty^{n_u}(\mathcal{I})$. The dependence on the vector w is indicated by the notation

$$u_M(t) := u_M(t; w) := u_M(t; w_1, \ldots, w_{N+k-1}).$$

The coefficients w_i, $i = 1, \ldots, N+k-1$, are known as *de Boor points*. Most authors prefer $k = 1$ (piecewise constant approximation) or $k = 2$ (continuous and piecewise linear approximation) in which cases the de Boor points w_i satisfy $w_{i+1} = u_M(t_i)$, $i = 0, \ldots, N-1$, and $w_{N+1} = u_M(t_N)$ if $k = 2$.

The choice of B-splines has two major advantages from the numerical point of view. First, it is easy to create approximations with prescribed smoothness properties. Second, the de Boor point w_i influences the function value $u_M(t)$ only for $t \in [\tau_i, \tau_{i+k}]$ due to the local support of B_i^k. This property leads to sparsity patterns in the Jacobian of the constraints. The exploitation of this sparsity reduces the computational effort for the numerical solution of the discretized problems considerably, see [112, 113].

Of course, alternative control parameterizations can also be constructed. However, the following example shows that choosing a control parameterization may lead to pitfalls. Suppose we would like to use the modified Euler scheme with Butcher array

$$\begin{array}{c|cc} 0 & 0 & 0 \\ 1/2 & 1/2 & 0 \\ \hline & 0 & 1 \end{array} \tag{5.22}$$

for the discretization of an ODE. This method requires function evaluations at the time points t_i and $t_i + h/2$. In particular, the method uses the control values $u_i := u_M(t_i)$ and $u_{i+1/2} := u_M(t_i + h/2)$. Instead of choosing a piecewise constant control approximation on $\mathbb{G}_N$, we might have the idea of treating u_i and $u_{i+1/2}$ as independent optimization variables in the discretized optimal control problem. Although this strategy seems reasonable as it provides more degrees of freedom than a piecewise constant control approximation, the following example shows that this strategy may fail.

Example 5.1.6 ([152, p. 272]). Consider the following optimal control problem:

Minimize

$$\frac{1}{2}\int_0^1 u(t)^2 + 2z(t)^2 dt$$

subject to the constraints

$$\dot{z}(t) = \frac{1}{2}z(t) + u(t), \quad z(0) = 1.$$

The optimal solution is

$$\hat{z}(t) = \frac{2\exp(3t) + \exp(3)}{\exp(3t/2)(2 + \exp(3))}, \qquad \hat{u}(t) = \frac{2(\exp(3t) - \exp(3))}{\exp(3t/2)(2 + \exp(3))}.$$

We consider the modified Euler method (5.22) and Heun's method defined by

$$\begin{array}{c|cc} 0 & 0 & 0 \\ 1 & 1 & 0 \\ \hline & 1/2 & 1/2 \end{array}$$

Table 5.1 shows the error in z in the norm $\|\cdot\|_\infty$ for Heun's method. The order of convergence is two.

Table 5.1: Error for Heun's method.

N	Error in z	Order
10	0.2960507253983891e-02	–
20	0.7225108094129906e-03	2.0347533
40	0.1783364646560370e-03	2.0184174
80	0.4342336372986644e-04	2.0380583
160	0.9861920395981549e-05	2.138531
320	0.2417855093361787e-05	2.0281408

Table 5.2 shows the error in z in the norm $\|\cdot\|_\infty$ for modified Euler's method, if the control is discretized at t_i and $t_i + h/2$ with independent optimization variables u_i and $u_{i+1/2}$.

Table 5.2: Error for the modified Euler method with independent optimization variables at t_i and $t_{i+1/2}$.

N	Error in z	Order
10	0.4254224673693650e+00	–
20	0.4258159920666613e+00	−0.0013339
40	0.4260329453139864e+00	−0.0007349
80	0.4260267362368171e+00	0.0000210
160	0.4261445411996390e+00	−0.0003989
320	0.4260148465889140e+00	0.0004391

There is no convergence at all. Figure 5.2 shows the numerical solution for $N = 40$. The control u oscillates strongly between 0 at $t_i + h/2$ and approximately $-1/(2h)$ at t_i.

Now we use a piecewise constant control approximation in the modified Euler method and obtain the error in Table 5.3. For this control approximation, the modified Euler method converges of second order! □

5.2 A brief introduction to sequential quadratic programming

The full discretization method in Problem 5.1.1 and the reduced discretization method in Problem 5.1.4 lead to constrained nonlinear optimization problems that have to be

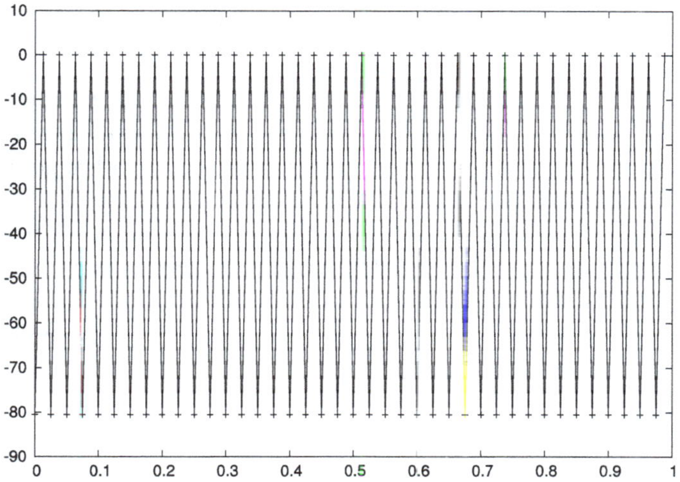

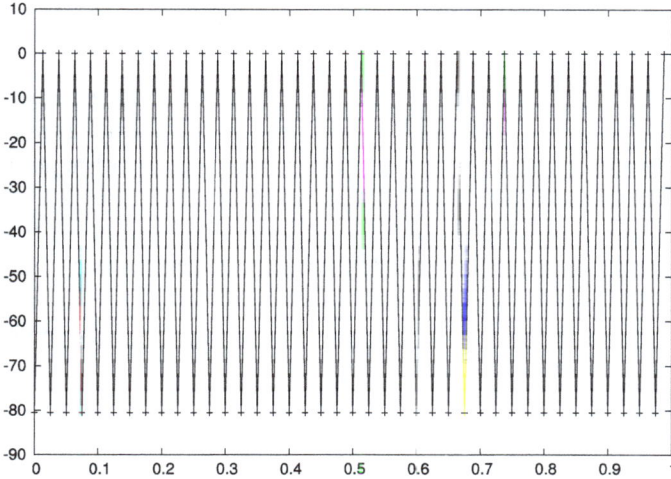

Figure 5.2: Oscillating control for the modified Euler method and $N = 40$.

Table 5.3: Error for the modified Euler method with piecewise constant control approximation.

N	Error in z	Order
10	0.3358800781952942e-03	–
20	0.8930396513584515e-04	1.9111501
40	0.2273822819465199e-04	1.9736044
80	0.5366500129055929e-05	2.0830664
160	0.1250729642299220e-05	2.1012115
320	0.7884779272826492e-06	0.6656277

solved numerically. In essence, any method for constrained nonlinear optimization problems can be applied. The most popular methods are sequential quadratic programming (SQP), interior-point methods, and multiplier-penalty methods, which exist in different versions and are described in most textbooks on nonlinear optimization methods, see [138, 95, 309, 21, 257, 64, 6, 111, 98].

Typically, these main algorithmic frameworks are enriched by globalization techniques to achieve convergence from arbitrary starting points, such as line-search methods using merit functions, trust-region methods, or filter methods. All of these approaches are well-investigated in view of their mathematical convergence properties. Most of them show at least global convergence to first-order stationary points and locally superlinear convergence under appropriate assumptions.

It is important to distinguish between the mathematical method and the implementation of the method. From a purely mathematical point of view, none of the approaches can be considered superior to the remaining ones on a sufficiently large test-set. So, the practical performance of a method depends to a large extent on the specific implementation and its strategies to deal with ill-conditioning, rank deficiencies, bad scaling, remote starting points, inaccurate derivatives, and large-scale and sparse problems.

Problems 5.1.1 and 5.1.4 require different implementations of methods due to their structural differences. While Problem 5.1.1 is large-scale but sparse with potentially millions of variables and constraints, Problem 5.1.4 is medium-scale but dense with at most 5000, say, variables and constraints in most applications.

Among the many different approaches, we focus on the Lagrange–Newton and the SQP methods. The SQP method has shown its capability in many applications and is discussed in, e. g., [156, 275, 138, 312, 299, 300, 6, 111, 90]. The SQP method exists in several implementations, e. g. [301, 183, 137, 136], and its special adaptations to discretized optimal control problems are described in [135, 302, 310, 28, 91].

A recent implementation of the SQP method for large-scale and sparse nonlinear optimization problems called WORHP can be obtained under www.worhp.de.

5.2.1 Lagrange–Newton method

Consider the equality constrained nonlinear optimization problem:

Problem 5.2.1. Let $J, H_j : \mathbb{R}^{n_{\bar{z}}} \longrightarrow \mathbb{R}$, $j = 1, \dots, n_H$, be twice continuously differentiable and $H := (H_1, \dots, H_{n_H})^\top$.

Minimize $J(\bar{z})$ with respect to $\bar{z} \in \mathbb{R}^{n_{\bar{z}}}$ subject to the constraints

$$H_j(\bar{z}) = 0, \quad j = 1, \dots, n_H.$$

Let $\hat{\bar{z}}$ be a local minimum of Problem 5.2.1, and let the gradients $\nabla H_j(\hat{\bar{z}})$, $j = 1, \dots, n_H$, be linearly independent, compare Definition 2.3.34. Then according to Corollary 2.3.35, there exist multipliers $\ell_0 = 1$ and $\hat{\lambda} = (\hat{\lambda}_1, \dots, \hat{\lambda}_{n_H})^\top \in \mathbb{R}^{n_H}$ such that

$$\begin{aligned} \nabla_{\bar{z}} L(\hat{\bar{z}}, \ell_0, \hat{\lambda}) &= 0_{\mathbb{R}^{n_{\bar{z}}}}, \\ H(\hat{\bar{z}}) &= 0_{\mathbb{R}^{n_H}}, \end{aligned}$$

where $L(\bar{z}, \ell_0, \lambda) = \ell_0 J(\bar{z}) + \lambda^\top H(\bar{z})$ denotes the Lagrange function of Problem 5.2.1. This is a nonlinear equation for $\hat{\bar{z}}$, and $\hat{\lambda}$ and we can rewrite it as

$$T(\hat{\bar{z}}, \hat{\lambda}) = 0_{\mathbb{R}^{n_{\bar{z}}+n_H}}, \tag{5.23}$$

where $T : \mathbb{R}^{n_{\bar{z}}} \times \mathbb{R}^{n_H} \longrightarrow \mathbb{R}^{n_{\bar{z}}+n_H}$ (with $\ell_0 = 1$) is defined by

$$T(\bar{z}, \lambda) := \begin{pmatrix} \nabla_{\bar{z}} L(\bar{z}, \ell_0, \lambda) \\ H(z) \end{pmatrix}.$$

Application of Newton's method to (5.23) yields

Algorithm 5.2.2 (Lagrange–Newton method).

(0) Choose $\bar{z}^{(0)} \in \mathbb{R}^{n_{\bar{z}}}$ and $\lambda^{(0)} \in \mathbb{R}^{n_H}$ and set $i := 0$.

(1) If $T(\bar{z}^{(i)}, \lambda^{(i)}) = 0_{\mathbb{R}^{n_{\bar{z}}+n_H}}$, STOP.

(2) Solve the linear equation (with $\ell_0 = 1$)

$$\begin{pmatrix} L''_{\bar{z}\bar{z}}(\bar{z}^{(i)}, \ell_0, \lambda^{(i)}) & H'(\bar{z}^{(i)})^\top \\ H'(\bar{z}^{(i)}) & \Theta \end{pmatrix} \begin{pmatrix} d \\ v \end{pmatrix} = - \begin{pmatrix} \nabla_{\bar{z}} L(\bar{z}^{(i)}, \ell_0, \lambda^{(i)}) \\ H(\bar{z}^{(i)}) \end{pmatrix} \tag{5.24}$$

and set

$$\bar{z}^{(i+1)} := \bar{z}^{(i)} + d, \quad \lambda^{(i+1)} := \lambda^{(i)} + v. \tag{5.25}$$

(3) Set $i \leftarrow i + 1$ and go to (1). □

The well-known convergence results for Newton's method yield

Theorem 5.2.3 (Local convergence of Lagrange–Newton method). *Assumptions:*

(a) *Let* $(\hat{\bar{z}}, \hat{\lambda})$ *be a KKT point.*

(b) *Let* $J, H_j, j = 1, \dots, n_H$, *be twice continuously differentiable with Lipschitz continuous second derivatives* J'' *and* H_j'', $j = 1, \dots, n_H$.

(c) *Let the* KKT matrix *(with* $\ell_0 = 1$*)*

$$\begin{pmatrix} L''_{\bar{z}\bar{z}}(\hat{\bar{z}}, \ell_0, \hat{\lambda}) & H'(\hat{\bar{z}})^\top \\ H'(\hat{\bar{z}}) & \Theta \end{pmatrix} \tag{5.26}$$

be non-singular.

Then there exists $\varepsilon > 0$ *such that the Lagrange–Newton method converges for all* $(\bar{z}^{(0)}, \lambda^{(0)}) \in B_\varepsilon(\hat{\bar{z}}, \hat{\lambda})$. *Furthermore, there exists a constant* $C \geq 0$ *such that*

$$\|(\bar{z}^{(i+1)}, \lambda^{(i+1)}) - (\hat{\bar{z}}, \hat{\lambda})\| \leq C \|(\bar{z}^{(i)}, \lambda^{(i)}) - (\hat{\bar{z}}, \hat{\lambda})\|^2$$

for all sufficiently large i, i. e. the convergence is quadratic. □

The so-called *KKT matrix* in (5.26) is non-singular, if the following conditions are satisfied:

(a) The gradients $\nabla H_j(\hat{\bar{z}}), j = 1, \dots, n_H$, are linearly independent.

(b) It holds

$$v^\top L''_{\bar{z}\bar{z}}(\hat{\bar{z}}, \ell_0, \hat{\lambda}) v > 0 \quad \text{for all } v \in \mathbb{R}^{n_{\bar{z}}}, v \neq 0_{\mathbb{R}^{n_{\bar{z}}}}, \text{ with } H'(\hat{\bar{z}}) v = 0_{\mathbb{R}^{n_H}}.$$

These conditions are actually sufficient for local minimality of $\hat{\bar{z}}$, compare Theorem 6.1.3 in Section 6.1.1.

5.2.2 Sequential Quadratic Programming (SQP)

The SQP method can be considered an extension of the Lagrange–Newton method for optimization problems with inequality constraints. We start with an observation regarding the linear equation (5.24) in step (2) of Algorithm 5.2.2. This equation can be obtained in a different way. To this end, consider the equality constrained optimization problem 5.2.1 and approximate it locally at $(\bar{z}^{(i)}, \lambda^{(i)})$ by the following quadratic optimization problem with $\ell_0 = 1$:

Minimize

$$\frac{1}{2} d^\top L''_{\bar{z}\bar{z}}(\bar{z}^{(i)}, \ell_0, \lambda^{(i)}) d + \nabla J(\bar{z}^{(i)})^\top d$$

with respect to $d \in \mathbb{R}^{n_{\bar{z}}}$ subject to the constraint

$$H(\bar{z}^{(i)}) + H'(\bar{z}^{(i)}) d = 0_{\mathbb{R}^{n_H}}.$$

Evaluation of necessary KKT conditions for the quadratic optimization problem yields

$$\begin{pmatrix} L''_{\bar{z}\bar{z}}(\bar{z}^{(i)}, \ell_0, \lambda^{(i)}) & H'(\bar{z}^{(i)})^\top \\ H'(\bar{z}^{(i)}) & \Theta \end{pmatrix} \begin{pmatrix} d \\ \eta \end{pmatrix} = - \begin{pmatrix} \nabla J(\bar{z}^{(i)}) \\ H(\bar{z}^{(i)}) \end{pmatrix}.$$

Subtracting $H'(\bar{z}^{(i)})^\top \lambda^{(i)}$ on both sides of the first equation yields

$$\begin{pmatrix} L''_{\bar{z}\bar{z}}(\bar{z}^{(i)}, \ell_0, \lambda^{(i)}) & H'(\bar{z}^{(i)})^\top \\ H'(\bar{z}^{(i)}) & \Theta \end{pmatrix} \begin{pmatrix} d \\ \eta - \lambda^{(i)} \end{pmatrix} = - \begin{pmatrix} \nabla_{\bar{z}} L(\bar{z}^{(i)}, \ell_0, \lambda^{(i)}) \\ H(\bar{z}^{(i)}) \end{pmatrix}.$$

A comparison with (5.24) reveals that these two linear equations are identical for $\nu := \eta - \lambda^{(i)}$. The new iterates in (5.25) are then given by

$$\bar{z}^{(i+1)} := \bar{z}^{(i)} + d, \quad \lambda^{(i+1)} := \lambda^{(i)} + \nu = \eta,$$

where η denotes the multiplier of the quadratic optimization problem.

We permit inequality constraints and consider

Problem 5.2.4. Let $J, G_i, H_j : \mathbb{R}^{n_{\bar{z}}} \longrightarrow \mathbb{R}$, $i = 1, \ldots, n_G$, $j = 1, \ldots, n_H$, be twice continuously differentiable and $G := (G_1, \ldots, G_{n_G})^\top$, $H := (H_1, \ldots, H_{n_H})^\top$.

Minimize $J(\bar{z})$ with respect to $\bar{z} \in \mathbb{R}^{n_{\bar{z}}}$ subject to the constraints

$$G_i(\bar{z}) \le 0, \quad i = 1, \ldots, n_G,$$
$$H_j(\bar{z}) = 0, \quad j = 1, \ldots, n_H.$$

At $(\bar{z}^{(i)}, \mu^{(i)}, \lambda^{(i)})$ a local approximation (with $\ell_0 = 1$) is given by

Problem 5.2.5 (Quadratic optimization problem).
Minimize

$$\frac{1}{2} d^\top L''_{\bar{z}\bar{z}}(\bar{z}^{(i)}, \ell_0, \mu^{(i)}, \lambda^{(i)}) d + \nabla J(\bar{z}^{(i)})^\top d$$

with respect to $d \in \mathbb{R}^{n_{\bar{z}}}$ *subject to the constraints*

$$G(\bar{z}^{(i)}) + G'(\bar{z}^{(i)}) d \leq 0_{\mathbb{R}^{n_G}},$$
$$H(\bar{z}^{(i)}) + H'(\bar{z}^{(i)}) d = 0_{\mathbb{R}^{n_H}}.$$

Sequential quadratic approximation yields

Algorithm 5.2.6 (Local SQP method).

(0) Choose $(\bar{z}^{(0)}, \mu^{(0)}, \lambda^{(0)}) \in \mathbb{R}^{n_{\bar{z}}} \times \mathbb{R}^{n_G} \times \mathbb{R}^{n_H}$ and set $i := 0$.
(1) If $(\bar{z}^{(i)}, \mu^{(i)}, \lambda^{(i)})$ is a KKT point of Problem 5.2.4, STOP.
(2) Compute a KKT point $(d^{(i)}, \mu^{(i+1)}, \lambda^{(i+1)}) \in \mathbb{R}^{n_{\bar{z}}} \times \mathbb{R}^{n_G} \times \mathbb{R}^{n_H}$ of the quadratic optimization problem 5.2.5.
(3) Set $\bar{z}^{(i+1)} := z^{(i)} + d^{(i)}$, $i \leftarrow i + 1$, and go to (1). □

It is not necessary to know the index set $A(\hat{\bar{z}})$ of active inequality constraints at a local minimum $\hat{\bar{z}}$ in advance.

The iterates $\bar{z}^{(i)}$ are not necessarily admissible for Problem 5.2.4.

There are powerful algorithms for the numerical solution of quadratic optimization problems using primal or dual active set methods, see [133, 140, 134, 309], or interior-point methods, see [321, 132]. The method in [132] is designed for large-scale and sparse problems. In Section 5.2.3, details of a semi-smooth Newton method for quadratic optimization problems are presented.

The local convergence of the SQP method is established in the following theorem, see [131, Theorem 7.5.4].

Theorem 5.2.7 (Local convergence of SQP method). *Assumptions:*

(a) *Let $\hat{\bar{z}}$ be a local minimum of Problem 5.2.4.*
(b) *Let J, G_i, $i = 1, \ldots, n_G$, and H_j, $j = 1, \ldots, n_H$, be twice continuously differentiable with Lipschitz continuous second derivatives J'', G_i'', $i = 1, \ldots, n_G$, and H_j'', $j = 1, \ldots, n_H$.*
(c) *Let the linear independence constraint qualification 2.3.34 hold at $\hat{\bar{z}}$.*
(d) *Let the strict complementarity condition $\hat{\mu}_i - G_i(\hat{\bar{z}}) > 0$ hold for $i \in A(\hat{\bar{z}})$, where $\hat{\mu}_i$ denotes the Lagrange multiplier for the ith inequality constraint.*
(e) *Let (with $\ell_0 = 1$)*

$$v^\top L''_{\bar{z}\bar{z}}(\hat{\bar{z}}, \ell_0, \hat{\mu}, \hat{\lambda}) v > 0$$

hold for all $v \in \mathbb{R}^{n_{\bar{z}}}$, $v \neq 0_{\mathbb{R}^{n_{\bar{z}}}}$, *with*

$$G_i'(\hat{\bar{z}})v = 0, \quad i \in A(\hat{\bar{z}}), \quad H_j'(\hat{\bar{z}})v = 0, \quad j = 1, \ldots, n_H.$$

Then there exist neighborhoods U *of* $(\hat{\bar{z}}, \hat{\mu}, \hat{\lambda})$ *and* V *of* $(0_{\mathbb{R}^{n_{\bar{z}}}}, \hat{\mu}, \hat{\lambda})$ *such that all quadratic optimization problems* 5.2.5 *have a unique local solution* $d^{(i)}$ *with unique multipliers* $\mu^{(i+1)}$ *and* $\lambda^{(i+1)}$ *in* V *for every* $(\bar{z}^{(0)}, \mu^{(0)}, \lambda^{(0)})$ *in* U.

Moreover, the sequence $\{(\bar{z}^{(i)}, \mu^{(i)}, \lambda^{(i)})\}_{i \in \mathbb{N}_0}$ *converges locally quadratically to* $(\hat{\bar{z}}, \hat{\mu}, \hat{\lambda})$. □

The convergence result shows that the SQP method converges for starting values within some neighborhood of a local minimum of Problem 5.2.4. Unfortunately, in practice, this neighborhood is not known, and it cannot be guaranteed that the starting values are within this neighborhood. Fortunately, the SQP method can be globalized in the sense that it converges for arbitrary starting values (under suitable conditions). The idea is to introduce a step-length parameter (or damping parameter) $t_i > 0$ in step (3) of Algorithm 5.2.6 and use

$$\bar{z}^{(i+1)} = \bar{z}^{(i)} + t_i d^{(i)}$$

as new iterate. The step-length t_i is obtained by performing a so-called *line-search* in direction $d^{(i)}$ for a suitable *merit function* (or *penalty function*). The merit function allows to decide whether the new iterate $\bar{z}^{(i+1)}$ is in some sense better than the old iterate $\bar{z}^{(i)}$. The new iterate is considered to be better than the old one, if either a sufficient decrease in the objective function J or an improvement of the total constraint violation is achieved, while the respective other value is not substantially declined. Improvement of the iterates is typically measured by one of the following merit functions, which depend on a *penalty parameter* $\alpha > 0$:

(a) The non-differentiable ℓ_1-merit function

$$\ell_1(\bar{z}; \alpha) := J(\bar{z}) + \alpha \sum_{i=1}^{n_G} \max\{0, G_i(\bar{z})\} + \alpha \sum_{j=1}^{n_H} |H_j(\bar{z})|$$

was used in [275].

(b) A differentiable merit function is the *augmented Lagrange function*

$$L_\alpha(\bar{z}, \mu, \lambda; \alpha) := J(\bar{z}) + \lambda^\top H(\bar{z}) + \frac{\alpha}{2}\|H(\bar{z})\|^2 + \frac{1}{2\alpha}\sum_{i=1}^{n_G}\left((\max\{0, \mu_i + \alpha G_i(\bar{z})\})^2 - \mu_i^2\right).$$

Both functions are exact under suitable assumptions, i. e. there exists a finite parameter $\hat{\alpha} > 0$, such that every local minimum $\hat{\bar{z}}$ of Problem 5.2.4 is also a local minimum of the merit function for all $\alpha \geq \hat{\alpha}$.

Without specifying all details, a globalized prototype version of the SQP method employing the ℓ_1-merit function and an Armijo rule for step-length determination is summarized in the following algorithm. The algorithm uses symmetric and positive definite matrices B_i, $i = 0, 1, \ldots$, instead of the Hessian matrix $L''_{\bar{z}\bar{z}}$, since the true Hessian matrix may be indefinite leading to a non-convex quadratic optimization problem, which is considerably more difficult to handle than a convex one.

Algorithm 5.2.8 (Globalized SQP method).

(0) Choose $(\bar{z}^{(0)}, \mu^{(0)}, \lambda^{(0)}) \in \mathbb{R}^{n_{\bar{z}}} \times \mathbb{R}^{n_G} \times \mathbb{R}^{n_H}$, $B_0 \in \mathbb{R}^{n_{\bar{z}} \times n_{\bar{z}}}$ symmetric and positive definite, $\alpha > 0$, $\beta \in (0,1)$, $\sigma \in (0,1)$, and set $i := 0$.

(1) If $(\bar{z}^{(i)}, \mu^{(i)}, \lambda^{(i)})$ is a KKT point of Problem 5.2.4, STOP.

(2) Compute a KKT point $(d^{(i)}, \mu^{(i+1)}, \lambda^{(i+1)}) \in \mathbb{R}^{n_{\bar{z}}} \times \mathbb{R}^{n_G} \times \mathbb{R}^{n_H}$ of the quadratic programming problem 5.2.5 with $L''_{\bar{z}\bar{z}}$ replaced by B_i.

(3) Adapt α appropriately.

(4) *Armijo line-search*:
Determine a step-size

$$t_i := \max \left\{ \beta^j \;\middle|\; \begin{array}{l} j \in \{0, 1, 2, \ldots\} \text{ and} \\ \ell_1(\bar{z}^{(i)} + \beta^j d^{(i)}; \alpha) \le \ell_1(\bar{z}^{(i)}; \alpha) + \sigma \beta^j \ell_1'(\bar{z}^{(i)}; d^{(i)}; \alpha) \end{array} \right\}.$$

(5) *Matrix update:*
Compute a suitable symmetric and positive definite matrix B_{i+1}.

(6) Set $\bar{z}^{(i+1)} := \bar{z}^{(i)} + t_i d^{(i)}$, $i \leftarrow i + 1$, and go to (1). □

In practical applications, a suitable value for the penalty parameter α is not known a priori. Strategies for adapting α in step (3) of Algorithm 5.2.8 iteratively and individually for each constraint can be found in [275, 300]. **i**

The quadratic subproblem 5.2.5 may become infeasible owing to the linearization of the nonlinear constraints. In this case, the constraints are typically relaxed using artificial slack variables, see [275, 44]. A convergence analysis for an SQP method using the augmented Lagrange function can be found in [299, 300]. **i**

The line-search procedure using a merit function can be replaced by a *filtering technique*, see [100, 99]. Herein, objective function and constraint violation are decoupled, and progress is measured in a multi-criteria fashion for these criteria. **i**

Depending on the structure of the nonlinear optimization problem 5.2.4, the matrices B_i, $i = 0, 1, \ldots$, in step (5) of Algorithm 5.2.8 are subject to further requirements.

Dense problems

If Problem 5.2.4 is small to medium-scale and dense, where dense refers to the fraction of non-zero elements in the Jacobian and Hessian matrices, then B_i can be updated by the modified BFGS update rule

$$B_{i+1} := B_i + \frac{q^{(i)}(q^{(i)})^\top}{(q^{(i)})^\top s^{(i)}} - \frac{B_i s^{(i)}(s^{(i)})^\top B_i}{(s^{(i)})^\top B_i s^{(i)}}, \tag{5.27}$$

where

$$\begin{aligned} s^{(i)} &= \bar{z}^{(i+1)} - \bar{z}^{(i)}, \\ q^{(i)} &= \theta_i \eta^{(i)} + (1-\theta_i) B_i s^{(i)}, \\ \eta^{(i)} &= \nabla_{\bar{z}} L(\bar{z}^{(i+1)}, \ell_0, \mu^{(i)}, \lambda^{(i)}) - \nabla_{\bar{z}} L(\bar{z}^{(i)}, \ell_0, \mu^{(i)}, \lambda^{(i)}), \\ \theta_i &= \begin{cases} 1, & \text{if } (s^{(i)})^\top \eta^{(i)} \geq 0.2 (s^{(i)})^\top B_i s^{(i)}, \\ \frac{0.8 (s^{(i)})^\top B_i s^{(i)}}{(s^{(i)})^\top B_i s^{(i)} - (s^{(i)})^\top \eta^{(i)}}, & \text{otherwise}. \end{cases} \end{aligned}$$

see [275]. This update formula guarantees that B_{i+1} remains symmetric and positive definite, if B_i was symmetric and positive definite. For $\theta_i = 1$, the well-known BFGS update formula arises, which is used in variable metric methods (or quasi Newton methods) for unconstrained optimization.

Large-scale and sparse problems

If Problem 5.2.4 is large-scale and sparse, then the modified BFGS update is not suitable as it tends to produce dense matrices. One of the following strategies can be applied instead:

(a) Set $B_i := I_{n_{\bar{z}}}$ for $i = 0, 1, \ldots$. The resulting method is a sequential linear programming (SLP) method. Only a linear convergence rate can be achieved with this choice.

(b) Use the regularized exact Hessian

$$B_i := L''_{\bar{z}\bar{z}}(\bar{z}^{(i)}, \ell_0, \mu^{(i)}, \lambda^{(i)}) + \nu_i I_{n_{\bar{z}}}$$

with $\nu_i > 0$ sufficiently large, such that positive definiteness of B_i is guaranteed. This requires efficient techniques to estimate eigenvalues of the sparse Hessian matrix, see [63, 26], and a procedure to choose ν_i appropriately. More precisely, ν_i has to be larger than the modulus of the smallest negative eigenvalue of the Hessian matrix. An estimate of the smallest eigenvalue, which often is not very accurate, though, can be obtained from Gerschgorin circles. Alternatively, ν_i can be adapted iteratively in a trial-and-error fashion until B_i turns out to be positive definite. Notice that modern sparse decomposition methods like MA57 provide information on the inertia of a matrix, i. e., the number of negative, zero and positive eigenvalues, and thus allow checking positive definiteness without computing eigenvalues explicitly.

(c) Use a sparse positive definite update formula, see [97, 101]. Such methods are expensive to realize, though.
(d) Use limited memory BFGS (L-BFGS) in combination with iterative solution methods, see [257, Section 7.2].

Sparsity also requires tailored methods to solve linear equations that arise during the solution of the quadratic subproblem 5.2.5. Common iterative techniques like interior-point methods and active set-methods lead to linear equations with indefinite saddle-point matrices of type

$$\begin{pmatrix} B & A^\top \\ A & -D \end{pmatrix},$$

where A and B are large-scale and sparse matrices, and D is a positive definite diagonal matrix. Linear equations with such matrices can be solved either iteratively, see [288, 298], or directly using sparse decomposition methods like MA57, PARDISO, or SUPERLU, see [85, 69, 296, 297, 295]. Sparse decomposition methods avoid a fill-in in the LU factors by suitable re-ordering strategies, which use permutations of columns and rows for the matrix to be decomposed. Figure 5.3 illustrates the effect using the so-called minimum-degree heuristic. If the standard Gaussian LU decomposition method is applied directly to a sparse matrix, then the LU factors tend to be dense matrices. If re-ordering strategies are applied to the matrix first, then the resulting LU factors still tend to be sparse, allowing for an efficient solution of sparse linear equations.

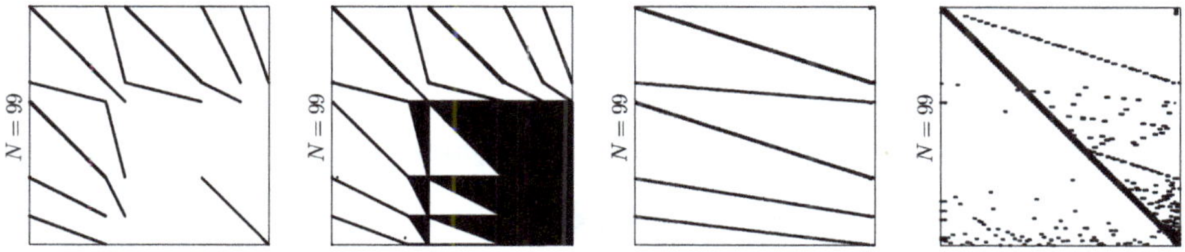

Figure 5.3: From left to right: Matrix A to be decomposed into $A = LU$, factors L (lower triangular part in matrix) and U (upper triangular part in matrix), re-ordered matrix $\tilde{A}$ to be decomposed into $\tilde{A} = \tilde{L}\tilde{U}$, factors $\tilde{L}$ (lower triangular part in matrix) and $\tilde{U}$ (upper triangular part in matrix), see [129].

The following example illustrates the application of an SQP method to a discretized optimal control problem subject to a partial differential equation.

Example 5.2.9 (Boundary control of the 1D wave equation, see [127]). Consider the one-dimensional wave equation

$$u_{tt}(x,t) = c^2 u_{xx}(x,t) \quad \text{in } \Omega = (0,L) \times (0,T) \tag{5.28}$$

$$u(x,0) = u_0(x) \quad \text{on } [0,L] \tag{5.29}$$

$$u_t(x,0) = u_1(x) \quad \text{on } [0,L] \tag{5.30}$$

$$u(0,t) = y_L(t) \quad \text{on } [0,T] \tag{5.31}$$

$$u(L,t) = y_R(t) \quad \text{on } [0,T] \tag{5.32}$$

with constants c, $T > 0$, $L > 0$ and given functions u_0 and u_1. The time-dependent functions y_L and y_R are supposed to be control functions that are applied on the left and right boundary of the domain at $x = 0$ and $x = L$, respectively, and allow to influence the wave equation. The wave equation is a hyperbolic partial differential equation, which models, for instance, a vibrating string.

The task of controlling the wave equation with minimal effort such that the string is at rest at time T leads to the following optimal control problem:

Minimize

$$\int_0^T y_L(t)^2 + y_R(t)^2 dt$$

subject to the wave equation (5.28)–(5.32) and the terminal constraints

$$u(x,T) = u_t(x,T) = 0 \quad \text{on } (0,L).$$

The problem is discretized using equidistant grids in time and space given by $x_i := ih$, $i = 0,\dots,M$, $h = L/M$, and $t_j = jk$, $j = 0,\dots,N$, $k = T/N$ with $M, N \in \mathbb{N}$. The second derivatives in (5.28) are approximated by the second-order finite difference schemes

$$u_{xx}(x_i,t_j) \approx \frac{1}{h^2}(u_{i+1}^j - 2u_i^j + u_{i-1}^j), \quad u_{tt}(x_i,t_j) \approx \frac{1}{k^2}(u_i^{j+1} - 2u_i^j + u_i^{j-1})$$

with $u_i^j \approx u(x_i,t_j)$ for $i = 1,\dots,M-1$, $j = 1,\dots,N-1$.

Introducing these approximations into (5.28) leads to

$$u_i^{j+1} = 2(1-\alpha^2)u_i^j + \alpha^2(u_{i+1}^j + u_{i-1}^j) - u_i^{j-1} \tag{5.33}$$

for $i = 1,\dots,M-1$, $j = 1,\dots,N-1$, where $\alpha := ck/h$.

Condition (5.30) is approximated by second-order central finite differences

$$u_1(x_i) = u_t(x_i,t_0) \approx \frac{1}{2k}(u_i^1 - u_i^{-1}), \quad i = 1,\dots,M-1,$$

which involves the approximation u_i^{-1} at an artificial time point t_{-1}. Combining this with (5.33) and elimination of u_i^{-1} using the combined relation leads to the conditions

$$u_i^1 = (1-\alpha^2)u_i^0 + \frac{\alpha^2}{2}(u_{i+1}^0 + u_{i-1}^0) + ku_1(x_i), \quad i = 1,\dots,M-1. \tag{5.34}$$

With these approximations, we obtain the following fully discretized problem:

Minimize

$$k \sum_{j=0}^{N-1} (y_L^j)^2 + (y_R^j)^2$$

with respect to u_i^j, $i = 0, \dots, M$, $j = 0, \dots, N$, *and* y_L^j, y_R^j, $j = 0, \dots, N$, *subject to the constraints* (5.33), (5.34), *and*

$$\begin{aligned} u_i^0 &= u_0(x_i), && i = 1, \dots, M-1, \\ u_0^j &= y_L^j, && j = 0, \dots, N, \\ u_M^j &= y_R^j, && j = 0, \dots, N, \\ u_i^N &= 0, && i = 1, \dots, M-1, \\ \frac{1}{k}\left(u_i^N - u_i^{N-1}\right) &= 0, && i = 1, \dots, M-1. \end{aligned}$$

We choose the particular data $T = 2\pi$, $L = \pi$, $c = 1$, $u_0(x) = \sin(x)$, $u_1(x) = \cos(x)$, compare [127, Example 2].

Figure 5.5 shows the numerical solution of the discretized problem for $h = k = \pi/140$, which coincides with the exact solution of this problem except at initial and final time and at the point of non-differentiability of y_L and y_R. Figure 5.4 shows the controls y_L and y_R.

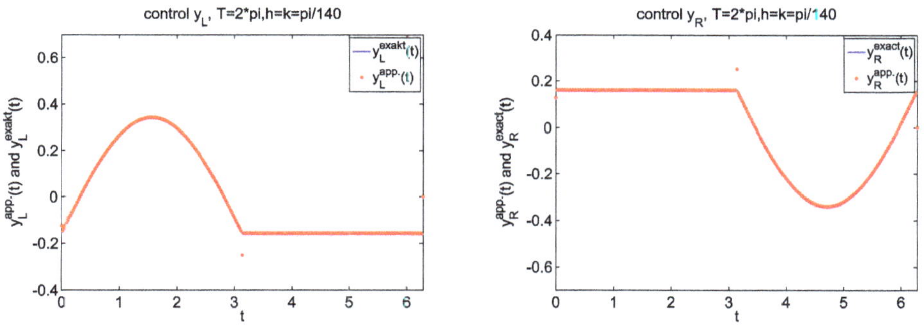

Figure 5.4: Comparison of exact and approximate boundary controls.

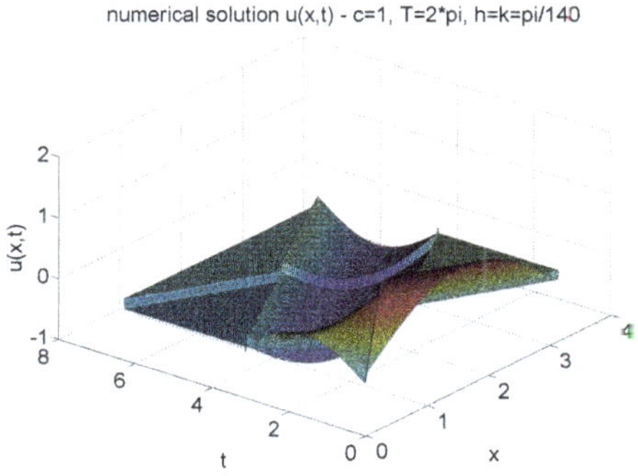

Figure 5.5: Controlling the wave equation to rest using boundary controls.

The SQP method with modified BFGS update, which has been used to solve the discretized problem, generates the following output:

Iter	$J(\bar{z})$	feasibility	step-size	optimality
0	5.620000e+002	2.328e+001		4.756e+001
1	5.824771e+002	1.107e-004	1.000e+000	2.342e+000
2	2.088953e+001	9.561e-005	1.000e+000	3.190e-001
3	2.087705e+001	2.536e-013	1.000e+000	1.051e-005
4	2.087705e+001	1.299e-013	7.000e-001	4.182e-006
5	2.087705e+001	1.373e-013	1.000e+000	4.143e-007

□

5.2.3 A semi-smooth Newton method for quadratic programs

The quadratic optimization problem 5.2.5 is at the core of the SQP method and needs to be solved efficiently. Amongst the various methods for quadratic programs, compare [68, Chapter 3] for an overview, we choose a *semi-smooth Newton method*, see [187, 188, 277, 278, 96, 174, 129]. The semi-smooth Newton method is appealing because it is suitable for small and dense problems and large-scale and sparse problems. It exhibits a locally quadratic convergence rate under suitable assumptions, and it has excellent warm-start properties. The latter is beneficial if a sequence of similar problems has to be solved, e. g., within a model-predictive control framework.

For the data $Q \in \mathbb{R}^{n_d \times n_d}$, $c \in \mathbb{R}^{n_d}$, $A \in \mathbb{R}^{n_A \times n_d}$, $\alpha \in \mathbb{R}^{n_A}$, $B \in \mathbb{R}^{n_B \times n_d}$, $\beta \in \mathbb{R}^{n_B}$, $n_d \in \mathbb{N}$, $n_A, n_B \in \mathbb{N}_0$, we consider a standard quadratic optimization problem:

Problem 5.2.10 (Quadratic program (QP)).
Minimize

$$\frac{1}{2} d^\top Q d + c^\top d$$

with respect to $d \in \mathbb{R}^{n_d}$ *subject to the constraints*

$$Ad = \alpha, \quad Bd \leq \beta.$$

Throughout Q is supposed to be symmetric and positive definite. Hence QP is strictly convex, and the KKT conditions in Corollary 2.3.25 with $\ell_0 = 1$ are necessary and sufficient for a minimum $\hat{d}$ of QP according to Theorem 2.3.36:

$$0 = Q\hat{d} + c + A^\top \lambda + B^\top \mu, \tag{5.35}$$

$$0 = A\hat{d} - \alpha, \tag{5.36}$$

$$0 \leq \mu, \quad \mu^\top (B\hat{d} - \beta) = 0, \quad \beta - B\hat{d} \geq 0. \tag{5.37}$$

The idea of the *semi-smooth Newton method* is to transform these optimality conditions into an equivalent system of nonlinear equations. This can be achieved by applying a so-called *nonlinear complementarity (NCP) function* to the complementarity conditions (5.37). A particular NCP function is the *Fischer–Burmeister function* $\phi_{FB} : \mathbb{R}^2 \longrightarrow \mathbb{R}$, which is defined by

$$\phi_{FB}(a,b) := \sqrt{a^2+b^2} - a - b, \tag{5.38}$$

see [96]. The function ϕ_{FB} has the important property

$$\phi_{FB}(a,b) = 0 \quad \Longleftrightarrow \quad 0 \le a, \quad a \cdot b = 0, \quad b \ge 0.$$

The vector-valued extension $\phi : \mathbb{R}^{n_B} \times \mathbb{R}^{n_B} \longrightarrow \mathbb{R}^{n_B}$, $\phi(a,b) := (\phi_{FB}(a_i,b_i))_{i=1,\dots,n_B}$ allows rewriting the optimality system (5.35)–(5.37) equivalently as

$$0 = T(\hat{d},\lambda,\mu) := \begin{pmatrix} Q\hat{d} + c + A^\top\lambda + B^\top\mu \\ A\hat{d} - \alpha \\ \phi(\beta - B\hat{d}, \mu) \end{pmatrix}. \tag{5.39}$$

Please observe ϕ_{FB} is not continuously differentiable at the origin $(a,b) = (0,0)$, and thus, the system of nonlinear equations (5.39) is non-smooth. However, ϕ_{FB} is Lipschitz continuous and so are ϕ and T. Rademacher's theorem implies that ϕ_{FB}, ϕ, and T are differentiable almost everywhere. This allows us to define the *B(ouligand)-differential* of T by

$$\partial_B T(d,\lambda,\mu) := \left\{ V \mid V = \lim_{\substack{(d_i,\lambda_i,\mu_i)\in D_T \\ (d_i,\lambda_i,\mu_i)\longrightarrow(d,\lambda,\mu)}} T'(d_i,\lambda_i,\mu_i) \right\},$$

where D_T denotes the set of points at which the Jacobian T' exists. The convex hull of the B-differential yields *Clarke's generalized Jacobian*

$$\partial T(d,\lambda,\mu) := \operatorname{conv}(\partial_B T(d,\lambda,\mu)),$$

see [62]. Please note that $\partial_B T$ and ∂T are set-valued mappings. The B-differential and Clarke's generalized Jacobian of ϕ_{FB} and ϕ can be defined accordingly.

Replacing the Jacobian in the classic Newton method by Clarke's generalized Jacobian yields the semi-smooth Newton method.

Algorithm 5.2.11 ((Local) semi-smooth Newton method).

(0) Choose $\eta^{(0)} = (d^{(0)}, \lambda^{(0)}, \mu^{(0)})^\top$ and set $i := 0$.

(1) If $T(\eta^{(i)}) = 0$, STOP.

(2) Choose $V_i \in \partial T(\eta^{(i)})$ and compute a search direction $\delta^{(i)}$ with

$$V_i \delta^{(i)} = -T(\eta^{(i)}).$$

(3) Set $\eta^{(i+1)} := \eta^{(i)} + \delta^{(i)}$, set $i \leftarrow i+1$, and go to (1). □

Under suitable assumptions, the algorithm has the same nice local convergence properties as the classic Newton method, i. e., a quadratic convergence rate. We refer the reader to [187, 188, 277, 278, 96, 129] for a detailed convergence analysis.

Algorithm 5.2.11 can be globalized by adding an Armijo-type line-search for the merit function $\frac{1}{2}\|T(d,\lambda,\mu)\|^2$, see [174].

Algorithm 5.2.12 (Globalized semi-smooth Newton method).

(0) Choose $\eta^{(0)} = (d^{(0)}, \lambda^{(0)}, \mu^{(0)})^\top$, $\sigma \in (0, 1/2)$, $\beta \in (0, 1)$, and set $i := 0$.

(1) If $T(\eta^{(i)}) = 0$, STOP.

(2) Choose $V_i \in \partial T(\eta^{(i)})$ and compute a search direction $\delta^{(i)}$ with

$$V_i \delta^{(i)} = -T(\eta^{(i)}). \tag{5.40}$$

(3) Find step-size $t_i = \max\{\beta^j \mid j \in \mathbb{N}_0\}$, with

$$\frac{1}{2}\|T(\eta^{(i)} + t_i\delta^{(i)})\|^2 \le \frac{1}{2}\|T(\eta^{(i)})\|^2 - \sigma t_i \|T(\eta^{(i)})\|^2 = \frac{1}{2}(1 - 2\sigma t_i)\|T(\eta^{(i)})\|^2.$$

(4) Set $\eta^{(i+1)} := \eta^{(i)} + t_i\delta^{(i)}$, set $i \leftarrow i + 1$, and go to (1). □

It remains to compute Clarke's generalized Jacobian of T. The chain rule for non-differentiable functions in [62, Theorem 2.6.6, Proposition 2.6.2] yields the relation

$$\begin{aligned} &\partial T(d,\lambda,\mu) \\ &\subseteq \left\{ \begin{pmatrix} Q & A^\top & B^\top \\ A & \Theta & \Theta \\ -S(d,\mu)B & \Theta & R(d,\mu) \end{pmatrix} \;\middle|\; \begin{array}{l} S(d,\mu) = \operatorname{diag}(s_j(d,\mu))_{j=1,\ldots,n_B}, \\ R(d,\mu) = \operatorname{diag}(r_j(d,\mu))_{j=1,\ldots,n_B}, \\ (s_j(d,\mu), r_j(d,\mu)) \in \partial\phi_{FB}((\beta - Bd)_j, \mu_j), \\ j = 1,\ldots,n_B \end{array} \right\}. \end{aligned} \tag{5.41}$$

Herein, the B-differential and Clarke's generalized Jacobian of ϕ_{FB} are given by

$$\partial_B \phi_{FB}(a,b) = \begin{cases} \{(\frac{a}{\sqrt{a^2+b^2}} - 1, \frac{b}{\sqrt{a^2+b^2}} - 1)\}, & \text{if } (a,b) \neq (0,0), \\ \{(s,r) \mid (r+1)^2 + (s+1)^2 = 1\}, & \text{if } (a,b) = (0,0), \end{cases}$$

and

$$\partial \phi_{FB}(a,b) = \begin{cases} \{(\frac{a}{\sqrt{a^2+b^2}} - 1, \frac{b}{\sqrt{a^2+b^2}} - 1)\}, & \text{if } (a,b) \neq (0,0), \\ \{(s,r) \mid (r+1)^2 + (s+1)^2 \le 1\}, & \text{if } (a,b) = (0,0), \end{cases}$$

respectively, see Figure 5.6.

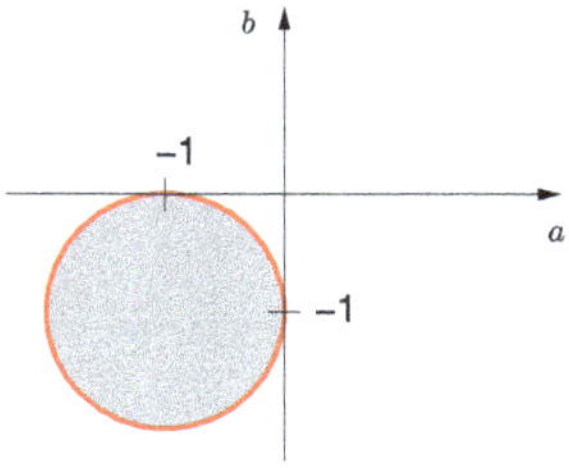

Figure 5.6: B-differential $\partial_B \phi_{FB}$ (red circle) and Clarke's generalized Jacobian $\partial \phi_{FB}$ (shaded disc).

We illustrate the semi-smooth Newton method with an example.

Example 5.2.13. Consider the QP 5.2.10 with $n_d = 2$, $n_A = 0$, $n_B = 6$ and the data

$$Q = \begin{pmatrix} 1 & 0 \\ 0 & 1 \end{pmatrix}, \quad c = \begin{pmatrix} 2 \\ 1 \end{pmatrix}, \quad B = \begin{pmatrix} 1 & 1 \\ -1 & -1 \\ -1 & 1 \\ 1 & 0 \\ 0 & 1 \\ 0 & -1 \end{pmatrix}, \quad \beta = \begin{pmatrix} 5 \\ 0 \\ 2 \\ 5 \\ 2 \\ 1 \end{pmatrix}.$$

The globalized semi-smooth Newton method with initial guess $d^{(0)} = (1,0)^\top$, $\mu^{(0)} = (0,0,0,0,0,0)^\top$, $\sigma = 0.01$, and $\beta = 0.9$ yields the output in Table 5.4. The solution is $\hat{d} = (-0.5, 0.5)^\top$ with Lagrange multiplier $\mu = (0, 1.5, 0, 0, 0, 0)^\top$. The local quadratic convergence rate is nicely visible in the last 3 to 4 iterations.

Table 5.4: Output of the semi-smooth Newton method with globalization: iteration i, residual norm $\|T(\eta^{(i)})\|$, violation in complementarity conditions (comp), constraint violation (con), and step-size t_i.

i	$\|T(\eta^{(i)})\|$	comp	con	t_i
0	3.1622777e+00	0.0000000e+00	0.0000000e+00	–
1	3.4259850e+00	0.0000000e+00	1.6244000e+00	6.5610000e-01
2	7.4164079e-01	7.2000000e-01	6.0000000e-01	1.0000000e+00
3	1.0941842e-01	1.5278640e-01	1.0557281e-01	1.0000000e+00
4	3.7056995e-03	5.5448424e-03	3.7011277e-03	1.0000000e+00
5	4.5661015e-06	6.8491314e-06	4.5660946e-06	1.0000000e+00
6	6.9497741e-12	1.0424578e-11	6.9497186e-12	1.0000000e+00

□

5.3 Exploitation of sparsity in the full discretization approach

The exploitation of sparsity structures is a must within the numerical solution of fully discretized optimal control problems since, otherwise, the computational effort on the linear algebra level for solving large-scale and sparse linear equations using dense LU factorizations would be prohibitively large. In [168], a full discretization approach for

DAE optimal control problems with structure exploitation for an interior point method is investigated. A semi-smooth Newton method with structure exploitation for fully discretized linear-quadratic optimal control problems is used in [39] within a model-predictive control scheme.

For the upcoming structural analysis, it is convenient to investigate structured nonlinear programs of the following type.

Problem 5.3.1 (Structured nonlinear program).
Minimize

$$J(\bar{z}) := \varphi(\bar{z}_0, \bar{z}_N, p)$$

with respect to $\bar{z} = (\bar{z}_0, \dots, \bar{z}_N, p)^\top$ *subject to the constraints*

$$\begin{aligned} H(j, \bar{z}_{j+1}, \bar{z}_j, p) &= 0, \quad j = 0, \dots, N-1, \\ \Psi(\bar{z}_0, \bar{z}_N, p) &= 0, \\ G(j, \bar{z}_j, p) &\leq 0, \quad j = 0, \dots, N. \end{aligned}$$

Problem 5.3.1 has to be understood as a general container. It covers many optimization problems that arise in the full discretization approach, for instance, Problem 5.1.1 and the problems in Examples 5.1.2 and 5.1.3. The vectors $\bar{z}_j, j = 0, \dots, N$, in Problem 5.3.1 are supposed to contain the state vectors *and* the control vectors (or its parameterization) at the grid points, compare Example 5.3.2. For the sake of increased generality, we added the optimization parameter vector p, which may describe, e. g., a free final time or other parameters to be optimized.

Example 5.3.2 (Full discretization with trapezoidal rule). Let us revisit Example 5.1.3 and the full discretization using the trapezoidal rule, where $f_j = f(t_j, x_j, u_j)$ and $f_{j+1} = f(t_{j+1}, x_{j+1}, u_{j+1})$:

Minimize

$$\varphi(x_0, x_N)$$

with respect to $x_0, \dots, x_N, u_0, \dots, u_N$ *subject to the constraints*

$$\begin{aligned} x_j + \frac{h_j}{2}(f_j + f_{j+1}) - x_{j+1} &= 0_{\mathbb{R}^{n_x}}, \quad j = 0, \dots, N-1, \\ \psi(x_0, x_N) &= 0_{\mathbb{R}^{n_\psi}}, \\ c(t_j, x_j, u_j) &\leq 0_{\mathbb{R}^{n_c}}, \quad j = 0, \dots, N, \\ s(t_j, x_j) &\leq 0_{\mathbb{R}^{n_s}}, \quad j = 0, \dots, N. \end{aligned}$$

With the settings $\bar{z}_j := (x_j, u_j)^\top, j = 0, \dots, N, \bar{z} = (\bar{z}_0, \dots, \bar{z}_N)^\top$, and

$$\begin{aligned} J(\bar{z}, p) &= \varphi(x_0, x_N), \\ H(j, \bar{z}_{j+1}, \bar{z}_j, p) &= x_j + \frac{h_j}{2}\big(f(t_j, x_j, u_j) + f(t_{j+1}, x_{j+1}, u_{j+1})\big) - x_{j+1}, \end{aligned}$$

$$\Psi(\bar{z}_0, \bar{z}_N, p) = \psi(x_0, x_N),$$
$$G(j, \bar{z}_j, p) = \begin{pmatrix} c(t_j, x_j, u_j) \\ s(t_j, x_j) \end{pmatrix},$$

a problem occurs, which fits in to class of Problem 5.3.1. □

The Lagrange function (with $\ell_0 = 1$) of Problem 5.3.1 is given by

$$L(\bar{z}, \ell_0, \mu, \lambda, \sigma) = \varphi(\bar{z}_0, \bar{z}_N, p) + \sigma^\top \Psi(\bar{z}_0, \bar{z}_N, p) + \sum_{j=0}^{N-1} \lambda_j^\top H(j, \bar{z}_{j+1}, \bar{z}_j, p) + \sum_{j=0}^{N} \mu_j^\top G(j, \bar{z}_j, p).$$

At a given $(\bar{z}, \lambda, \mu, \sigma)$, define

$$\begin{aligned}
L_{j,j} &:= L''_{\bar{z}_j,\bar{z}_j}(\bar{z}, \ell_0, \mu, \lambda, \sigma), & L_{j,p} &:= L''_{\bar{z}_j,p}(\bar{z}, \ell_0, \mu, \lambda, \sigma), & j &= 0, \dots, N,\\
G_z[j] &:= G'_{\bar{z}_j}(j, \bar{z}_j, p), & G_p[j] &:= G'_p(j, \bar{z}_j, p), & j &= 0, \dots, N,\\
L_{j,j+1} &:= L''_{\bar{z}_j,\bar{z}_{j+1}}(\bar{z}, \ell_0, \mu, \lambda, \sigma), & H_{z_j}[j] &:= H'_{\bar{z}_j}(j, \bar{z}_{j+1}, \bar{z}_j, p), & j &= 0, \dots, N-1,\\
H_{z_{j+1}}[j] &:= H'_{\bar{z}_{j+1}}(j, \bar{z}_{j+1}, \bar{z}_j, p), & H_p[j] &:= H'_p(j, \bar{z}_{j+1}, \bar{z}_j, p), & j &= 0, \dots, N-1,\\
L_{0,N} &:= L''_{\bar{z}_0,\bar{z}_N}(\bar{z}, \ell_0, \mu, \lambda, \sigma), & L_{p,p} &:= L''_{p,p}(\bar{z}, \ell_0, \mu, \lambda, \sigma), & &\\
\Psi_0 &:= \Psi'_{\bar{z}_0}(\bar{z}_0, \bar{z}_N, p), & \Psi_N &:= \Psi'_{\bar{z}_N}(\bar{z}_0, \bar{z}_N, p), & \Psi_p &:= \Psi'_p(\bar{z}_0, \bar{z}_N, p),
\end{aligned}$$

and

$$\begin{aligned}
\alpha_N &:= -\Psi(\bar{z}_0, \bar{z}_N, p), & \alpha_j &:= -H(j, \bar{z}_{j+1}, \bar{z}_j, p), & j &= 0, \dots, N-1,\\
\beta_j &:= -G(j, \bar{z}_j, p), & c_j &:= \nabla_{\bar{z}_j} \varphi(\bar{z}_0, \bar{z}_N, p), & j &= 0, \dots, N,\\
c_p &:= \nabla_p \varphi(\bar{z}_0, \bar{z}_N, p). & & & &
\end{aligned}$$

The Hessian matrix of the Lagrange function is a large-scale and sparse matrix:

$$L''_{\bar{z},\bar{z}}(\bar{z}, \ell_0, \mu, \lambda, \sigma) = \left(\begin{array}{cccc|c}
L_{0,0} & L_{0,1} & & L_{0,N} & L_{0,p}\\
L_{0,1}^\top & \ddots & \ddots & & L_{1,p}\\
 & \ddots & \ddots & L_{N-1,N} & \vdots\\
L_{0,N}^\top & & L_{N-1,N}^\top & L_{N,N} & L_{N,p}\\
\hline
L_{0,p}^\top & L_{1,p}^\top & \cdots & L_{N,p}^\top & L_{p,p}
\end{array}\right).$$

With these settings, the quadratic optimization problem 5.2.5 within the SQP method 5.2.6 for Problem 5.3.1 reads as follows.

Problem 5.3.3 (Structured quadratic program).
Minimize

$$\frac{1}{2}d_p^\top L_{p,p}d_p + d_0^\top L_{0,N}d_N + c_p^\top d_p + \sum_{j=0}^{N} \frac{1}{2}d_j^\top L_{j,j}d_j + d_j^\top L_{j,p}d_p + c_j^\top d_j + \sum_{j=0}^{N-1} d_j^\top L_{j,j+1}d_{j+1}$$

with respect to $d = (d_0, \dots, d_N, d_p)^\top$ *subject to the constraints*

$$\begin{aligned} H_{z_j}[j]d_j + H_{z_{j+1}}[j]d_{j+1} + H_p[j]d_p &= a_j, \quad j = 0, \dots, N-1, \\ \Psi_0 d_0 + \Psi_N d_N + \Psi_p d_p &= a_N, \\ G_z[j]d_j + G_p[j]d_p &\le \beta_j, \quad j = 0, \dots, N. \end{aligned}$$

The structured QP 5.3.3 is a special case of Problem 5.2.10, and the semi-smooth Newton method from Section 5.2.3 can be applied. In each iteration, the linear equation in (5.40) with an element of the generalized Jacobian in (5.41) has to be solved. This could be achieved using suitable sparse LU factorization software packages like MA48 or SuperLU without further structure exploitation. Please note that the matrix in (5.41) is non-symmetric.

For Problem 5.3.3, however, one can further exploit the structure. To this end, let the multipliers for the equality and inequality constraints in Problem 5.3.3 be denoted by $\tilde{\lambda}_j$, $j = 0, \dots, N-1$, $\tilde{\sigma}$, and $\tilde{\mu}_j$, $j = 0, \dots, N$, respectively. A thorough inspection of the linear equation system in (5.40) reveals that the linear equation system can be re-arranged using column and row permutations. The resulting equivalent linear equation system for the (permuted) search direction $\delta = (\delta_0, \dots, \delta_N, \delta_{\tilde{\sigma}}, \delta_{d_p})^\top$ with $\delta_j = (\delta_{d_j}, \delta_{\tilde{\mu}_j}, \delta_{\tilde{\lambda}_j})^\top$, $j = 0, \dots, N-1$, $\delta_N = (\delta_{d_N}, \delta_{\tilde{\mu}_N})^\top$, reads as follows:

$$\left(\begin{array}{cccc|c|c} \Gamma_0 & \Omega_0 & & L_{0,N} & \Delta_0^\top & \tilde{\Sigma}_0 \\ \Omega_0^\top & \Gamma_1 & \ddots & & & \tilde{\Sigma}_1 \\ & \ddots & \ddots & \Omega_{N-1} & & \vdots \\ L_{0,N}^\top & & \Omega_{N-1}^\top & \Gamma_N & \Delta_N^\top & \tilde{\Sigma}_N \\ \hline \Delta_0 & & & \Delta_N & & \Psi_p \\ \hline \Sigma_0 & \Sigma_1 & \cdots & \Sigma_N & \Psi_p^\top & L_{p,p} \end{array}\right) \left(\begin{array}{c} \delta_0 \\ \delta_1 \\ \vdots \\ \delta_N \\ \hline \delta_{\tilde{\sigma}} \\ \hline \delta_{d_p} \end{array}\right) = -\left(\begin{array}{c} r_0 \\ r_1 \\ \vdots \\ r_N \\ \hline r_\Psi \\ \hline r_p \end{array}\right). \tag{5.42}$$

Herein, we have $\Delta_0 = (\Psi_0 \quad \Theta \quad \Theta)$, $\Delta_N = (\Psi_N \quad \Theta)$, and

$$\Gamma_j = \begin{pmatrix} L_{j,j} & G_z[j]^\top & H_{z_j}[j]^\top \\ -S_j G_z[j] & R_j & \Theta \\ H_{z_j}[j] & \Theta & \Theta \end{pmatrix}, \quad j = 0, \dots, N-1,$$

$$\Omega_j = \begin{pmatrix} L_{j,j+1} & \Theta & \Theta \\ \Theta & \Theta & \Theta \\ H_{z_{j+1}}[j] & \Theta & \Theta \end{pmatrix}, \quad j = 0, \dots, N-2,$$

$$\Sigma_j = \begin{pmatrix} L_{j,p}^\top & G_p[j]^\top & H_p[j]^\top \end{pmatrix}, \qquad j = 0, \dots, N-1,$$

$$\tilde{\Sigma}_j = \begin{pmatrix} L_{j,p} \\ -S_j G_p[j] \\ H_p[j] \end{pmatrix}, \qquad j = 0, \dots, N-1,$$

and

$$\Gamma_N = \begin{pmatrix} L_{N,N} & G_z[N]^\top \\ -S_N G_z[N] & R_N \end{pmatrix}, \quad \Omega_{N-1} = \begin{pmatrix} L_{N-1,N} & \Theta \\ \Theta & \Theta \\ H_{z_N}[N-1] & \Theta \end{pmatrix},$$

$$\Sigma_N = \begin{pmatrix} L_{N,p}^\top & G_p[N]^\top \end{pmatrix}, \qquad \tilde{\Sigma}_N = \begin{pmatrix} L_{N,p} \\ -S_N G_p[N] \end{pmatrix},$$

where

$$(S_j, R_j) \in \partial\phi(\beta_j - G_z[j]d_j^{(i)} - G_p[j]d_p^{(i)}, \tilde{\mu}_j^{(i)}), \quad j = 0, \dots, N.$$

The right-hand side in (5.42) is given by

$$r_j = \begin{pmatrix} \nabla_{d_j} \tilde{L}(d^{(i)}, \ell_0, \tilde{\mu}^{(i)}, \tilde{\lambda}^{(i)}, \tilde{\sigma}^{(i)}) \\ \phi(\beta_j - G_z[j]d_j^{(i)} - G_p[j]d_p^{(i)}, \tilde{\mu}_j^{(i)}) \\ H_{z_j}[j]d_j^{(i)} + H_{z_{j+1}}[j]d_{j+1}^{(i)} + H_p[j]d_p^{(i)} - \alpha_j \end{pmatrix}, \quad j = 0, \dots, N-1,$$

$$r_N = \begin{pmatrix} \nabla_{d_N} \tilde{L}(d^{(i)}, \ell_0, \tilde{\mu}^{(i)}, \tilde{\lambda}^{(i)}, \tilde{\sigma}^{(i)}) \\ \phi(\beta_N - G_z[N]d_N^{(i)} - G_p[N]d_p^{(i)}, \tilde{\mu}_N^{(i)}) \end{pmatrix},$$

$$r_\Psi = \Psi_0 d_0^{(i)} + \Psi_N d_N^{(i)} + \Psi_p d_p^{(i)} - \alpha_N,$$

$$r_p = \nabla_{d_p} \tilde{L}(d^{(i)}, \ell_0, \tilde{\mu}^{(i)}, \tilde{\lambda}^{(i)}, \tilde{\sigma}^{(i)}),$$

where $\tilde{L}$ (with $\ell_0 = 1$) denotes the Lagrange function of Problem 5.3.3, and $d^{(i)} = (d_0^{(i)}, \dots, d_N^{(i)}, d_p^{(i)})^\top$, $\tilde{\mu}^{(i)}$, $\tilde{\lambda}^{(i)}$, $\tilde{\sigma}^{(i)}$ denotes the ith iterate of the semi-smooth Newton method for Problem 5.3.3.

5.3.1 Special cases

In many applications, objective function φ and boundary conditions are separated, that is,

$$\varphi(\bar{z}_0, \bar{z}_N, p) = \varphi_0(\bar{z}_0, p) + \varphi_N(\bar{z}_N, p), \quad \Psi(\bar{z}_0, \bar{z}_N, p) = \begin{pmatrix} \Psi_a(\bar{z}_0, p) \\ \Psi_b(\bar{z}_N, p) \end{pmatrix}.$$

As a consequence, the matrix $L_{0,N}$ is equal to zero, and the Jacobians of Ψ with respect to $\bar{z}_0$, $\bar{z}_N$, and p are

$$\Psi_0 = \begin{pmatrix} \Psi'_{a,\bar{z}_0}(\bar{z}_0, p) \\ \Theta \end{pmatrix}, \quad \Psi_N = \begin{pmatrix} \Theta \\ \Psi'_{b,\bar{z}_N}(\bar{z}_N, p) \end{pmatrix}, \quad \Psi_p = \begin{pmatrix} \Psi_{a,p} \\ \Psi_{b,p} \end{pmatrix}.$$

Let $\delta_{\bar{\sigma}} = (\delta_{\bar{\sigma}_a}, \delta_{\bar{\sigma}_b})^\top$, $\Delta_a = (\Psi'_{a,\bar{z}_0}(\bar{z}_0, p) \quad \Theta \quad \Theta)$, $\Delta_b = (\Psi'_{b,\bar{z}_N}(\bar{z}_N, p) \quad \Theta)$, and $r_\Psi = \begin{pmatrix} r_{\Psi_a} \\ r_{\Psi_b} \end{pmatrix}$. Exploiting this special case and performing a row and column permutation for the row with Δ_a, and the column corresponding to δ_{σ_a} in (5.42) yields the linear equation system

$$\left(\begin{array}{c|cccc|c|c} & \Delta_a & & & & & \Psi_{a,p} \\ \hline \Delta_a^\top & \Gamma_0 & \Omega_0 & & & & \tilde{\Sigma}_0 \\ & \Omega_0^\top & \Gamma_1 & \ddots & & & \tilde{\Sigma}_1 \\ & & \ddots & \ddots & \Omega_{N-1} & & \vdots \\ & & & \Omega_{N-1}^\top & \Gamma_N & \Delta_b^\top & \tilde{\Sigma}_N \\ \hline & & & & \Delta_b & & \Psi_{b,p} \\ \hline \Psi_{a,p}^\top & \Sigma_0 & \Sigma_1 & \cdots & \Sigma_N & \Psi_{b,p}^\top & L_{p,p} \end{array}\right) \begin{pmatrix} \delta_{\bar{\sigma}_a} \\ \hline \delta_0 \\ \delta_1 \\ \vdots \\ \delta_N \\ \hline \delta_{\bar{\sigma}_b} \\ \hline \delta_{d_p} \end{pmatrix} = - \begin{pmatrix} r_{\Psi_a} \\ \hline r_0 \\ r_1 \\ \vdots \\ r_N \\ \hline r_{\Psi_b} \\ \hline r_p \end{pmatrix}. \tag{5.43}$$

Moreover, if there are no parameters present in the optimization problems, then (5.43) further reduces to a linear equation system with a *block-banded matrix*:

$$\left(\begin{array}{c|cccc|c} & \Delta_a & & & & \\ \hline \Delta_a^\top & \Gamma_0 & \Omega_0 & & & \\ & \Omega_0^\top & \Gamma_1 & \ddots & & \\ & & \ddots & \ddots & \Omega_{N-1} & \\ & & & \Omega_{N-1}^\top & \Gamma_N & \Delta_b^\top \\ \hline & & & & \Delta_b & \end{array}\right) \begin{pmatrix} \delta_{\bar{\sigma}_a} \\ \hline \delta_0 \\ \delta_1 \\ \vdots \\ \delta_N \\ \hline \delta_{\bar{\sigma}_b} \end{pmatrix} = - \begin{pmatrix} r_{\Psi_a} \\ \hline r_0 \\ r_1 \\ \vdots \\ r_N \\ \hline r_{\Psi_b} \end{pmatrix}. \tag{5.44}$$

The linear equation system in (5.44) can be solved efficiently by a banded LU factorization, e. g., the LAPACK routine DGBTRF from https://www.netlib.org/lapack/. The bandwidth of the matrix merely depends on the number of states, controls, and constraints.

5.3.2 Solving structured linear equation systems

The linear equation systems (5.42) and (5.43) can be partitioned as follows:

$$\begin{pmatrix} M & D \\ C & W \end{pmatrix} \begin{pmatrix} x \\ y \end{pmatrix} = \begin{pmatrix} a \\ b \end{pmatrix}.$$

We assume:

(i) The matrix M is non-singular, large-scale, and banded with moderate bandwidth.

(ii) The matrix W is low-dimensional and dense.

(iii) The matrix $W - CM^{-1}D$ is non-singular.

The first assumption implies that M can be factorized (and inverted) efficiently. Then, solving the first equation yields

$$x = M^{-1}(a - Dy).$$

Introducing x into the second equation yields

$$(W - CM^{-1}D)y = b - CM^{-1}a.$$

The matrix $W - CM^{-1}D$ is low-dimensional and can be factorized efficiently.

Example 5.3.4 (Compare [87]). We illustrate the semi-smooth Newton method for a structured QP of type 5.3.3, which arises in a path tracking task with obstacle avoidance for the state $x_j \in \mathbb{R}^4$, the control $u_j \in \mathbb{R}$, $j = 0, \ldots, N$, and a discretization by the trapezoidal rule with step-size $h = t_f/N$, $N \in \mathbb{N}$. The resulting structured QP reads as follows:

Minimize

$$\frac{h}{4}\left(x_0^\top Q x_0 + \frac{1}{2}u_0^2\right) + \frac{h}{2}\sum_{j=1}^{N-1}\left(x_j^\top Q x_j + \frac{1}{2}u_j^2\right) + \frac{h}{4}\left(x_N^\top Q x_N + \frac{1}{2}u_N^2\right)$$

with respect to (x_j, u_j), $j = 0, \ldots, N$, *subject to the constraints*

$$\begin{pmatrix} I - \frac{h}{2}A & -\frac{h}{2}B \end{pmatrix}\begin{pmatrix} x_{j+1} \\ u_{j+1} \end{pmatrix} - \begin{pmatrix} I + \frac{h}{2}A & \frac{h}{2}B \end{pmatrix}\begin{pmatrix} x_j \\ u_j \end{pmatrix} = a$$

for $j = 0, \ldots, N-1$ *and*

$$Gx_j \le g_j, \quad |u_j| \le u_{\max}, \quad j = 0, \ldots, N, \quad x_0 = 0_{\mathbb{R}^4}.$$

Herein

$$A = \begin{pmatrix} 0 & V & 0 & -V \\ 0 & 0 & V & 0 \\ 0 & 0 & 0 & 0 \\ 0 & 0 & 0 & 0 \end{pmatrix}, \quad B = \begin{pmatrix} 0 \\ 0 \\ 1 \\ 0 \end{pmatrix}, \quad a = \begin{pmatrix} 0 \\ 0 \\ 0 \\ h \cdot V \cdot \kappa \end{pmatrix}$$

with constants V and κ, and

$$Q = \begin{pmatrix} 1 & 0 & 0 & 0 \\ 0 & 1 & 0 & -1 \\ 0 & 0 & 0 & 0 \\ 0 & -1 & 0 & 1 \end{pmatrix}, \quad G = \begin{pmatrix} 1 & 0 & 0 & 0 \\ -1 & 0 & 0 & 0 \\ 0 & 0 & 1 & 0 \\ 0 & 0 & -1 & 0 \end{pmatrix}, \quad g_j = \begin{pmatrix} r_{\max,j} \\ -r_{\min,j} \\ \kappa_{\max} \\ -\kappa_{\min} \end{pmatrix}.$$

Linear equation systems of type (5.44) have to be solved within the semi-smooth Newton method. Figure 5.7 shows the results for $t_f = 10$, $V = 15$, $\kappa = 0.02$, $u_{\max} = 0.3$, $\kappa_{\max} = -\kappa_{\min} = 0.1$, and

$$r_{\max,j} = \begin{cases} -1, & \text{if } j \cdot h \cdot V \in [100, 110], \\ 4, & \text{otherwise,} \end{cases} \qquad r_{\min,j} = \begin{cases} 1, & \text{if } j \cdot h \cdot V \in [40, 50], \\ -4, & \text{otherwise} \end{cases}$$

for $j = 0, \ldots, N$.

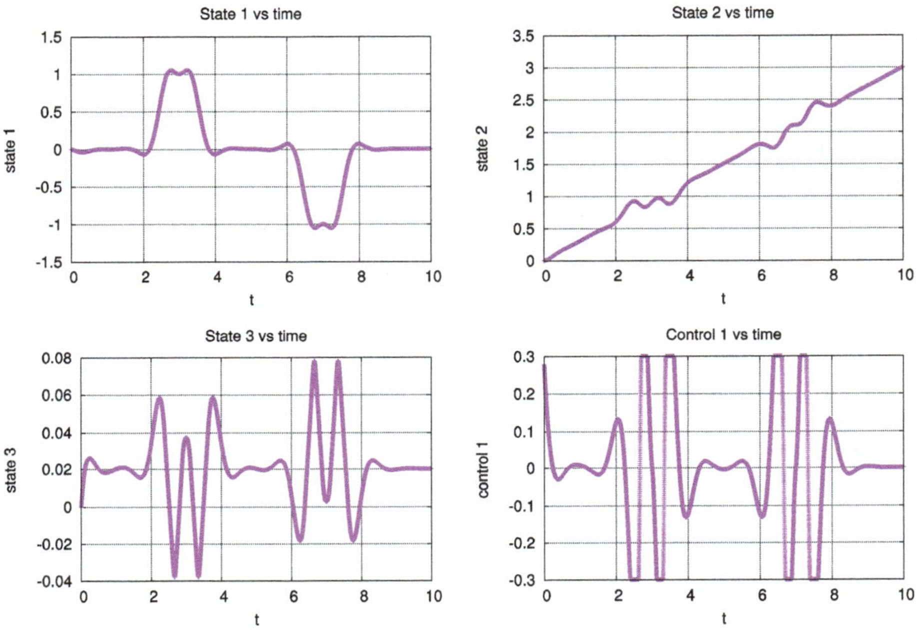

Figure 5.7: States one to three and control for $N = 16000$.

Table 5.5 shows the CPU times and the number of iterations until convergence for the semi-smooth Newton method. □

5.4 Calculation of derivatives for reduced discretization

The application of the SQP method to the discretized optimal control problems 5.1.1 and 5.1.4, respectively, is straightforward provided the derivatives J', G', H', and $L''_{\bar{z}\bar{z}}$ are available. Efficient procedures for the calculation of the Hessian matrix $L''_{\bar{z}\bar{z}}$ for a large-scale and sparse problem such as Problem 5.1.1 are beyond the scope of this book and can be found in [63, 26]. For dense problems such as Problem 5.1.4 it is assumed that the Hessian is replaced by the modified BFGS update formula in (5.27).

Table 5.5: CPU times for the semi-smooth Newton method on a PC with Intel®Core©i7-7700K CPU with 4.2 GHz.

N	CPU time in [s]	#iterations
30	0.001958	7
60	0.004341	8
125	0.013193	9
250	0.056735	11
500	0.168870	16
1000	0.411074	20
2000	1.116120	23
4000	2.949900	24
8000	9.280530	33
16000	41.61100	44

Hence, we focus on the calculation of J', G', H'. In the case of the full discretization approach 5.1.1, the derivatives J' G', H' are easily computed according to (5.9)–(5.11), but for the reduced discretization approach 5.1.4, it is more involved to compute these derivatives, since sensitivity information of the numerical solution with respect to parameters is required, compare [224, 59, 58, 93, 177, 117].

Different approaches exist:

(a) The *sensitivity differential equation approach* is advantageous if the number of constraints is (much) larger than the number of variables in the discretized problem.

(b) The *adjoint equation approach* is preferable if the number of constraints is less than the number of variables in the discretized problem.

(c) A powerful tool for the evaluation of derivatives is *algorithmic differentiation*. This approach assumes that the evaluation of a function is performed by a FORTRAN or C procedure (or any other supported programming language). Algorithmic differentiation means that the complete procedure is differentiated step-by-step using, roughly speaking, chain and product rules. The result is again a FORTRAN or C procedure that provides the derivative of the function. Essentially, the so-called *forward mode* in algorithmic differentiation corresponds to the sensitivity equation approach, while the *backward mode* corresponds to the adjoint approach. Further details can be found on the web page www.autodiff.org and in [146, 147].

(d) The approximation by *finite differences* is straightforward but has the drawback of being computationally expensive and often suffers from low accuracy. Nevertheless, this approach is often used, if some solver depending on optimization variables is used as a black box inside an algorithm.

The first two approaches for calculating derivatives in the reduced discretization approach are discussed in detail subsequently. For both approaches, there is a *discrete version* dealing with the discretized dynamics only and a *continuous version* dealing with the dynamics as a function of time directly. The continuous version eventually requires to solve a *sensitivity DAE* or an *adjoint DAE*, respectively. The discrete version, in turn, can be viewed as a discretization of the continuous sensitivity or adjoint equation.

5.4.1 Sensitivity equation approach

Consider the one-step method

$$Z_{i+1}(\bar{z}) = Z_i(\bar{z}) + h_i \Phi(t_i, Z_i(\bar{z}), w, h_i), \quad i = 0, 1, \dots, N-1. \tag{5.45}$$

Recall that $\bar{z} = (z_0, w)^\top \in \mathbb{R}^{n_z+M}$ denotes the variables in the optimization problem 5.1.4. We intend to compute the sensitivities

$$S_i := Z_i'(\bar{z}), \quad i = 0, 1, \dots, N.$$

For $i = 0$, it holds

$$S_0 = Z_0'(\bar{z}) \in \mathbb{R}^{n_z \times (n_z+M)}, \tag{5.46}$$

where $Z_0(\bar{z})$ is supposed to be a continuously differentiable function that provides a consistent initial value, compare Section 4.5.1. Differentiation of (5.45) with respect to $\bar{z}$ yields the relationship

$$S_{i+1} = S_i + h_i \left(\Phi_z'(t_i, Z_i(\bar{z}), w, h_i) \cdot S_i + \Phi_w'(t_i, Z_i(\bar{z}), w, h_i) \cdot \frac{\partial w}{\partial \bar{z}} \right) \tag{5.47}$$

for $i = 0, 1, \dots, N-1$ with $\frac{\partial w}{\partial \bar{z}} := (0_{\mathbb{R}^{M \times n_z}} | I_M \ )$. This approach is known as *internal numerical differentiation (IND)*, see [30]. The IND-approach is based on the differentiation of the discretization scheme (the one-step method) with respect to $\bar{z}$. Computing the derivatives for the reduced problem 5.1.4 essentially amounts to solving one initial value problem of size $n_z(1 + n_z + M)$. It is worth pointing out that the size of the sensitivity equation in (5.47) depends on the number of unknowns in the optimization problem, but it does not depend on the number of constraints.

The computation of the derivatives Φ_z' and Φ_w' can be non-trivial and will be explained in detail at the end of Subsection 5.4.2.

Next, we point out the relation of the IND approach to the continuous sensitivity DAE, which has been used in Section 4.6 already for shooting techniques. To this end, we use the implicit Euler method and consider the integration step $t_i \longrightarrow t_{i+1} = t_i + h_i$. Discretization of (5.1) with $Z_i \approx z(t_i)$, $Z_{i+1} \approx z(t_{i+1})$ leads to the nonlinear equation

$$F\left(t_{i+1}, Z_{i+1}, \frac{Z_{i+1} - Z_i}{h_i}, u_M(t_{i+1}; w) \right) = 0_{\mathbb{R}^{n_z}}. \tag{5.48}$$

Provided that this nonlinear equation possesses a solution and we may apply the implicit function theorem, this solution will depend on $\bar{z}$, i. e. $Z_{i+1} = Z_{i+1}(\bar{z})$. Now, we need the derivative of the numerical solution $Z_{i+1}(\bar{z})$ with respect to $\bar{z}$, i. e. S_{i+1}. Differentiation of (5.48) with respect to $\bar{z}$ yields the linear equation

$$\left(F'_z + \frac{1}{h_i}F'_{\dot{z}}\right)\bigg|_{Z_{i+1}} \cdot S_{i+1} = -F'_u|_{Z_{i+1}} \cdot u'_{M,\bar{z}}(t_{i+1}; w) + \frac{1}{h_i}F'_{\dot{z}}\bigg|_{Z_{i+1}} \cdot S_i \tag{5.49}$$

by the implicit function theorem.

This formula can be obtained in a different way. Let $z(t;\bar{z})$ denote the solution of the DAE (5.1) for given $\bar{z}$. If F is sufficiently smooth, z is continuously differentiable with respect to t and $\bar{z}$, and if

$$\frac{\partial}{\partial \bar{z}} \frac{dz(t;\bar{z})}{dt} = \frac{d}{dt} \frac{\partial z(t;\bar{z})}{\partial \bar{z}},$$

then differentiation of (5.1) with respect to $\bar{z}$ results in a linear matrix DAE – the *sensitivity DAE*

$$F'_z[t] \cdot S(t) + F'_{\dot{z}}[t] \cdot \dot{S}(t) + F'_u[t] \cdot u'_{M,\bar{z}}(t; w) = 0_{\mathbb{R}^{n_z \times n_z}} \tag{5.50}$$

for the *sensitivity matrix* $S(t) := \partial z(t;\bar{z})/\partial \bar{z}$. Discretization of the sensitivity DAE with the same method as in (5.48) and for the same time step leads again to the linear equation (5.49). Hence, both approaches coincide, provided that Z_{i+1} solves (5.48) exactly. Equation (5.48) is solved numerically by Newton's method

$$\left(F'_z + \frac{1}{h_i}F'_{\dot{z}}\right)\bigg|_{Z^{(k)}_{i+1}} \cdot \Delta Z^{(k)}_{i+1} = -F|_{Z^{(k)}_{i+1}}, \quad Z^{(k+1)}_{i+1} = Z^{(k)}_{i+1} + \Delta Z^{(k)}_{i+1}, \quad k = 0, 1, 2, \ldots.$$

Notice that the iteration matrix $F'_z + \frac{1}{h_i}F'_{\dot{z}}$ at the last iterate can be re-used for computing S_{i+1}.

In practice, deviations in the sensitivities in (5.47), which are obtained by the IND approach, and the sensitivities obtained by solving the sensitivity DAE (5.50) directly using a state-of-the-art DAE integrator may occur, because state-of-the-art DAE integrators use automatic step-size and order selection algorithms, only perform few Newton steps, and keep the iteration matrix within the Newton method constant for several Newton steps or even several integration steps to speed up the procedure. The overall integration procedure then typically becomes non-differentiable with respect to parameters. Hence, the so-computed sensitivity S_{i+1} is only an approximation of the correct derivative $Z'_{i+1}(\bar{z})$ from the IND approach. Numerical experiments show that the accuracy of this sensitivity matrix approximation is often too low to use in a gradient-based optimization procedure.

In order to increase accuracy, it is important to re-evaluate the iteration matrix in the Newton method in each integration step prior to the calculation of the sensitivity matrix S_{i+1}.

---

i The IND approach is also applicable for multi-step method such as BDF. Similar strategies using sensitivity equations are discussed in [59, 224, 35, 160, 177]. A comparison of strategies can be found in [93].

5.4.2 Adjoint equation approach: the discrete case

The adjoint method avoids the calculation of the sensitivities S_i. We demonstrate the method for a prototype function of type

$$\Gamma(\bar{z}) := \gamma(Z_0(\bar{z}), Z_N(\bar{z}), \bar{z}).$$

Obviously, φ and ψ in (5.13) and (5.15) are of this type. The components of G in (5.14) also are of this type, if $Z_0(\bar{z})$ is neglected and $Z_i(\bar{z})$ is used instead of $Z_N(\bar{z})$.

We intend to derive a procedure for calculating $\Gamma'(\bar{z})$ subject to the difference equations

$$Z_{i+1}(\bar{z}) - Z_i(\bar{z}) - h_i\Phi(t_i, Z_i(\bar{z}), w, h_i) = 0_{\mathbb{R}^{n_z}}, \quad i = 0, \dots, N-1. \tag{5.51}$$

The initial state $Z_0(\bar{z}) \in \mathbb{R}^{n_z}$ is assumed to be sufficiently smooth as a function of $\bar{z}$ and consistent with the underlying DAE for all $\bar{z}$.

Consider the auxiliary functional

$$\Gamma_a(\bar{z}) := \Gamma(\bar{z}) + \sum_{i=0}^{N-1} \lambda_{i+1}^\top (Z_{i+1}(\bar{z}) - Z_i(\bar{z}) - h_i\Phi(t_i, Z_i(\bar{z}), w, h_i))$$

with multipliers λ_i, $i = 1, \dots, N$. Differentiating Γ_a with respect to $\bar{z}$ yields the expression

$$\begin{aligned}
\Gamma_a'(\bar{z}) &= \gamma_{z_0}' \cdot S_0 + \gamma_{z_N}' \cdot S_N + \gamma_{\bar{z}}' \\
&\quad + \sum_{i=0}^{N-1} \lambda_{i+1}^\top \left(S_{i+1} - S_i - h_i\Phi_z'[t_i] \cdot S_i - h_i\Phi_w'[t_i] \cdot \frac{\partial w}{\partial \bar{z}} \right) \\
&= \gamma_{z_0}' \cdot S_0 + \gamma_{z_N}' \cdot S_N + \gamma_{\bar{z}}' \\
&\quad + \sum_{i=1}^{N} \lambda_i^\top S_i - \sum_{i=0}^{N-1} \lambda_{i+1}^\top \left(S_i + h_i\Phi_z'[t_i] \cdot S_i + h_i\Phi_w'[t_i] \cdot \frac{\partial w}{\partial \bar{z}} \right) \\
&= (\gamma_{z_0}' - \lambda_1^\top - h_0\lambda_1^\top \Phi_z'[t_0]) \cdot S_0 + (\gamma_{z_N}' + \lambda_N^\top) \cdot S_N + \gamma_{\bar{z}}' \\
&\quad + \sum_{i=1}^{N-1} (\lambda_i^\top - \lambda_{i+1}^\top - h_i\lambda_{i+1}^\top \Phi_z'[t_i]) \cdot S_i - \sum_{i=0}^{N-1} h_i\lambda_{i+1}^\top \Phi_w'[t_i] \cdot \frac{\partial w}{\partial \bar{z}}.
\end{aligned}$$

The terms $S_i = \partial Z_i(\bar{z})/\partial \bar{z}$ are just the sensitivities in the sensitivity equation approach, which we do not (!) want to compute here. Hence, we have to ensure that the expressions involving S_i are eliminated. This leads to the *discrete adjoint equation*

$$\lambda_i^\top - \lambda_{i+1}^\top - h_i \lambda_{i+1}^\top \Phi_z'[t_i] = 0_{\mathbb{R}^{n_z}}^\top, \quad i = 0, \dots, N-1, \tag{5.52}$$

and the *transversality condition*

$$\lambda_N^\top = -\gamma_{z_N}'(Z_0(\bar{z}), Z_N(\bar{z}), \bar{z}). \tag{5.53}$$

Notice that the adjoint equation is solved backwards in time. With these expressions the derivative of Γ_a reduces to

$$\Gamma_a'(z) = (\gamma_{z_0}' - \lambda_0^\top) \cdot S_0 + \gamma_{\bar{z}}' - \sum_{i=0}^{N-1} h_i \lambda_{i+1}^\top \Phi_w'[t_i] \cdot \frac{\partial w}{\partial \bar{z}}. \tag{5.54}$$

Herein, the sensitivity matrix S_0 is given by

$$S_0 = Z_0'(\bar{z}). \tag{5.55}$$

It remains to show that $\Gamma_a'(\bar{z}) = \Gamma'(\bar{z})$:

Theorem 5.4.1. *It holds*

$$\Gamma'(\bar{z}) = \Gamma_a'(\bar{z}) = (\gamma_{z_0}' - \lambda_0^\top) \cdot S_0 + \gamma_{\bar{z}}' - \sum_{i=0}^{N-1} h_i \lambda_{i+1}^\top \Phi_w'[t_i] \cdot \frac{\partial w}{\partial \bar{z}}. \tag{5.56}$$

Proof. Multiplication of the sensitivity equation (5.47) by $\lambda_{i+1}^\top$ from the left yields

$$-h_i \lambda_{i+1}^\top \Phi_w'[t_i] \cdot \frac{\partial w}{\partial \bar{z}} = -\lambda_{i+1}^\top S_{i+1} + \lambda_{i+1}^\top S_i + h_i \lambda_{i+1}^\top \Phi_z'[t_i] S_i$$

for $i = 0, \dots, N-1$ and hence

$$\begin{aligned}
\Gamma_a'(\bar{z}) &= (\gamma_{z_0}' - \lambda_0^\top) \cdot S_0 + \gamma_{\bar{z}}' + \sum_{i=0}^{N-1} \lambda_{i+1}^\top S_i + h_i \lambda_{i+1}^\top \Phi_z'[t_i] S_i - \lambda_{i+1}^\top S_{i+1} \\
&\overset{(5.52)}{=} (\gamma_{z_0}' - \lambda_0^\top) \cdot S_0 + \gamma_{\bar{z}}' + \sum_{i=0}^{N-1} (\lambda_i^\top S_i - \lambda_{i+1}^\top S_{i+1}) \\
&= (\gamma_{z_0}' - \lambda_0^\top) \cdot S_0 + \gamma_{\bar{z}}' + \lambda_0^\top S_0 - \lambda_N^\top S_N \\
&\overset{(5.53)}{=} \gamma_{z_0}' S_0 + \gamma_{\bar{z}}' + \gamma_{z_N}' S_N \\
&= \Gamma'(\bar{z}).
\end{aligned}$$

□

With $\Gamma'(\bar{z}) = \Gamma_a'(\bar{z})$ we finally found a formula for the gradient of Γ. Γ itself is a placeholder for the functions J, $G = (G_1, \dots, G_{n_G})^\top$, $H = (H_1, \dots, H_{n_H})^\top$ in (5.13)–(5.15).

In order to compute J', G', H' of (5.13)–(5.15) for each (!) component of J, G, H an adjoint equation with appropriate transversality condition has to be solved. This essentially corresponds to solving an initial value problem of dimension $n_z(2+n_G+n_H)$. The trajectory $Z_i(\bar{z})$, $i=0,\dots,N$, has to be stored. It is important to mention that the effort for solving the adjoint equations does not depend on the number M of control parameters! The method is particularly efficient if no or very few constraints are present.

5.4.2.1 Symplecticity

The combined integration scheme consisting of adjoint equation (5.52) and one-step method (5.51) has a nice additional property – *symplecticity*, see also [201] for ODE optimal control problems. The combined integration scheme reads as

$$\begin{aligned}\lambda_{i+1} &= \lambda_i - h_i\Phi_z'(t_i,Z_i,w,h_i)^\top\lambda_{i+1},\\ Z_{i+1} &= Z_i + h_i\Phi(t_i,Z_i,w,h_i)\end{aligned}$$

or in Hamiltonian form as

$$\begin{aligned}\lambda_{i+1} &= \lambda_i - h_iH_z'(t_i,Z_i,\lambda_{i+1},w,h_i)^\top,\\ Z_{i+1} &= Z_i + h_iH_\lambda'(t_i,Z_i,\lambda_{i+1},w,h_i)^\top\end{aligned}$$

with the auxiliary function

$$H(t,z,\lambda,w,h) := \lambda^\top\Phi(t,z,w,h).$$

Solving this equation leads to

$$\begin{pmatrix}\lambda_{i+1}\\ Z_{i+1}\end{pmatrix} = \begin{pmatrix}(I_{n_z}+h_i\Phi_z'(t_i,Z_i,w,h_i)^\top)^{-1}\lambda_i\\ Z_i+h_i\Phi(t_i,Z_i,w,h_i)\end{pmatrix} =: \begin{pmatrix}\Psi_1(\lambda_i,Z_i)\\ \Psi_2(\lambda_i,Z_i)\end{pmatrix} =: \Psi(\lambda_i,Z_i).$$

Hence, we obtain the integration scheme

$$\begin{aligned}\Psi_1(\lambda_i,Z_i) + h_iH_z'(t_i,Z_i,\Psi_1(\lambda_i,Z_i),w,h_i)^\top &= \lambda_i,\\ \Psi_2(\lambda_i,Z_i) - h_iH_\lambda'(t_i,Z_i,\Psi_1(\lambda_i,Z_i),w,h_i)^\top &= Z_i.\end{aligned}$$

Differentiation with respect to (λ_i,Z_i) leads to

$$\begin{pmatrix}I_{n_z}+h_i(H_{z\lambda}'')^\top & \Theta\\ -h_iH_{\lambda\lambda}'' & I_{n_z}\end{pmatrix}\begin{pmatrix}\Psi_{1,\lambda_i}' & \Psi_{1,z_i}'\\ \Psi_{2,\lambda_i}' & \Psi_{2,z_i}'\end{pmatrix} = \begin{pmatrix}I_{n_z} & -h_iH_{zz}''\\ \Theta & I_{n_z}+h_i(H_{\lambda z}'')^\top\end{pmatrix}.$$

Exploiting $H_{\lambda\lambda}'' = \Theta$ yields the Jacobian of Ψ to be

$$\Psi' = \begin{pmatrix}\Psi_{1,\lambda_i}' & \Psi_{1,z_i}'\\ \Psi_{2,\lambda_i}' & \Psi_{2,z_i}'\end{pmatrix} = \begin{pmatrix}(I_{n_z}+h_i(H_{z\lambda}'')^\top)^{-1} & -h_i(I_{n_z}+h_i(H_{z\lambda}'')^\top)^{-1}H_{zz}''\\ \Theta & I_{n_z}+h_iH_{z\lambda}''\end{pmatrix}.$$

It is straightforward to show $(\Psi')^\top \Lambda \Psi' = \Lambda$, where

$$\Lambda = \begin{pmatrix} \Theta & I_{n_z} \\ -I_{n_z} & \Theta \end{pmatrix}.$$

Thus, we proved

Theorem 5.4.2. *Ψ is symplectic, i. e. it holds $(\Psi')^\top \Lambda \Psi' = \Lambda$.* □

5.4.2.2 Application to Runge–Kutta methods

The increment function Φ and its derivatives Φ'_z and Φ'_w in the preceding sections are to be specified for implicit Runge–Kutta methods and the DAE

$$F(t, z(t), \dot{z}(t), u_M(t; w)) = 0_{\mathbb{R}^{n_z}}.$$

One step of the s-stage implicit Runge–Kutta method in Definition 4.1.9 from t_i to $t_i + h$ starting at Z_i is given by

$$Z_{i+1} = Z_i + h\Phi(t_i, Z_i, w, h)$$

with the increment function

$$\Phi(t, z, w, h) := \sum_{j=1}^{s} b_j k_j(t, z, w, h) \tag{5.57}$$

and the stage derivatives $k = (k_1, \dots, k_s)$, which are implicitly defined by the nonlinear equation

$$G(k, t, z, w, h) := \begin{pmatrix} F(\tau_1, z + h\sum_{j=1}^{s} a_{1j}k_j, k_1, u_M(\tau_1; w)) \\ F(\tau_2, z + h\sum_{j=1}^{s} a_{2j}k_j, k_2, u_M(\tau_2; w)) \\ \vdots \\ F(\tau_s, z + h\sum_{j=1}^{s} a_{sj}k_j, k_s, u_M(\tau_s; w)) \end{pmatrix} = 0_{\mathbb{R}^{sn_z}}, \tag{5.58}$$

where the abbreviations $\tau_i := t + c_i h$, $i = 1, \dots, s$, are used.

The nonlinear equation (5.58) is solved numerically by Newton's method:

$$\begin{aligned} G'_k(k^{(j)}, t, z, w, h)\Delta k^{(j)} &= -G(k^{(j)}, t, z, w, h), \\ k^{(j+1)} &= k^{(j)} + \Delta k^{(j)}, \quad j = 0, 1, 2, \dots. \end{aligned}$$

Under the assumptions that a solution $\hat{k}$ for (t, z, w, h) exists and the derivative G'_k is non-singular at $\hat{k}$, the implicit function theorem yields the existence of the function $k = k(t, z, w, h)$ satisfying

$$G(k(t, z, w, h), t, z, w, h) = 0_{\mathbb{R}^{sn_z}}$$

in appropriate neighborhoods. Differentiation of this identity with respect to z and w yields

$$k'_z(t,z,w,h) = -G'_k(k,t,z,w,h)^{-1}G'_z(k,t,z,w,h),$$
$$k'_w(t,z,w,h) = -G'_k(k,t,z,w,h)^{-1}G'_w(k,t,z,w,h),$$

where

$$G'_k(k,t,z,w,h) = \begin{pmatrix} ha_{11}M_1 + T_1 & ha_{12}M_1 & \cdots & ha_{1s}M_1 \\ ha_{21}M_2 & ha_{22}M_2 + T_2 & \ddots & \vdots \\ \vdots & \ddots & \ddots & ha_{s-1,s}M_{s-1} \\ ha_{s1}M_s & \cdots & ha_{s,s-1}M_s & ha_{ss}M_s + T_s \end{pmatrix},$$

$$M_j := F'_z\left(\tau_j, z + h\sum_{\ell=1}^{s} a_{jl}k_\ell, k_j, u_M(\tau_j; w)\right), \quad j = 1,\dots,s,$$

$$T_j := F'_{\dot z}\left(\tau_j, z + h\sum_{\ell=1}^{s} a_{j\ell}k_\ell, k_j, u_M(\tau_j; w)\right), \quad j = 1,\dots,s,$$

$$G'_z(k,t,z,w,h) = \begin{pmatrix} F'_z(\tau_1, z + h\sum_{j=1}^{s} a_{1j}k_j, k_1, u_M(\tau_1; w)) \\ \vdots \\ F'_z(\tau_s, z + h\sum_{j=1}^{s} a_{sj}k_j, k_s, u_M(\tau_s; w)) \end{pmatrix},$$

$$G'_w(k,t,z,w,h) = \begin{pmatrix} F'_u(\tau_1, z + h\sum_{j=1}^{s} a_{1j}k_j, k_1, u_M(\tau_1; w)) \cdot u'_{M,w}(\tau_1; w) \\ \vdots \\ F'_u(\tau_s, z + h\sum_{j=1}^{s} a_{sj}k_j, k_s, u_M(\tau_s; w)) \cdot u'_{M,w}(\tau_s; w) \end{pmatrix}.$$

Notice that (5.52) and (5.54) require the partial derivatives $\Phi'_z(t,z,w,h)$ and $\Phi'_w(t,z,w,h)$. According to (5.57), these values are given by

$$\Phi'_z(t,z,w,h) = \sum_{j=1}^{s} b_j k'_{j,z}(t,z,w,h),$$

$$\Phi'_w(t,z,w,h) = \sum_{j=1}^{s} b_j k'_{j,w}(t,z,w,h).$$

Example 5.4.3 (Implicit Euler method). For the implicit Euler method, we find

$$G(k,t,z,w,h) = F(t+h, z+hk, k, u_M(t+h; w))$$

and

$$\begin{aligned}\Phi'_z(t,z,w,h) &= k'_z(t,z,w,h) \\ &= -\left(hF'_z[t+h] + F'_{\dot z}[t+h]\right)^{-1} \cdot F'_z[t+h],\end{aligned}$$

$$
\begin{aligned}
\Phi'_w(t,z,w,h) &= k'_w(t,z,w,h) \\
&= -\left(hF'_z[t+h] + F'_{\dot z}[t+h]\right)^{-1} \cdot F'_u[t+h] \cdot u'_{M,w}(t+h;w).
\end{aligned}
$$
□

Example 5.4.4 (Explicit Runge–Kutta methods) We briefly discuss the important subclass of ODEs

$$F(t,z(t),\dot z(t),u_M(t;w)) := \dot z(t) - f(t,z(t),u_M(t;w))$$

in combination with an s-stage explicit Runge–Kutta method

$$Z_{i+1} = Z_i + h\sum_{j=1}^{s} b_j k_j(t_i, Z_i, w, h),$$

where the stage derivatives $k_j = k_j(t,z,w,h)$ for $j = 1,\dots,s$ are recursively defined by

$$k_j(t,z,w,h) := f\left(t + c_j h, z + h\sum_{\ell=1}^{j-1} a_{j\ell}k_\ell, u_M(t+c_j h;w)\right). \tag{5.59}$$

Differentiation with respect to z and w yields

$$
\begin{aligned}
k'_{j,z} &= f'_z\left(t + c_j h, z + h\sum_{\ell=1}^{j-1} a_{j\ell}k_\ell, u_M(t+c_j h;w)\right)\left(I + h\sum_{\ell=1}^{j-1} a_{j\ell}k'_{\ell,z}\right),\\
k'_{j,w} &= f'_z\left(t + c_j h, z + h\sum_{\ell=1}^{j-1} a_{j\ell}k_\ell, u_M(t+c_j h;w)\right)\left(h\sum_{\ell=1}^{j-1} a_{j\ell}k'_{\ell,w}\right)\\
&\quad + f'_u\left(t + c_j h, z + h\sum_{\ell=1}^{j-1} a_{j\ell}k_\ell, u_M(t+c_j h;w)\right)\cdot u'_{M,w}(t+c_j h;w)
\end{aligned}
$$

for $j = 1,\dots,s$, and

$$
\begin{aligned}
\Phi'_z(t,z,w,h) &= \sum_{j=1}^{s} b_j k'_{j,z}(t,z,w,h),\\
\Phi'_w(t,z,w,h) &= \sum_{j=1}^{s} b_j k'_{j,w}(t,z,w,h).
\end{aligned}
$$
□

5.4.3 Adjoint equation approach: the continuous case

If the DAE is solved by a state-of-the-art integrator with automatic step-size and order selection, then the adjoint equation approach as described in Subsection 5.4.2 becomes costly in terms of storage requirements as the approximation Z_i at every time point t_i generated by the step-size selection algorithm has to be stored. This is a potentially large number. In this case, it is more convenient to consider the DAE (5.1) with the discretized control $u_M(t; w)$ as an *infinite* DAE constraint for every $t \in \mathcal{I}$:

$$F(t, z(t; \bar{z}), \dot{z}(t; \bar{z}), u_M(t; w)) = 0_{\mathbb{R}^{n_z}}, \quad z(t_0; \bar{z}) = Z_0(\bar{z}). \tag{5.60}$$

This leads to a state approximation $z(t; \bar{z})$ for *every* $t \in \mathcal{I}$ that depends on $\bar{z}$. The initial state $Z_0(\bar{z}) \in \mathbb{R}^{n_z}$ is assumed to be sufficiently smooth as a function of $\bar{z}$ and consistent with the DAE for all $\bar{z}$.

As in the discrete case in Subsection 5.4.2, we aim at computing the gradient of the function

$$\Gamma(\bar{z}) := \gamma(z(t_0; \bar{z}), z(t_f; \bar{z}), \bar{z}) \tag{5.61}$$

with respect to $\bar{z} \in \mathbb{R}^{n_{\bar{z}}}$. Consider the auxiliary functional

$$\Gamma_a(\bar{z}) := \Gamma(\bar{z}) + \int_{t_0}^{t_f} \lambda(t)^\top F(t, z(t; \bar{z}), \dot{z}(t; \bar{z}), u_M(t; w)) dt,$$

where λ is a function to be defined later. Differentiating Γ_a with respect to $\bar{z}$ yields

$$\begin{aligned}
\Gamma_a'(\bar{z}) &= \Gamma'(\bar{z}) + \int_{t_0}^{t_f} \lambda(t)^\top (F_z'[t] \cdot S(t) + F_{\dot{z}}'[t] \cdot \dot{S}(t) + F_u'[t] \cdot u_{M,\bar{z}}'(t; w)) dt \\
&= \gamma_{z_0}' \cdot S(t_0) + \gamma_{z_f}' \cdot S(t_f) + \gamma_{\bar{z}}' + \int_{t_0}^{t_f} \lambda(t)^\top \cdot F_u'[t] \cdot u_{M,\bar{z}}'(t; w) dt \\
&\quad + \int_{t_0}^{t_f} \lambda(t)^\top \cdot F_z'[t] \cdot S(t) + \lambda(t)^\top \cdot F_{\dot{z}}'[t] \cdot \dot{S}(t) dt.
\end{aligned}$$

Integration by parts of the latter term yields

$$\begin{aligned}
\Gamma_a'(\bar{z}) &= (\gamma_{z_f}' + \lambda(t_f)^\top \cdot F_{\dot{z}}'[t_f]) \cdot S(t_f) + (\gamma_{z_0}' - \lambda(t_0)^\top \cdot F_{\dot{z}}'[t_0]) \cdot S(t_0) + \gamma_{\bar{z}}' \\
&\quad + \int_{t_0}^{t_f} \lambda(t)^\top \cdot F_u'[t] \cdot u_{M,\bar{z}}'(t; w) dt \\
&\quad + \int_{t_0}^{t_f} \left(\lambda(t)^\top \cdot F_z'[t] - \frac{d}{dt} (\lambda(t)^\top \cdot F_{\dot{z}}'[t]) \right) \cdot S(t) dt.
\end{aligned}$$

λ is chosen such that the adjoint DAE

$$\lambda(t)^\top \cdot F_z'[t] - \frac{d}{dt} (\lambda(t)^\top \cdot F_{\dot{z}}'[t]) = 0_{\mathbb{R}^{n_z}}^\top \tag{5.62}$$

is satisfied, provided that such a function λ exists. Then the derivative of Γ_a reduces to

$$\Gamma_a'(\bar{z}) = (\gamma_{z_f}' + \lambda(t_f)^\top \cdot F_{\dot{z}}'[t_f]) \cdot S(t_f) + (\gamma_{z_0}' - \lambda(t_0)^\top \cdot F_{\dot{z}}'[t_0]) \cdot S(t_0) + \gamma_{\bar{z}}' + \int_{t_0}^{t_f} \lambda(t)^\top \cdot F_u'[t] \cdot u_{M,\bar{z}}'(t;w)dt.$$

A connection with the sensitivity DAE (5.50) arises as follows. With (5.62) and (5.50) almost everywhere, it holds

$$\begin{aligned}
&\frac{d}{dt}(\lambda(t)^\top \cdot F_{\dot{z}}'[t] \cdot S(t)) \\
&= \frac{d}{dt}(\lambda(t)^\top \cdot F_{\dot{z}}'[t]) \cdot S(t) + \lambda(t)^\top \cdot F_{\dot{z}}'(t) \cdot \dot{S}(t) \\
&= \lambda(t)^\top F_z'[t] \cdot S(t) + \lambda(t)^\top(-F_z'[t] \cdot S(t) - F_u'[t] \cdot u_{M,\bar{z}}'(t;w)) \\
&= -\lambda(t)^\top F_u'[t] \cdot u_{M,\bar{z}}'(t;w).
\end{aligned}$$

Using the fundamental theorem of calculus, we obtain

$$[\lambda(t)^\top \cdot F_{\dot{z}}'[t] \cdot S(t)]_{t_0}^{t_f} = -\int_{t_0}^{t_f} \lambda(t)^\top F_u'[t] \cdot u_{M,\bar{z}}'(t;w)dt. \tag{5.63}$$

The gradient then becomes

$$\begin{aligned}
\Gamma_a'(\bar{z}) &= (\gamma_{z_f}' + \lambda(t_f)^\top \cdot F_{\dot{z}}'[t_f]) \cdot S(t_f) \\
&\quad + (\gamma_{z_0}' - \lambda(t_0)^\top \cdot F_{\dot{z}}'[t_0]) \cdot S(t_0) + \gamma_{\bar{z}}' \\
&\quad + \int_{t_0}^{t_f} \lambda(t)^\top \cdot F_u'[t] \cdot u_{M,\bar{z}}'(t;w)dt \\
&\overset{(5.63)}{=} \gamma_{z_f}' \cdot S(t_f) + \gamma_{z_0}' \cdot S(t_0) + \gamma_{\bar{z}}' \\
&= \Gamma'(\bar{z}).
\end{aligned}$$

Hence, $\Gamma_a'(\bar{z})$ and $\Gamma'(\bar{z})$ coincide provided that λ satisfies (5.62).

While it is cheap to compute the sensitivity matrix

$$S(t_0) = Z_0'(\bar{z}), \tag{5.64}$$

the sensitivity $S(t_f)$ in the expression for $\Gamma_a'(\bar{z})$ has to be eliminated by a proper choice of $\lambda(t_f)$ as we intend to avoid the explicit computation of $S(t_f)$. Moreover, $\lambda(t_f)$ has to be chosen consistently with the adjoint DAE (5.62). We discuss the procedure of defining appropriate conditions for $\lambda(t_f)$ for some common cases, see [58]:

The index-zero case

Let $F'_{\dot{z}}$ be non-singular almost everywhere in $\mathcal{I}$. Then the DAE has index zero, and $\lambda(t_f)$ is defined by the linear equation

$$\gamma'_{z_f} + \lambda(t_f)^\top \cdot F'_{\dot{z}}[t_f] = 0^\top_{\mathbb{R}^{n_z}} \tag{5.65}$$

and we find

$$\lambda(t_f)^\top = -\gamma'_{z_f} F'_{\dot{z}}[t_f]^{-1}$$

and thus

$$\begin{aligned}\Gamma'_a(\bar{z}) &= (\gamma'_{z_0} - \lambda(t_0)^\top \cdot F'_{\dot{z}}[t_0]) \cdot S(t_0) + \gamma'_{\bar{z}} \\ &\quad + \int_{t_0}^{t_f} \lambda(t)^\top \cdot F'_u[t] \cdot u'_{M,\bar{z}}(t; w) dt.\end{aligned} \tag{5.66}$$

Example 5.4.5. In the special case of an explicit ODE

$$F(t, z, \dot{z}, u) = f(t, z, u) - \dot{z},$$

equations (5.62) and (5.65) reduce to

$$\dot{\lambda}(t)^\top = -\lambda(t)^\top f'_x[t], \quad \lambda(t_f)^\top = \gamma'_{z_f}.$$

□

Semi-explicit DAEs

Consider a semi-explicit DAE with $z := (x, y)^\top \in \mathbb{R}^{n_x+n_y}$ and

$$F(t, z, \dot{z}, u) := \begin{pmatrix} f(t, x, y, u) - \dot{x} \\ g(t, x, y, u) \end{pmatrix} \in \mathbb{R}^{n_x+n_y}.$$

The corresponding adjoint equation (5.62) for $\lambda := (\lambda_f, \lambda_g)^\top \in \mathbb{R}^{n_x+n_y}$ reads as

$$\lambda_f(t)^\top f'_x[t] + \lambda_g(t)^\top g'_x[t] + \dot{\lambda}_f(t)^\top = 0^\top_{\mathbb{R}^{n_x}}, \tag{5.67}$$

$$\lambda_f(t)^\top f'_y[t] + \lambda_g(t)^\top g'_y[t] = 0^\top_{\mathbb{R}^{n_y}}. \tag{5.68}$$

With $S := (S^x, S^y)^\top$, we find

$$\begin{aligned}&(\gamma'_{z_f} + \lambda(t_f)^\top \cdot F'_{\dot{z}}[t_f]) \cdot S(t_f) \\ &\quad = (\gamma'_{x_f} - \lambda_f(t_f)^\top) S^x(t_f) + \gamma'_{y_f} S^y(t_f).\end{aligned} \tag{5.69}$$

A consistent value $\lambda(t_f) = (\lambda_f(t_f), \lambda_g(t_f))^\top$ is sought for (5.67)–(5.68), such that the expression in (5.69) does not depend explicitly on $S(t_f)$.

(i) Let g_y' be non-singular almost everywhere in $\mathcal{I}$. Then the DAE has index one and

$$\lambda_g(t_f)^\top = -\lambda_f(t_f)^\top f_y'[t_f] g_y'[t_f]^{-1}$$

is consistent with the algebraic constraint (5.68) of the adjoint system for any $\lambda_f(t_f)$. The expression (5.69) vanishes if

$$\lambda_f(t_f)^\top = \gamma_{x_f}' \quad \text{and} \quad \gamma_{y_f}' = 0_{\mathbb{R}^{n_y}}^\top.$$

With these settings $\Gamma_a'(\bar{z})$ is given by (5.66).
The latter assumption is not as restrictive as it seems as it fits well into the theoretical investigations in Chapter 3. For, the algebraic component y is an L_∞-function and thus it makes no sense to allow a pointwise evaluation of y at t_0 or t_f. Consequently, it is natural to prohibit γ depending explicitly on $y(t_f)$ (and $y(t_0)$).

(ii) Let $g_y' \equiv 0$ and let $g_x' f_y'$ be non-singular almost everywhere in $\mathcal{I}$. Then the DAE has index two and

$$\lambda_g(t_f)^\top = -\lambda_f(t_f)^\top \left(f_x'[t_f] f_y'[t_f] - \frac{d}{dt} f_y'[t_f] \right) (g_x'[t_f] f_y'[t_f])^{-1}$$

is consistent with the derivative of the algebraic constraint (5.68) for any $\lambda_f(t_f)$, provided that the time derivative of f_y' at t_f exists. Not any $\lambda_f(t_f)$, however, is consistent with the constraint (5.68). We choose the ansatz

$$\gamma_{x_f}' - \lambda_f(t_f)^\top = \xi^\top g_x'[t_f]$$

with $\xi \in \mathbb{R}^{n_y}$ to be determined such that $\lambda_f(t_f)$ is consistent with (5.68), i. e.

$$0_{\mathbb{R}^{n_y}}^\top = \lambda_f(t_f)^\top f_y'[t_f] = \gamma_{x_f}' f_y'[t_f] - \xi^\top g_x'[t_f] f_y'[t_f],$$

and thus

$$\xi^\top = \gamma_{x_f}' f_y'[t_f] (g_x'[t_f] f_y'[t_f])^{-1}.$$

Moreover, the sensitivity matrix S^x satisfies almost everywhere the algebraic equation

$$g_x'[t] S^x(t) + g_u'[t] u_{M,\bar{z}}'(t; w) = 0_{\mathbb{R}^{n_y}}$$

and thus the first term of the right-hand side in (5.69) computes to

$$(\gamma_{x_f}' - \lambda_f(t_f)^\top) S^x(t_f) = \xi^\top g_x'[t_f] S^x(t_f) = -\xi^\top g_u'[t_f] u_{M,\bar{z}}'(t_f; w).$$

Notice that the expression on the right does not depend on $S(t_f)$ anymore. Finally, if, as above, we assume $\gamma'_{y_f} = 0^\top_{\mathbb{R}^{n_y}}$, then $\Gamma'_a(\bar{z})$ can be computed without computing $S(t_f)$ according to

$$\begin{aligned}\Gamma'_a(\bar{z}) &= -\gamma'_{x_f} f'_y[t_f](g'_x[t_f] f'_y[t_f])^{-1} g'_u[t_f] u'_{M,\bar{z}}(t_f; w) \\ &\quad + (\gamma'_{z_0} - \lambda(t_0)^\top \cdot F'_{\dot{z}}[t_0]) \cdot S(t_0) + \gamma'_{\bar{z}} \\ &\quad + \int_{t_0}^{t_f} \lambda(t)^\top \cdot F'_u[t] \cdot u'_{M,\bar{z}}(t; w) dt.\end{aligned}$$

Stability results for the forward and the adjoint system are derived in [58]. It is shown for explicit ODEs and semi-explicit index-one and index-two Hessenberg DAEs that stability is preserved for the adjoint DAE.

5.4.3.1 Numerical adjoint approximation by BDF

We discuss the application of the BDF method to the adjoint DAE (5.62). First, the implicit DAE

$$F(t, z(t), \dot{z}(t), u_M(t; w)) = 0_{\mathbb{R}^{n_z}} \tag{5.70}$$

is transformed formally to semi-explicit form:

$$\begin{aligned}0_{\mathbb{R}^{n_z}} &= \dot{z}(t) - v(t), \\ 0_{\mathbb{R}^{n_z}} &= F(t, z(t), v(t), u_M(t; w)).\end{aligned}$$

Using the function

$$\begin{aligned}\hat{H}(t, z, \dot{z}, v, u, \lambda_v, \lambda_z) &:= (\lambda_v^\top, \lambda_z^\top) \begin{pmatrix} \dot{z} - v \\ F(t, z, v, u) \end{pmatrix} \\ &= \lambda_v^\top (\dot{z} - v) + \lambda_z^\top F(t, z, v, u),\end{aligned}$$

the adjoint DAE (5.62) for $y = (z, v)^\top$ is given by

$$\hat{H}'_y[t] - \frac{d}{dt} \hat{H}'_{\dot{y}}[t] = 0^\top_{\mathbb{R}^{2n_z}}.$$

Transposition yields

$$\begin{pmatrix} F'_z[t]^\top \lambda_z(t) - \dot{\lambda}_v(t) \\ -\lambda_v(t) + F'_{\dot{z}}[t]^\top \lambda_z(t) \end{pmatrix} = \begin{pmatrix} 0_{\mathbb{R}^{n_z}} \\ 0_{\mathbb{R}^{n_z}} \end{pmatrix}. \tag{5.71}$$

This is a linear DAE in $\lambda = (\lambda_v, \lambda_z)^\top$. Employing the BDF discretization method for the integration step $t_{m+k-1} \longrightarrow t_{m+k}$ and using the approximation

$$\dot{\lambda}_v(t_{m+k}) \approx \frac{1}{h}\sum_{i=0}^{k} \alpha_i \lambda_v(t_{m+i})$$

yields the linear equation

$$\begin{pmatrix} F_z'[t_{m+k}]^\top \lambda_z(t_{m+k}) - \frac{1}{h}\sum_{i=0}^{k} \alpha_i \lambda_v(t_{m+i}) \\ -\lambda_v(t_{m+k}) + F_{\dot z}'[t_{m+k}]^\top \lambda_z(t_{m+k}) \end{pmatrix} = \begin{pmatrix} 0_{\mathbb{R}^{n_z}} \\ 0_{\mathbb{R}^{n_z}} \end{pmatrix}$$

for $\lambda_z(t_{m+k})$ and $\lambda_v(t_{m+k})$. The size of the equation can be reduced by introducing the second equation

$$\lambda_v(t_{m+k}) = F_{\dot z}'[t_{m+k}]^\top \lambda_z(t_{m+k}),$$

into the first equation. Herein, the relations

$$\lambda_v(t_{m+i}) = F_{\dot z}'[t_{m+i}]^\top \lambda_z(t_{m+i}), \quad i = 0, 1, \ldots, k,$$

are exploited. It remains to solve the linear equation

$$\left(F_z'[t_{m+k}]^\top - \frac{\alpha_k}{h} F_{\dot z}'[t_{m+k}]^\top\right)\lambda_z(t_{m+k}) = \frac{1}{h}\sum_{i=0}^{k-1} \alpha_i F_{\dot z}'[t_{m+i}]^\top \lambda_z(t_{m+i}).$$

Recall that λ_z is the desired adjoint λ of the original system in (5.70), whose Hamilton function is given by

$$H(t, z, \dot z, u, \lambda) = \lambda^\top F(t, z, \dot z, u).$$

For this, the adjoint DAE is given by

$$F_z'[t]^\top \lambda(t) - \frac{d}{dt}(F_{\dot z}'[t]^\top \lambda(t)) = 0_{\mathbb{R}^{n_z}}$$

Notice that this equation arises, if the second equation in (5.71) is introduced into the first one.

For fully implicit index-zero and index-one DAEs, the augmented adjoint system in (5.71) is stable, if the original DAE was stable, see [58, Theorem 4.3].

Example 5.4.6 (See Example 1.1.10). We revisit Example 1.1.10 without dynamic pressure constraint and compare the sensitivity equation approach and the adjoint equation approach in view of CPU time. We used the fourth-order classic Runge–Kutta method with fixed step-size for time-integration of the ODE. The control is approximated by a continuous and piecewise linear function.

Figure 5.8 summarizes computational results obtained for the sensitivity equation approach and the adjoint equation approach. While the sensitivity approach grows nonlinearly with the number N of equidistant time intervals in the control grid, the adjoint approach grows at a linear rate. Hence, in this case, the adjoint approach is more efficient than the sensitivity approach. This is the expected behavior since the number of constraints is significantly smaller than the number of optimization variables. □

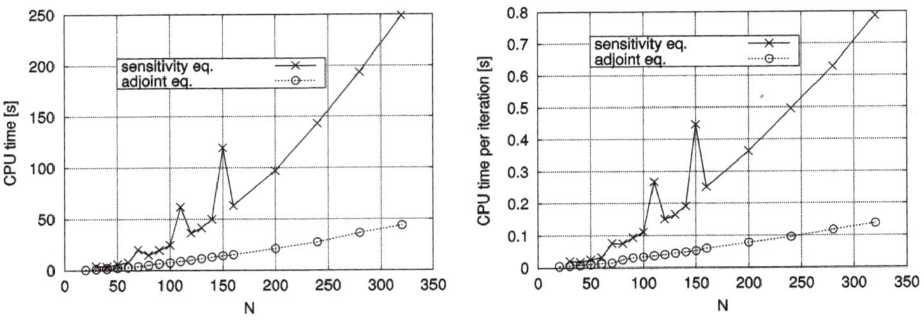

Figure 5.8: Computational results for the emergency landing maneuver without dynamic pressure constraint: Performance of the sensitivity equation approach compared to the adjoint equation approach with NPSOL.

5.5 Discrete minimum principle and approximation of adjoints

The aim of this section is to state a local minimum principle for the full discretization of Problem 3.4.1 using the implicit Euler method. It turns out that this discrete local minimum principle can be interpreted as discretization of the local minimum principles 3.4.3 and 3.4.4, respectively. This observation allows constructing approximations of the multipliers λ_f, λ_g, η, μ, and σ of Section 3.4 to which we refer to as *continuous* multipliers for brevity (which does not mean that these multipliers are actually continuous). We consider Problem 3.4.1 with $f_0 \equiv 0$:

Problem 5.5.1 (DAE optimal control problem). Let $\mathcal{I} := [t_0, t_f] \subset \mathbb{R}$ be a non-empty compact time interval with $t_0 < t_f$ fixed. Let

$$
\begin{aligned}
\varphi &: \mathbb{R}^{n_x} \times \mathbb{R}^{n_x} \longrightarrow \mathbb{R},\\
f &: \mathcal{I} \times \mathbb{R}^{n_x} \times \mathbb{R}^{n_y} \times \mathbb{R}^{n_u} \longrightarrow \mathbb{R}^{n_x},\\
g &: \mathcal{I} \times \mathbb{R}^{n_x} \times \mathbb{R}^{n_y} \times \mathbb{R}^{n_u} \longrightarrow \mathbb{R}^{n_y},\\
c &: \mathcal{I} \times \mathbb{R}^{n_x} \times \mathbb{R}^{n_y} \times \mathbb{R}^{n_u} \longrightarrow \mathbb{R}^{n_c},\\
s &: \mathcal{I} \times \mathbb{R}^{n_x} \longrightarrow \mathbb{R}^{n_s},\\
\psi &: \mathbb{R}^{n_x} \times \mathbb{R}^{n_x} \longrightarrow \mathbb{R}^{n_\psi}
\end{aligned}
$$

be sufficiently smooth functions and $\mathcal{U} \subseteq \mathbb{R}^{n_u}$ a set.

Minimize

$$\varphi(x(t_0), x(t_f)) \tag{5.72}$$

with respect to $x \in W^{n_x}_{1,\infty}(\mathcal{I})$, $y \in L^{n_y}_{\infty}(\mathcal{I})$, $u \in L^{n_u}_{\infty}(\mathcal{I})$ *subject to the constraints*

$$\dot{x}(t) = f(t, x(t), y(t), u(t)), \tag{5.73}$$

$$0_{\mathbb{R}^{n_y}} = g(t, x(t), y(t), u(t)), \tag{5.74}$$

$$\psi(x(t_0), x(t_f)) = 0_{\mathbb{R}^{n_\psi}}, \tag{5.75}$$

$$c(t, x(t), y(t), u(t)) \leq 0_{\mathbb{R}^{n_c}}, \tag{5.76}$$

$$s(t, x(t)) \leq 0_{\mathbb{R}^{n_s}} \tag{5.77}$$

$$u(t) \in \mathcal{U}. \tag{5.78}$$

The Hamilton function and the augmented Hamilton function are defined as usual (with $f_0 \equiv 0$):

$$\mathcal{H}(t, x, y, u, \lambda_f, \lambda_g) = \lambda_f^\top f(t, x, y, u) + \lambda_g^\top g(t, x, y, u),$$

$$\hat{\mathcal{H}}(t, x, y, u, \lambda_f, \lambda_g, \eta) = \mathcal{H}(t, x, y, u, \lambda_f, \lambda_g) + \eta^\top c(t, x, y, u).$$

Problem 5.5.1 is discretized using the *implicit* Euler method on the not necessarily equidistant grid $\mathbb{G}_N$ in (5.5). The resulting finite-dimensional nonlinear program reads as follows:

Problem 5.5.2 (Discretized DAE optimal control problem).
Minimize

$$\varphi(x_0, x_N) \tag{5.79}$$

with respect to grid functions

$$x_N : \mathbb{G}_N \longrightarrow \mathbb{R}^{n_x}, \quad x_i := x_N(t_i),$$

$$y_N : \mathbb{G}_N \longrightarrow \mathbb{R}^{n_y}, \quad y_i := y_N(t_i),$$

$$u_N : \mathbb{G}_N \longrightarrow \mathbb{R}^{n_u}, \quad u_i := u_N(t_i),$$

subject to the constraints

$$f(t_i, x_i, y_i, u_i) - \frac{x_i - x_{i-1}}{h_{i-1}} = 0_{\mathbb{R}^{n_x}}, \quad i = 1, \ldots, N, \tag{5.80}$$

$$g(t_i, x_i, y_i, u_i) = 0_{\mathbb{R}^{n_y}}, \quad i = 1, \ldots, N, \tag{5.81}$$

$$\psi(x_0, x_N) = 0_{\mathbb{R}^{n_\psi}}, \tag{5.82}$$

$$c(t_i, x_i, y_i, u_i) \leq 0_{\mathbb{R}^{n_c}}, \quad i = 1, \ldots, N, \tag{5.83}$$

$$s(t_i, x_i) \leq 0_{\mathbb{R}^{n_s}}, \quad i = 0, \ldots, N, \tag{5.84}$$

$$u_i \in \mathcal{U}, \quad i = 1, \ldots, N.$$

Equations (5.80)–(5.81) result from the implicit Euler method applied to the DAE (5.73)–(5.74). Notice that we can dispense with the algebraic constraint (5.74) and the mixed control-state constraint (5.76) at $t = t_0$ in the discretized problem. As u_0 and y_0 do not enter the objective function, it would be superfluous to impose the constraints $g(t_0, x_0, y_0, u_0) = 0_{\mathbb{R}^{n_y}}$ and $c(t_0, x_0, y_0, u_0) \leq 0_{\mathbb{R}^{n_c}}$ in Problem 5.5.2, provided feasible u_0 and y_0 exist for given t_0 and x_0. This can be guaranteed under similar assumptions as in Theorem 3.4.4.

Evaluation of the Fritz John conditions in Theorem 2.3.25 yields the following necessary optimality conditions.

Theorem 5.5.3 (Discrete local minimum principle). *Let the following Assumptions hold:*
(i) *The functions $\varphi, f, g, c, s, \psi$ are continuously differentiable with respect to x, y, and u.*
(ii) *$\mathcal{U} \subseteq \mathbb{R}^{n_u}$ is a closed and convex set with non-empty interior.*
(iii) *$(\hat{x}_N, \hat{y}_N, \hat{u}_N)$ is a local minimum of Problem* 5.5.2.

Then there exist multipliers $\kappa_0 \in \mathbb{R}$, $\kappa \in \mathbb{R}^{n_\psi}$,

$$
\begin{aligned}
\lambda_{f,N} &: \mathbb{G}_N \longrightarrow \mathbb{R}^{n_x}, & \lambda_{f,i} &:= \lambda_{f,N}(t_i),\\
\lambda_{g,N} &: \mathbb{G}_N \longrightarrow \mathbb{R}^{n_y}, & \lambda_{g,i} &:= \lambda_{g,N}(t_i),\\
\zeta_N &: \mathbb{G}_N \longrightarrow \mathbb{R}^{n_c}, & \zeta_i &:= \zeta_N(t_i),\\
\nu_N &: \mathbb{G}_N \longrightarrow \mathbb{R}^{n_s}, & \nu_i &:= \nu_N(t_i),
\end{aligned}
$$

such that the following conditions are satisfied:
(a) $\kappa_0 \geq 0$, $(\kappa_0, \kappa, \lambda_{f,N}, \lambda_{g,N}, \zeta_N, \nu_N) \neq \Theta$.
(b) Discrete adjoint equations: *For $i = 1, \dots, N$, it holds*

$$
\begin{aligned}
\lambda_{f,i-1}^\top &= \lambda_{f,i}^\top + h_{i-1}\hat{\mathcal{H}}_x'\left(t_i, \hat{x}_i, \hat{y}_i, \hat{u}_i, \lambda_{f,i-1}, \lambda_{g,i-1}, \frac{\zeta_i}{h_{i-1}}\right) + \nu_i^\top s_x'(t_i, \hat{x}_i)\\
&= \lambda_{f,N}^\top + \sum_{j=i}^{N} h_{j-1}\hat{\mathcal{H}}_x'\left(t_j, \hat{x}_j, \hat{y}_j, \hat{u}_j, \lambda_{f,j-1}, \lambda_{g,j-1}, \frac{\zeta_j}{h_{j-1}}\right) + \sum_{j=i}^{N} \nu_j^\top s_x'(t_j, \hat{x}_j),
\end{aligned}
\tag{5.85}
$$

$$
0_{\mathbb{R}^{n_y}} = \hat{\mathcal{H}}_y'\left(t_i, \hat{x}_i, \hat{y}_i, \hat{u}_i, \lambda_{f,i-1}, \lambda_{g,i-1}, \frac{\zeta_i}{h_{i-1}}\right)^\top. \tag{5.86}
$$

(c) Discrete transversality conditions:

$$
\lambda_{f,0}^\top = -(\kappa_0 \varphi_{x_0}'(\hat{x}_0, \hat{x}_N) + \kappa^\top \psi_{x_0}'(\hat{x}_0, \hat{x}_N) + \nu_0^\top s_x'(t_0, \hat{x}_0)), \tag{5.87}
$$

$$
\lambda_{f,N}^\top = \kappa_0 \varphi_{x_N}'(\hat{x}_0, \hat{x}_N) + \kappa^\top \psi_{x_N}'(\hat{x}_0, \hat{x}_N). \tag{5.88}
$$

(d) Discrete optimality conditions: *For all $i = 1, \dots, N$ and all $u \in \mathcal{U}$, it holds*

$$
\hat{\mathcal{H}}_u'\left(t_i, \hat{x}_i, \hat{y}_i, \hat{u}_i, \lambda_{f,i-1}, \lambda_{g,i-1}, \frac{\zeta_i}{h_{i-1}}\right)(u - \hat{u}_i) \geq 0. \tag{5.89}
$$

(e) Discrete complementarity conditions: *It holds*

$$\zeta_i \geq 0_{\mathbb{R}^{n_c}}, \quad i = 1, \dots, N, \tag{5.90}$$

$$\nu_i \geq 0_{\mathbb{R}^{n_s}}, \quad i = 0, \dots, N, \tag{5.91}$$

$$\zeta_i^\top c(t_i, \hat{x}_i, \hat{y}_i, \hat{u}_i) = 0, \quad i = 1, \dots, N, \tag{5.92}$$

$$\nu_i^\top s(t_i, \hat{x}_i) = 0, \quad i = 0, \dots, N. \tag{5.93}$$

Proof. Let $\kappa_0 \in \mathbb{R}$, $\kappa \in \mathbb{R}^{n_\psi}$,

$$\begin{aligned}
x &= (\hat{x}_0, \dots, \hat{x}_N)^\top \in \mathbb{R}^{n_x(N+1)}, \\
y &= (\hat{y}_1, \dots, \hat{y}_N)^\top \in \mathbb{R}^{n_y N}, \\
u &= (\hat{u}_1, \dots, \hat{u}_N)^\top \in \mathbb{R}^{n_u N}, \\
\tilde{\lambda}_f &= (\tilde{\lambda}_{f,0}, \dots, \tilde{\lambda}_{f,N-1})^\top \in \mathbb{R}^{n_x N}, \\
\tilde{\lambda}_g &= (\tilde{\lambda}_{g,0}, \dots, \tilde{\lambda}_{g,N-1})^\top \in \mathbb{R}^{n_y N}, \\
\zeta &= (\zeta_1, \dots, \zeta_N)^\top \in \mathbb{R}^{n_c N}, \\
\nu &= (\nu_0, \dots, \nu_N)^\top \in \mathbb{R}^{n_s(N+1)}.
\end{aligned}$$

The Lagrange function of Problem 5.5.2 is given by

$$\begin{aligned}
&L(x, y, u, \tilde{\lambda}_f, \tilde{\lambda}_g, \zeta, \nu, \kappa, \kappa_0) \\
&= \kappa_0 \varphi(x_0, x_N) + \kappa^\top \psi(x_0, x_N) \\
&\quad + \sum_{i=1}^{N} \hat{\mathcal{H}}(t, x_i, y_i, u_i, \tilde{\lambda}_{f,i-1}, \tilde{\lambda}_{g,i-1}, \zeta_i) + \sum_{i=1}^{N} \tilde{\lambda}_{f,i-1}^\top \frac{x_{i-1} - x_i}{h_{i-1}} + \sum_{i=0}^{N} \nu_i^\top s(t_i, x_i).
\end{aligned}$$

Application of Theorem 2.3.25 results in the following equations:

1. $L'_{u_i}(u - \hat{u}_i) \geq 0$ for all $u \in \mathcal{U}$: For $i = 1, \dots, N$, it holds for all $u \in \mathcal{U}$

$$\hat{\mathcal{H}}'_u(t_i, \hat{x}_i, \hat{y}_i, \hat{u}_i, \tilde{\lambda}_{f,i-1}, \tilde{\lambda}_{g,i-1}, \zeta_i)(u - \hat{u}_i) \geq 0.$$

2. $L'_{y_i} = 0^\top_{\mathbb{R}^{n_y}}$: For $i = 1, \dots, N$, it holds

$$\hat{\mathcal{H}}'_y(t_i, \hat{x}_i, \hat{y}_i, \hat{u}_i, \tilde{\lambda}_{f,i-1}, \tilde{\lambda}_{g,i-1}, \zeta_i) = 0^\top_{\mathbb{R}^{n_y}}.$$

3. $L'_{x_i} = 0^\top_{\mathbb{R}^{n_x}}$: For $i = 0$, it holds

$$(\kappa_0 \varphi'_{x_0}(\hat{x}_0, \hat{x}_N) + \kappa^\top \psi'_{x_0}(\hat{x}_0, \hat{x}_N) + \nu_0^\top s'_x(t_0, \hat{x}_0)) + \frac{1}{h_0} \tilde{\lambda}_{f,0}^\top = 0^\top_{\mathbb{R}^{n_x}}.$$

   For $i = 1, \dots, N-1$, it holds

$$\hat{\mathcal{H}}'_x(t_i, \hat{x}_i, \hat{y}_i, \hat{u}_i, \tilde{\lambda}_{f,i-1}, \tilde{\lambda}_{g,i-1}, \zeta_i) - \frac{1}{h_{i-1}} \tilde{\lambda}_{f,i-1}^\top + \frac{1}{h_i} \tilde{\lambda}_{f,i}^\top + \nu_i^\top s'_x(t_i, \hat{x}_i) = 0^\top_{\mathbb{R}^{n_x}}.$$

For $i = N$, it holds

$$\kappa_0 \varphi'_{x_N}(\hat{x}_0, \hat{x}_N) + \kappa^\top \psi'_{x_N}(\hat{x}_0, \hat{x}_N) + \nu_N^\top s'_x(t_N, \hat{x}_N) + \hat{\mathcal{H}}'_x(t_N, \hat{x}_N, \hat{y}_N, \hat{u}_N, \tilde{\lambda}_{f,N-1}, \tilde{\lambda}_{g,N-1}, \zeta_N) - \frac{1}{h_{N-1}} \tilde{\lambda}_{f,N-1}^\top = 0_{\mathbb{R}^{n_x}}^\top.$$

With the definitions

$$\lambda_{f,i} := \frac{1}{h_i} \tilde{\lambda}_{f,i}, \quad \lambda_{g,i} := \frac{1}{h_i} \tilde{\lambda}_{g,i}, \quad i = 0, \dots, N-1,$$

and

$$\lambda_{f,N}^\top := \kappa_0 \varphi'_{x_N}(\hat{x}_0, \hat{x}_N) + \kappa^\top \psi'_{x_N}(\hat{x}_0, \hat{x}_N)$$

we obtain the discrete optimality conditions, the discrete adjoint equations, and the discrete transversality conditions. □

---

i The local minimum principle for DAE optimal control problems can be extended to a global minimum principle, see Section 7.1. The question arises, whether a similar result holds for the discretized optimal control problem 5.5.2. This is not the case in general, but an approximate minimum principle holds, see [251]. With additional convexity-like conditions, a discrete minimum principle also holds, see [169, Section 6.4, p. 277].

We compare the necessary optimality conditions in Theorem 5.5.3 with those in Theorems 3.4.3 and 3.4.4. Our intention is to provide an interpretation of the *discrete* multipliers in Theorem 5.5.3 and put them into relation to the *continuous* multipliers in Theorem 3.4.3. Following this interpretation, it becomes possible to construct estimates of the continuous multipliers using the discrete multipliers only. Of course, the discrete multipliers can be computed numerically by solving the discretized optimal control problem 5.5.2 by SQP.

In line with the situation of Theorem 3.4.3, we discuss Problem 5.5.1 without mixed control-state constraints, i. e. $n_c = 0$ and $c \equiv 0_{\mathbb{R}^{n_c}}$.

The discrete optimality conditions (5.89) and the discrete adjoint equations (5.86) are easily being recognized as pointwise discretizations of the optimality condition (3.75) and the algebraic equation in (3.72), respectively, provided we assume for all i:

$$\hat{x}_i \approx \hat{x}(t_i), \quad \hat{y}_i \approx \hat{y}(t_i), \quad \hat{u}_i \approx \hat{u}(t_i), \quad \lambda_{f,i} \approx \lambda_f(t_i), \quad \lambda_{g,i} \approx \lambda_g(t_i).$$

Actually, there will be an exceptional interpretation for $\lambda_{f,0}$ later on.

Comparing the discrete transversality condition (5.88)

$$\lambda_{f,N}^\top = \kappa_0 \varphi'_{x_N}(\hat{x}_0, \hat{x}_N) + \kappa^\top \psi'_{x_f}(\hat{x}_0, \hat{x}_N)$$

and the transversality condition (3.74)

$$\lambda_f(t_f)^\top = \ell_0 \varphi'_{x_f}(\hat{x}(t_0), \hat{x}(t_f)) + \sigma^\top \psi'_{x_f}(\hat{x}(t_0), \hat{x}(t_f)),$$

it is natural to presume

$$\kappa_0 \approx \ell_0, \quad \kappa \approx \sigma.$$

The adjoint equation (3.71) for $t \in \mathcal{I}$ reads as

$$\lambda_f(t) = \lambda_f(t_f) + \int_t^{t_f} \mathcal{H}'_x(\tau, \hat{x}(\tau), \hat{y}(\tau), \hat{u}(\tau), \lambda_f(\tau), \lambda_g(\tau))^\top d\tau + \int_t^{t_f} s'_x(\tau, \hat{x}(\tau))^\top d\mu(\tau).$$

The discrete adjoint equation (5.85) for $i = 1, \dots, N$ is given by

$$\lambda_{f,i-1} = \lambda_{f,N} + \sum_{j=i}^{N} h_{j-1} \mathcal{H}'_x(t_j, \hat{x}_j, \hat{y}_j, \hat{u}_j, \lambda_{f,j-1}, \lambda_{g,j-1})^\top + \sum_{j=i}^{N} s'_x(t_j, \hat{x}_j)^\top \nu_j.$$

The first sum is easily being recognized to be a Riemann sum on $\mathbb{G}_N$, and thus

$$\sum_{j=i}^{N} h_{j-1} \mathcal{H}'_x(t_j, \hat{x}_j, \hat{y}_j, \hat{u}_j, \lambda_{f,j-1}, \lambda_{g,j-1})^\top \approx \int_{t_{i-1}}^{t_f} \mathcal{H}'_x(\tau, \hat{x}(\tau), \hat{y}(\tau), \hat{u}(\tau), \lambda_f(\tau), \lambda_g(\tau))^\top d\tau \tag{5.94}$$

for $i = 1, \dots, N$. This observation encourages us to expect that for $i = 1, \dots, N$

$$\sum_{j=i}^{N} s'_x(t_j, \hat{x}_j)^\top \nu_j \approx \int_{t_{i-1}}^{t_f} s'_x(\tau, \hat{x}(\tau))^\top d\mu(\tau). \tag{5.95}$$

In order to interpret the approximation in (5.95), we have to recall the definition of a Riemann–Stieltjes integral, see Definition 2.1.24. The Riemann–Stieltjes integral in (5.95) is defined to be the limit of the sum

$$\sum_{j=i}^{m} s'_x(\xi_j, \hat{x}(\xi_j))^\top (\mu(t_j) - \mu(t_{j-1}))$$

where $t_{i-1} < t_i < \cdots < t_m = t_f$ is an arbitrary partition of $[t_{i-1}, t_f]$ and $\xi_j \in [t_{j-1}, t_j]$ are arbitrary points. If we choose the particular grid $\mathbb{G}_N$ and the points $\xi_j := t_j$, we obtain the approximation

$$\int_{t_{i-1}}^{t_f} s'_x(\tau, \hat{x}(\tau))^\top d\mu(\tau) \approx \sum_{j=i}^{N} s'_x(t_j, \hat{x}(t_j))^\top (\mu(t_j) - \mu(t_{j-1})).$$

Together with (5.95) we draw the conclusion that the relationship of the discrete multipliers ν_i and the continuous counterpart μ must be given by

$$\nu_i \approx \mu(t_i) - \mu(t_{i-1}), \quad i = 1, \dots, N.$$

Now, we address the remaining transversality condition (3.73):

$$\lambda_f(t_0)^\top = -(\ell_0 \varphi'_{x_0}(\hat{x}(t_0), \hat{x}(t_f)) + \sigma^\top \psi'_{x_0}(\hat{x}(t_0), \hat{x}(t_f)))$$

and the discrete counterpart (5.87)

$$\lambda_{f,0}^\top = -(\kappa_0 \varphi'_{x_0}(\hat{x}_0, \hat{x}_N) + \kappa^\top \psi'_{x_0}(\hat{x}_0, \hat{x}_N) + \nu_0^\top s'_x(t_0, \hat{x}_0)).$$

These two conditions can be brought into accordance as follows. Recall that the multiplier μ is normalized, i. e. $\mu(t_0) = 0_{\mathbb{R}^{n_s}}$ and μ is continuous from the right in the open interval (t_0, t_f). Hence, μ may jump at t_0 with $0 \le \mu(t_0+) - \mu(t_0) = \mu(t_0+)$, where $\mu(t_0+)$ denotes the right-sided limit of μ at t_0. Similarly, λ_f may jump at t_0, too. Similar to the derivation of the jump conditions (3.76), it follows

$$\lambda_f(t_0+) - \lambda_f(t_0) = \lim_{\varepsilon \downarrow 0} \lambda_f(t_0 + \varepsilon) - \lambda_f(t_0) = -s'_x(t_0, \hat{x}(t_0))^\top (\mu(t_0+) - \mu(t_0))$$

and thus

$$\begin{aligned} \lambda_f(t_0+)^\top &= \lambda_f(t_0)^\top - (\mu(t_0+) - \mu(t_0))^\top s'_x(t_0, \hat{x}(t_0)) \\ &= -(\ell_0 \varphi'_{x_0}(\hat{x}(t_0), \hat{x}(t_f)) + \sigma^\top \psi'_{x_0}(\hat{x}(t_0), \hat{x}(t_f)) + (\mu(t_0+) - \mu(t_0))^\top s'_x(t_0, \hat{x}(t_0))). \end{aligned}$$

A comparison with (5.87) shows that both are in accordance, if ν_0 is interpreted as the jump height of μ at $t = t_0$, i. e.

$$\nu_0 \approx \mu(t_0+) - \mu(t_0)$$

and if $\lambda_{f,0}$ is interpreted as the right-sided limit of λ_f at t_0, i. e.

$$\lambda_{f,0} \approx \lambda_f(t_0+).$$

As an approximation of the value $\lambda_f(t_0)$, we then use the value

$$-(\kappa_0 \varphi'_{x_0}(\hat{x}_0, \hat{x}_N) + \kappa^\top \psi'_{x_0}(\hat{x}_0, \hat{x}_N)).$$

The above interpretations cope well with the complementarity conditions, if we recall that ν_i denotes the multiplier for the constraint $s(t_i, x_i) \le 0_{\mathbb{R}^{n_s}}$. Then the complementarity conditions yield

$$0_{\mathbb{R}^{n_s}} \le \nu_i \approx \mu(t_i) - \mu(t_{i-1}),$$

which reflects the fact that μ is non-decreasing. The condition $\nu_i^\top s(t_i, x_i) = 0$ implies

$$s(t_i, x_i) < 0_{\mathbb{R}^{n_s}} \quad \Longrightarrow \quad 0_{\mathbb{R}^{n_s}} = \nu_i \approx \mu(t_i) - \mu(t_{i-1}),$$

which reflects that μ is constant on inactive arcs.

The above interpretations remain valid for problems with additional mixed control-state constraints and $\mathcal{U} = \mathbb{R}^{n_u}$, see Theorem 3.4.4. A comparison of the respective necessary conditions in Theorem 3.4.4 and Theorem 5.5.3 yields the additional relationship

$$\frac{\zeta_i}{h_{i-1}} \approx \eta(t_i), \quad i = 1, \dots, N,$$

for the multiplier η. The complementarity conditions for ζ_i are discrete versions of the complementarity condition in Theorem 3.4.4.

5.5.1 Example

The subsequent example shows that the above interpretations are meaningful. Since the problem is an ODE optimal control problem, we use the explicit Euler method for discretization instead of the implicit Euler method. The above interpretations can be adapted accordingly. Though the example at first glance is very simple, it has the nice feature that one of the two-state constraints becomes active only at the final time point. This causes the corresponding multiplier to jump only at the final time point. Correspondingly, the adjoint also jumps. It turns out that the above interpretations allow constructing approximations for the multipliers that reflect this behavior for the numerical solution.

Consider a system of two water boxes, where $x(t)$ and $y(t)$ denote the volume of water in the two boxes and $u(t)$ and $v(t)$ denote the outflow rate of water for the respective boxes at time t, see Figure 5.9.

An optimal control problem is given by

Problem 5.5.4. Let $\mathcal{I} := [0, 10]$.

Minimize

$$-\int_0^{10} (10 - t)u(t) + tv(t)dt$$

subject to the constraints

$$\begin{aligned} \dot{x}(t) &= -u(t), & x(0) &= 4, \\ \dot{y}(t) &= u(t) - v(t), & y(0) &= 4, \end{aligned}$$

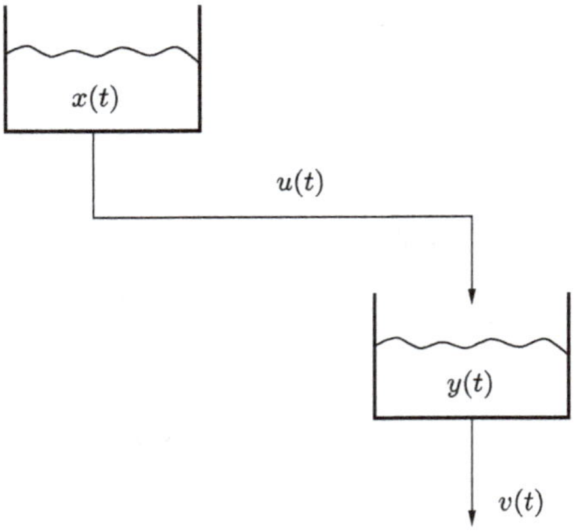

Figure 5.9: System of two water boxes with controllable outflow rates.

$$x(t) \geq 0,$$
$$y(t) \geq 0,$$
$$u(t) \in [0,1],$$
$$v(t) \in [0,1].$$

Evaluation of the local minimum principle 3.4.3 or 3.2.7, respectively, yields the optimal solution in

Theorem 5.5.5. *The following functions satisfy the local minimum principle* 3.4.3 *(and* 3.2.7*) with $\ell_0 = 1$ for the linear optimal control problem* 5.5.4 *and hence are optimal according to Theorem* 2.3.36.

The optimal control variables are given by

$$\hat{u}(t) = \begin{cases} 1, & \text{if } 0 \leq t < 4, \\ 0, & \text{if } 4 \leq t \leq 10, \end{cases} \qquad \hat{v}(t) = \begin{cases} 0, & \text{if } 0 \leq t < 2, \\ 1, & \text{if } 2 \leq t \leq 10. \end{cases}$$

The optimal state variables are given by

$$\hat{x}(t) = \begin{cases} 4 - t, & \text{if } 0 \leq t < 4, \\ 0, & \text{if } 4 \leq t \leq 10, \end{cases} \qquad \hat{y}(t) = \begin{cases} 4 + t, & \text{if } 0 \leq t < 2, \\ 6, & \text{if } 2 \leq t < 4, \\ 10 - t, & \text{if } 4 \leq t \leq 10. \end{cases}$$

The adjoints are given by

$$\hat{\lambda}_x(t) = \mu_x(t) - \mu_x(10), \quad \hat{\lambda}_y(t) = \mu_y(t) - \mu_y(10) = \begin{cases} -2, & \text{if } 0 \leq t < 10, \\ 0, & \text{if } t = 10, \end{cases}$$

where the multiplier μ_x is non-decreasing and satisfies $\mu_x(t) = 0$ for $t \in [0,4)$ and

$$\int_4^{10} d\mu_x(\tau) = \mu_x(10) - \mu_x(4) = 8, \quad \mu_x(10) - \mu_x(t) > 12 - t, \quad t \in (4, 10).$$

The multiplier μ_y is given by

$$\hat{\mu}_y(t) = \begin{cases} 0, & \text{if } 0 \le t < 10, \\ 2, & \text{if } t = 10. \end{cases}$$

Proof. We apply Theorem 3.4.3. The Hamilton function for Problem 5.5.4 is given by

$$\mathcal{H}(t, x, y, u, v, \lambda_x, \lambda_y, \ell_0) = -\ell_0((10 - t)u + tv) - \lambda_x u + \lambda_y(u - v).$$

Let $(\hat{x}, \hat{y}, \hat{u}, \hat{v})$ be a local minimum of Problem 5.5.4. Then by Theorem 3.4.3, there exist multipliers $\ell_0 \ge 0$, $\sigma = (\sigma_x, \sigma_y)^\top \in \mathbb{R}^2$, $\lambda = (\lambda_x, \lambda_y)^\top \in BV^2(\mathcal{I})$, and $\mu = (\mu_x, \mu_y)^\top \in NBV^2(\mathcal{I})$ with $(\ell_0, \sigma, \lambda, \mu) \ne \Theta$ and

$$\begin{aligned}
\lambda_x(10) = \lambda_y(10) &= 0, \\
\lambda_x(0) &= -\sigma_x, \\
\lambda_y(0) &= -\sigma_y, \\
\lambda_x(t) &= \lambda_x(10) + \int_t^{10} \mathcal{H}_x'[\tau]d\tau - \int_t^{10} d\mu_x(\tau) = \mu_x(t) - \mu_x(10), \\
\lambda_y(t) &= \lambda_y(10) + \int_t^{10} \mathcal{H}_y'[\tau]d\tau - \int_t^{10} d\mu_y(\tau) = \mu_y(t) - \mu_y(10).
\end{aligned}$$

Notice that the case $\ell_0 = 0$ can be excluded, since the Mangasarian–Fromowitz conditions are easily being checked to be satisfied for this problem. Hence, we set $\ell_0 = 1$.

Almost everywhere in $\mathcal{I}$, it holds for every $u, v \in [0, 1]$:

$$\begin{aligned}
0 &\le \mathcal{H}_u'[t](u - \hat{u}(t)) = (-(10 - t) - \lambda_x(t) + \lambda_y(t))(u - \hat{u}(t)) \\
&= (-(10 - t) - \mu_x(t) + \mu_x(10) + \mu_y(t) - \mu_y(10))(u - \hat{u}(t)), \\
0 &\le \mathcal{H}_v'[t](v - \hat{v}(t)) = (-t - \lambda_y(t))(v - \hat{v}(t)) \\
&= (-t - \mu_y(t) + \mu_y(10))(v - \hat{v}(t)).
\end{aligned}$$

Finally, μ_x, μ_y are non-decreasing functions with $\mu_x(0) = \mu_y(0) = 0$. μ_x and μ_y are constant on intervals with measure greater than zero and $x(t) > 0$ and $y(t) > 0$, respectively.

From the optimality conditions, we conclude

$$\hat{u}(t) = \begin{cases} 0, & \text{if } \lambda_y(t) - \lambda_x(t) > 10 - t, \\ 1, & \text{if } \lambda_y(t) - \lambda_x(t) < 10 - t, \end{cases}$$

$$\hat{v}(t) = \begin{cases} 1, & \text{if } -\lambda_y(t) < t, \\ 0, & \text{if } -\lambda_y(t) > t. \end{cases}$$

The monotonicity properties of μ_x and μ_y imply that $\lambda_x(t) = \mu_x(t) - \mu_x(10) \le 0$ and $\lambda_y(t) = \mu_y(t) - \mu_y(10) \le 0$ are non-decreasing functions. Since the function t is strictly increasing and $-\lambda_y(t) \ge 0$ is monotonically decreasing, there exists at most one point $\hat{t} \in [0, 10]$ with $-\lambda_y(\hat{t}) = \hat{t}$. Hence, the control $\hat{v}$ is a bang-bang control with at most one switching point $\hat{t}$. This defines the structure of the control $\hat{v}$.

The first differential equation yields

$$\hat{x}(t) = 4 - \int_0^t \hat{u}(\tau)d\tau, \quad 0 \le t \le 10.$$

Let us assume for a moment that $\hat{x}(t) > 0$ holds for all $t \in [0, 10]$, i. e. the state constraint $x(t) \ge 0$ is inactive. Then, $\mu_x(t) \equiv 0$ and $\lambda_x(t) = \mu_x(t) - \mu_x(10) \equiv 0$. The optimality condition yields $\hat{u}(t) \equiv 1$, because $\lambda_y(t) - \lambda_x(t) = \lambda_y(t) \le 0 \le 10 - t$. But this contradicts $\hat{x}(t) > 0$ for all $t \in [0, 10]$. Hence, there exists a first point $\tilde{t} \in (0, 10]$ with $\hat{x}(\tilde{t}) = 0$. Once $\hat{x}(\tilde{t}) = 0$ is fulfilled, it holds $\hat{x}(t) = 0$ and $\hat{u}(t) = 0$ for all $t \in [\tilde{t}, 10]$ because of

$$0 \le \hat{x}(t) = \hat{x}(\tilde{t}) - \int_{\tilde{t}}^t \hat{u}(\tau)d\tau = -\int_{\tilde{t}}^t \hat{u}(\tau)d\tau \le 0.$$

Due to the optimality conditions this implies $\lambda_y(t) - \lambda_x(t) \ge 10 - t$ in $(\tilde{t}, 10]$.

Since $\tilde{t}$ is the first point satisfying $\hat{x}(\tilde{t}) = 0$, it holds $\hat{x}(t) > 0$ in $[0, \tilde{t})$ and thus $\mu_x(t) \equiv 0$, respectively, $\lambda_x(t) \equiv -\mu_x(10)$ in $[0, \tilde{t})$. Hence, $\lambda_y(t) - \lambda_x(t)$ is non-decreasing in $[0, \tilde{t})$. Since $10 - t$ is strictly decreasing, there is at most one point $t_1 \in [0, \tilde{t})$ with $\lambda_y(t_1) - \lambda_x(t_1) = 10 - t_1$. Assume $t_1 < \tilde{t}$. Then, necessarily, $\hat{u}(t) = 0$ in $(t_1, \tilde{t})$ and thus $\hat{u}(t) = 0$ in $(t_1, 10]$ and $\hat{x}(t) \equiv \hat{x}(t_1)$ in $[t_1, 10]$. Then, either $\hat{x}(t_1) > 0$, which contradicts the existence of $\tilde{t}$, or $\hat{x}(t_1) = 0$, which contradicts the minimality of $\tilde{t}$. Consequently, there is no point t_1 in $[0, \tilde{t})$ with $\lambda_y(t_1) - \lambda_x(t_1) = 10 - t_1$. Therefore, either $\lambda_y(t) - \lambda_x(t) > 10 - t$ in $[0, \tilde{t})$, which implies $\hat{u}(t) = 0$ and contradicts the existence of $\tilde{t}$, or $\lambda_y(t) - \lambda_x(t) < 10 - t$ in $[0, \tilde{t})$, which implies $\hat{u}(t) = 1$ in $[0, \tilde{t})$. Summarizing, these considerations yield

$$\begin{aligned} \tilde{t} &= 4, \\ \hat{u}(t) &= \begin{cases} 1, & \text{if } t \in [0, 4), \\ 0, & \text{if } t \in [4, 10], \end{cases} \\ \hat{x}(t) &= \begin{cases} 4 - t, & \text{if } t \in [0, 4), \\ 0, & \text{if } t \in [4, 10], \end{cases} \\ \mu_x(t) &= 0, \quad t \in [0, 4), \\ \lambda_x(t) = \mu_x(t) - \mu_x(10) &= -\mu_x(10), \quad t \in [0, 4), \\ \lambda_y(\tilde{t}) - \lambda_x(\tilde{t}) &= 10 - \tilde{t} = 6. \end{aligned}$$

Now, we have to determine the switching point $\hat{t}$ of $\hat{v}$. It turns out that $\hat{t} = 2$ is the only possible choice. All other cases ($\hat{t}$ does not occur in $[0, 10]$, $\hat{t} \neq 2$) will lead to contradictions. Hence, we have

$$\hat{t} = 2, \qquad \hat{v}(t) = \begin{cases} 0, & \text{if } t \in [0, 2), \\ 1, & \text{if } t \in [2, 10]. \end{cases}$$

Using these controls, it is easy to check that $\hat{y}$ becomes active exactly at $t = 10$. Hence,

$$\mu_y(t) = \begin{cases} 0, & \text{in } [0, 10), \\ \mu_y(10) \geq 0, & \text{if } t = 10, \end{cases}$$

$$\lambda_y(t) = \mu_y(t) - \mu_y(10) = \begin{cases} -\mu_y(10), & \text{in } [0, 10), \\ 0, & \text{if } t = 10. \end{cases}$$

On the other hand, we know already $\hat{t} = -\lambda_y(\hat{t})$ and hence $\mu_y(10) = 2$.

Moreover, $\mu_x(t) = 0$ in $[0, 4)$ and hence

$$\lambda_x(t) = \mu_x(t) - \mu_x(10) = \begin{cases} -\mu_x(10), & \text{if } 0 \leq t < 4, \\ \leq 0, & \text{if } 4 \leq t < 10, \\ 0, & \text{if } t = 10. \end{cases}$$

On the other hand, we know already $6 = \lambda_y(\tilde{t}) - \lambda_x(\tilde{t}) = -2 - \lambda_x(\tilde{t})$ and hence

$$\lambda_x(\tilde{t}) = \mu_x(\tilde{t}) - \mu_x(10) = -\int_{\tilde{t}}^{10} d\mu_x(\tau) = -8.$$

Finally, since $\hat{u}(t) \equiv 0$ in $[4, 10]$, it follows

$$\lambda_y(t) - \lambda_x(t) = -2 - \lambda_x(t) > 10 - t$$

in $(4, 10)$, i. e. $\mu_x(10) - \mu_x(t) > 12 - t$ in $(0, 4)$. □

Next, the discretization of Problem 5.5.4 by the explicit Euler method is investigated in detail. Application of the explicit Euler method with constant step-size $h = (t_f - t_0)/N$, and equidistant grid points $t_i = t_0 + ih$, $i = 0, \dots, N$, to Problem 5.5.4 leads to

Problem 5.5.6 (Discretized problem).

Minimize

$$-h \sum_{i=0}^{N-1} (10 - t_i)u_i + t_i v_i$$

subject to the constraints

$$
\begin{aligned}
x_0 &= 4, \\
x_{i+1} &= x_i - hu_i, && i = 0,1,\dots,N-1, \\
y_0 &= 4, \\
y_{i+1} &= y_i + hu_i - hv_i, && i = 0,1,\dots,N-1, \\
x_i &\geq 0, && i = 0,1,\dots,N, \\
y_i &\geq 0, && i = 0,1,\dots,N, \\
u_i &\in [0,1], && i = 0,1,\dots,N-1, \\
v_i &\in [0,1], && i = 0,1,\dots,N-1.
\end{aligned}
$$

The Lagrange function of Problem 5.5.6 with

$$
\begin{aligned}
x &= (x_1,\dots,x_N)^\top, & y &= (y_1,\dots,y_N)^\top, \\
u &= (u_0,\dots,u_{N-1})^\top, & v &= (v_0,\dots,v_{N-1})^\top, \\
\lambda^x &= (\lambda_1^x,\dots,\lambda_N^x)^\top, & \lambda^y &= (\lambda_1^y,\dots,\lambda_N^y)^\top, \\
\mu^x &= (\mu_1^x,\dots,\mu_N^x)^\top, & \mu^y &= (\mu_1^y,\dots,\mu_N^y)^\top
\end{aligned}
$$

is given by

$$
\begin{aligned}
L(x,y,u,v,\lambda^x,\lambda^y,\mu^x,\mu^y) = {} & -h \sum_{i=0}^{N-1} [(10 - t_i)u_i + t_i v_i] \\
& + \sum_{i=0}^{N-1} \lambda_{i+1}^x (x_i - hu_i - x_{i+1}) \\
& + \sum_{i=0}^{N-1} \lambda_{i+1}^y (y_i + hu_i - hv_i - y_{i+1}) \\
& + \sum_{i=0}^{N} \mu_i^x (-x_i) + \sum_{i=0}^{N} \mu_i^y (-y_i).
\end{aligned}
$$

Notice that $x_0 = 4$ and $y_0 = 4$ are not considered as constraints in Problem 5.5.6.

We intend to evaluate the Fritz John conditions in Theorem 2.3.25 for Problem 5.5.6. Notice that we can choose $\ell_0 = 1$, since Problem 5.5.6 is linear. Theorem 2.3.25 with $\ell_0 = 1$ yields the following necessary (and in this case also sufficient) optimality conditions for an optimal solution $\hat{x}, \hat{y}, \hat{u}, \hat{v}$ with multipliers $\lambda^x, \lambda^y, \mu^x, \mu^y$:

(i) For $i = 1,\dots,N$, it holds $0 = L'_{x_i}(\hat{x},\hat{y},\hat{u},\hat{v},\lambda^x,\lambda^y,\mu^x,\mu^y)$, i.e.

$$
\begin{aligned}
0 &= \lambda_{i+1}^x - \lambda_i^x - \mu_i^x, \quad i = 1,\dots,N-1, \\
0 &= \lambda_N^x - \mu_N^x.
\end{aligned}
$$

Recursive evaluation leads to

$$
\lambda_i^x = -\sum_{j=i}^{N} \mu_j^x, \quad i = 1,\dots,N.
$$

(ii) For $i = 1, \ldots, N$, it holds $0 = L'_{y_i}(\hat{x}, \hat{y}, \hat{u}, \hat{v}, \lambda^x, \lambda^y, \mu^x, \mu^y)$, i. e.

$$\begin{aligned} 0 &= \lambda^y_{i+1} - \lambda^y_i - \mu^y_i, \quad i = 1, \ldots, N-1, \\ 0 &= \lambda^y_N - \mu^y_N. \end{aligned}$$

Recursive evaluation leads to

$$\lambda^y_i = -\sum_{j=i}^{N} \mu^y_j, \quad i = 1, \ldots, N.$$

(iii) For $i = 0, \ldots, N$, it holds $L'_{u_i}(\hat{x}, \hat{y}, \hat{u}, \hat{v}, \lambda^x, \lambda^y, \mu^x, \mu^y)(u - \hat{u}_i) \geq 0$ for all $u \in [0, 1]$, i. e.

$$(-h(10 - t_i) - h\lambda^x_{i+1} + h\lambda^y_{i+1})(u - \hat{u}_i) \geq 0.$$

This implies

$$\hat{u}_i = \begin{cases} 1, & \text{if } -(10 - t_i) + \sum_{j=i+1}^{N}(\mu^x_j - \mu^y_j) < 0, \\ 0, & \text{if } -(10 - t_i) + \sum_{j=i+1}^{N}(\mu^x_j - \mu^y_j) > 0, \\ \text{undefined}, & \text{otherwise}. \end{cases} \tag{5.96}$$

(iv) For $i = 0, \ldots, N$, it holds $L'_{v_i}(\hat{x}, \hat{y}, \hat{u}, \hat{v}, \lambda^x, \lambda^y, \mu^x, \mu^y)(v - \hat{v}_i) \geq 0$ for all $v \in [0, 1]$, i. e.

$$(-ht_i - h\lambda^y_{i+1})(v - \hat{v}_i) \geq 0.$$

This implies

$$\hat{v}_i = \begin{cases} 1, & \text{if } -t_i + \sum_{j=i+1}^{N}\mu^y_j < 0, \\ 0, & \text{if } -t_i + \sum_{j=i+1}^{N}\mu^y_j > 0, \\ \text{undefined}, & \text{otherwise}. \end{cases} \tag{5.97}$$

(v) It holds $\mu^x_i x_i = 0$, $\mu^x_i \geq 0$, $i = 0, \ldots, N$, and $\mu^y_i y_i = 0$, $\mu^y_i \geq 0$, $i = 0, \ldots, N$.

We verify that

$$\hat{u}_i = \begin{cases} 1, & \text{for } i = 0, \ldots, m-1, \\ 4/h - m, & \text{for } i = m, \\ 0, & \text{for } i = m+1, \ldots, N-1, \end{cases} \tag{5.98}$$

$$\hat{v}_i = \begin{cases} 0, & \text{for } i = 0, \ldots, q-1, \\ q + 1 - 2/h, & \text{for } i = q, \\ 1, & \text{for } i = q+1, \ldots, N-1, \end{cases} \tag{5.99}$$

with

$$m = \left\lfloor \frac{4}{h} \right\rfloor, \quad q = \left\lfloor \frac{2}{h} \right\rfloor,$$

is an optimal control for the discretized problem 5.5.6. To this end, by application of these controls, we find

$$\begin{aligned}
\hat{x}_i &= 4 - h\sum_{j=0}^{i-1} \hat{u}_j = 4 - ih > 0, \quad i = 0, \dots, m, \\
\hat{x}_{m+1} &= x_m - h\hat{u}_m = 4 - mh - h(4/h - m) = 0, \\
\hat{x}_i &= 0, \quad i = m+2, \dots, N.
\end{aligned}$$

Similarly, we have

$$\begin{aligned}
\hat{y}_i &= 4 + h\sum_{j=0}^{i-1}(\hat{u}_j - \hat{v}_j) = 4 + ih, \quad i = 0, \dots, q, \\
\hat{y}_{q+1} &= \hat{y}_q + h(\hat{u}_q - \hat{v}_q) = 4 + qh + h(1 - (q + 1 - 2/h)) = 6, \\
\hat{y}_i &= 6, \quad i = q+2, \dots, m, \\
\hat{y}_{m+1} &= \hat{y}_m + h(\hat{u}_m - \hat{v}_m) = 6 + h(4/h - m - 1) = 10 - h(m+1), \\
\hat{y}_i &= \hat{y}_{m+1} + h\sum_{j=m+1}^{i-1}(\hat{u}_j - \hat{v}_j) \\
&= 10 - h(m+1) - h(i - 1 - m) = 10 - ih, \quad i = m+2, \dots, N.
\end{aligned}$$

Notice that $\hat{y}_i > 0$ for $i = 0, \dots, N-1$ and $\hat{y}_N = 0$. Hence, according to the complementarity conditions (v), the multipliers μ_i^y must satisfy $\mu_i^y = 0$ for $i = 1, \dots, N-1$. With (5.97) and (5.99), necessarily it holds

$$\mu_N^y \in (t_{q-1}, t_q).$$

Taking the limit $h \longrightarrow 0$, we find $t_q = qh = \lfloor 2/h \rfloor h \longrightarrow 2$ and $t_{q-1} \longrightarrow 2$ and thus $\mu_N^y \longrightarrow 2$.

According to the complementarity conditions (v), the multipliers μ_i^x must satisfy $\mu_i^x = 0$ for $i = 1, \dots, m$. With (5.96) and (5.98), necessarily it holds

$$\begin{aligned}
-(10 - t_{m-1}) + \sum_{j=m}^{N}(\mu_j^x - \mu_j^y) &< 0, \\
-(10 - t_m) + \sum_{j=m+1}^{N}(\mu_j^x - \mu_j^y) &> 0,
\end{aligned}$$

and thus

$$\mu_N^y + (10 - t_m) < \sum_{j=m}^{N} \mu_j^x < \mu_N^y + (10 - t_{m-1}).$$

Taking the limit $h \longrightarrow 0$ yields $t_m = mh = \lfloor 4/h \rfloor h \longrightarrow 4$ and

$$\sum_{j=m}^{N} \mu_j^x \longrightarrow 8.$$

Summarizing, the above considerations showed that (5.98), (5.99), and the resulting discrete state variables are optimal solutions of the discretized problem provided the multipliers μ^x and μ^y are chosen accordingly. Furthermore, the switching points t_q and t_m converge at a linear rate to the switching points of the continuous problem 5.5.4. The discrete states (viewed as continuous, piecewise linear functions) converge for the norm $\|\cdot\|_{1,\infty}$, whereas the discrete controls (viewed as piecewise constant functions) do not converge to the continuous controls for the norm $\|\cdot\|_\infty$. The discrete controls, however, do converge for the norm $\|\cdot\|_1$. Due to the non-uniqueness of the continuous and discrete multipliers, it is hard to observe convergence for these quantities.

Figure 5.10 shows the numerical solution for $N = 999$ and piecewise constant control approximation. Figures 5.11, 5.12, 5.13, and 5.14 depict the multiplier estimates for μ^x, μ^y, λ^x, and λ^y, respectively. The numerical solution was computed by the software package OCPID-DAE1 [123].

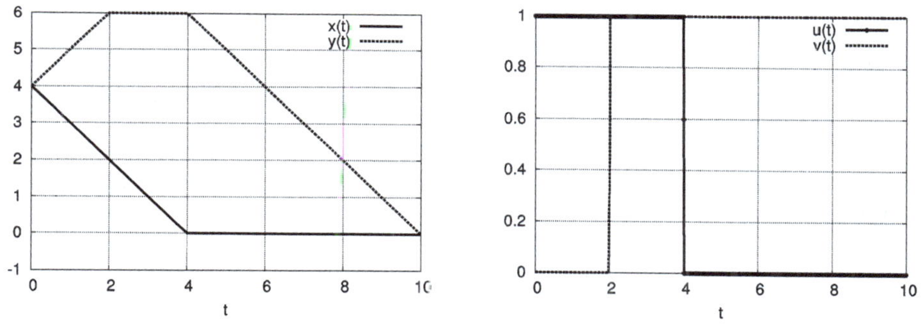

Figure 5.10: Optimal approximate state and control for $N = 999$ grid points and piecewise constant control approximation.

5.6 An overview on convergence results

The convergence of discretized optimal control problems is a current field of research. Only a few results are available for ODE and DAE optimal control problems. We summarize the existing results without proving them. All results assume that the optimal control is at least continuous. A convergence result for linear optimal control problems with discontinuous controls can be found in [2].

5.6.1 Convergence of the Euler discretization for ODE optimal control problems

The proof of convergence for the full Euler discretization can be found in [223]. More specifically, the authors investigate the following optimal control problem on $\mathcal{I} := [0, t_f]$ with $\xi \in \mathbb{R}^{n_x}$ given:

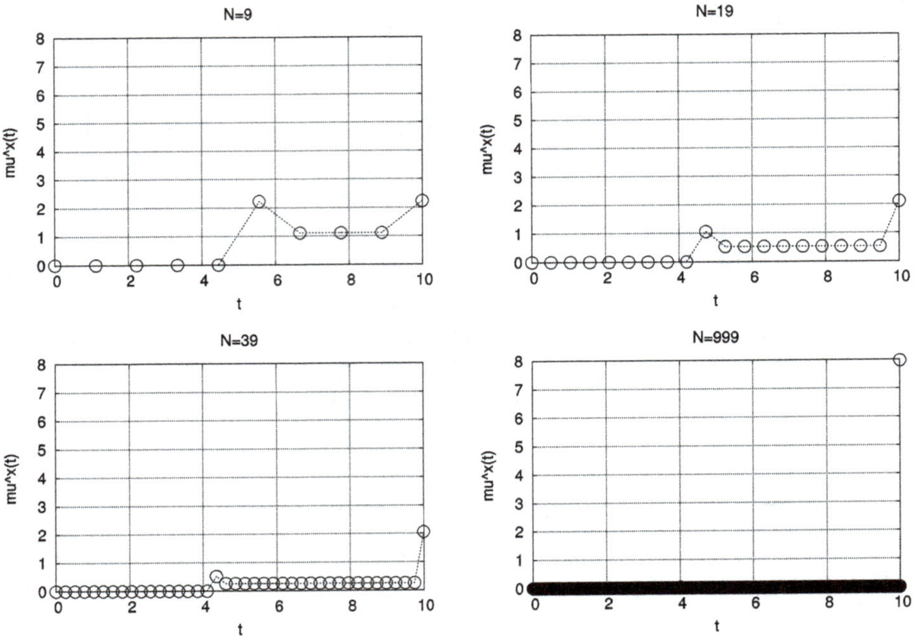

Figure 5.11: Lagrange multiplier μ^x for the discretized state constraint $x_i \geq 0$ in Problem 5.5.6 for $N = 9, 19, 39, 999$.

Minimize

$$\varphi\big(x(t_f)\big) + \int_0^{t_f} f_0\big(x(t), u(t)\big)dt$$

subject to the constraints

$$\begin{aligned} \dot{x}(t) - f\big(x(t), u(t)\big) &= 0_{\mathbb{R}^{n_x}}, \\ x(0) - \xi &= 0_{\mathbb{R}^{n_x}}, \\ \psi\big(x(t_f)\big) &= 0_{\mathbb{R}^{n_\psi}}, \\ c\big(x(t), u(t)\big) &\leq 0_{\mathbb{R}^{n_c}}. \end{aligned}$$

Its discretization by the explicit Euler method looks as follows:

Minimize

$$\varphi(x_N) + h \sum_{j=0}^{N-1} f_0(x_j, u_j)$$

subject to the constraints

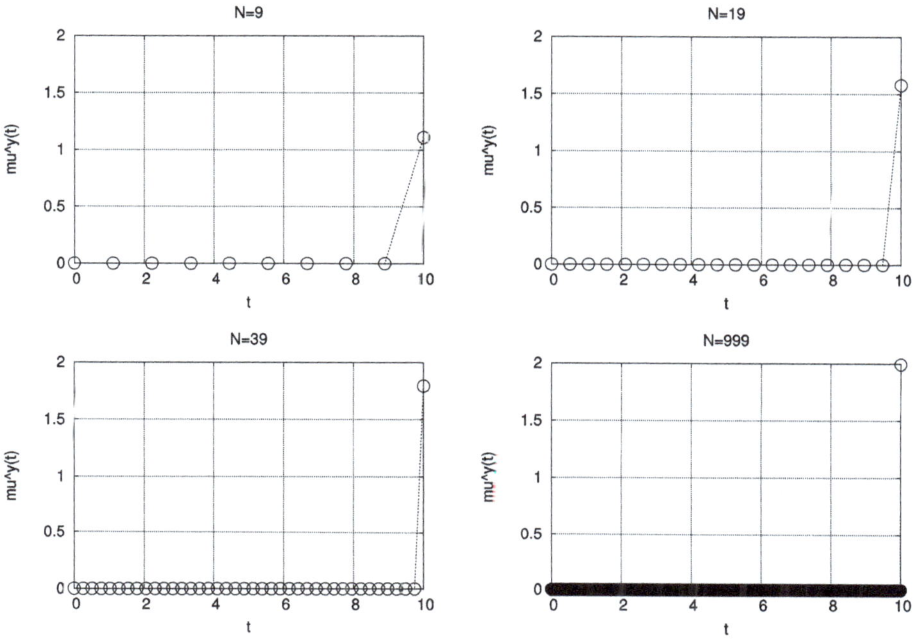

Figure 5.12: Lagrange multiplier μ^y for the discretized state constraint $y_j \geq 0$ in Problem 5.5.6 for $N = 9, 19, 39, 999$.

$$
\begin{aligned}
\frac{x_{j+1} - x_j}{h} - f(x_j, u_j) &= 0_{\mathbb{R}^{n_x}}, \quad j = 0, \dots, N-1, \\
x_C - \xi &= 0_{\mathbb{R}^{n_x}}, \\
\psi(x_N) &= 0_{\mathbb{R}^{n_\psi}}, \\
c(x_j, u_j) &\leq 0_{\mathbb{R}^{n_c}}, \quad j = 0, \dots, N-1.
\end{aligned}
$$

We merely summarize the assumptions needed to prove a convergence result. $\hat{\mathcal{H}}$ denotes the augmented Hamilton function as usual.

Assumption 5.6.1. (a) f_0, f, c, φ, ψ are differentiable with locally Lipschitz continuous derivatives.

(b) There exists a local solution $(\hat{x}, \hat{u}) \in C_1^{n_x}(\mathcal{I}) \times C^{n_u}(\mathcal{I})$ of the optimal control problem.

(c) *Uniform rank condition for c*: Let

$$
\mathcal{J}(t) := \{i \in \{1, \dots, n_c\} \mid c_i(\hat{x}(t), \hat{u}(t)) \geq -\alpha\} \quad \text{for some } \alpha > 0
$$

denote the index set of active mixed control-state constraints at t and $c_{\mathcal{J}(t)}[t]$ the active constraints at t. There exists a constant $\beta > 0$ with

$$
\|c'_{\mathcal{J}(t),u}[t]^\top d\| \geq \beta \|d\| \quad \text{for all } d \in \mathbb{R}^{|\mathcal{J}(t)|} \text{ almost everywhere in } \mathcal{I}.
$$

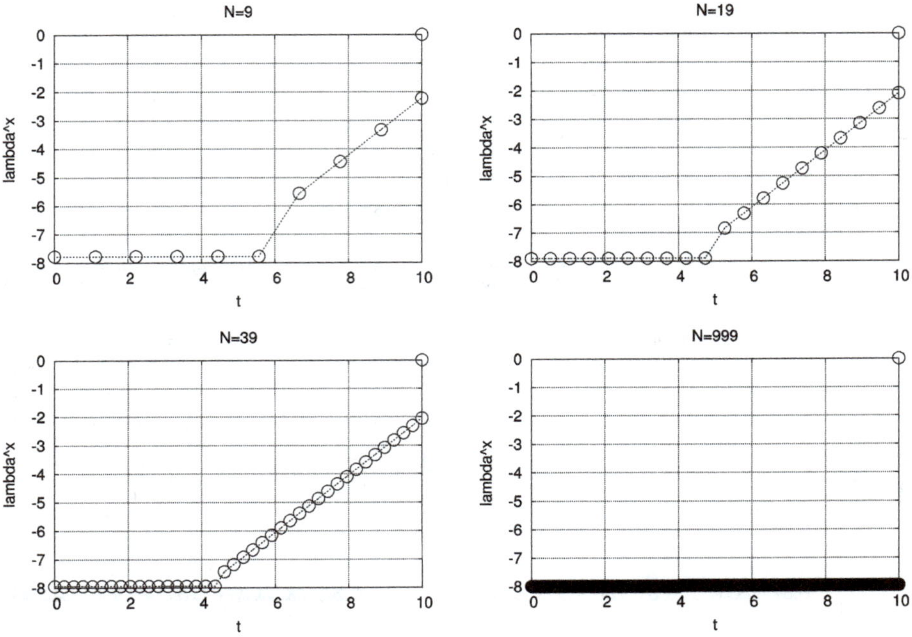

Figure 5.13: Adjoint estimation λ^x for the discretized differential equation in Problem 5.5.6 for $N = 9, 19, 39, 999$.

(d) *Surjectivity of the linearized equality constraints:* The boundary value problem

$$\dot{y}(t) - \tilde{A}(t)y(t) - \tilde{B}(t)v(t) = 0_{\mathbb{R}^{n_x}}, \quad y(0) = 0_{\mathbb{R}^{n_x}}, \quad \psi'_{x_f}(\hat{x}(t_f))y(t_f) = h$$

possesses a solution for any $h \in \mathbb{R}^{n_\psi}$, where

$$\begin{aligned}\tilde{A}(t) &= f'_x[t] - f'_u[t]c'_{\mathcal{J}(t),u}[t]^\top \left(c'_{\mathcal{J}(t),u}[t]c'_{\mathcal{J}(t),u}[t]^\top\right)^{-1} c'_{\mathcal{J}(t),x}[t],\\ \tilde{B}(t) &= f'_u[t]\left(I_{n_u} - c'_{\mathcal{J}(t),u}[t]^\top \left(c'_{\mathcal{J}(t),u}[t]c'_{\mathcal{J}(t),u}[t]^\top\right)^{-1} c'_{\mathcal{J}(t),x}[t]\right).\end{aligned}$$

(e) *Coercivity:* Define $\mathcal{J}^+(t) := \{i \in \mathcal{J}(t) \mid \eta_i(t) > \delta\}$ for some $\delta \geq 0$, where η denotes the multiplier for the mixed control-state constraint c.
There exists $\bar{\beta} > 0$ such that

$$d^\top \hat{\mathcal{H}}''_{uu}[t]d \geq \bar{\beta}\|d\|^2$$

holds for all $d \in \mathbb{R}^{n_u}$ with

$$c'_{J^+(t),u}[t]d = 0_{\mathbb{R}^{|\mathcal{J}^+(t)|}}.$$

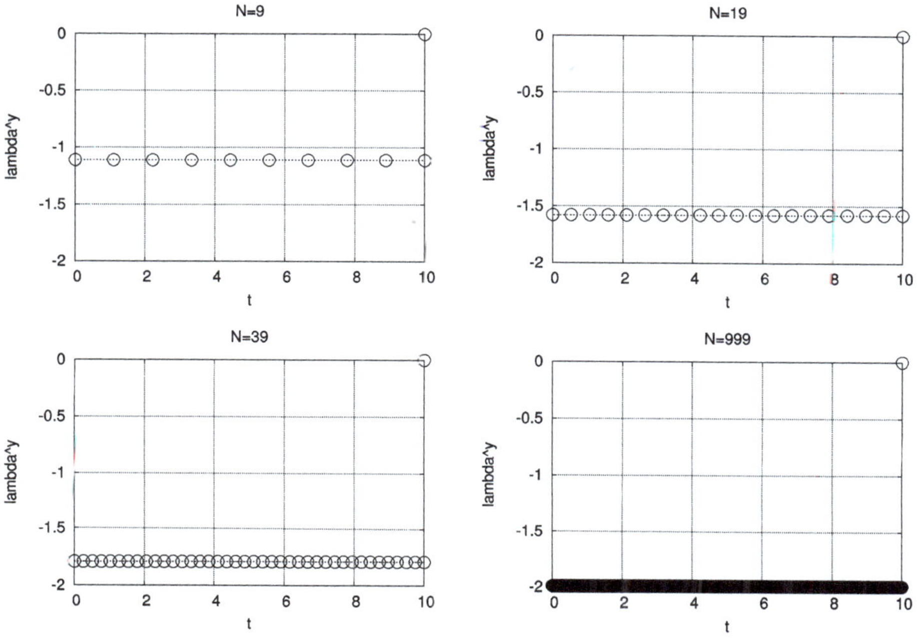

Figure 5.14: Adjoint estimation λ^y for the discretized differential equation in Problem 5.5.6 for $N = 9, 19, 39, 999$.

(f) *Riccati equation:* Let the Riccati equation

$$\begin{aligned}\dot{Q}(t) &= -Q(t)f_x'[t] - f_x'[t]^\top Q(t) - \hat{\mathcal{H}}_{xx}''[t] \\ &\quad + \left[\begin{pmatrix}\hat{\mathcal{H}}_{ux}''[t]\\ c_{J^+(t),x}'[t]\end{pmatrix}^\top + Q(t)\begin{pmatrix} f_u'[t]^\top \\ \Theta\end{pmatrix}^\top\right]\begin{pmatrix}\hat{\mathcal{H}}_{uu}''[t] & c_{J^+(t),u}'[t]^\top\\ c_{J^+(t),u}'[t] & \Theta\end{pmatrix} \\ &\quad \cdot \left[\begin{pmatrix} f_u'[t]^\top\\ \Theta\end{pmatrix} Q(t) + \begin{pmatrix}\hat{\mathcal{H}}_{ux}''[t]\\ c_{J^+(t),x}'[t]\end{pmatrix}\right]\end{aligned}$$

possess a bounded solution Q on $\mathcal{I}$ that satisfies the rank condition

$$d^\top(\Gamma - Q(t_f))d \geq 0 \quad \text{for all } d \in \mathbb{R}^{n_x} \text{ with } \psi_{x_f}'(\hat{x}(t_f))d = 0_{\mathbb{R}^{n_\psi}},$$

where

$$\Gamma := (\varphi(\hat{x}(t_f)) + \sigma_f^\top \psi(\hat{x}(t_f)))_{xx}''.$$

□

All function evaluations are at the optimal solution $(\hat{x}, \hat{u})$. The following convergence result was obtained in [223]:

Theorem 5.6.2. *Let Assumption* 5.6.1 *hold. Then for sufficiently small step-sizes* $h > 0$, *there exist a locally unique KKT point* $(x_h, u_h, \lambda_h, \eta_h, \kappa_0, \kappa_f)$ *of the discretized problem satisfying*

$$\max\{\|x_h - \hat{x}\|_{1,\infty}, \|u_h - \hat{u}\|_\infty, \|\lambda_h - \lambda\|_{1,\infty}, \|\kappa_0 - \sigma_0\|, \|\kappa_f - \sigma_f\|, \|\eta_h - \eta\|_\infty\} = \mathcal{O}(h),$$

where λ_h *denotes the discrete adjoint,* η_h *the discrete multiplier for the mixed control-state constraint,* κ_0 *the discrete multiplier for the initial condition, and* κ_f *the discrete multiplier for the final condition.* □

The assumptions (e) and (f) together are sufficient for local optimality of $(\hat{x}, \hat{u})$.

Similar convergence results can be found in [81, 80].

5.6.2 Higher order of convergence for Runge–Kutta discretizations for ODE optimal control problems

Hager [152] investigates optimal control problems on $\mathcal{I} := [0,1]$ with $\xi \in \mathbb{R}^{n_x}$ given:

Minimize $\varphi(1)$ *subject to the constraints*

$$\dot{x}(t) = f\big(x(t), u(t)\big), \quad x(0) = \xi.$$

Its discretization by an s-stage Runge–Kutta method with fixed step-size $h = 1/N$ looks as follows:

Minimize $\varphi(x_N)$ *subject to the constraints*

$$\begin{aligned} \frac{x_{j+1} - x_j}{h} &= \sum_{i=1}^{s} b_i f(\eta_i, u_{ji}), && j = 0, \ldots, N-1, \\ \eta_i &= x_j + h \sum_{\ell=1}^{s} a_{i\ell} f(\eta_\ell, u_{j\ell}), && i = 1, \ldots, s, \\ x_0 &= \xi. \end{aligned}$$

Notice that for each function evaluation of f an independent optimization variable u_{ji} is introduced, compare Example 5.1.6. Hence, there is no coupling through, e. g., piecewise constant interpolation or linear interpolation.

Assumption 5.6.3. (a) *Smoothness:* The optimal control problem possesses a solution

$$(\hat{x}, \hat{u}) \in W^{n_x}_{p,\infty}(\mathcal{I}) \times W^{n_u}_{p-1,\infty}(\mathcal{I}) \quad \text{with } p \geq 2.$$

The first p derivatives of f and φ are supposed to be locally Lipschitz continuous in some neighborhood of $(\hat{x}, \hat{u})$.

(b) *Coercivity:* There exists some $\alpha > 0$ with

$$\mathcal{B}(x, u) \geq \alpha \|u\|_2^2 \quad \text{for all } (x, u) \in \mathcal{M},$$

where

$$\mathcal{B}(x, u) := \frac{1}{2}\left(x(1)^\top V x(1) + \int_0^1 x(t)^\top Q(t) x(t) + 2x(t)^\top S(t) u(t) + u(t)^\top R(t) u(t) dt \right)$$

and

$$\begin{aligned} A(t) &:= f_x'(\hat{x}(t), \hat{u}(t)), & B(t) &:= f_u'(\hat{x}(t), \hat{u}(t)), \\ V &:= \varphi''(\hat{x}(1)), & Q(t) &:= \mathcal{H}_{xx}''(\hat{x}(t), \hat{u}(t), \lambda(t)), \\ R(t) &:= \mathcal{H}_{uu}''(\hat{x}(t), \hat{u}(t), \lambda(t)), & S(t) &:= \mathcal{H}_{xu}''(\hat{x}(t), \hat{u}(t), \lambda(t)), \end{aligned}$$

and

$$\mathcal{M} := \{(x, u) \in W_{1,2}^{n_x}(\mathcal{I}) \times L_2^{n_u}(\mathcal{I}) \mid \dot{x} = Ax + Bu,\ x(0) = 0_{\mathbb{R}^{n_x}}\}. \qquad \square$$

The smoothness property in (a) implies that the optimal control $\hat{u}$ is at least continuous.

It turns out that the well-known order conditions for Runge–Kutta methods for initial value problems are not enough to ensure higher-order convergence for discretized optimal control problems. To ensure the latter, the Runge–Kutta method has to satisfy the stronger conditions in Table 5.6. To distinguish these conditions from the ordinary conditions for initial value problems, we refer to the conditions in Table 5.6 as *optimal control order conditions*. A closer investigation yields that they are identical with the initial value problem conditions only up to order $p = 2$.

Table 5.6: Optimal control order conditions for Runge–Kutta methods up to order 4 (higher-order conditions require the validity of all conditions of lower order).

Order	Conditions ($c_i = \sum a_{ij}$, $d_j = \sum b_i a_{ij}$)
$p = 1$	$\sum b_i = 1$
$p = 2$	$\sum d_i = \frac{1}{2}$
$p = 3$	$\sum c_i d_i = \frac{1}{6}, \sum b_i c_i^2 = \frac{1}{3}, \sum \frac{d_i^2}{b_i} = \frac{1}{3}$
$p = 4$	$\sum b_i c_i^3 = \frac{1}{4}, \sum b_i c_i a_{ij} c_j = \frac{1}{8}, \sum d_i c_i^2 = \frac{1}{12}, \sum d_i a_{ij} c_j = \frac{1}{24},$ $\sum \frac{c_i d_i^2}{b_i} = \frac{1}{12}, \sum \frac{d_i^3}{b_i^2} = \frac{1}{4}, \sum \frac{b_i c_i a_{ij} d_j}{b_j} = \frac{5}{24}, \sum \frac{d_i a_{ij} d_j}{b_j} = \frac{1}{8}$

The following convergence result holds, see [152]:

Theorem 5.6.4. *Let Assumption 5.6.3 hold. Let $b_i > 0$, $i = 1, \dots, s$, hold for the coefficients of the Runge–Kutta method. Let the Runge–Kutta method be of optimal control order p, see Table 5.6.*

Then, for any sufficiently small step-size $h > 0$, there exists a strict local minimum of the discretized optimal control problem.

If $d^{p-1}\hat{u}/dt^{p-1}$ is of bounded variation, then

$$\max_{0 \le j \le N} \{\|x_j - \hat{x}(t_j)\| + \|\lambda_j - \lambda(t_j)\| + \|u^*(x_j, \lambda_j) - \hat{u}(t_j)\|\} = \mathcal{O}(h^p).$$

If $d^{p-1}\hat{u}/dt^{p-1}$ is Riemann-integrable, then

$$\max_{0 \le j \le N} \{\|x_j - \hat{x}(t_j)\| + \|\lambda_j - \lambda(t_j)\| + \|u^*(x_j, \lambda_j) - \hat{u}(t_j)\|\} = o(h^{p-1}).$$

Herein, $u^*(x_k, \lambda_k)$ denotes a local minimum of the Hamilton function $\mathcal{H}(x_j, u, \lambda_j)$ with respect to u. □

---

i The assumption $b_i > 0$, $i = 1, \dots, s$, is essential as Example 5.1.6 shows.

5.6.3 Convergence results for DAE optimal control problems

The following summary of convergence results is based on the publications [225, 229, 228, 227, 226]. The index-1 case with an implicit Euler discretization is discussed in [229]. Higher-order convergence results for Runge–Kutta methods in the index-1 case are proven in [226]. For sufficient conditions, we refer the reader to [225, Chapter 3].

Next, we focus on semi-explicit DAEs of index two in order to illustrate the difficulties which arise in proving the convergence of an implicit Euler discretization. The major difficulty that needs to be overcome is a *structural discrepancy* of necessary conditions of the infinite optimal control problem and the necessary conditions of its discretization. To illustrate this discrepancy, consider the following autonomous optimal control problem:

Minimize

$$\varphi(x(t_0), x(t_f))$$

subject to the constraints

$$\begin{aligned} 0_{\mathbb{R}^{n_x}} &= f(x(t), y(t), u(t)) - \dot{x}(t), \\ 0_{\mathbb{R}^{n_y}} &= g(x(t)), \\ 0_{\mathbb{R}^{n_\psi}} &= \psi(x(t_0), x(t_f)). \end{aligned}$$

We assume the DAE is of index two, and the assumptions of Theorem 3.1.9 are fulfilled for a local minimum $(\hat{x}, \hat{y}, \hat{u})$. Then the following conditions are necessary:

(i) *Adjoint DAE:* Almost everywhere in $[t_0, t_f]$, we have

$$\begin{aligned}\dot{\lambda}_f(t) &= -\nabla_x \mathcal{H}(\hat{x}(t), \hat{y}(t), \hat{u}(t), \lambda_f(t), \lambda_g(t)),\\ 0_{\mathbb{R}^{n_y}} &= \nabla_y \mathcal{H}(\hat{x}(t), \hat{y}(t), \hat{u}(t), \lambda_f(t), \lambda_g(t)).\end{aligned}$$

(ii) The *transversality conditions* (3.24), (3.25) apply.

(iii) *Stationarity of Hamilton function:* Almost everywhere in $[t_0, t_f]$, we have

$$\nabla_u \mathcal{H}(\hat{x}(t), \hat{y}(t), \hat{u}(t), \lambda_f(t), \lambda_g(t)) = 0_{\mathbb{R}^{n_u}}.$$

Herein, the Hamilton function is defined by

$$\mathcal{H}(x, y, u, \lambda_f, \lambda_g) = \lambda_f^\top f(x, y, u) + \lambda_g^\top g'(x) f(x, y, u).$$

Please notice that the Hamilton function $\mathcal{H}$ includes the term $g'(x)f(x,y,u)$ and not $g(x)$. The former term defines the hidden algebraic equation and arises, if an index reduction is performed for the DAE by differentiating the algebraic constraint $g(x) = 0_{\mathbb{R}^{n_y}}$. Thus, an internal index reduction takes place under the necessary conditions.

Now consider a direct discretization of the DAE optimal control problem using the implicit Euler method on an equidistant grid with step-size $h > 0$:

Minimize

$$\varphi(x_0, x_N)$$

with respect to

$$x_h = (x_0, \ldots, x_N)^\top, \quad y_h = (y_1, \ldots, y_N)^\top, \quad u_h = (u_1, \ldots, u_N)^\top$$

and subject to the constraints

$$\begin{aligned}0_{\mathbb{R}^{n_x}} &= hf(x_j, y_j, u_j) - (x_j - x_{j-1}), \quad && j = 1, \ldots, N,\\ 0_{\mathbb{R}^{n_y}} &= g(x_j), && j = 0, \ldots, N,\\ 0_{\mathbb{R}^{n_\psi}} &= \psi(x_0, x_N).\end{aligned}$$

The discrete local minimum principle in Theorem 5.5.3 with control constraints and state constraints neglected (i. e. $c \equiv 0$ and $s \equiv 0$) is valid for the higher index case as well (with a minor modification in the transversality conditions) and yields the following necessary conditions for a local minimum $(\hat{x}_h, \hat{y}_h, \hat{u}_h)$ of the discretized problem:

(i) *Discrete adjoint equations:*

$$\frac{\lambda_{f,j} - \lambda_{f,j-1}}{h} = -\nabla_x \mathcal{H}(\hat{x}_j, \hat{y}_j, \hat{u}_j, \lambda_{f,j-1}, \lambda_{g,j-1}), \quad j = 1, \ldots, N,$$

$$0_{\mathbb{R}^{n_y}} = \nabla_y \tilde{\mathcal{H}}(\hat{x}_j, \hat{y}_j, \hat{u}_j, \lambda_{f,j-1}, \lambda_{g,j-1}), \quad j = 1, \dots, N.$$

(ii) The *discrete transversality conditions* (5.88) and

$$\lambda_{f,0}^\top = -(\kappa_0 \varphi'_{x_0}(\hat{x}_0, \hat{x}_N) + \kappa^\top \psi'_{x_0}(\hat{x}_0, \hat{x}_N) + \eta^\top g'_x(\hat{x}_0))$$

apply.

(iii) *Discrete optimality conditions:*

$$\nabla_u \tilde{\mathcal{H}}(\hat{x}_j, \hat{y}_j, \hat{u}_j, \lambda_{f,j-1}, \lambda_{g,j-1}) = 0_{\mathbb{R}^{n_u}}, \quad j = 1, \dots, N.$$

Herein, the Hamilton function $\tilde{\mathcal{H}}$ of the discretized problem is defined by

$$\tilde{\mathcal{H}}(x, y, u, \lambda_f, \lambda_g) = \lambda_f^\top f(x, y, u) + \lambda_g^\top g(x).$$

Please notice that the Hamilton function $\mathcal{H}$ of the infinite optimal control problem and the Hamilton function $\tilde{\mathcal{H}}$ of its discretization differ! This is the source of the above mentioned *structural discrepancy* between the necessary conditions of the infinite problem and the necessary conditions of its discretization!

The structural discrepancy between the sets of necessary conditions does not allow proving convergence, in general, owing to a missing consistency of the conditions. In [225, 228], however, an approach was found to overcome this inconsistency. The key idea is to reformulate the algebraic constraints for the discretized problem. The constraints

$$0_{\mathbb{R}^{n_y}} = g(x_j), \quad j = 0, 1, \dots, N,$$

can be equivalently rewritten as

$$\begin{aligned} 0_{\mathbb{R}^{n_y}} &= \frac{g(x_j) - g(x_{j-1})}{h} = \frac{g(x_j) - g(x_j - hf(x_j, y_j, u_j))}{h}, \quad j = 1, \dots, N, \\ 0_{\mathbb{R}^{n_y}} &= g(x_0). \end{aligned}$$

By this reformulation, which resembles an index reduction on discrete level, optimal solutions remain optimal, but the multipliers will change. For this reformulation, a convergence result for problems with mixed control-state constraints was obtained in [228] following the procedures in [223]. An extension, including pure state constraints, can be found in [227].

5.7 Numerical examples

The reduced discretization approach is illustrated for some examples using the software package OCPID-DAE1 [123]. The first example revisits Goddard's problem 1.0.2, [144].

Example 5.7.1 (Goddard problem). Consider the following optimal control problem.

Maximize $h(t_f)$ *subject to the differential equations*

$$\dot{h}(t) = v(t),$$
$$\dot{v}(t) = \frac{1}{m(t)}\big(u(t) - D(v(t), h(t))\big) - \frac{1}{h(t)^2},$$
$$\dot{m}(t) = -2u(t)$$

with $D(v,h) = 310v^2 \exp(500(1-h))$, *the boundary conditions*

$$h(0) = 1, \quad v(0) = 0, \quad m(0) = 1, \quad m(t_f) = 0.6,$$

and the control constraints

$$0 \le u(t) \le 3.5.$$

Figure 5.15 shows the result of a reduced discretization approach using the classic Runge–Kutta method with fixed step-size and $N = 200$. The final time computes to approximately 0.198858. The control structure is bang-singular-bang. □

The next example illustrates a practical strategy – the homotopy strategy – to overcome potential convergence problems. The idea is to solve a sequence of problems by relaxing the constraints and tightening them step-by-step while using the solution of the previous step as an initial guess for the successor.

Example 5.7.2 (Obstacle avoidance and homotopy). We consider a path planning task with fixed obstacles in the (x, y)-plane. The obstacles are assumed to be circles with centers and radii given by

$$(X_k, Y_k, R_k) \in \mathbb{R}^3 \qquad k = 1, \dots, M.$$

We aim to find an optimal path avoiding the obstacles and consider the following optimal control problem with free final time subject to pure state constraints and control constraints:

Minimize

$$a_1 t_f + a_2 \int_0^{t_f} u(t)^2 dt$$

subject to

$$\dot{x}(t) = \cos\psi(t), \qquad x(0) = 0, x(t_f) = 100,$$
$$\dot{y}(t) = \sin\psi(t), \qquad y(0) = 0, y(t_f) = 50,$$
$$\dot{\psi}(t) = u(t), \qquad \psi(0) = 0, \psi(t_f) = -\pi,$$

state constraints

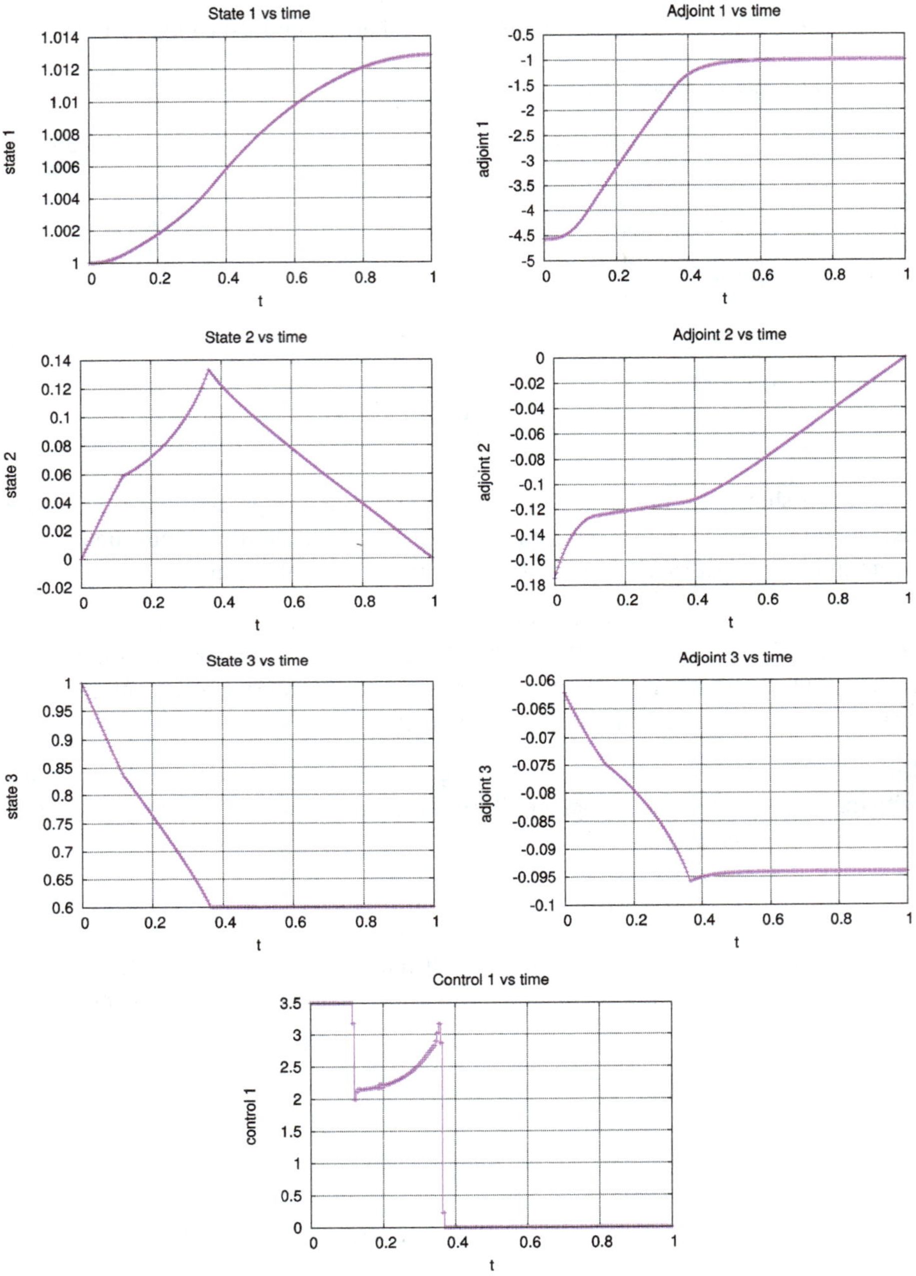

Figure 5.15: Numerical results for the Goddard problem: States h, v, m (left column), adjoints (right column), and control (bottom row) as functions of normalized time.

$$\left(x(t) - X_k\right)^2 + \left(y(t) - Y_k\right)^2 \geq R_k^2, \quad k = 1, \ldots, M,$$

and control constraint

$$u(t) \in [-0.1, 0.1].$$

Herein, a_1 and a_2 are given non-negative weights. Please note the curve $t \mapsto (x(t), y(t))^\top$ is parameterized with respect to its arclength, that is, the "time" t is actually the arclength and u is the curvature.

Problems with many state constraints typically require a good initial guess. We apply a *homotopy strategy* to solve the obstacle avoidance problem. To this end, consider the following parametric optimal control problem OCP(δ) with a *homotopy parameter* $\delta \in [0, 1]$:

Minimize

$$a_1 t_f + a_2 \int_0^{t_f} u(t)^2 dt$$

subject to the dynamics

$$\begin{aligned}
\dot{x}(t) &= \cos \psi(t), & x(0) &= 0, x(t_f) = 100, \\
\dot{y}(t) &= \sin \psi(t), & y(0) &= 0, y(t_f) = 50, \\
\dot{\psi}(t) &= u(t), & \psi(0) &= 0, \psi(t_f) = -\pi,
\end{aligned}$$

state constraints

$$\left(x(t) - X_k\right)^2 + \left(y(t) - Y_k\right)^2 \geq (\delta R_k)^2, \quad k = 1, \ldots, M,$$

and control constraint

$$u(t) \in [-0.1, 0.1].$$

The homotopy parameter $\delta \in [0, 1]$ is used to relax the state constraints, and a solution of the obstacle avoidance problem can be obtained by the following homotopy strategy:

(0) Choose numbers $m \in \mathbb{N}_0$ and $0 = \delta_0 < \delta_1 < \cdots < \delta_m = 1$, e. g., $\delta_k = k/m, k = 0, \ldots, m$.

(1) For $k = 0, \ldots, m$, solve OCP(δ_k) and use the solution of OCP(δ_{k-1}) as initial guess. For $k = 0$, use some suitable initial guess.

Figure 5.16 shows three homotopy steps with $\delta_k = k/3$, $k = 0, 1, 2, 3$, for the $M = 5$ obstacles given by

$$\begin{aligned}
(X_1, Y_1, R_1) &= (10, 5, 6.25), \\
(X_2, Y_2, R_2) &= (58, 30, 7.5), \\
(X_3, Y_3, R_3) &= (50, 50, 12.5), \\
(X_4, Y_4, R_4) &= (75, 50, 9.375),
\end{aligned}$$

$$(X_5, Y_5, R_5) = (30, 25, 15),$$

and $\alpha_1 = 0.01$, $\alpha_2 = 10^{-4}$. □

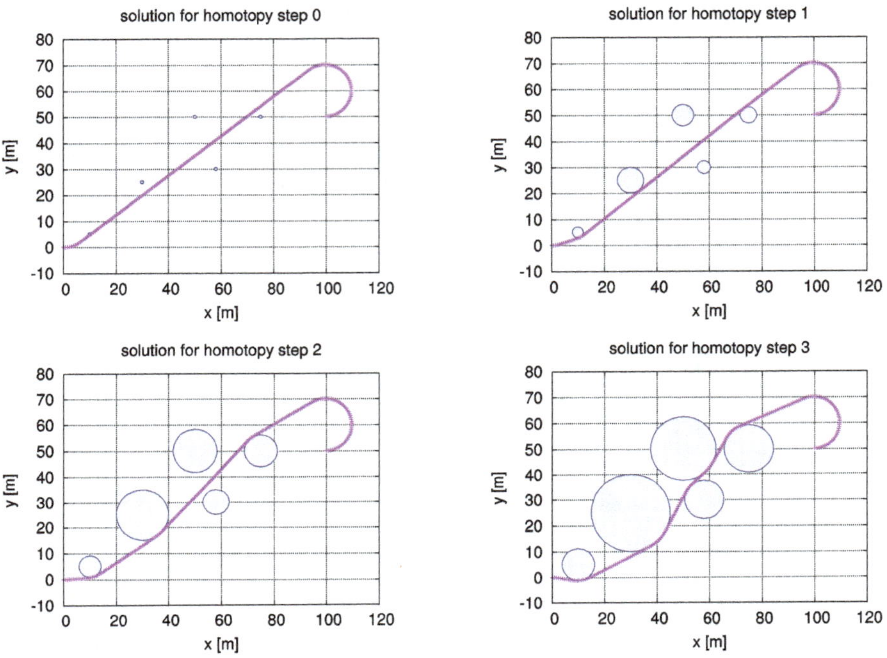

Figure 5.16: Solutions of three homotopy steps with $\delta_k = k/3$, $k = 0, 1, 2, 3$.

Example 5.7.3 (Ski driver). A ski driver of mass m moves in a plane with elevation angle ψ. The plane is given by the equation $0 = z - x\tan(\psi)$, see Figure 5.17. The ski driver can control its motion by the steering angle velocity $\dot{\delta}(t) = u(t)$ and its downhill force $F(t)$. The equations of motion for the ski driver with center of gravity $S = (x, y, z)^\top \in \mathbb{R}^3$ without air resistance and friction read as follows:

$$\begin{aligned} \dot{x}(t) &= v_x(t) \\ \dot{y}(t) &= v_y(t) \\ \dot{z}(t) &= v_z(t) \\ m\,\dot{v}_x(t) &= \tan(\psi)\lambda(t) + F(t)\cos(\delta(t))\cos(\psi) \\ m\,\dot{v}_y(t) &= F(t)\sin(\delta(t)) \\ m\,\dot{v}_z(t) &= -mg - \lambda(t) + F(t)\cos(\delta(t))\sin(\psi) \\ \dot{\delta}(t) &= u(t) \\ 0 &= z(t) - x(t)\tan(\psi). \end{aligned} \tag{5.100}$$

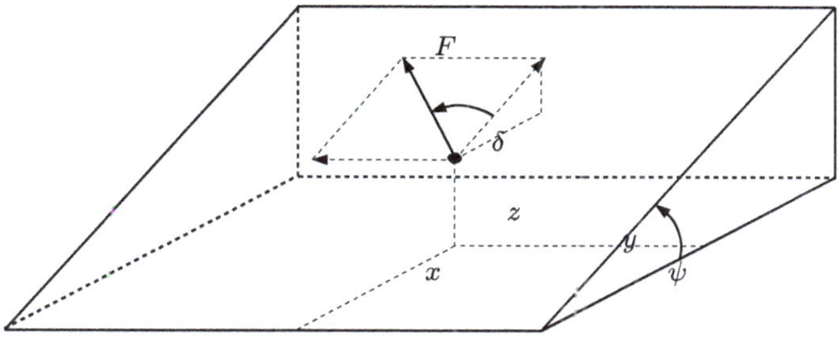

Figure 5.17: Motion in a plane.

Equation (5.100) is a DAE of index three. Initial conditions are given by

$$\begin{aligned} &x(0) = y(0) = z(0) = \delta(0) = 0, \\ &v_x(0) = -5, \quad v_y(0) = 5, \quad v_z(0) = v_x(0)\tan(\psi). \end{aligned} \tag{5.101}$$

The aim of the skier is to reach a given final position in minimal time:

Minimize t_f subject to (5.100), (5.101), and

$$\begin{aligned} x(t_f) + 100 &= 0, \\ y(t_f) + 20 &= 0, \\ v_y(t_f) &= 0, \\ u(t) &\in [-10, 10], \\ F(t) &\in [-200, 200]. \end{aligned}$$

The control appears linearly in the optimal control problem. The solution of the reduced discretization approach with $t_f = 8.426448607129171$ [s], $m = 80$ [kg], $\psi = 10°$, and 401 equidistant discretization points is depicted in Figure 5.18. The control $u(t)$ has two so-called singular arcs, where the control is in between the control bounds, and two boundary arcs, where $u(t)$ is at the boundary of the control constraints. One can observe oscillations at the entry and leaving points of the singular arcs, which is typical for direct discretization methods. The control $F(t)$ stays at the lower boundary of the control constraints with the exception of a small interval at normalized time 0.6, i. e. the ski driver basically accelerates in a downhill direction at all time. □

Example 5.7.4. We revisit Example 1.1.17 with an additional smoothing of the control u_2:

Minimize

$$\int_0^3 2x_1(t) + \frac{1}{2}y(t)^2 dt$$

subject to the DAE

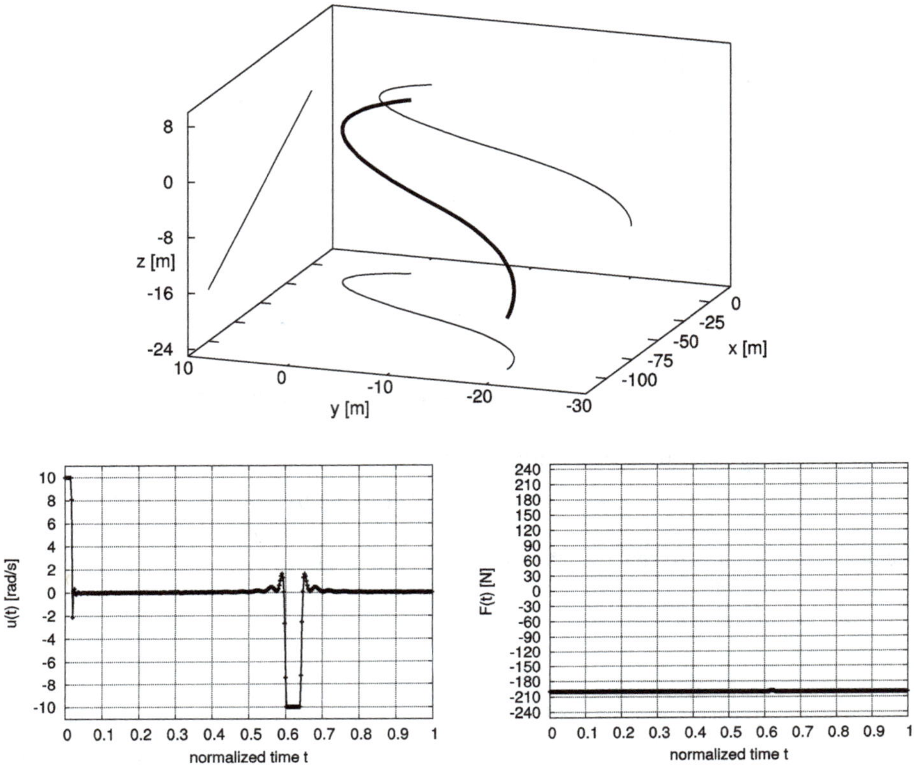

Figure 5.18: Downhill drive of a ski driver: Track and controls (normalized time).

$$
\begin{aligned}
\dot{x}_1(t) &= x_2(t), & x_1(0) &= 1,\\
\dot{x}_2(t) &= y(t) + u_1(t), & x_2(0) &= \frac{2}{3},\\
\dot{x}_3(t) &= x_4(t), & x_3(0) &= 1,\\
\dot{x}_4(t) &= u_2(t), & x_4(0) &= \frac{2}{3},\\
0 &= x_2(t) - x_4(t),
\end{aligned}
$$

the control constraint

$$-1 \le u_1(t) \le 1,$$

and the state constraint

$$0 \le x_3(t) \le 200.$$

Figures 5.19 and 5.20 depict the numerical solution of the problem obtained by the reduced discretization approach. The numerical solution shows small oscillations in the controls beyond the time point $t = 2$. □

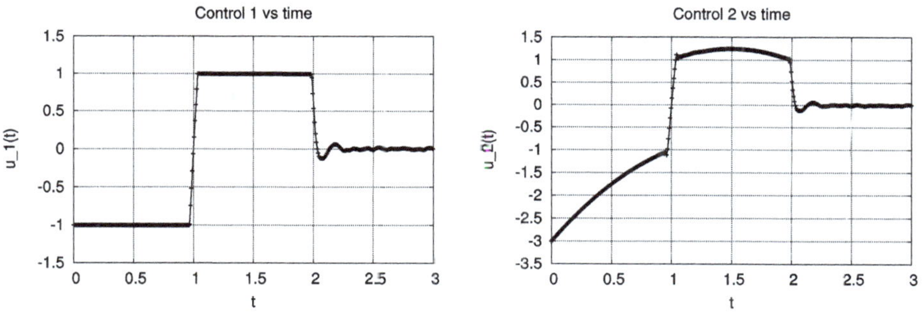

Figure 5.19: Control variables: Numerical solution with $N = 400$, linearized Runge–Kutta method and continuous, piecewise linear control approximation.

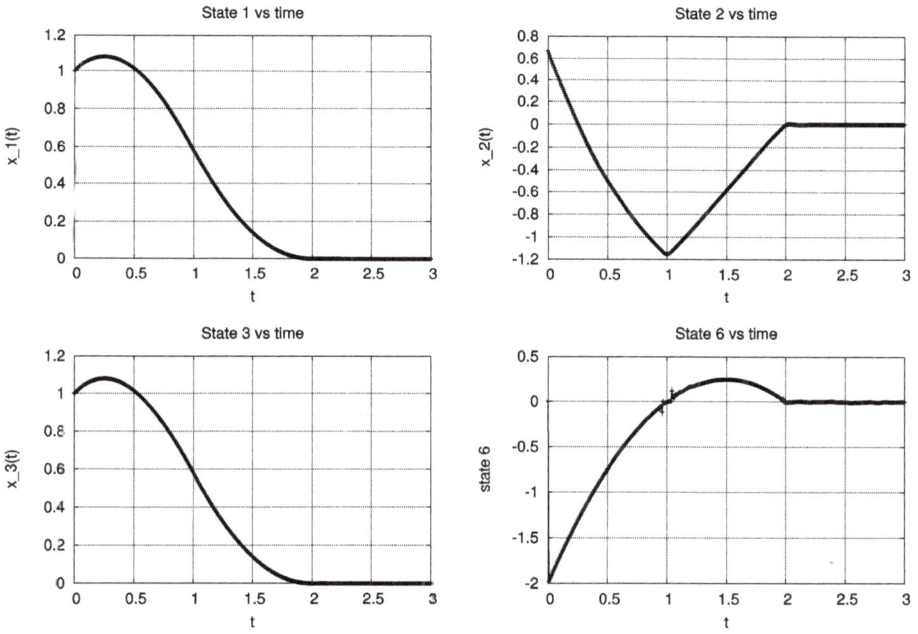

Figure 5.20: State variables: Numerical solution with $N = 400$, linearized Runge–Kutta method and continuous, piecewise linear control approximation.

Example 5.7.5 (Robot control). We consider an optimal control problem for a manipulator robot in three space dimensions, see Figure 5.21. Herein, $q = (q_1, q_2, q_3)^\top$ denotes the vector of joint angles at the joints of the robot, and $\dot{q} = (\dot{q}_1, \dot{q}_2, \dot{q}_3)^\top$ denotes the vector of joint angle velocities.

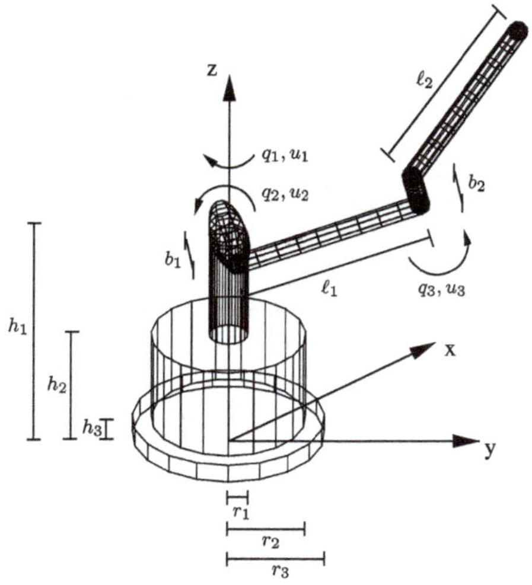

Figure 5.21: Configuration of the manipulator robot.

Let the rotation matrices be defined as

$$S_1(\alpha) := \begin{pmatrix} \cos\alpha & -\sin\alpha & 0 \\ \sin\alpha & \cos\alpha & 0 \\ 0 & 0 & 1 \end{pmatrix}, \quad S_2(\beta) := S_3(\beta) := \begin{pmatrix} 1 & 0 & 0 \\ 0 & \cos\beta & -\sin\beta \\ 0 & \sin\beta & \cos\beta \end{pmatrix},$$

and

$$S_{12}(\alpha,\beta) := S_1(\alpha)S_2(\beta), \quad S_{123}(\alpha,\beta,\gamma) := S_1(\alpha)S_2(\beta)S_3(\gamma).$$

Then the mount points of the first and second link, respectively, are given by

$$P_1 := S_1(q_1)\begin{pmatrix} -b_1 \\ 0 \\ h_1 \end{pmatrix}, \quad P_2 := P_1 + S_{12}(q_1,q_2)\begin{pmatrix} b_2 \\ \ell_1 \\ 0 \end{pmatrix}.$$

The centers of gravity of the three links of the manipulator in the reference system are given by

$$R_1 := \begin{pmatrix} 0 \\ 0 \\ \frac{h_1}{2} \end{pmatrix},$$

$$R_2 := P_1 + S_{12}(q_1,q_2)\begin{pmatrix} 0 \\ \frac{\ell_1}{2} \\ 0 \end{pmatrix} = \begin{pmatrix} -b_1\cos q_1 - \frac{\ell_1}{2}\sin q_1\cos q_2 \\ -b_1\sin q_1 + \frac{\ell_1}{2}\cos q_1\cos q_2 \\ h_1 + \frac{\ell_1}{2}\sin q_2 \end{pmatrix},$$

and

$$R_3 := P_2 + S_{123}(q_1, q_2, q_3)\begin{pmatrix} 0 \\ \frac{\ell_2}{2} \\ 0 \end{pmatrix}$$

$$= \begin{pmatrix} (b_2 - b_1)\cos q_1 - \sin q_1(\ell_1 \cos q_2 + \frac{\ell_2}{2}\cos(q_2 + q_3)) \\ (b_2 - b_1)\sin q_1 + \cos q_1(\ell_1 \cos q_2 + \frac{\ell_2}{2}\cos(q_2 + q_3)) \\ h_1 + \ell_1 \sin q_2 - \frac{\ell_2}{2}\sin(q_2 + q_3) \end{pmatrix}.$$

The center of gravity of the load is given by

$$R_4 := \begin{pmatrix} (b_2 - b_1)\cos q_1 - \sin q_1(\ell_1 \cos q_2 + \ell_2 \cos(q_2 + q_3)) \\ (b_2 - b_1)\sin q_1 + \cos q_1(\ell_1 \cos q_2 + \ell_2 \cos(q_2 + q_3)) \\ h_1 + \ell_1 \sin q_2 + \ell_2 \sin(q_2 + q_3) \end{pmatrix}. \tag{5.102}$$

The angular velocities of the respective body coordinate systems with respect to the reference coordinate system expressed in the body reference coordinate system are given by

$$\omega_1 = \begin{pmatrix} 0 \\ 0 \\ \dot{q}_1 \end{pmatrix}, \quad \omega_2 = \begin{pmatrix} \dot{q}_2 \\ \dot{q}_1 \sin q_2 \\ \dot{q}_1 \cos q_2 \end{pmatrix}, \quad \omega_3 = \omega_4 = \begin{pmatrix} \dot{q}_2 + \dot{q}_3 \\ \dot{q}_1 \sin(q_2 + q_3) \\ \dot{q}_1 \cos(q_2 + q_3) \end{pmatrix}.$$

The kinetic energy of the manipulator robot is

$$T(q, \dot{q}) := \frac{1}{2}\sum_{i=1}^{4}(m_i\|\dot{R}_i\|_2^2 + \omega_i^\top J_i \omega_i),$$

where m_i and $J_i = \operatorname{diag}(J_{x,i}, J_{y,i}, J_{z,i})$, $i = 1, 2, 3, 4$, denote the masses of the robot links and the load and the moments of inertia, respectively.

Application of the torques u_1, u_2, and u_3 at the centers of gravity of the robot links allow controlling the robot. The equations of motion are then given by

$$\ddot{q} = M(q)^{-1}(G(q, \dot{q}) + F(q)), \tag{5.103}$$

where $M(q) := T''_{\dot{q},\dot{q}}(q, \dot{q})$ denotes the symmetric and positive definite mass matrix, $G(q, \dot{q})^\top := T'_q(q, \dot{q}) - T''_{\dot{q},q}(q, \dot{q})\dot{q}$ denotes the generalized Coriolis forces and

$$F(q) = \begin{pmatrix} u_1 \\ u_2 - g\ell_1 \cos q_2(\frac{m_2}{2} + m_3 + m_4) - g\ell_2 \cos(q_2 + q_3)(\frac{m_3}{2} + m_4) \\ u_3 - g\ell_2 \cos(q_2 + q_3)(\frac{m_3}{2} + m_4) \end{pmatrix}$$

denotes the vector of applied joint torques and gravity forces.

An optimal control problem is obtained by the task to move the robot from some initial configuration to some terminal configuration minimizing a linear combination of final time and control effort:

Minimize

$$w_1 t_f + w_2 \int_0^{t_f} u_1(t)^2 + u_2(t)^2 + u_3(t)^2 \, dt$$

subject to the equations of motion (5.103), the initial position

$$q(0) = q_0, \quad \dot{q}(0) = 0_{\mathbb{R}^3},$$

the terminal condition

$$q(t_f) = q_f, \quad \dot{q}(t_f) = 0_{\mathbb{R}^3},$$

and the control constraints

$$-u_{i,\max} \leq u_i \leq u_{i,\max}, \quad i = 1, 2, 3.$$

Herein, the terminal time $t_f > 0$ is free, $w_1, w_2 \geq 0$ are user-defined weights, $q_0, q_f \in \mathbb{R}^3$ are given vectors, and $u_{i,\max} > 0$, $i = 1, 2, 3$, are given control bounds.

Instead of using the terminal condition $q(t_f) = q_f$, which defines the terminal angles, and thus the terminal position in a unique way, it might be better to allow additional degrees of freedom in the final position of the robot. This can be achieved by replacing the terminal condition $q(t_f) = q_f$ by the nonlinear boundary condition

$$R_4(t_f) - R_f = 0_{\mathbb{R}^3}, \tag{5.104}$$

where $R_f = (x_f, y_f, z_f)^\top \in \mathbb{R}^3$ is a user defined target position of the load and $R_4(t_f)$ is given by (5.102) with $q = q(t_f)$.

The moments of inertia for the cylindric links of radius r and the spherical load of radius r_4 are computed as follows:

$$\begin{aligned} J_{z,1} &= \frac{1}{2} m_1 r^2, \quad J_{x,4} = J_{y,4} = J_{z,4} = \frac{2}{5} m_4 r_4^2, \\ J_{x,2} = J_{z,2} &= \frac{1}{4} m_2 r^2 + \frac{1}{12} m_2 \ell_1, \quad J_{y,2} = \frac{1}{2} m_2 r^2, \\ J_{x,3} = J_{z,3} &= \frac{1}{4} m_3 r^2 + \frac{1}{12} m_3 \ell_2, \quad J_{y,3} = \frac{1}{2} m_3 r^2. \end{aligned}$$

The parameters in Table 5.7 were used for numerical computations.

Figures 5.22, 5.23, 5.24, and 5.25 show numerical results for the above optimal control problem with $w_1 = 1$ and $w_2 = 0$ (minimum time problem), terminal condition (5.104) with

$$R_f = \begin{pmatrix} (b_2 - b_1)\cos\alpha - \sin\alpha(\ell_1 \cos\beta + \ell_2 \cos(\beta + \gamma)) \\ (b_2 - b_1)\sin\alpha + \cos\alpha(\ell_1 \cos\beta + \ell_2 \cos(\beta + \gamma)) \\ h_1 + \ell_1 \sin\beta + \ell_2 \sin(\beta + \gamma) \end{pmatrix}$$

and $\alpha = -\pi/4$, $\beta = \pi/4$, $\gamma = \pi/4$.

Table 5.7: Parameters for manipulator robot.

Parameter	Value	Description
m_1	10 [kg]	mass of socket
m_2	2 [kg]	mass of link 1
m_3	2 [kg]	mass of link 2
m_4	1 [kg]	mass of load
h_1	1 [m]	height of socket
h_2	0.5 [m]	height of 2nd socket
h_3	0.1 [m]	height of platform
r_1	0 [m]	radius of socket
r_2	0.3 [m]	radius of 2nd socket
r_3	0.5 [m]	radius of platform
ℓ_2	1 [m]	length link 1
ℓ_3	1 [m]	length link 2
r_4	0.1 [m]	radius of load
r	0.1 [m]	radius of cylindric links
$u_{i,\max}$, $i = 1, 2, 3$	100 [N]	control bound
g	9.81 [m/s^2]	earth acceleration

The numerical results were obtained using $N = 51$ grid points, piecewise linear control approximations, and classic Runge–Kutta integrator for the equations of motion. The final time computes to $t_f \approx 0.417674$. □

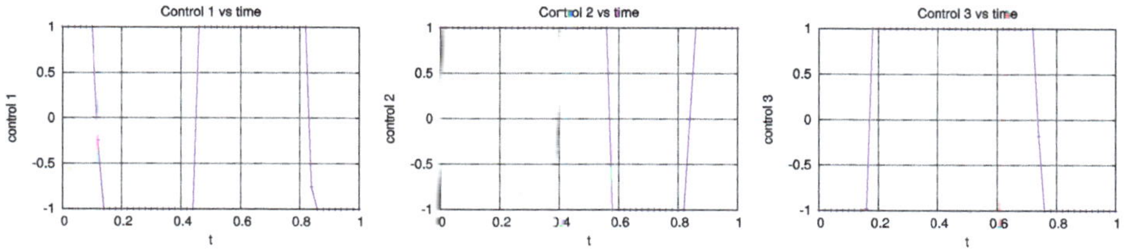

Figure 5.22: Control variables $(u_1, u_2, u_3)^\top$ for the manipulator robot (scaled to $[-1, 1]$, normalized time).

Example 5.7.6 (Semi-active chassis control). We aim to optimize the ride comfort and handling properties of a vehicle by actively controlling the damper forces in the suspension of the vehicle. For simplicity, a half-car model is used, see Figure 5.26.

The vehicle is supposed to travel with constant longitudinal velocity v. The road excitation is modeled by a time-dependent excitation $\xi_f(t)$ of the front wheel. The excitation arrives at the rear wheel with a delay $\delta(t) = \ell \cos \phi(t)/v$, where ϕ denotes the pitch angle. The excitation of the rear wheel is thus given by $\xi_r(t) = \xi_f(t - \delta(t))$. Herein $\ell = \ell_f + \ell_r$, where ℓ_f and ℓ_r are the distances from the chassis' center of gravity to the front and rear suspension mount points, respectively. The motion of the vehicle is

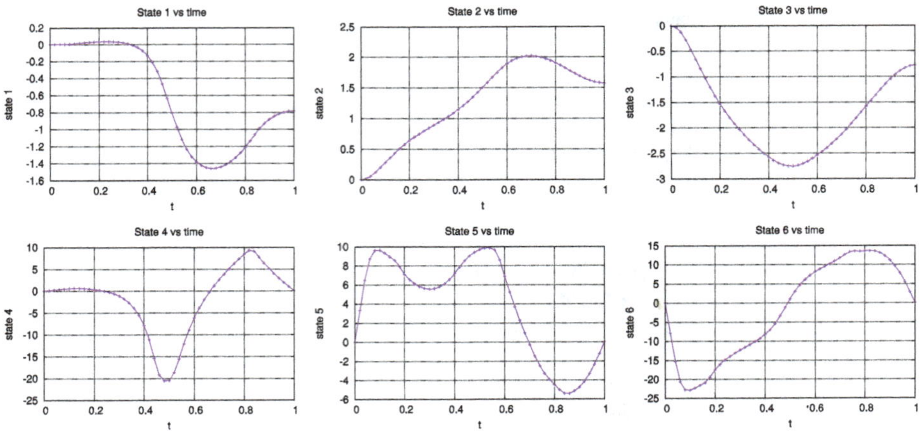

Figure 5.23: State variables $(q_1, q_2, q_3, \dot{q}_1, \dot{q}_2, \dot{q}_3)^\top$ for the manipulator robot (normalized time).

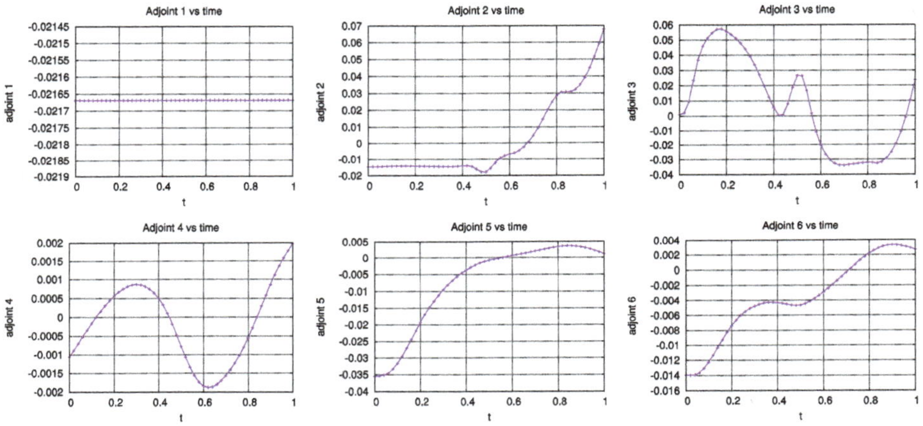

Figure 5.24: Adjoint variables associated to $(q_1, q_2, q_3, \dot{q}_1, \dot{q}_2, \dot{q}_3)^\top$ for the manipulator robot (normalized time).

modeled by the following second-order index-three DAE, where the dependence on t is suppressed for notational convenience:

$$m_c \ddot{z}_c = -m_c g + F_f + F_r, \tag{5.105}$$

$$m_f \ddot{z}_f = -m_f g - F_f - \lambda_f, \tag{5.106}$$

$$m_r \ddot{z}_r = -m_r g - F_r - \lambda_r, \tag{5.107}$$

$$J_c \ddot{\phi} = \cos\phi(\ell_r F_r - \ell_f F_f), \tag{5.108}$$

$$0 = z_f - \xi_f, \tag{5.109}$$

$$0 = z_r - \xi_r. \tag{5.110}$$

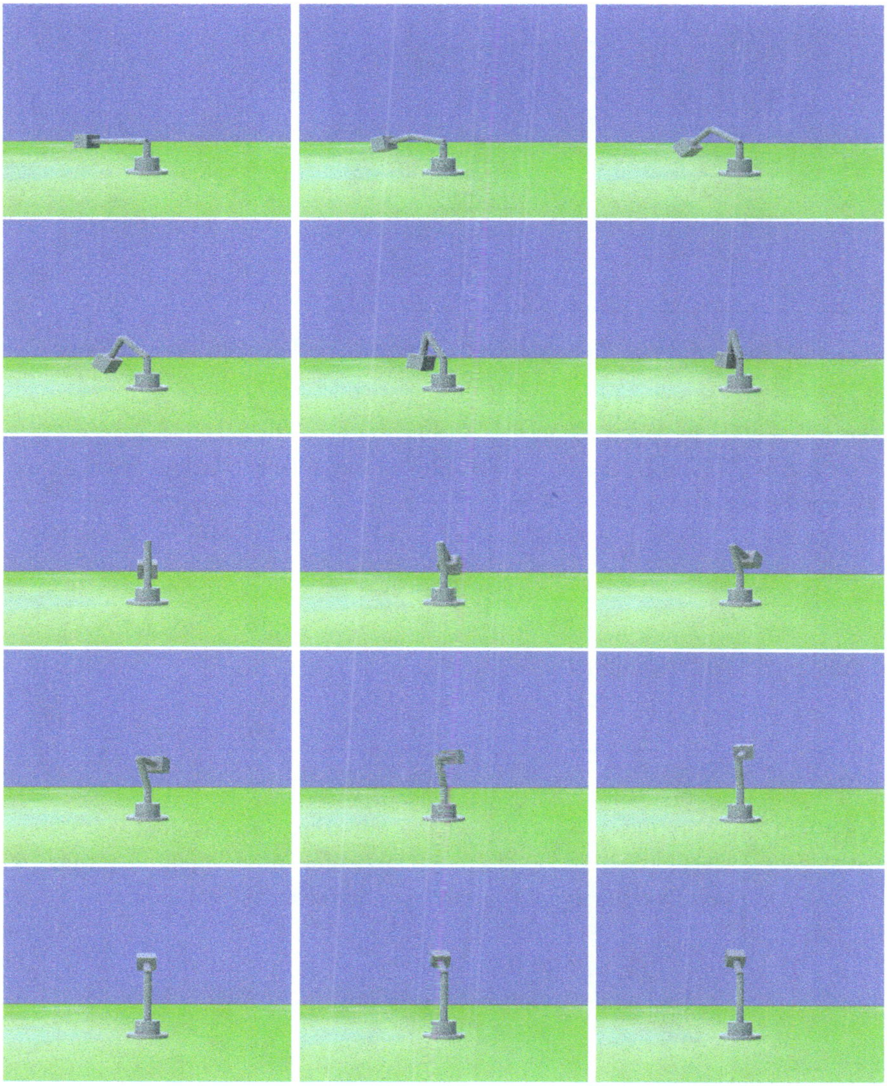

Figure 5.25: Snapshots of the motion of the manipulator robot.

The spring and damper forces are modeled by linear force laws

$$F_f = c_f^0 - c_f(z_c - \ell_f \sin\phi - z_f) - u_f d_f(\dot{z}_c - \ell_f\dot{\phi}\cos\phi - \dot{z}_f),$$
$$F_r = c_r^0 - c_r(z_c + \ell_r \sin\phi - z_r) - u_r d_r(\dot{z}_c + \ell_r\dot{\phi}\cos\phi - \dot{z}_r)$$

with offset parameters c_f^0, c_r^0, spring coefficients c_f, c_r and damper coefficients $u_f d_f, u_r d_r$. Herein $u_f \in [0,1]$ and $u_r \in [0,1]$ serve as controls and allow adjusting the dampers at front and rear.

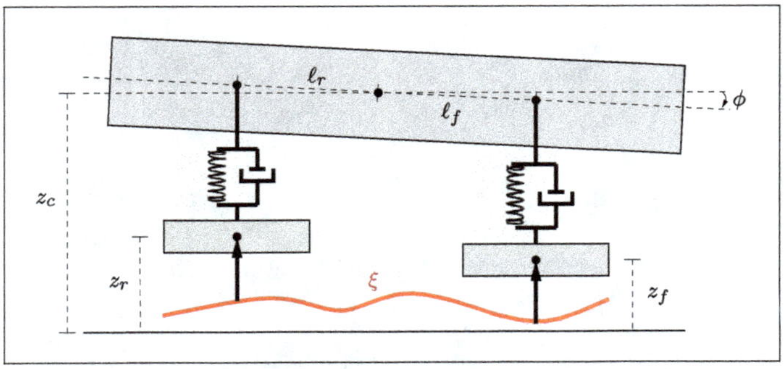

Figure 5.26: Configuration of the half-car model for the investigation of the vertical dynamics of a car.

Let the initial state be given by an equilibrium state on the plane, that is, $\xi_f(0) = \xi_r(0) = \dot{\xi}_f(0) = \dot{\xi}_r(0) = \ddot{\xi}_f(0) = \ddot{\xi}_r(0) = 0$, and all initial velocities and accelerations are assumed to be zero. Thus, $z_f(0) = z_r(0) = \dot{z}_f(0) = \dot{z}_r(0) = 0$, $\lambda_f(0) = -m_f g - F_f$, $\lambda_r(0) = -m_r g - F_r$, $F_f + F_r = m_c g$, and $\ell_r F_r - \ell_f F_f = 0$ (assuming $\cos\phi \neq 0$). This yields

$$F_f = \frac{\ell_r m_c g}{\ell}, \quad F_r = \frac{\ell_f m_c g}{\ell}$$

and, after some lengthly computations,

$$z_c(0) = \frac{\ell_f c_r^0}{\ell c_r} + \frac{\ell_r c_f^0}{\ell c_f} - \frac{m_c g}{\ell^2}\left(\frac{\ell_f^2}{c_r} + \frac{\ell_r^2}{c_f}\right), \quad \sin\phi(0) = \frac{1}{\ell}\left(\frac{c_r^0}{c_r} - \frac{c_f^0}{c_f} - \frac{m_c g}{\ell}\left(\frac{\ell_f}{c_r} - \frac{\ell_r}{c_f}\right)\right).$$

Together with $\dot{z}_c(0) = \dot{z}_f(0) = \dot{z}_r(0) = \dot{\phi}(0) = 0$ this initial state is consistent.

The algebraic variables λ_f and λ_r are found by twofold differentiation of the algebraic constraints, provided the signal ξ_f is twice continuously differentiable:

$$\begin{aligned}\lambda_f &= m_f g + F_f + m_f \ddot{\xi}_f,\\ \lambda_r &= m_r g + F_r + m_r \ddot{\xi}_r\end{aligned}$$

with

$$\ddot{\xi}_r(t) = \ddot{\xi}_f(t - \delta(t))\left(1 + \frac{\ell}{v}\dot{\phi}(t)\sin\phi(t)\right) + \frac{\ell}{v}\dot{\xi}_f(t - \delta(t))(\ddot{\phi}(t)\sin\phi(t) + \dot{\phi}(t)^2\cos\phi(t)).$$

The following optimal control problem aims to minimize a linear combination of a comfort criterion, a handling criterion, and a regularization term.

Minimize

$$a_1 \int_0^{t_f} \ddot{z}_c(t)^2 + \ddot{\phi}(t)^2\,dt + a_2 \int_0^{t_f} \lambda_f(t)^2 + \lambda_r(t)^2\,dt + a_3 \int_0^{t_f} \left(u_f(t) - 0.5\right)^2 + \left(u_r(t) - 0.5\right)^2\,dt$$

subject to (5.105)–(5.110) with initial equilibrium state and the control constraints

$$u_f(t) \in [0,1], \quad u_r(t) \in [0,1].$$

Figure 5.27 shows the numerical solution using a reduced discretization approach and a BDF method with GGL stabilization and $N = 101$, $\alpha_1 = 1$, $\alpha_2 = 10^{-9}$, $\alpha_3 = 10^{-2}$, $\ell_f = 1.1$ [m], $\ell_r = 1.4$ [m], $m_c = 1300$ [kg], $m_f = m_r = 40$ [kg], $J_c = 1659.97$ [kg m^2], $c_f = c_r = 25000$ [N/m], $d_f = d_r = 2500$ [Ns/m], $v = 20$ [m/s], $t_f = 3$ [s], $g = 9.81$ [m/s^2], and

$$\xi_f(t) = \begin{cases} 0.02(\cos(\frac{2\pi(t-0.3)}{0.2}) - 1), & \text{if } t \in [0.3, 0.5], \\ 0, & \text{else.} \end{cases}$$

Please not that ξ_f is only once continuously differentiable with the Lipschitz continuous first derivative. This causes discontinuities in the algebraic states λ_f and λ_r. □

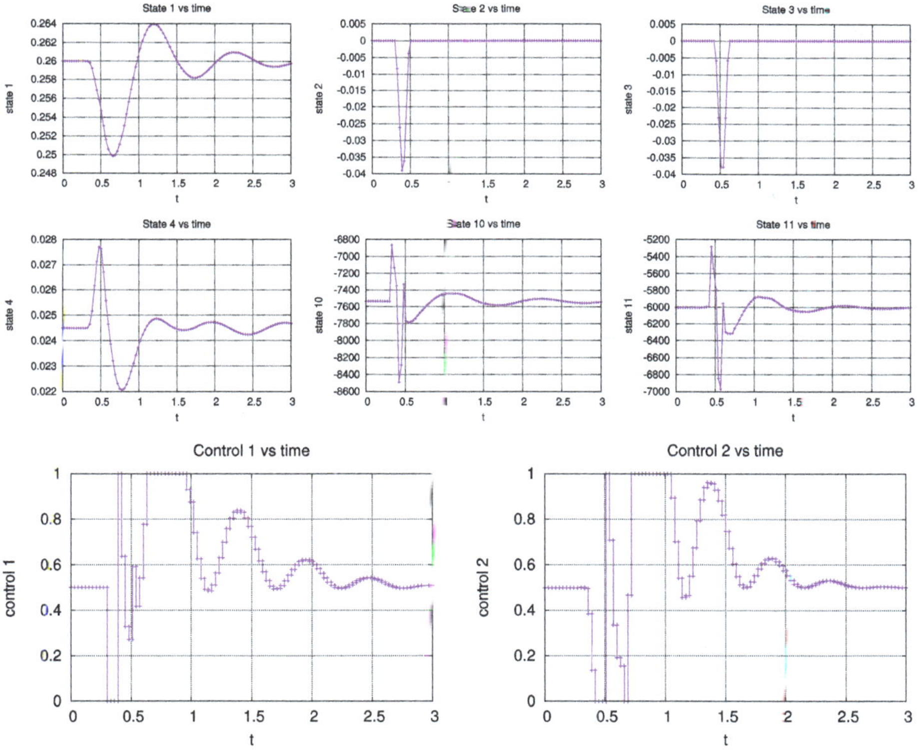

Figure 5.27: States z_c, z_f, z_r, ϕ, λ_f, and λ_r (from top left to middle right), and controls u_f and u_r (bottom row, left to right).

5.8 Exercises

Consider the nonlinear optimization problem

$$\begin{aligned} \textit{Minimize} \quad & J(z_1, z_2) = z_1 + z_2 \\ \textit{subject to} \quad & G_1(z_1, z_2) = z_1^2 - z_2 = 0. \end{aligned}$$

Solve the problem by hand using the Lagrange–Newton method with $z^{(0)} = 0$ and $\lambda^{(0)} \neq 0$. How does the initial guess of $\lambda^{(0)}$ influence the algorithm?

Let $J : \mathbb{R}^{n_z} \longrightarrow \mathbb{R}$ and $H : \mathbb{R}^{n_z} \longrightarrow \mathbb{R}^{n_H}$ be twice continuously differentiable.

(a) Implement the Lagrange–Newton method in Algorithm 5.2.2 to solve the nonlinear optimization problem

$$\textit{Minimize} \quad J(z) \quad \textit{subject to} \quad H(z) = 0_{\mathbb{R}^{n_z}}.$$

(b) Test the program for the following example:

$$\begin{aligned} \textit{Minimize} \quad & 2z_1^4 + z_2^4 + 4z_1^2 - z_1 z_2 + 6z_2^2 \\ \textit{subject to} \quad & 2z_1 - z_2 = -4, \quad z_1, z_2 \in \mathbb{R}. \end{aligned}$$

Implement the full discretization approach in Problem 5.1.1 using the implicit Euler method.

The resulting nonlinear optimization problem can be solved, for instance, by the MATLAB routine `fmincon` or by the code WORHP, which is available at www.worhp.de.

Test the program for the constrained minimum energy problem 3.2.13.

(Sensitivity Analysis by Adjoint Equation) Let $f : [t_0, t_f] \times \mathbb{R}^{n_z} \times \mathbb{R}^{n_p} \longrightarrow \mathbb{R}^{n_z}$ and $Z_0 : \mathbb{R}^{n_p} \longrightarrow \mathbb{R}^{n_z}$ be sufficiently smooth functions and

$$\begin{aligned} z(t_0) &= Z_0(p), \\ \dot{z}(t) &= f\big(t, z(t), p\big), \quad t \in [t_0, t_f], \end{aligned}$$

a parametric initial value problem with solution $z(t; p)$.

(a) Implement the adjoint equation approach in Subsection 5.4.2 for the classic 4-stage Runge–Kutta method with Butcher array

$$\begin{array}{c|cccc} 0 & & & & \\ 1/2 & 1/2 & & & \\ 1/2 & 0 & 1/2 & & \\ 1 & 0 & 0 & 1 & \\ \hline & 1/6 & 1/3 & 1/3 & 1/6 \end{array}$$

to compute the derivative

$$\frac{d}{dp}\varphi\big(z(t_f;p),p\big)$$

of the function $\varphi : \mathbb{R}^{n_z} \times \mathbb{R}^{n_p} \longrightarrow \mathbb{R}$, $(z,p) \mapsto \varphi(z,p)$.

(b) Test the program for $[t_0, t_f] = [0, 10]$, step-size $h = 0.1$, $\varphi(z_1, z_2) := z_1^2 - z_2$, and

$$\begin{aligned} z_1(0) &= p_1, \\ z_2(0) &= p_2 \\ \dot{z}_1(t) &= z_2(t), \\ \dot{z}_2(t) &= -\frac{p_3}{p_4}\sin\big(z_1(t)\big) \end{aligned}$$

with $p_1 = \frac{\pi}{2}$, $p_2 = 0$, $p_3 = 9.81$, $p_4 = 1$.

Implement the reduced discretization method in Subsection 5.1.2 using the classic 4-stage Runge–Kutta method with a fixed step-size $h = (t_f - t_0)/N$, $N \in \mathbb{N}$, and a piecewise constant control approximation to solve the following optimal control problem:

Minimize

$$\int_0^{4.5} u(t)^2 + x_1(t)^2 dt$$

subject to

$$\begin{aligned} \dot{x}_1(t) &= x_2(t), \\ \dot{x}_2(t) &= -x_1(t) + x_2(t)\big(1.4 - 0.14x_2(t)^2\big) + 4u(t), \\ x_1(0) &= x_2(0) = -5, \\ x_1(4.5) &= x_2(4.5) = 0. \end{aligned}$$

You may use the Lagrange–Newton method to solve the nonlinear optimization problem and either the sensitivity equation approach or the adjoint equation approach to compute the first derivatives of the objective function and the constraints. Replace the Hessian of the Lagrange function in the Lagrange–Newton method by the modified BFGS update in (5.27).

Solve the following discrete dynamic optimization problem for $k = 1$, $c = 1$, $b = 0.6$, and $N = 5$ using the discrete minimum principle in Theorem 5.5.3:

Minimize $-x_2(N)$ subject to the constraints

$$\begin{aligned} x_1(j+1) &= x_1(j)\big(0.9 + bu(j)\big), && j = 0, 1, \ldots, N-1, \\ x_2(j+1) &= x_2(j) + c\big(1 - u(j)\big)x_1(j), && j = 0, 1, \ldots, N-1, \\ x_1(0) &= k, \\ x_2(0) &= 0, \\ u(j) &\in [0, 1], && j = 0, 1, \ldots, N-1. \end{aligned}$$

Solve the following discrete dynamic optimization problem for $N = 5$ using the discrete minimum principle in Theorem 5.5.3:

Minimize

$$\sum_{j=0}^{N-1} u(j)^2$$

subject to the constraints

$$\begin{aligned} x_1(j+1) &= x_1(j) + 2x_2(j), \quad j = 0, 1, \ldots, N-1, \\ x_2(j+1) &= 2u(j) - x_2(j), \quad j = 0, 1, \ldots, N-1, \\ x_1(0) &= 0, \\ x_1(N) &= 4, \\ x_2(0) &= 0, \\ x_2(N) &= 0. \end{aligned}$$

Let the following optimal control problem be given:

Minimize

$$\|u\|_\infty$$

with respect to $x \in W_{2,\infty}([0, t_f])$ *and* $u \in L_\infty([0, t_f])$ *subject to the constraints*

$$\begin{aligned} \ddot{x}(t) &= u(t), \\ x(0) &= x(t_f) = \dot{x}(t_f) = 0, \\ \dot{x}(0) &= 1, \\ |u(t)| &\le 1. \end{aligned}$$

Discretize the problem for $N = 10$ and $t_f = 3$ on the grid

$$t_i = ih, \quad i = 0, \ldots, N, \quad h := t_f/N.$$

Formulate the discretized problem as a linear optimization problem and solve it.

Investigate numerically the order of convergence of the state x and the control u with respect to the norms $\|\cdot\|_\infty$, $\|\cdot\|_1$, and $\|\cdot\|_2$ for the following optimal control problems. To this end, use a full discretization method with the explicit Euler method and an equidistant grid with step-size $h = t_f/N$ and $N = 5, 10, 20, 40, 80, 160, 320, 640, 1280$.

(a) (Compare [2, Example 2.8])

Minimize

$$\int_0^2 -2x(t) + 3u(t)dt$$

subject to

$$\dot{x}(t) = x(t) + u(t), \quad x(0) = 5,$$
$$u(t) \in [0, 2].$$

True solution ($t_f = 2$):

$$\hat{x}(t) = \begin{cases} 7\exp(t) - 2, & \text{if } 0 \le t < 2 - \ln(5/2), \\ 7\exp(t) - 5\exp(t-2), & \text{if } 2 - \ln(5/2) \le t \le t_f, \end{cases}$$

$$\hat{u}(t) = \begin{cases} 2, & \text{if } 0 \le t < 2 - \ln(5/2), \\ 0, & \text{if } 2 - \ln(5/2) \le t \le t_f. \end{cases}$$

(b) *Minimize*

$$\frac{1}{2}\int_0^3 x(t)^2 dt$$

subject to

$$\dot{x}(t) = u(t), \quad x(0) = 1,$$
$$u(t) \in [-1, 1].$$

True solution ($t_f = 3$):

$$\hat{x}(t) = \begin{cases} 1 - t, & \text{if } 0 \le t < 1, \\ 0, & \text{if } 1 \le t \le t_f, \end{cases} \qquad \hat{u}(t) = \begin{cases} -1, & \text{if } 0 \le t < 1, \\ 0, & \text{if } 1 \le t \le t_f. \end{cases}$$

How can the Lagrange multipliers in the discrete local minimum principle in Theorem 5.5.3 be related to the adjoints and the multipliers in Theorems 3.1.4, 3.2.2, and 3.3.2 for the special case that g in Problem 5.5.1 does not depend on y and u (this is the higher index case)?

---

6 Real-time control

The direct method in Chapter 5, the indirect method in Chapters 3 and 4, and the function space methods in Chapter 8 allow solving a given optimal control problem. The outcome of all of these methods is an (approximate) optimal control $\hat{u}$ as a function of time in some interval $\mathcal{I}$ – a so-called *open loop control*. Implementing the open loop control directly into the underlying real-world process often does not yield the desired outcome owing to simplifying assumptions in the mathematical process model, deviations in parameters, or other perturbations. As an open loop control does not consider deviations between the desired and actual state of the process, control techniques are required to provide such a *feedback*. For time-crucial processes, *real-time capability* of such feedback control laws is essential.

In this chapter, we will discuss different approaches. Section 6.1 addresses parametric (discretized) optimal control problems and online update techniques for approximate solutions of perturbed problems. If enough time is available, then the simplest approach is to resolve the perturbed problem using suitable solution techniques. For time-critical processes, however, the re-computation of an optimal solution for the perturbed problem may not be fast enough to provide an optimal solution in real-time. Therefore, alternative methods are needed that are capable of providing at least an approximation of the optimal solution for the perturbed problem in real-time. Such techniques have been developed in [49, 51, 50] for ODE optimal control problems and in [124] optimal control problems subject to index-one DAEs. The idea is to perform a *sensitivity analysis* of the optimal solution with respect to parameters entering the optimization problem. Under suitable assumptions, it is possible to prove solution differentiability with respect to these parameters. A simple update rule for perturbed solutions is then obtained locally by a Taylor expansion. This approach is particularly effective, if only small perturbations in parameters occur. The parametric sensitivity analysis is illustrated for the emergency landing maneuver and a problem from test-driving.

The construction of feedback control laws, which provide the control as a linear or nonlinear function of the deviation measured in the state trajectory, is the subject of Sections 6.2 and 6.3. Section 6.2 briefly summarizes the basic construction principles of linear-quadratic regulators (LQR). Such an LQR controller exploits the first-order necessary optimality conditions for linear-quadratic optimal control problems and requires offline solving a Riccati differential equation. However, it is incapable of taking into account control or state constraints.

The model-predictive control paradigm in Section 6.3 can overcome the shortcomings of the LQR controller, since control or state constraints can be incorporated at the cost of a higher computational effort. In this book, it is impossible to cover all aspects of controller design and model-predictive control, such as stability, observability, and controllability, so the presentation is restricted to the basic working principles only. For a detailed discussion of nonlinear model-predictive control theory and algorithms, please refer to [76, 77, 78, 150].

https://doi.org/10.1515/9783110797893-006

6.1 Parametric sensitivity analysis and open-loop real-time control

In the sequel, a method based on parametric sensitivity analysis of the underlying discretized optimal control problem is suggested to calculate real-time approximations of optimal solutions. The discussion is restricted to general finite-dimensional nonlinear optimization problems. Discretized optimal control problems define a subclass with a special structure. Extensions to general infinite optimization problems can be found in [211, 34] and the literature cited therein. The solution differentiability of optimal control problems subject to ODEs in appropriate Banach spaces is investigated in detail in [242, 243, 244, 220, 221, 34, 222, 16, 241]. Numerical approaches based on a linearization of the necessary optimality conditions of the optimal control problem are discussed in [265, 266, 267, 268].

6.1.1 Parametric sensitivity analysis of nonlinear optimization problems

Consider the following parametric optimization problem:

Problem 6.1.1 (Parametric nonlinear optimization problem NLP(p)). Let $p \in \mathbb{R}^{n_p}$ be a given parameter and $J, G_i, H_j : \mathbb{R}^{n_z} \times \mathbb{R}^{n_p} \longrightarrow \mathbb{R}$, $i = 1, \dots, n_G$, $j = 1, \dots, n_H$, sufficiently smooth functions.

Minimize $J(z,p)$ with respect to $z \in \mathbb{R}^{n_z}$ subject to the constraints

$$\begin{aligned} G_i(z,p) &\le 0, \quad i = 1, \dots, n_G, \\ H_j(z,p) &= 0, \quad j = 1, \dots, n_H. \end{aligned}$$

Let $\hat{p}$ denote a fixed *nominal parameter*. Let us assume that Problem 6.1.1 possesses an optimal solution $\hat{z} = z(\hat{p})$ for the nominal parameter $\hat{p}$. We are interested in the following questions:

(a) Does Problem 6.1.1 possess a solution for perturbed parameters p in some neighborhood of the nominal parameter $\hat{p}$?

(b) Which properties has the solution mapping $p \mapsto z(p)$ that assigns to each parameter p in some neighborhood of $\hat{p}$ an optimal solution of NLP(p)?

The admissible set of *NLP*(p) is defined by

$$\Sigma(p) := \{z \in \mathbb{R}^{n_z} \mid G_i(z,p) \le 0,\ i = 1, \dots, n_G,\ H_j(z,p) = 0,\ j = 1, \dots, n_H\}.$$

The index set of active inequality constraints is given by

$$A(z,p) := \{i \in \{1, \dots, n_G\} \mid G_i(z,p) = 0\}.$$

Definition 6.1.2 (Strongly regular local solution). A local minimum $\hat{z}$ of $NLP(\hat{p})$ is called *strongly regular*, if the following properties hold:

(a) $\hat{z}$ is admissible: $\hat{z} \in \Sigma(\hat{p})$.

(b) $\hat{z}$ fulfills the linear independence constraint qualification: The gradients

$$\nabla_z G_i(\hat{z}, \hat{p}), \quad i \in A(\hat{z}, \hat{p}), \quad \nabla_z H_j(\hat{z}, \hat{p}), \quad j = 1, \dots, n_H,$$

are linearly independent.

(c) The KKT conditions hold at $(\hat{z}, \hat{\mu}, \hat{\lambda})$, where $\hat{\mu}$ and $\hat{\lambda}$ denote the Lagrange multipliers for the inequality and equality constraints, respectively.

(d) The strict complementarity condition holds:

$$\hat{\mu}_i - G_i(\hat{z}, \hat{p}) > 0 \quad \text{for all } i = 1, \dots, n_G.$$

(e) Let

$$L''_{zz}(\hat{z}, \ell_0, \hat{\mu}, \hat{\lambda}, \hat{p})(d, d) > 0$$

for all $d \in T_C(\hat{z}, \hat{p})$ with $d \neq 0_{\mathbb{R}^{n_z}}$ and $\ell_0 = 1$, where

$$L(z, \ell_0, \mu, \lambda, p) := \ell_0 J(z, p) + \mu^\top G(z, p) + \lambda^\top H(z, p)$$

denotes the Lagrange function of NLP(p) and

$$T_C(z, p) := \left\{ d \in \mathbb{R}^{n_z} \;\middle|\; \begin{array}{l} G'_{i,z}(z, p)(d) \le 0, \; i \in A(z, p), \mu_i = 0, \\ G'_{i,z}(z, p)(d) = 0, \; i \in A(z, p), \mu_i > 0, \\ H'_{j,z}(z, p)(d) = 0, \; j = 1, \dots, n_H \end{array} \right\}$$

denotes the *critical cone* of NLP(p). □

In fact, condition (e) in Definition 6.1.2 turns out to be *sufficient* for optimality of a KKT point $\hat{z}$ for a fixed parameter $\hat{p}$. To see this, we have a closer look at the critical cone.

The term *critical cone* is due to the following reasoning. Directions d with

$$G'_{i,z}(\hat{z}, \hat{p})(d) > 0 \quad \text{for some } i \in A(\hat{z}, \hat{p})$$

or

$$H'_{j,z}(\hat{z}, \hat{p})(d) \neq 0 \quad \text{for some } j \in \{1, \dots, n_H\}$$

are infeasible directions. So, consider only feasible directions d with

$$G'_{i,z}(\hat{z}, \hat{p})(d) \le 0 \quad \text{for } i \in A(\hat{z}, \hat{p})$$

and

$$H'_{j,z}(\hat{z},\hat{p})(d) = 0 \quad \text{for } j = 1,\dots,n_H.$$

For such directions, the KKT conditions yield

$$J'_z(\hat{z},\hat{p})(d) + \sum_{i\in A(\hat{z},\hat{p})} \underbrace{\mu_i G'_{i,z}(\hat{z},\hat{p})(d)}_{\leq 0} + \sum_{j=1}^{n_H} \lambda_j \underbrace{H'_{j,z}(\hat{z},\hat{p})(d)}_{=0} = 0,$$

and thus $J'_z(\hat{z},\hat{p})(d) \geq 0$. If even $J'_z(\hat{z},\hat{p})(d) > 0$ holds, then d is a direction of ascent, and the direction d is not interesting for the investigation of sufficient conditions. So, let $J'_z(\hat{z},\hat{p})(d) = 0$. This is the critical case. In this critical case, it holds

$$\sum_{i\in A(\hat{z},\hat{p})} \mu_i G'_{i,z}(\hat{z},\hat{p})(d) = \sum_{i\in A(\hat{z},\hat{p}),\mu_i>0} \mu_i G'_{i,z}(\hat{z},\hat{p})(d) = 0,$$

and thus $G'_{i,z}(\hat{z},\hat{p})(d) = 0$ for all $i \in A(\hat{z},\hat{p})$ with $\mu_i > 0$. Hence, $d \in T_C(\hat{z},\hat{p})$ and for such directions, we need additional assumptions about the curvature (2nd derivative!). The following theorem can be found in [111, Theorem 2.55, p. 67], [6, Theorem 7.3.1, p. 281], [21, Theorem 4.4.2, p. 169]:

Theorem 6.1.3 (Second order sufficient optimality condition). *Let J, G_i, $i = 1,\dots,n_G$, and H_j, $j = 1,\dots,n_H$, in Problem 6.1.1 be twice continuously differentiable with respect to the argument z with $\hat{p}$ fixed. Let $(\hat{z},\hat{\mu},\hat{\lambda})$ be a KKT point of Problem 6.1.1 with*

$$L''_{zz}(\hat{z},\ell_0,\hat{\mu},\hat{\lambda},\hat{p})(d,d) > 0$$

for every $d \in T_C(\hat{z},\hat{p})$ with $d \neq 0_{\mathbb{R}^{n_z}}$ and $\ell_0 = 1$.

Then there exists a neighborhood $B_\varepsilon(\hat{z})$ of $\hat{z}$ and some $\alpha > 0$ such that

$$J(z,\hat{p}) \geq J(\hat{z},\hat{p}) + \alpha\|z - \hat{z}\|^2$$

for every $z \in \Sigma(\hat{p}) \cap B_\varepsilon(\hat{z})$.

Proof. (a) Let $d \in T(\Sigma(\hat{p}),\hat{z})$ with $d \neq 0_{\mathbb{R}^{n_z}}$. $T(\Sigma(\hat{p}),\hat{z})$ denotes the tangent cone of $\Sigma(\hat{p})$ at $\hat{z}$.

Then there exist sequences $z_k \in \Sigma(\hat{p})$, $z_k \longrightarrow \hat{z}$, and $\alpha_k \downarrow 0$ with

$$\lim_{k\to\infty} \frac{z_k - \hat{z}}{\alpha_k} = d.$$

For $i \in A(\hat{z},\hat{p})$, we have

$$0 \geq \frac{G_i(z_k,\hat{p}) - G_i(\hat{z},\hat{p})}{\alpha_k} = G'_{i,z}(\xi_k,\hat{p})\left(\frac{z_k - \hat{z}}{\alpha_k}\right) \longrightarrow G'_{i,z}(\hat{z},\hat{p})(d)$$

by the mean-value theorem. Similarly, we show $H'_{j,z}(\hat{z},\hat{p})(d) = 0$ for $j = 1,\dots,n_H$. Since $(\hat{z},\hat{\mu},\hat{\lambda})$ is a KKT point with $\hat{\mu}_i = 0$, if $G_i(\hat{z},\hat{p}) < 0$, we obtain

$$J'_z(\hat{z},\hat{p})(d) = -\sum_{i=1}^{n_G} \hat{\mu}_i G'_{i,z}(\hat{z},\hat{p})(d) - \sum_{j=1}^{n_H} \hat{\lambda}_j H'_{j,z}(\hat{z},\hat{p})(d) \geq 0.$$

Hence, $\hat{z}$ fulfills the first-order necessary condition $J'_z(\hat{z},\hat{p})(d) \geq 0$ for all $d \in T(\Sigma(\hat{p}),\hat{z})$.

(b) Assume that the statement of the theorem is wrong. Then for any ball around $\hat{z}$ with radius $1/i$, there exists a point $z_i \in \Sigma(\hat{p})$ with $z_i \neq \hat{z}$ and

$$J(z_i,\hat{p}) - J(\hat{z},\hat{p}) < \frac{1}{i}\|z_i - \hat{z}\|^2, \quad \|z_i - \hat{z}\| \leq \frac{1}{i} \quad \text{for all } i \in \mathbb{N}. \tag{6.1}$$

Since the unit ball with respect to $\|\cdot\|$ is compact in $\mathbb{R}^{n_z}$, there exists a convergent subsequence $\{z_{i_k}\}$ with

$$\lim_{k\to\infty} \frac{z_{i_k} - \hat{z}}{\|z_{i_k} - \hat{z}\|} = d, \quad \lim_{k\to\infty} \|z_{i_k} - \hat{z}\| = 0.$$

Hence, $d \in T(\Sigma(\hat{p}),\hat{z}) \setminus \{0_{\mathbb{R}^{n_z}}\}$. Taking the limit in (6.1) yields

$$J'_z(\hat{z},\hat{p})(d) = \lim_{k\to\infty} \frac{J(z_{i_k},\hat{p}) - J(\hat{z},\hat{p})}{\|z_{i_k} - \hat{z}\|} \leq 0.$$

Together with (a), we have

$$J'_z(\hat{z},\hat{p})(d) = 0.$$

(c) Since $\hat{z}$ is a KKT point, it follows

$$J'_z(\hat{z},\hat{p})(d) = -\sum_{i\in A(\hat{z},\hat{p})} \underbrace{\hat{\mu}_i}_{\geq 0} \underbrace{G'_{i,z}(\hat{z},\hat{p})(d)}_{\leq 0} - \sum_{j=1}^{n_H} \hat{\lambda}_j \underbrace{H'_{j,z}(\hat{z},\hat{p})(d)}_{=0} = 0.$$

Thus, it is $G'_{i,z}(\hat{z},\hat{p})(d) = 0$, if $\hat{\mu}_i > 0$, and hence $d \in T_C(\hat{z},\hat{p})$.
According to (6.1), it holds

$$\lim_{k\to\infty} \frac{J(z_{i_k},\hat{p}) - J(\hat{z},\hat{p})}{\|z_{i_k} - \hat{z}\|^2} \leq \lim_{k\to\infty} \frac{1}{i_k} = 0 \tag{6.2}$$

for the direction d. Furthermore, it is $(\ell_0 = 1)$

$$L(z_{i_k},\ell_0,\hat{\mu},\hat{\lambda},\hat{p}) = J(z_{i_k},\hat{p}) + \sum_{i=1}^{n_G} \hat{\mu}_i G_i(z_{i_k},\hat{p}) + \sum_{j=1}^{n_H} \hat{\lambda}_j H_j(z_{i_k},\hat{p})$$

$$
\begin{aligned}
&\leq J(z_{i_k}, \hat{p}), \\
L(\hat{z}, \ell_0, \hat{\mu}, \hat{\lambda}, \hat{p}) &= J(\hat{z}, \hat{p}) + \sum_{i=1}^{n_G} \hat{\mu}_i G_i(\hat{z}, \hat{p}) + \sum_{j=1}^{n_H} \hat{\lambda}_j H_j(\hat{z}, \hat{p}) \\
&= J(\hat{z}, \hat{p}), \\
L_z'(\hat{z}, \ell_0, \hat{\mu}, \hat{\lambda}, \hat{p}) &= J_z'(\hat{z}, \hat{p}) + \sum_{i=1}^{n_G} \hat{\mu}_i G_{i,z}'(\hat{z}, \hat{p}) + \sum_{j=1}^{n_H} \hat{\lambda}_j H_{j,z}'(\hat{z}, \hat{p}) \\
&= 0_{\mathbb{R}^{n_z}}^\top .
\end{aligned}
$$

Taylor expansion of L with $\ell_0 = 1$ with respect to z at $\hat{z}$ yields

$$
\begin{aligned}
J(z_{i_k}, \hat{p}) &\geq L(z_{i_k}, \ell_0, \hat{\mu}, \hat{\lambda}, \hat{p}) \\
&= L(\hat{z}, \ell_0, \hat{\mu}, \hat{\lambda}, \hat{p}) + L_z'(\hat{z}, \ell_0, \hat{\mu}, \hat{\lambda}, \hat{p})(z_{i_k} - \hat{z}) + \frac{1}{2} L_{zz}''(\xi_k, \ell_0, \hat{\mu}, \hat{\lambda}, \hat{p})(z_{i_k} - \hat{z}, z_{i_k} - \hat{z}) \\
&= J(\hat{z}, \hat{p}) + \frac{1}{2} L_{zz}''(\xi_k, \ell_0, \hat{u}, \hat{\lambda}, \hat{p})(z_{i_k} - \hat{z}, z_{i_k} - \hat{z}),
\end{aligned}
$$

where ξ_k is some point between $\hat{z}$ and z_{i_k}. Division by $\|z_{i_k} - \hat{z}\|^2$ and taking the limit yields together with (6.2)

$$
0 \geq \frac{1}{2} L_{zz}''(\hat{z}, \ell_0, \hat{\mu}, \hat{\lambda}, \hat{p})(d, d).
$$

This contradicts the assumption $L_{zz}''(\hat{z}, \ell_0, \hat{\mu}, \hat{\lambda}, \hat{p})(d, d) > 0$ for all $d \in T_C(\hat{z}, \hat{p})$ with $d \neq 0_{\mathbb{R}^{n_z}}$. □

A second-order *necessary* optimality condition involving the critical cone can be found in [111, Theorem 2.54, p. 65].

The following result is based on [94, 95, 309]. Extensions can be found in [18, 180].

Theorem 6.1.4 (Sensitivity theorem). *Let $J, G_1, \ldots, G_{n_G}, H_1, \ldots, H_{n_H} : \mathbb{R}^{n_z} \times \mathbb{R}^{n_p} \longrightarrow \mathbb{R}$ be twice continuously differentiable and $\hat{p}$ a nominal parameter. Let $\hat{z}$ be a strongly regular local minimum of NLP($\hat{p}$) with Lagrange multipliers $\hat{\lambda}$ and $\hat{\mu}$.*

Then there exist neighborhoods $B_\epsilon(\hat{p})$ and $B_\delta(\hat{z}, \hat{\mu}, \hat{\lambda})$, such that NLP(p) has a unique strongly regular local minimum

$$
(z(p), \mu(p), \lambda(p)) \in B_\delta(\hat{z}, \hat{\mu}, \hat{\lambda})
$$

for each $p \in B_\epsilon(\hat{p})$, and

$$
A(\hat{z}, \hat{p}) = A(z(p), p).
$$

In addition, $(z(p), \mu(p), \lambda(p))$ is continuously differentiable with respect to p with

$$\begin{pmatrix} \frac{dz}{dp}(\hat{p}) \\ \frac{d\mu}{dp}(\hat{p}) \\ \frac{d\lambda}{dp}(\hat{p}) \end{pmatrix} = -\begin{pmatrix} L''_{zz} & (G'_z)^\top & (H'_z)^\top \\ \hat{\Xi}\cdot G'_z & \hat{\Gamma} & \Theta \\ H'_z & \Theta & \Theta \end{pmatrix}^{-1} \cdot \begin{pmatrix} L''_{zp} \\ \hat{\Xi}\cdot G'_p \\ H'_p \end{pmatrix} \tag{6.3}$$

where

$$\hat{\Xi} := \operatorname{diag}(\hat{\mu}_1, \dots, \hat{\mu}_{n_G}), \quad \hat{\Gamma} := \operatorname{diag}(G_1, \dots, G_{n_G}).$$

All functions and their derivatives are evaluated at $(\hat{z}, \hat{\mu}, \hat{\lambda}, \hat{p})$.

Proof. Consider the nonlinear equation

$$T(z,\mu,\lambda,p) := \begin{pmatrix} L'_z(z,\ell_0,\mu,\lambda,p)^\top \\ \Xi\cdot G(z,p) \\ H(z,p) \end{pmatrix} = 0_{\mathbb{R}^{n_z+n_G+n_H}}, \tag{6.4}$$

where $\ell_0 = 1$ and $\Xi := \operatorname{diag}(\mu_1, \dots, \mu_{n_G})$. T is continuously differentiable, and it holds

$$T(\hat{z},\hat{\mu},\hat{\lambda},\hat{p}) = 0_{\mathbb{R}^{n_z+n_G+n_H}}.$$

We intend to apply the implicit function theorem. Hence, we have to show the non-singularity of

$$T'_{(z,\mu,\lambda)}(\hat{z},\hat{\mu},\hat{\lambda},\hat{p}) = \begin{pmatrix} L''_{zz}(\hat{z},\ell_0,\hat{\mu},\hat{\lambda},\hat{p}) & G'_z(\hat{z},\hat{p})^\top & H'_z(\hat{z},\hat{p})^\top \\ \hat{\Xi}\cdot G'_z(\hat{z},\hat{p}) & \hat{\Gamma} & \Theta \\ H'_z(\hat{z},\hat{p}) & \Theta & \Theta \end{pmatrix}.$$

To this end, we assume without loss of generality, that the index set of active inequality constraints is given by $A(\hat{z},\hat{p}) = \{\ell+1, \dots, n_G\}$, where ℓ denotes the number of inactive inequality constraints. The strict complementarity condition implies

$$\hat{\Xi} = \begin{pmatrix} \Theta & \Theta \\ \Theta & \hat{\Xi}_2 \end{pmatrix}, \quad \hat{\Gamma} = \begin{pmatrix} \hat{\Gamma}_1 & \Theta \\ \Theta & \Theta \end{pmatrix},$$

with non-singular matrices

$$\hat{\Xi}_2 := \operatorname{diag}(\hat{\mu}_{\ell+1}, \dots, \hat{\mu}_{n_G}) \quad \text{and} \quad \hat{\Gamma}_1 := \operatorname{diag}(G_1(\hat{z},\hat{p}), \dots, G_\ell(\hat{z},\hat{p})).$$

Consider the linear equation

$$\begin{pmatrix} L''_{zz}(\hat{z},\ell_0,\hat{\mu},\hat{\lambda},\hat{p}) & G'_z(\hat{z},\hat{p})^\top & H'_z(\hat{z},\hat{p})^\top \\ \hat{\Xi}\cdot G'_z(\hat{z},\hat{p}) & \hat{\Gamma} & \Theta \\ H'_z(\hat{z},\hat{p}) & \Theta & \Theta \end{pmatrix} \begin{pmatrix} v_1 \\ v_2 \\ v_3 \end{pmatrix} = \begin{pmatrix} 0_{\mathbb{R}^{n_z}} \\ 0_{\mathbb{R}^{n_G}} \\ 0_{\mathbb{R}^{n_H}} \end{pmatrix}$$

for $v_1 \in \mathbb{R}^{n_z}$, $v_2 = (v_{21}, v_{22})^\top \in \mathbb{R}^{\ell+(n_G-\ell)}$, and $v_3 \in \mathbb{R}^{n_H}$. Exploitation of the special structure of $\hat{\Xi}$ and $\hat{\Gamma}$ yields $\hat{\Gamma}_1 v_{21} = 0_{\mathbb{R}^\ell}$ and, since $\hat{\Gamma}_1$ is non-singular, it follows $v_{21} = 0_{\mathbb{R}^\ell}$. With this, it remains to investigate the reduced system

$$\begin{pmatrix} A & B^\top & C^\top \\ B & \Theta & \Theta \\ C & \Theta & \Theta \end{pmatrix} \begin{pmatrix} v_1 \\ v_{22} \\ v_3 \end{pmatrix} = \begin{pmatrix} 0_{\mathbb{R}^{n_z}} \\ 0_{\mathbb{R}^{n_G-\ell}} \\ 0_{\mathbb{R}^{n_H}} \end{pmatrix}$$

with $A := L''_{zz}(\hat{z}, \ell_0, \hat{\mu}, \hat{\lambda}, \hat{p})$, $B := (G'_{i,z}(\hat{z}, \hat{p}))_{i=\ell-1,\dots,n_G}$, and $C := H'_z(\hat{z}, \hat{p})$. Notice that the second block equation has been multiplied with $\hat{\Xi}_2^{-1}$. The last two block equations yield $Bv_1 = 0_{\mathbb{R}^{n_G-\ell}}$ and $Cv_1 = 0_{\mathbb{R}^{n_H}}$. Multiplication of the first block equation from the left with $v_1^\top$ yields

$$0 = v_1^\top A v_1 + (Bv_1)^\top v_{22} + (Cv_1)^\top v_3 = v_1^\top A v_1.$$

Since A is positive definite on $T_C(\hat{z}, \hat{p}) \setminus \{0_{\mathbb{R}^{n_z}}\}$, i. e. it holds $d^\top A d > 0$ for all $d \neq 0_{\mathbb{R}^{n_z}}$ with $Bd = 0_{\mathbb{R}^{n_G-\ell}}$ and $Cd = 0_{\mathbb{R}^{n_H}}$, it follows $v_1 = 0_{\mathbb{R}^{n_z}}$. Taking this property into account, the first block equation reduces to

$$B^\top v_{22} + C^\top v_3 = 0_{\mathbb{R}^{n_z}}.$$

By the linear independence of the gradients

$$\nabla_z G_i(\hat{z}, \hat{p}), \quad i \in A(\hat{z}, \hat{p}), \quad \text{and} \quad \nabla_z H_j(\hat{z}, \hat{p}), \quad j = 1, \dots, n_H,$$

we obtain $v_{22} = 0_{\mathbb{R}^{n_G-\ell}}$, $v_3 = 0_{\mathbb{R}^{n_H}}$. In summary, the above linear equation has the unique solution $v_1 = 0_{\mathbb{R}^{n_z}}$, $v_2 = 0_{\mathbb{R}^{n_G}}$, $v_3 = 0_{\mathbb{R}^{n_H}}$, which implies that the matrix $T'_{(z,\mu,\lambda)}$ is non-singular and the implicit function theorem is applicable.

By the implicit function theorem, there exist neighborhoods $B_\varepsilon(\hat{p})$ and $B_\delta(\hat{z}, \hat{\mu}, \hat{\lambda})$, and uniquely defined functions

$$(z(\cdot), \mu(\cdot), \lambda(\cdot)) : B_\varepsilon(\hat{p}) \longrightarrow B_\delta(\hat{z}, \hat{\mu}, \hat{\lambda})$$

satisfying

$$T(z(p), \mu(p), \lambda(p), p) = 0_{\mathbb{R}^{n_z+n_G+n_H}} \tag{6.5}$$

for all $p \in B_\varepsilon(\hat{p})$. Furthermore, these functions are continuously differentiable, and (6.3) arises by differentiation of the identity (6.5) with respect to p.

It remains to verify that $z(p)$ actually is a strongly regular local minimum of NLP(p). The continuity of the functions $z(p)$, $\mu(p)$, and G together with $\mu_i(\hat{p}) = \hat{\mu}_i > 0$, $i = \ell + 1, \dots, n_G$, and $G_i(z(\hat{p}), \hat{p}) = G_i(\hat{z}, \hat{p}) < 0$, $i = 1, \dots, \ell$, guarantees $\mu_i(p) > 0$, $i = \ell + 1, \dots, n_G$, and $G_i(z(p), p) < 0$, $i = 1, \dots, \ell$, for all $p \in B_\varepsilon(\hat{p})$ (after diminishing $\varepsilon > 0$ if necessary).

From (6.5), it follows $G_i(z(p),p) = 0, i = \ell+1,\dots,n_G$, and $H_j(z(p),p) = 0, j = 1,\dots,n_H$. Thus, $z(p) \in \Sigma(p)$ and the KKT conditions are satisfied in $B_\varepsilon(\hat{p})$. Furthermore, the index set $A(z(p),p) = A(\hat{z},\hat{p})$ remains unchanged in $B_\varepsilon(\hat{p})$.

Finally, we have to show that $L''_{zz}(z(p),\ell_0,\mu(p),\lambda(p),p)$ remains positive definite on $T_C(z(p),p)$ for p sufficiently close to $\hat{p}$. Notice that the critical cone $T_C(z(p),p)$ varies with p. By now, we only know that $L''_{zz}(\hat{p}) := L''_{zz}(z(\hat{p}),\ell_0,\mu(\hat{p}),\lambda(\hat{p}),\hat{p})$ is positive definite on $T_C(z(\hat{p}),\hat{p})$, which is equivalent to the existence of some $\alpha > 0$ with $d^\top L''_{zz}(\hat{p})d \geq \alpha\|d\|^2$ for all $d \in T_C(z(\hat{p}),\hat{p})$. Owing to the strict complementarity in the neighborhood $B_\varepsilon(\hat{p})$, it holds

$$T_C(z(p),p) = \left\{ d \in \mathbb{R}^{n_z} \;\middle|\; \begin{array}{l} G'_{i,z}(z(p),p)(d) = 0,\ i \in A(\hat{z},\hat{p}), \\ H'_{j,z}(z(p),p)(d) = 0,\ j = 1,\dots,n_H, \end{array} \right\}$$

for $p \in B_\varepsilon(\hat{p})$. Assume that for every $i \in \mathbb{N}$, there exists some $p^i \in \mathbb{R}^{n_p}$ with $\|p^i - \hat{p}\| \leq \frac{1}{i}$ such that for all $j \in \mathbb{N}$ there exists some $d^{i_j} \in T_C(z(p^i),p^i)$, $d^{i_j} \neq 0_{\mathbb{R}^{n_z}}$, with

$$(d^{i_j})^\top L''_{zz}(p^{i_j})d^{i_j} < \frac{1}{j}\|d^{i_j}\|^2.$$

Since the unit ball with respect to $\|\cdot\|$ is compact in $\mathbb{R}^{n_z}$, there exists a convergent subsequence $\{p^{i_{j_k}}\}$ with $\lim_{k\to\infty} p^{i_{j_k}} = \hat{p}$ and

$$\lim_{k\to\infty} \frac{d^{i_{j_k}}}{\|d^{i_{j_k}}\|} = \hat{d}, \quad \|\hat{d}\| = 1, \quad \hat{d} \in T_C(z(\hat{p}),\hat{p})$$

and

$$\hat{d}^\top L''_{zz}(\hat{p})\hat{d} \leq 0.$$

This contradicts the positive definiteness of $L''_{zz}(\hat{p})$. Hence, L''_{zz} remains positive definite on $T_C(z(p),p)$ for $p \in B_\varepsilon(\hat{p})$ (after diminishing $\varepsilon > 0$ if necessary).

Finally, the gradients of the active constraints remain linearly independent for $p \in B_\varepsilon(\hat{p})$ (after diminishing $\varepsilon > 0$ again if necessary) owing to the continuity of z and the first derivatives of G and H. □

The sensitivity Theorem 6.1.4 gives rise to a method that allows approximating the optimal solution for a perturbed parameter by linearization, see [51]. Under the assumptions of Theorem 6.1.4, the solution $z(p)$ of NLP(p) is continuously differentiable in some neighborhood of $\hat{p}$ with

$$z(p) = z(\hat{p}) + \frac{dz}{dp}(\hat{p})(p-\hat{p}) + o(\|p-\hat{p}\|).$$

Herein, the sensitivity dz/dp is given by the linear equation (6.3).

Algorithm 6.1.5 (Real-time approximation for NLP(p)).

(0) Let a nominal parameter $\hat{p}$ be given.

(1) *Offline computation:* Solve NLP($\hat{p}$) and the linear equation (6.3).

(2) *Online computation (real-time approximation):* For a perturbed parameter $p \neq \hat{p}$ compute

$$\tilde{z}(p) := z(\hat{p}) - \frac{dz}{dp}(\hat{p})(p - \hat{p}) \tag{6.6}$$

and use $\tilde{z}(p)$ as an approximation of $z(p)$. □

The computations in step (1) of Algorithm 6.1.5 can be very time-consuming depending on the size and complexity of NLP($\hat{p}$). In contrast, the computation of the real-time approximation $\tilde{z}(p)$ in step (2) only requires a matrix-vector product and two vector summations. The computation time for these operations is negligible, and the update rule (6.6) is real-time capable for most applications.

The downside of the approach in Algorithm 6.1.5 is that the linearization in (6.6) is only justified locally in some neighborhood of the nominal parameter $\hat{p}$. Theorem 6.1.4 unfortunately does not indicate how large this neighborhood is. In particular, if the index set of active inequality constraints changes, then Theorem 6.1.4 is not applicable anymore.

The real-time approximation $\tilde{z}(p)$ is not feasible in general owing to the linearization error. Feasibility can be achieved by projecting $\tilde{z}(p)$ onto the feasible set, e. g., by solving the following optimization problem:

Minimize

$$\frac{1}{2}\|z - \tilde{z}(p)\|^2$$

with respect to $z \in \mathbb{R}^{n_z}$ *subject to the constraints*

$$G_i(z,p) \leq 0, \quad i = 1,\ldots,n_G,$$
$$H_j(z,p) = 0, \quad j = 1,\ldots,n_H.$$

The sensitivity differential $dz/dp_i(\hat{p})$ for the ith component p_i of the parameter vector $p = (p_1,\ldots,p_{n_p})^\top$ can be approximated alternatively by the central finite difference scheme

$$\frac{dz}{dp_i}(\hat{p}) \approx \frac{z(\hat{p} + he_i) - z(\hat{p} - he_i)}{2h},$$

where $z(\hat{p} \pm he_i)$ denotes the optimal solution of NLP($\hat{p} \pm he_i$), and e_i is the ith unity vector. This approach requires to solve $2n_p$ nonlinear optimization problems but has the advantage that $z(\hat{p})$ may serve as an initial guess for NLP($\hat{p} \pm he_i$), which is usually very good. This approach may be preferable to solving (6.3), if the optimal solution cannot be calculated very accurately.

It remains to check the strong regularity condition in Definition 6.1.2 for a given local minimizer $\hat{z}$ numerically. Admissibility, strict complementarity, and KKT conditions are easy to check. If strict complementarity holds, then the second order sufficient optimality condition in Theorem 6.1.3 states that the Hessian matrix of the Lagrange function is positive definite on the null-space of the active constraints, that is

$$d^\top L''_{zz}(\hat{z}, \ell_0, \hat{\mu}, \hat{\lambda}, \hat{p})d > 0 \quad \text{for all } d \in \mathbb{R}^{n_z}, d \neq 0_{\mathbb{R}^{n_z}} \text{ with } Bd = \Theta,$$

where

$$B := \begin{pmatrix} G'_{i,z}(\hat{z}, \hat{p}), i \in A(\hat{z}, \hat{p}) \\ H'_z(\hat{z}, \hat{p}) \end{pmatrix}.$$

This condition can be checked numerically using an orthogonal decomposition of $B^\top$. To this end, compute a QR decomposition of $B^\top$ with an orthogonal matrix Q and an upper triangular matrix R, for instance by Householder reflections:

$$\begin{aligned} B^\top &= Q\begin{pmatrix} R \\ \Theta \end{pmatrix}, \\ Q &= (Q_1 \quad Q_2) \in \mathbb{R}^{n_z \times n_z}, \\ Q_1 &\in \mathbb{R}^{n_z \times (|A(\hat{z},\hat{p})|+n_H)}, \\ Q_2 &\in \mathbb{R}^{n_z \times (n_z - |A(\hat{z},\hat{p})| - n_H)}, \\ R &\in \mathbb{R}^{(|A(\hat{z},\hat{p})|+n_H) \times (|A(\hat{z},\hat{p})|+n_H)}. \end{aligned}$$

R is non-singular, if the columns of $B^\top$ are linearly independent. This allows checking the linear independence constraint qualification. Q is orthogonal, Q_1 contains an orthogonal basis of the image of $B^\top$, while Q_2 contains an orthogonal basis of the null-space of B. Every d in the null-space of B can thus be represented uniquely as $d = Q_2 d_2$ with some $d_2 \in \mathbb{R}^{n_z - |A(\hat{z},\hat{p})| - n_H}$.

The task of checking L''_{zz} for positive definiteness on the null-space of B is thus equivalent with checking the *reduced Hessian matrix*

$$(Q_2)^\top L''_{zz}(\hat{z}, \ell_0, \hat{\mu}, \hat{\lambda}, \hat{p})Q_2 \in \mathbb{R}^{(n_z - |A(\hat{z},\hat{p})| - n_H) \times (n_z - |A(\hat{z},\hat{p})| - n_H)}$$

for positive definiteness.

6.1.2 Open-loop real-time control via sensitivity analysis

The results of Subsection 6.1.1 are applied to discretized optimal control problems. To this end, consider

Problem 6.1.6 (Perturbed DAE optimal control problem OCP(p)). Let $\mathcal{I} := [t_0, t_f] \subset \mathbb{R}$ be compact with $t_0 < t_f$, and let

$$\begin{aligned} \varphi &: \mathbb{R}^{n_z} \times \mathbb{R}^{n_z} \times \mathbb{R}^{n_p} \longrightarrow \mathbb{R}, \\ F &: \mathbb{R}^{n_z} \times \mathbb{R}^{n_z} \times \mathbb{R}^{n_u} \times \mathbb{R}^{n_p} \longrightarrow \mathbb{R}^{n_z}, \\ c &: \mathbb{R}^{n_z} \times \mathbb{R}^{n_u} \times \mathbb{R}^{n_p} \longrightarrow \mathbb{R}^{n_c}, \\ \psi &: \mathbb{R}^{n_z} \times \mathbb{R}^{n_z} \times \mathbb{R}^{n_p} \longrightarrow \mathbb{R}^{n_\psi} \end{aligned}$$

be sufficiently smooth functions.

Minimize

$$\varphi\big(z(t_0), z(t_f), p\big)$$

with respect to z and u subject to the constraints

$$\begin{aligned} F\big(z(t), \dot{z}(t), u(t), p\big) &= 0_{\mathbb{R}^{n_z}}, \quad \text{a. e. in } \mathcal{I}, \\ \psi\big(z(t_0), z(t_f), p\big) &= 0_{\mathbb{R}^{n_\psi}}, \\ c\big(z(t), u(t), p\big) &\leq 0_{\mathbb{R}^{n_c}}, \quad \text{a. e. in } \mathcal{I}. \end{aligned}$$

Notice that $p \in \mathbb{R}^{n_p}$ is not an optimization variable. Let $\hat{p} \in \mathbb{R}^{n_p}$ denote a fixed *nominal parameter*. The corresponding optimal control problem OCP($\hat{p}$) is called *nominal problem*.

The reduced discretization method of Subsection 5.1.2 is applied to Problem 6.1.6 for its numerical solution. To this end, let Z_0 be a function that provides a consistent initial value for the optimization vector $\bar{z} = (z_0, w)^\top$ and parameter p. The values Z_j, $j = 1, \ldots, N$, are given by a suitable integration scheme, e. g., a one-step method:

$$Z_{j+1}(\bar{z}, p) = Z_j(\bar{z}, p) + h_j \Phi(t_j, Z_j(\bar{z}, p), w, p, h_j), \quad j = 0, 1, \ldots, N-1.$$

According to (5.6) the discretized control is given by

$$u_M(\cdot) = \sum_{i=1}^{M} w_i B_i(\cdot)$$

with basis functions B_i, $i = 1, \ldots, M$, for instance B-Splines. With these approximations, we obtain

Problem 6.1.7 (Reduced discretization DOCP(p)).
Minimize

$$\varphi\big(Z_0(\bar{z}, p), Z_N(\bar{z}, p), p\big)$$

with respect to $\bar{z} = (z_0, w) \in \mathbb{R}^{n_z + M}$ *subject to the constraints*

$$\begin{aligned} \psi\big(Z_0(\bar{z}, p), Z_N(\bar{z}, p), p\big) &= 0_{\mathbb{R}^{n_\psi}}, \\ c\big(Z_j(\bar{z}, p), u_M(t_j; w), p\big) &\leq 0_{\mathbb{R}^{n_c}}, \quad j = 0, 1, \ldots, N, \end{aligned}$$

Problem 6.1.7 is a special case of Problem 6.1.1, and the sensitivity Theorem 6.1.4 is applicable, if all assumptions hold. Theorem 6.1.4 then guarantees the differentiability of the solution mapping $p \mapsto \bar{z}(p)$ in some neighborhood of $\hat{p}$. Furthermore, a solution of the linear equation (6.3) yields the sensitivity differential $\frac{d\bar{z}}{dp}(\hat{p})$ at the solution $\hat{\bar{z}}$. Taking into account the structure of $\bar{z}$ in DOCP(p), that is $\bar{z}(p) = (z_0(p), w(p))$, the optimal discretized control for DOCP(p) is given by

$$u_M(\cdot) = \sum_{i=1}^{M} w_i(p)B_i(\cdot).$$

Application of Algorithm 6.1.5 to DOCP(p) yields

Algorithm 6.1.8 (Real-time approximation for DOCP(p)).

(0) Let a nominal parameter $\hat{p}$ be given.

(1) *Offline computation:* Solve DOCP($\hat{p}$) and the linear equation (6.3).

(2) *Online computation (real-time approximation):* For a perturbed parameter $p \neq \hat{p}$ compute

$$\tilde{z}_0(p) := z_0(\hat{p}) + \frac{dz_0}{dp}(\hat{p})(p - \hat{p}),$$
$$\tilde{w}(p) := w(\hat{p}) + \frac{dw}{dp}(\hat{p})(p - \hat{p}).$$

Use $\tilde{z}_0(p)$ as an approximation of the initial value $z_0(p)$ and

$$\tilde{u}_M(\cdot) := \sum_{i=1}^{M} \tilde{w}_i(p)B_i(\cdot) \tag{6.7}$$

as an approximate optimal control for the parameter p. □

Again, only matrix-vector products and vector-vector summations are needed to evaluate the real-time update formulae in Algorithm 6.1.8.

A consistent initial value for the real-time approximation $\tilde{z}_0(p)$ is given by

$$Z_0(\tilde{z}(p), p) = Z_0\left(\bar{z}(\hat{p}) + \frac{d\bar{z}}{dp}(\hat{p})(p - \hat{p}), p\right)$$

and an approximate consistent initial value is obtained by Taylor expansion:

$$\tilde{Z}_0(p) \approx Z_0(\bar{z}(\hat{p}), \hat{p}) + \frac{\partial Z_0}{\partial z}(\bar{z}(\hat{p}), \hat{p})\frac{d\bar{z}}{dp}(\hat{p})(p - \hat{p}) + \frac{\partial Z_0}{\partial p}(\bar{z}(\hat{p}), \hat{p})(p - \hat{p}).$$

It remains to compute the sensitivity differential $d\bar{z}(\hat{p})/dp$. The solution of (6.3) requires the second derivatives $L''_{\bar{z}\bar{z}}$ and $L''_{\bar{z}p}$ of the Lagrange function, the derivatives $H'_{\bar{z}}$, $G'_{\bar{z}}$ in (5.17)–(5.16), and the corresponding derivatives H'_p, G'_p at $\hat{\bar{z}}$. In fact, only the derivatives of the active constraints are needed. The computation of the derivatives $H'_{\bar{z}}$, $G'_{\bar{z}}$, H'_p, and G_p' can be done efficiently either by the sensitivity equation approach of Section 5.4.1 or by the adjoint equation approach of Sections 5.4.2 and 5.4.3.

If the derivatives cannot be obtained analytically or by algorithmic differentiation, then approximations of $L''_{\bar{z}\bar{z}}$ and $L''_{\bar{z}p}$ can be obtained by finite differences

$$\begin{aligned} L''_{\bar{z}_i\bar{z}_j}(\hat{\bar{z}}, \ell_0, \hat{\mu}, \hat{\lambda}, \hat{p}) &\approx \frac{1}{4h^2}(L(\hat{\bar{z}} + he_i + he_j, \ell_0, \hat{\mu}, \hat{\lambda}, \hat{p}) - L(\hat{\bar{z}} - he_i + he_j, \ell_0, \hat{\mu}, \hat{\lambda}, \hat{p}) \\ &\quad - L(\hat{\bar{z}} + he_i - he_j, \ell_0, \hat{\mu}, \hat{\lambda}, \hat{p}) + L(\hat{\bar{z}} - he_i - he_j, \ell_0, \hat{\mu}, \hat{\lambda}, \hat{p})), \\ L''_{\bar{z}_ip_j}(\hat{\bar{z}}, \ell_0, \hat{\mu}, \hat{\lambda}, \hat{p}) &\approx \frac{1}{4h^2}(L(\hat{\bar{z}} + he_i, \ell_0, \hat{\mu}, \hat{\lambda}, \hat{p} + he_j) - L(\hat{\bar{z}} - he_i, \ell_0, \hat{\mu}, \hat{\lambda}, \hat{p} + he_j) \\ &\quad - L(\hat{\bar{z}} + he_i, \ell_0, \hat{\mu}, \hat{\lambda}, \hat{p} - he_j) + L(\hat{\bar{z}} - he_i, \ell_0, \hat{\mu}, \hat{\lambda}, \hat{p} - he_j)) \end{aligned} \tag{6.8}$$

with appropriate unity vectors e_i, e_j. For each evaluation of the Lagrange function L, one DAE has to be solved numerically. Hence, this approach leads to a total number of $4\frac{n_{\bar{z}}(n_{\bar{z}}+1)}{2} + 4n_{\bar{z}}n_p$ DAE evaluations, if the symmetry of the Hessian $L''_{\bar{z}\bar{z}}$ is exploited.

An alternative approach is to exploit the information obtained by the sensitivity equation. Recall that one evaluation of the DAE and the corresponding sensitivity equation for the NLP variable $\bar{z}$ provides the gradient of the objective function and the Jacobian of the constraints, and with this information, the gradient of the Lagrangian is obtained easily. A finite difference approximation with these gradients yields

$$\begin{aligned} L''_{\bar{z}\bar{z}_i}(\hat{\bar{z}}, \ell_0, \hat{\mu}, \hat{\lambda}, \hat{p}) &\approx \frac{L'_{\bar{z}}(\hat{\bar{z}} + he_i, \ell_0, \hat{\mu}, \hat{\lambda}, \hat{p}) - L'_{\bar{z}}(\hat{\bar{z}} - he_i, \ell_0, \hat{\mu}, \hat{\lambda}, \hat{p})}{2h}, \\ L''_{\bar{z}p_j}(\hat{\bar{z}}, \ell_0, \hat{\mu}, \hat{\lambda}, \hat{p}) &\approx \frac{L'_{\bar{z}}(\hat{\bar{z}}, \ell_0, \hat{\mu}, \hat{\lambda}, \hat{p} + he_j) - L'_{\bar{z}}(\hat{\bar{z}}, \ell_0, \hat{\mu}, \hat{\lambda}, \hat{p} - he_j)}{2h}. \end{aligned}$$

This time only $2n_{\bar{z}} + 2n_p$ DAEs including the corresponding sensitivity equations (5.47) with respect to $\bar{z}$ have to be solved. Since the evaluation of $n_{\bar{z}}$ nonlinear DAEs in combination with the corresponding linear sensitivity DAE with $n_{\bar{z}}$ columns is usually much cheaper than the evaluation of $n_{\bar{z}}^2$ nonlinear DAEs, the second approach is found to be much faster than the finite difference approximation (6.8).

The choice of h in the finite difference schemes is crucial. As a rule of thumb, h should be in the range $\sqrt{\varepsilon}$ to $10\sqrt{\varepsilon}$, where ε denotes the tolerance for the KKT conditions in the numerical solution of DOCP($\hat{p}$). Typically DOCP(p) needs to be solved sufficiently accurately with $\varepsilon \approx \sqrt{eps}$, where eps is the machine precision, i. e. $eps \approx 2.22 \cdot 10^{-16}$ for double precision computations.

---

i Many implementations of methods for dense nonlinear optimization problems, e. g., SQP methods, use update formulas instead of the Hessian of the Lagrange function. The update matrices B_k, $k = 0, 1, \dots$, in the ideal case satisfy a condition of type

$$\lim_{k\to\infty} \frac{\|P(\bar{z}^{(k)}, \hat{p})(B_k - L''_{\bar{z}\bar{z}}(\hat{\bar{z}}, \ell_0, \hat{\mu}, \hat{\lambda}, \hat{p}))d^{(k)}\|}{\|d^{(k)}\|} = 0,$$

where P is a projector on the tangent space of the constraints, compare [31]. Hence, the update provides a good approximation of the Hessian only when projected on the linearized constraints at iterate $\bar{z}^{(k)}$ in direction $d^{(k)}$. This is not sufficient for sensitivity analysis, and it is necessary to evaluate the Hessian $L''_{\bar{z}\bar{z}}$ as well as the quantities $L''_{\bar{z}p}$, $H'_{\bar{z}}$, $G'_{\bar{z}}$, H'_p, and G'_p in the solution.

Several examples follow that illustrate the sensitivity analysis.

Example 6.1.9 (Pendulum). We revisit Example 1.1.9 with $\ell = 1$, but we add control constraints to the optimal control problem and consider the Gear-Gupta–Leimkuhler stabilized DAE of index two:

Minimize

$$\int_0^{4.5} u(t)^2 dt$$

with respect to $x = (x_1, \ldots, x_4)^\top$, $y = (y_1, y_2)^\top$, *and* u *subject to the constraints*

$$\begin{aligned}
\dot{x}_1(t) &= x_3(t) - 2x_1(t)y_2(t),\\
\dot{x}_2(t) &= x_4(t) - 2x_2(t)y_2(t),\\
m\dot{x}_3(t) &= \quad - 2x_1(t)y_1(t) + \frac{u(t)x_2(t)}{\ell},\\
m\dot{x}_4(t) &= -mg - 2x_2(t)y_1(t) - \frac{u(t)x_1(t)}{\ell},\\
0 &= x_1(t)^2 + x_2(t)^2 - \ell^2,\\
0 &= x_1(t)x_3(t) + x_2(t)x_4(t),
\end{aligned}$$

and

$$x_1(0) = -x_2(0) = \frac{1}{\sqrt{2}}, \quad x_3(0) = x_4(0) = 0, \quad x_1(4.5) = x_3(4.5) = 0$$

and

$$-1 \le u(t) \le 1.$$

The sensitivity analysis in Algorithm 6.1.8 is performed for the parameters m and g with nominal values $\hat{m} = 1$ and $\hat{g} = 9.81$. For the numerical computation, $N = 400$ grid intervals were used in combination with a piecewise constant control approximation and a linearized Runge–Kutta method for integration. Figure 6.1 shows the optimal nominal control $\hat{u}$ for the problem and its sensitivities with respect to the parameters m and g. Notice that the sensitivities are zero on active control intervals. The smallest eigenvalue of the reduced Hessian matrix is approximately 0.0217. □

The following example performs the sensitivity analysis for optimal flight paths in a space shuttle mission. Herein, a common model for the air density depends on the altitude and certain parameters describing a standard atmosphere. In reality, the air density differs from the model, and the actual air density can be viewed as a perturbation of the model. This, in turn, implies that the computed optimal flight path for the standard atmosphere is not optimal anymore for the actual air density. The flight path has to be adapted to the perturbed situation in an optimal way.

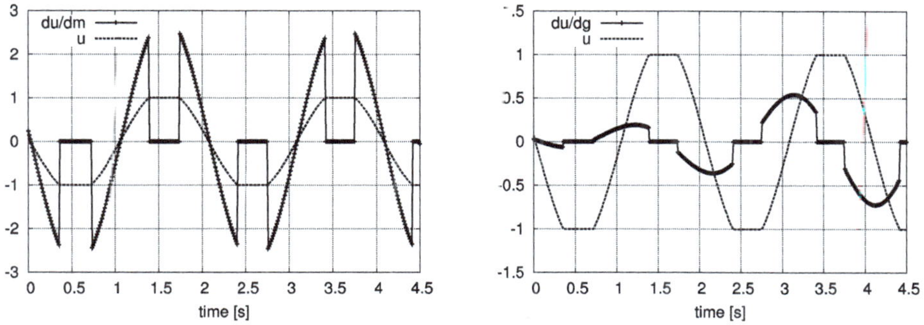

Figure 6.1: Optimal nominal control and sensitivities du/dm and du/dg for the pendulum problem with control constraints.

Example 6.1.10 (Emergency landing maneuver). We revisit the emergency landing maneuver in Example 1.1.10 and investigate the dependence of the optimal solution with respect to the following three parameters:

(i) The parameter $p_1 \in \mathbb{R}$ is used to model uncertainties in the air density $\rho(h) = \rho_0 \exp(-\beta h)$ by the relation

$$\rho_0 = 1.249512(1 + p_1).$$

The nominal parameter value is $\hat{p}_1 = 0$.

(ii) The parameter $p_2 \in \mathbb{R}$ is used to model uncertainties in the initial altitude $h(0)$ according to

$$h(0) = 33900 + 10^4 p_2.$$

Herein, the initial altitude is assumed to be free with the restriction $h(0) \geq 33900$. The nominal parameter value is $\hat{p}_2 = 0$.

(iii) The parameter $p_3 \in \mathbb{R}$ is used to model uncertainties in the terminal altitude $h(t_f)$ according to

$$h(t_f) = 500 - p_3.$$

The nominal parameter value is $\hat{p}_3 = 0$.

These three parameters influence the optimal control problem in different components. p_1 influences the dynamics and the dynamic pressure constraint. p_2 influences the initial value and p_3 influences the boundary condition.

The infinite-dimensional optimal control problem is discretized by the reduced discretization approach on an equidistant grid with 201 grid points. We used the classical fourth order Runge–Kutta scheme for time integration. The control is approximated by a piecewise constant function.

Figure 6.3 shows the numerical solution of the unperturbed problem with nominal parameter $\hat{p} := (\hat{p}_1, \hat{p}_2, \hat{p}_3)^\top = 0_{\mathbb{R}^3}$. The dynamic pressure constraint is active in the approximate normalized time interval $[0.185, 0.19]$.

The sensitivities of the free final time t_f with nominal value $t_f \approx 726.57$ compute to

$$\frac{dt_f}{dp_1}(\hat{p}) \approx 92.54, \quad \frac{dt_f}{dp_2}(\hat{p}) \approx 9.164, \quad \frac{dt_f}{dp_3}(\hat{p}) \approx 0.014.$$

The smallest eigenvalue of the reduced Hessian matrix of the Lagrange function amounts to $0.411 \cdot 10^{-4}$, and the second-order sufficient conditions are satisfied at the nominal solution.

The overall CPU time is 5 minutes and 17.45 seconds on a dual core CPU with 2 GHz. The approximate nominal objective function value is -0.765. Figures 6.2, 6.4, and 6.5 show the sensitivities of the nominal controls C_L and μ with respect to the perturbation parameter p. Note that all sensitivities jump at the time point where the state constraint gets active. Moreover, the sensitivities of C_L jump at the time point where the control constraint for C_L gets active.

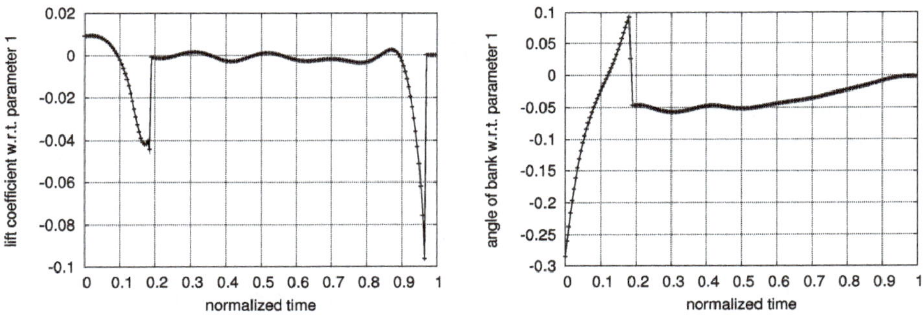

Figure 6.2: Sensitivities dC_L/dp_1 and $d\mu/dp_1$ at $\hat{p}$ (normalized time scale) for 201 grid points.

If a deviation $p = (p_1, p_2, p_3)^\top$ of the nominal parameter $\hat{p} = (0,0,0)^\top$ is detected, then an approximate solution for the perturbed problem is given by

$$\begin{aligned} t_f(p) &\approx t_f(\hat{p}) + \frac{dt_f}{dp}(\hat{p})(p - \hat{p}), \\ C_L(t;p) &\approx C_L(t;\hat{p}) + \frac{dC_L}{dp}(t;\hat{p})(p - \hat{p}), \quad t \in \mathcal{I}, \\ \mu(t;p) &\approx \mu(t;\hat{p}) + \frac{d\mu}{dp}(t;\hat{p})(p - \hat{p}), \quad t \in \mathcal{I}, \end{aligned}$$

and $h(0;p) = 33900 + 10^4(p_2 - \hat{p}_2)$. □

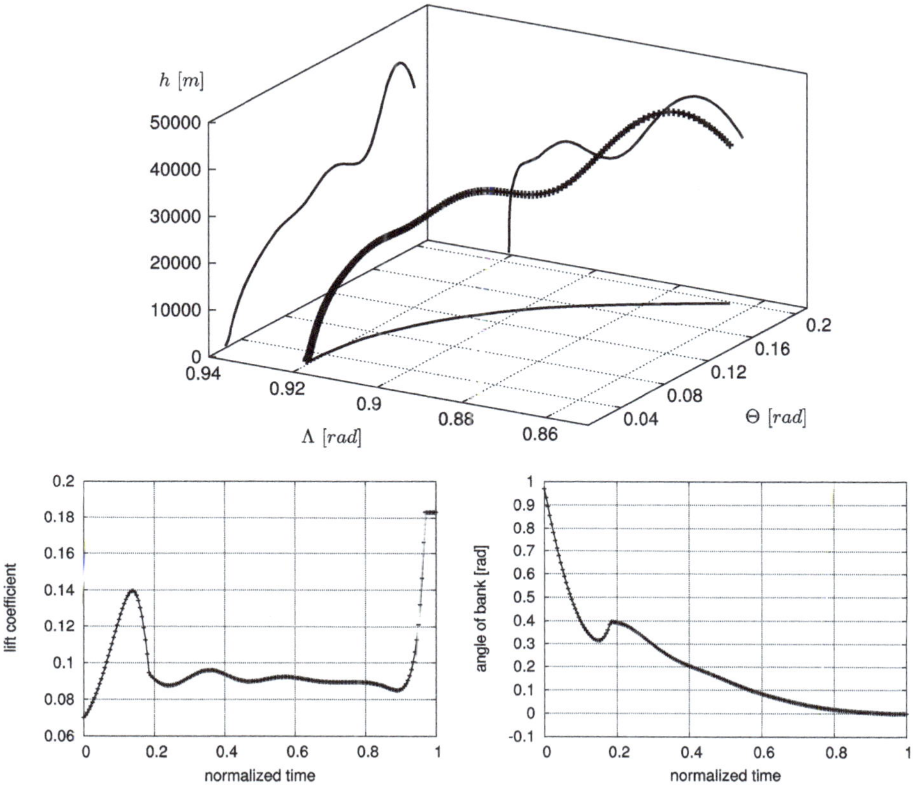

Figure 6.3: Numerical nominal solution for the problem with dynamic pressure constraint: 3D plot of the flight path (top) and the approximate optimal controls lift coefficient C_L and angle of bank μ (bottom, normalized time scale) for 201 grid points.

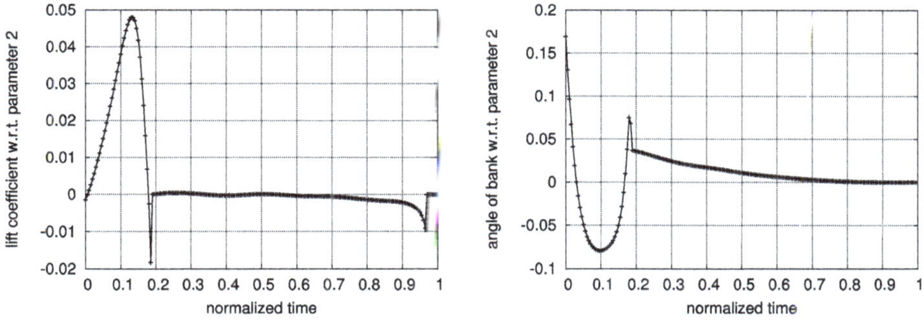

Figure 6.4: Sensitivities dC_L/dp_2 and $d\mu/dp_2$ at $\hat{p}$ (normalized time scale) for 201 grid points.

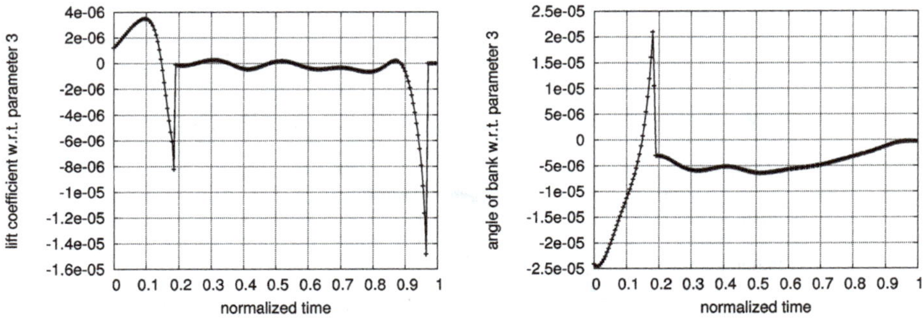

Figure 6.5: Sensitivities dC_L/dp_3 and $d\mu/dp_3$ at $\hat{p}$ (normalized time scale) for 201 grid points.

---

i A corrector iteration method for the reduction of constraint violations, which unavoidably occur due to the linearization in (6.6), was developed in [46], and it was applied to the emergency landing problem in [48].

Example 6.1.11 (Test-drive, compare [115]). We consider the simulation of a test-drive for the double-lane change maneuver, compare 7.2.4, and use a full car model of a BMW 1800/2000. The driver has to complete the test-course in Figure 6.6, which models a typical obstacle avoidance maneuver:

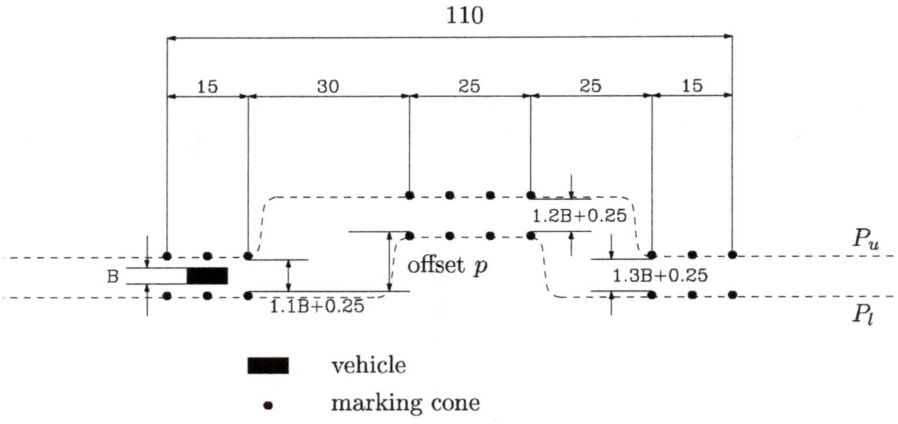

Figure 6.6: Measurements of the double-lane-change maneuver, compare [331].

The equations of motion form a semi-explicit index-one DAE (1.11)–(1.12) with $n_x = 40$ differential equations and $n_y = 4$ algebraic equations. The detailed model can be found in [323, 112, 114].

We assume that the double-lane change maneuver is driven at (almost) constant velocity $v(0) = 25$ [m/s], i. e. the braking force is set to zero, the acceleration is chosen such that it compensates the effect of rolling resistance, and the gear is fixed. The remaining control u denotes the steering wheel velocity. There are two real-time parameters p_1 and p_2 involved in the problem. The first one, p_1, denotes the offset of the test-course with nominal value $\hat{p}_1 = 3.5\,[\mathrm{m}]$, compare Figure 6.6. The second one, p_2, denotes the height of the car's center of gravity with nominal value $\hat{p}_2 = 0.56\,[\mathrm{m}]$.

The driver is modeled by formulation of an appropriate optimal control problem with free final time t_f similar as in Example 7.2.4. While p_2 influences the dynamics of the car, p_1 influences only the state constraints of the optimal control problem. The car's initial position on the course is fixed. At final time t_f, boundary conditions are given by prescribed x-position (140 [m]) and yaw angle (0 [rad]). In addition, the steering capability of the driver is restricted by $|u(t)| \leq 4.6\,[\mathrm{rad/s}]$. The objective is to minimize a linear combination of final time and steering effort, i. e.

$$40\, t_f + \int_0^{t_f} u(t)^2 dt \longrightarrow \min.$$

Figure 6.7 shows the nominal control $\hat{u} = u(\hat{p})$ and the sensitivity differentials for the control at 201 grid points obtained with a piecewise linear approximation of the control, i. e., a B-spline representation with $k = 2$, and the linearized RADAUIIA method of Section 4.3 for time integration.

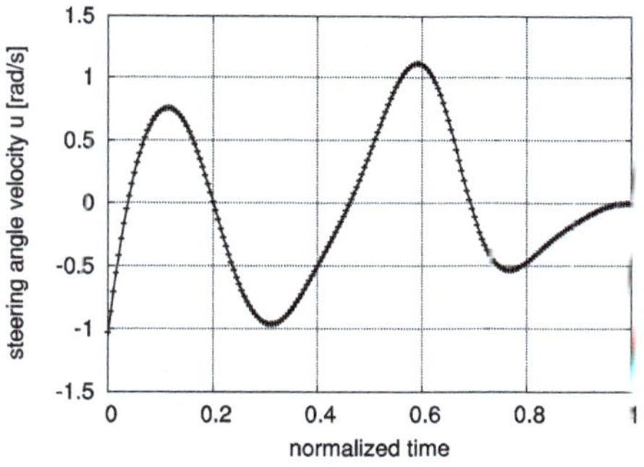

Figure 6.7: Nominal control for $\hat{p}_1 = 3.5\,[\mathrm{m}]$ and $\hat{p}_2 = 0.56\,[\mathrm{m}]$ for 201 grid points.

The nominal objective function value for $\hat{p} = (\hat{p}_1, \hat{p}_2)^\top$ is 274.221. The optimal final time is $t_f \approx 6.79784\,[\mathrm{s}]$, and its sensitivities are

$$\frac{dt_f}{dp_1}(\hat{p}) \approx 0.04638, \qquad \frac{dt_f}{dp_2}(\hat{p}) \approx 0.02715.$$

The second-order sufficient conditions in Theorem 6.1.3 are satisfied. The minimal eigenvalue of the reduced Hessian is approximately 0.0017.

The sensitivities are obtained by a sensitivity analysis of the discretized optimal control problem and are depicted in Figure 6.9.

Figure 6.8 shows snapshots of the nominal solution and the real-time approximation for a perturbation of +10 % in each component of the nominal parameter $\hat{p}$. □

$t_0 = 0$ [s]

$t_f = 6.79784$ [s]

Figure 6.8: Snapshots of the nominal solution (red car) and the real-time approximation (blue car) for perturbed parameters $p_1 = 3.85$ [m] and $p_2 = 0.616$ [m].

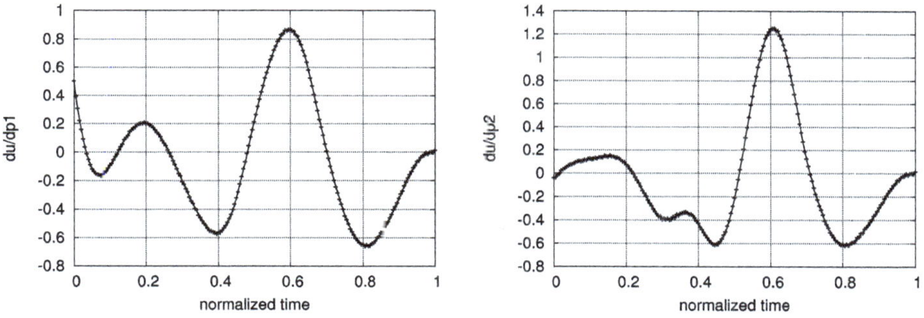

Figure 6.9: Sensitivities of the steering wheel velocity u with respect to the parameters p_1 (offset in double-lane-change maneuver, left) and p_2 (height of center of gravity, right) for 201 control grid points.

6.2 Feedback controller design by optimal control techniques – the linear-quadratic regulator

We briefly discuss approaches to control dynamic systems using feedback control laws. For simplicity, the discussion is restricted to ODEs. An extension to semi-explicit index-one DAEs is straightforward. Further results concerning more general DAEs can be found in [192, 198]. The latter uses properly stated leading terms and analyzes solvability of Riccati DAEs.

Problem 6.2.1 (Control problem). For $x_0 \in \mathbb{R}^{n_x}$ and $f : \mathbb{R}^{n_x} \times \mathbb{R}^{n_v} \longrightarrow \mathbb{R}^{n_x}$, let the following *control problem* be given:

$$\dot{x}(t) = f(x(t), v(t)),$$
$$x(0) = x_0.$$

In addition, let a reference state trajectory $x_{\text{ref}}(t)$ and a reference control $v_{\text{ref}}(t)$ for $t \in [0, t_f]$ be given. The task is to derive a *feedback control law* of type

$$v = v_{\text{ref}} + K(x - x_{\text{ref}})$$

with a suitable function K that tracks the reference state x_{ref} given the actual (measured) state x of the control problem.

The reference state x_{ref} and the reference control v_{ref} are often equilibrium solutions for the control problem or optimal solutions of an optimal control problem. Implementation of the reference control v_{ref} in a real process will usually not result in the reference state x_{ref} owing to model simplifications or disturbances influencing the dynamic behavior. Hence, a feedback control law is needed that controls the actual state towards the reference state. The working principle of feedback controllers is illustrated in Figure 6.10.

If no dynamic model is given for the process, one often uses one of the following approaches or a suitable combination thereof:

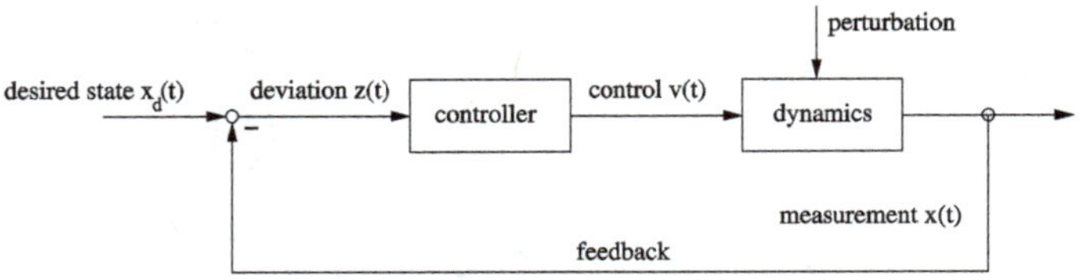

Figure 6.10: Basic scheme of a feedback control law.

(a) *Proportional controller (P-controller):*

$$v(t) = v_{\mathrm{ref}}(t) + c_P(x_{\mathrm{ref}}(t) - x(t)).$$

(b) *Integral controller (I-controller):*

$$v(t) = v_{\mathrm{ref}}(t) + c_I \int_{t-h}^{t} x_{\mathrm{ref}}(\tau) - x(\tau)d\tau, \quad h > 0.$$

(c) *Differential controller (D-controller):*

$$v(t) = v_{\mathrm{ref}}(t) + c_D(\dot{x}_{\mathrm{ref}}(t) - \dot{x}(t)).$$

(d) *Combination: PID-controller*

$$v(t) = v_{\mathrm{ref}}(t) + c_P(x_{\mathrm{ref}}(t) - x(t)) + c_I \int_{t-h}^{t} x_{\mathrm{ref}}(\tau) - x(\tau)d\tau + c_D(\dot{x}_{\mathrm{ref}}(t) - \dot{x}(t)).$$

These feedback controllers have in common that they solely rely on measurements of the actual state and they are easy to implement. Typical questions in constructing such controllers (and other controllers as well) address stability, controllability, and observability.

Example 6.2.2 (Control of a radiator). Let $x_1(t)$ denote the room temperature at time t and $x_2(t)$ the temperature of a radiator at time t, which can be changed by a thermostat with control $v(t)$. Herein, $v(t) > 0$ increases the temperature of the radiator, and $v(t) < 0$ decreases it according to

$$\dot{x}_2(t) = v(t).$$

Let us assume that the reference room temperature (and the temperature outside) is zero, say. A simple model for the room temperature is given by

$$\dot{x}_1(t) = -x_1(t) + x_2(t).$$

Now we use a proportional controller to control the room temperature to the target temperature. In a first attempt, we assume that the room temperature can be measured and obtain the feedback law

$$v(t) = -c_1 x_1(t) \quad \text{with } c_1 > 0,$$

i. e. if the room temperature x_1 is above the reference room temperature, then the radiator is cooled down proportionally to the deviation between room temperature and reference temperature and vice versa.

Simulation of the process with the following scilab program (see www.scilab.org) using Euler's method yields the result in Figure 6.11.

```
function radiator1(x10,x20,h,c,n)
  x1 = zeros(1,n);
  x2 = zeros(1,n);
  v = zeros(1,n-1);
  x1(1) = x10;
  x2(1) = x20;
  for i=1:n-1,
      x1(i+1) = x1(i) + h*(-x1(i)+x2(i));
      v(i) = -c*x1(i);
      x2(i+1) = x2(i) + h*v(i);
  end;
endfunction
```

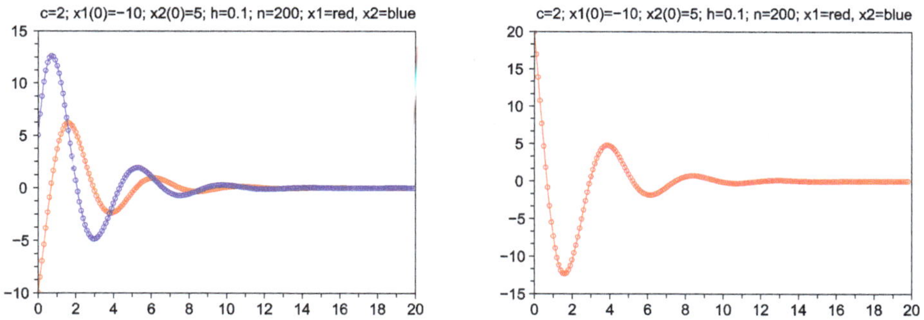

Figure 6.11: Control of a radiator with feedback law $v = -c_1 x_1$, $c_1 = 2$, $x_1(0) = -10$, $x_2(0) = 5$: Room temperature (left picture, red curve), radiator temperature (left picture, blue curve), and control v (right picture).

In a second attempt, it is assumed that only the radiator temperature can be measured and the resulting feedback law is

$$v(t) = -c_2 x_2(t) \quad \text{with } c_2 > 0,$$

i. e. if the radiator temperature x_2 is above the reference room temperature, then the radiator is cooled down proportionally to the deviation between radiator temperature and reference temperature and vice versa. The results are depicted in Figure 6.12.

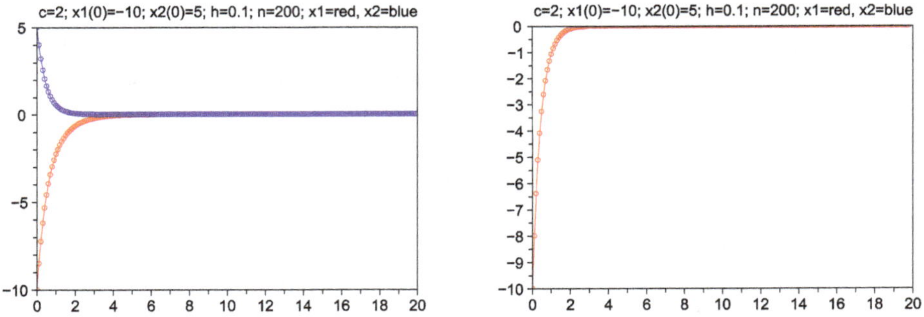

Figure 6.12: Control of a radiator with feedback law $v = -c_2 x_2$, $c_2 = 1$, $x_1(0) = -10$, $x_2(0) = 5$: Room temperature (left picture, red curve), radiator temperature (left picture, blue curve), and control v (right picture).

In terms of control effort (which is assumed to be proportional to heating costs), the second strategy appears to be more efficient. □

We are particularly interested in model-based control and particularly in optimal controllers, which are derived from suitably defined optimal control problems, and focus on the important class of linear-quadratic optimal control problems. To this end, we *linearize* the nonlinear dynamics

$$\dot{x}(t) = f(x(t), v(t)) \quad \text{for } t \in [0, t_f]$$

in the reference state x_{ref} and control v_{ref} as follows. Define the deviations

$$z(t) := x(t) - x_{\text{ref}}(t), \qquad u(t) := v(t) - v_{\text{ref}}(t).$$

Taylor expansion yields

$$\begin{aligned}
\dot{z}(t) &= \dot{x}(t) - \dot{x}_{\text{ref}}(t) \\
&= f(x(t), v(t)) - f(x_{\text{ref}}(t), v_{\text{ref}}(t)) \\
&\approx f'_x(x_{\text{ref}}(t), v_{\text{ref}}(t))(x(t) - x_{\text{ref}}(t)) + f'_v(x_{\text{ref}}(t), v_{\text{ref}}(t))(v(t) - v_{\text{ref}}(t)) \\
&=: A(t)z(t) + B(t)u(t)
\end{aligned}$$

with initial value $z(0) = x(0) - x_{\text{ref}}(0)$ and

$$A(t) := f'_x(x_{\text{ref}}(t), v_{\text{ref}}(t)), \qquad B(t) := f'_v(x_{\text{ref}}(t), v_{\text{ref}}(t)).$$

In case a deviation $z(0) = z_0 \neq 0$ is detected at $t = 0$, one can try to minimize the state deviation z and the control deviation by solving the following optimal control problem:

Problem 6.2.3 (Linear-quadratic optimal control problem). Let time-dependent and essentially bounded matrices $R(t) \in \mathbb{R}^{n_z \times n_z}$, $S(t) \in \mathbb{R}^{n_u \times n_u}$, $A(t) \in \mathbb{R}^{n_z \times n_z}$, and $B(t) \in \mathbb{R}^{n_z \times n_u}$ be given. Let R and S be symmetric, R positive semidefinite, and S uniformly positive definite for all $t \in [0, t_f]$.

Minimize

$$\frac{1}{2}\int_0^{t_f} z(t)^\top R(t)z(t) + u(t)^\top S(t)u(t)dt$$

with respect to $z \in W_{1,\infty}^{n_z}([0, t_f])$ and $u \in L_\infty^{n_u}([0, t_f])$ subject to the constraints

$$\begin{aligned}\dot{z}(t) &= A(t)z(t) + B(t)u(t),\\ z(0) &= z_0.\end{aligned}$$

Theorem 3.1.9 is applied to Problem 6.2.3. The Hamilton function for Problem 6.2.3 reads as

$$\mathcal{H}(t, z, u, \lambda) = \frac{1}{2}(z^\top R(t)z + u^\top S(t)u) + \lambda^\top (A(t)z + B(t)u).$$

The condition $\mathcal{H}_u' = 0_{\mathbb{R}^{n_u}}^\top$ yields

$$0_{\mathbb{R}^{n_u}} = Su + B^\top \lambda \quad \Longrightarrow \quad u = -S^{-1}B^\top \lambda.$$

Recall that S is symmetric and positive definite and thus non-singular. The adjoint equation is given by

$$\dot{\lambda}(t) = -R(t)z(t) - A(t)^\top \lambda(t), \quad \lambda(t_f) = 0_{\mathbb{R}^{n_z}}.$$

Summarizing, the two-point boundary value problem

$$\begin{aligned}\dot{z}(t) &= A(t)z(t) - B(t)S(t)^{-1}B(t)^\top \lambda(t),\\ \dot{\lambda}(t) &= -R(t)z(t) - A(t)^\top \lambda(t),\\ z(0) &= z_0,\\ \lambda(t_f) &= 0_{\mathbb{R}^{n_z}}\end{aligned}$$

has to be solved. These conditions are actually sufficient for optimality according to Theorem 2.3.36, since the objective function is convex and the constraints are affine linear.

We choose the ansatz

$$\lambda(t) := P(t)z(t), \quad P(t) \in \mathbb{R}^{n_z \times n_z},$$

with a matrix P to be determined. Differentiation with respect to t yields

$$\begin{aligned}\dot{\lambda}(t) &= \dot{P}(t)z(t) + P(t)\dot{z}(t)\\ &= \dot{P}(t)z(t) + P(t)\big(A(t)z(t) - B(t)S(t)^{-1}B(t)^\top\lambda(t)\big).\end{aligned}$$

Exploitation of the adjoint equation leads to the equation

$$0_{\mathbb{R}^{n_z}} = \big(R(t) + A(t)^\top P(t) + \dot{P}(t) + P(t)A(t) - P(t)B(t)S(t)^{-1}B(t)^\top P(t)\big)z(t).$$

P is chosen such that the expression in the brackets vanishes guaranteeing that the equation is satisfied for any z. This results in the *Riccati differential equation*

$$\begin{aligned}\dot{P}(t) &= -R(t) - A(t)^\top P(t) - P(t)A(t) + P(t)B(t)S(t)^{-1}B(t)^\top P(t),\\ P(t_f) &= 0_{\mathbb{R}^{n_z\times n_z}}\end{aligned}$$

that needs to be solved. Notice that the Riccati differential equation only depends on the data of the linear-quadratic optimal control problem. Hence, P can be determined once and in advance of the control process. Introducing the adjoint $\lambda = Pz$ into the control law yields the *linear feedback control law*

$$u(t) = -S(t)^{-1}B(t)^\top P(t)z(t).$$

With the feedback matrix $K := BS^{-1}B^\top P$, we obtain the controlled system

$$\dot{z}(t) = \big(A(t) - K(t)\big)z(t), \quad z(0) = z_0.$$

Using this feedback law for the nonlinear control problem yields

$$\begin{aligned}v(t) &= v_{\text{ref}}(t) + u(t)\\ &= v_{\text{ref}}(t) - S(t)^{-1}B(t)^\top P(t)z(t)\\ &= v_{\text{ref}}(t) - S(t)^{-1}B(t)^\top P(t)\big(x(t) - x_{\text{ref}}(t)\big),\end{aligned}$$

where $x(t)$ denotes the current (measured) state at time t.

---

i In the case of time-invariant matrices R, S, A, B, and $t_f = \infty$, steady state solutions of the Riccati equation with $\dot{P} = 0_{\mathbb{R}^{n_z\times n_z}}$ are of interest and lead to the nonlinear *algebraic Riccati equation*

$$0_{\mathbb{R}^{n_z\times n_z}} = -R - A^\top P - PA + PBS^{-1}B^\top P. \tag{6.9}$$

One can show: If (A, B) is stabilizable and S is positive definite, then the Riccati equation has a unique positive semidefinite solution and the feedback controlled linear system is asymptotically stable, see [213, Theorem 5.10] and also [307, Theorem 41, p. 384], where controllability was assumed.

We illustrate the single steps for a simple example.

Example 6.2.4. We construct an optimal controller with the aim to track a reference velocity v_{ref} and a reference position x_{ref} related by $\dot{x}_{\text{ref}}(t) = v_{\text{ref}}(t)$ within the time interval $[0, t_f]$. To this end, consider the following linear-quadratic optimal control problem with $c_1 \geq 0$ and $c_2 > 0$:

Minimize

$$\frac{1}{2}\int_0^{t_f} c_1\big(x(t) - x_{\text{ref}}(t)\big)^2 + c_2\big(v(t) - v_{\text{ref}}(t)\big)^2 dt$$

subject to the constraints

$$\dot{x}(t) = v(t), \quad x(0) = x_0.$$

Define

$$\begin{aligned} z(t) &:= x(t) - x_{\text{ref}}(t), \\ u(t) &:= v(t) - v_{\text{ref}}(t), \end{aligned}$$

such that

$$\dot{z}(t) = \dot{x}(t) - \dot{x}_{\text{ref}}(t) = v(t) - v_{\text{ref}}(t) = u(t).$$

The optimal control problem then reduces to

Minimize

$$\frac{1}{2}\int_0^{t_f} c_1 z(t)^2 + c_2 u(t)^2 dt$$

subject to the constraints

$$\dot{z}(t) = u(t), \quad z(0) = x_0 - x_{\text{ref}}(0).$$

The Hamilton function is given by

$$\mathcal{H}(z, u, \lambda) = \frac{1}{2}(c_1 z^2 + c_2 u^2) + \lambda u$$

and the necessary optimality conditions read as

$$\begin{aligned} 0 = \mathcal{H}'_u &= c_2 u(t) + \lambda(t) \quad \Longrightarrow \quad u(t) = -\frac{\lambda(t)}{c_2}, \\ \dot{\lambda}(t) &= -\mathcal{H}'_z = -c_1 z(t), \\ \lambda(t_f) &= 0. \end{aligned}$$

We intend to derive a feedback law and choose the ansatz $\lambda(t) = P(t)z(t)$ with a function P. Differentiation yields

$$\begin{aligned}\dot{\lambda}(t) &= \dot{P}(t)z(t) + P(t)\dot{z}(t)\\ &= \dot{P}(t)z(t) + P(t)u(t)\\ &= \dot{P}(t)z(t) - P(t)\frac{\lambda(t)}{c_2}\\ &= \dot{P}(t)z(t) - P(t)\frac{P(t)z(t)}{c_2}.\end{aligned}$$

With $\dot{\lambda} = -c_1 z$, it follows

$$0 = \left(\dot{P}(t) - \frac{1}{c_2}P(t)^2 + c_1\right)z(t).$$

P is chosen such that the term in the brackets vanishes:

$$\dot{P}(t) = \frac{1}{c_2}P(t)^2 - c_1.$$

This Riccati differential equation for P has the solution

$$P(t) = \sqrt{c_1 c_2}\tanh\left(\sqrt{\frac{c_1}{c_2}}(t_f - t)\right),$$

where the terminal condition $P(t_f) = 0$ was taken into account.

Summarizing, we obtain the feedback law

$$v(t) = v_{\text{ref}}(t) - \sqrt{\frac{c_1}{c_2}}\tanh\left(\sqrt{\frac{c_1}{c_2}}(t_f - t)\right)(x(t) - x_{\text{ref}}(t)),$$

where $x(t)$ is the current (measured) state at time t.

This feedback law can be used to track a given position. v plays the role of a velocity.

In the case $t_f = \infty$, we are interested in steady-state solutions of the Riccati equation with $\dot{P} = 0$ and hence the Riccati equation

$$0 = \frac{1}{c_2}P^2 - c_1$$

needs to be solved. This quadratic equation has the positive solution $P = \sqrt{c_1 c_2}$. The resulting feedback law is

$$v(t) = v_{\text{ref}}(t) - \sqrt{\frac{c_1}{c_2}}(x(t) - x_{\text{ref}}(t)).$$

Example 6.2.5 (Pendulum). Consider the pendulum equations with damping,

$$\begin{aligned}\dot{x}_1(t) &= x_2(t)\\ \dot{x}_2(t) &= \frac{1}{\ell}(-cx_2(t) - g\sin(x_1(t)) - v(t))\end{aligned}$$

with $g = 9.81$, $\ell = 1$, $c = 0.4$.

Linearization at the upper (unstable) equilibrium position

$$x_{\text{ref}} := (\pi, 0)^\top, \quad v_{\text{ref}}(t) := 0$$

yields the linear differential equation for $z(t) := x(t) - x_{\text{ref}}$:

$$\begin{pmatrix}\dot{z}_1(t)\\ \dot{z}_2(t)\end{pmatrix} = \underbrace{\begin{pmatrix}0 & 1\\ \frac{g}{\ell} & -\frac{c}{\ell}\end{pmatrix}}_{=:A}\begin{pmatrix}z_1(t)\\ z_2(t)\end{pmatrix} + \underbrace{\begin{pmatrix}0\\ -\frac{1}{\ell}\end{pmatrix}}_{=:B}u(t).$$

Choose in the linear-quadratic optimal control problem 6.2.3 $R := \beta I_2$, $S := \alpha$ with $\alpha, \beta > 0$.

The linear feedback control law with $z(t) = x(t) - x_{\text{ref}}$ is given by

$$v(t) = -S^{-1}B^\top P(t)(x(t) - x_{\text{ref}}) = -\frac{1}{\alpha}\begin{pmatrix}0 & -\frac{1}{\ell}\end{pmatrix}P(t)(x(t) - x_{\text{ref}}),$$

where P solves the Riccati differential equation

$$\dot{P}(t) = -\beta I - P(t)A - A^\top P(t) + \frac{1}{\alpha}P(t)BB^\top P(t)$$

with terminal value $P(t_f) = 0_{\mathbb{R}^{2\times 2}}$.

Figure 6.13 shows the solution P for $t_f = 5$, $\alpha = 1$, $\beta = 1000$, $\ell = 1$, $c = 0.4$, $g = 9.81$.

We apply the feedback law

$$v(t) = -\frac{1}{\alpha}\begin{pmatrix}0 & -\frac{1}{\ell}\end{pmatrix}P(t)(x(t) - x_{\text{ref}})$$

to the *nonlinear* control problem for the pendulum with a perturbation in $x_1(0)$ of $+10°$ from the equilibrium state $\hat{x}_1 = \pi$. Although the upper equilibrium is unstable, the feedback control law moves the pendulum back into the upright position, see Figure 6.14. □

6.2.1 Systems in discrete time

So far, we have dealt with the continuous-time case, but the main steps to design a linear-quadratic regulator can also be performed for dynamic systems in discrete time. This time-discrete viewpoint is often adopted in practice since the system can only be influenced at discrete (time) points t_k, $k \in \mathbb{N}_0$.

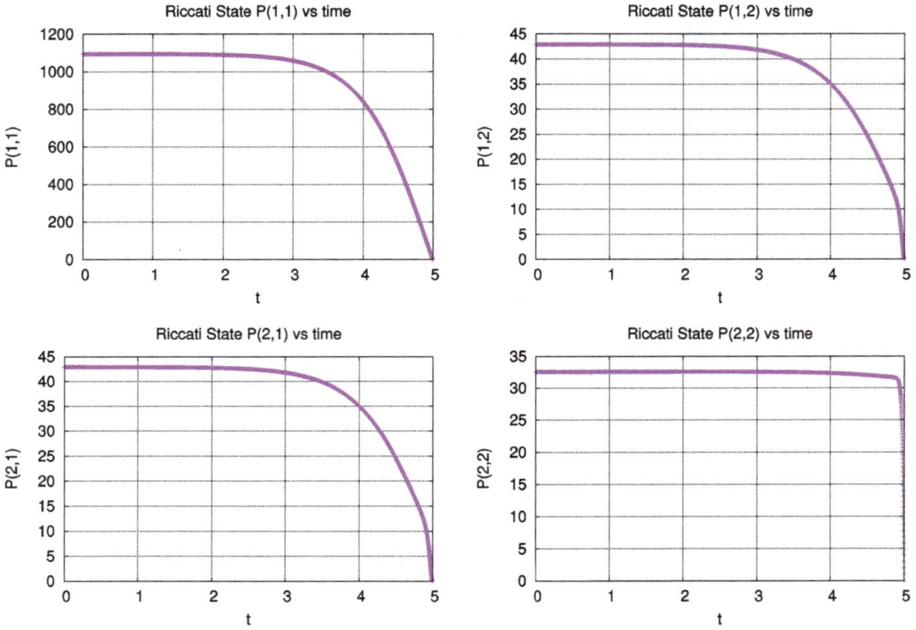

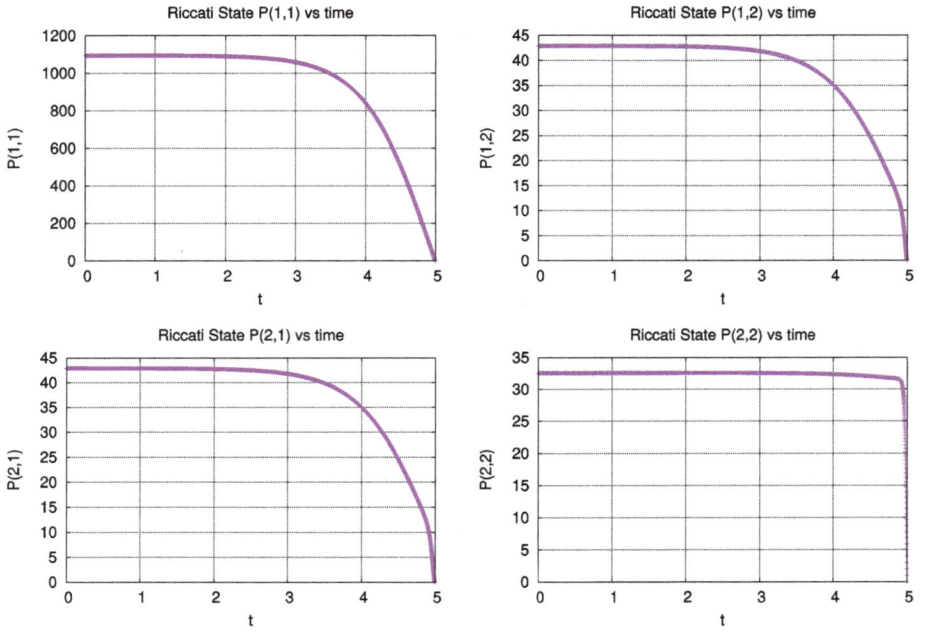

Figure 6.13: Solution of the Riccati equation.

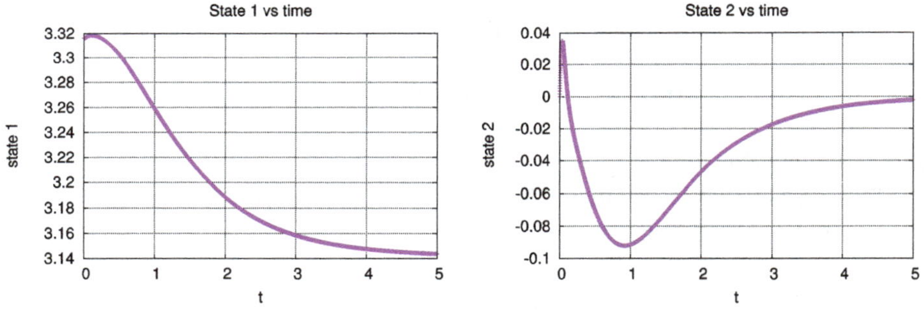

Figure 6.14: Controlled motion of pendulum.

Throughout, we identify the index $k \in \mathbb{N}_0$ with the time instance t_k, $k \in \mathbb{N}_0$, in order to simplify the notation.

Analogous to the continuous-time case, we consider a linear-quadratic optimal control problem in discrete time.

Problem 6.2.6 (Linear-quadratic optimal control problem in discrete time).
Minimize

$$\frac{1}{2}\sum_{k=0}^{N-1}\left(z(k)^\top R(k)z(k) + u(k)^\top S(k)u(k)\right)$$

subject to

$$\begin{aligned} z(k+1) &= A(k)z(k) + B(k)u(k) \quad && (k = 0,\dots,N-1) \\ z(0) &= z_0 && (z_0 \text{ given}) \end{aligned}$$

With the Lagrange function (with $\ell_0 = 1$)

$$\begin{aligned} L(z,u,\ell_0,\lambda,\sigma) := {} & \frac{1}{2}\sum_{k=0}^{N-1}(z(k)^\top R(k)z(k) - u(k)^\top S(k)u(k)) + \sigma^\top(z(0) - z_0) \\ & + \sum_{k=0}^{N-1}\lambda(k+1)^\top\big(A(k)z(k) + B(k)u(k) - z(k+1)\big) \end{aligned}$$

and by application of the KKT conditions, see Theorem 2.3.36, we obtain the following necessary and sufficient condition.

Theorem 6.2.7 (Necessary and sufficient condition). *Let the matrices $S(k)$ be symmetric and positive definite for all $k \in \mathbb{N}_0$, and let the matrices $R(k)$ be symmetric and positive semi-definite for all $k \in \mathbb{N}_0$.*

Then, Problem 6.2.6 is a convex linear-quadratic optimization problem, and the following conditions are necessary and sufficient for a minimum $(\hat{z},\hat{u})$:

(a) *The optimal control is given by*

$$\hat{u}(k) = -S(k)^{-1}B(k)^\top\lambda(k+1) \quad (k = 0,1,\dots,N-1).$$

(b) *The discrete adjoint equation holds:*

$$\begin{aligned} \lambda(N) &= 0_{\mathbb{R}^{n_z}} \\ \lambda(k) &= R(k)\hat{z}(k) + A(k)^\top\lambda(k+1) \quad (k = N-1,\dots,1). \end{aligned}$$

Proof. Under the assumptions of the theorem, the Hessian matrix of the objective function is positive semi-definite, and the constraints are linear. Hence, the problem is convex and the well-known KKT conditions are necessary and sufficient in this case. The KKT conditions yield

$$0_{\mathbb{R}^{n_z}} = \nabla_{z(0)}L = \sigma + R(0)\hat{z}(0) + A(0)^\top\lambda(1) \tag{6.10}$$

$$0_{\mathbb{R}^{n_z}} = \nabla_{z(k)}L = R(k)\hat{z}(k) + A(k)^\top\lambda(k+1) - \lambda(k) \quad (k = 1,\dots,N-1) \tag{6.11}$$

$$0_{\mathbb{R}^{n_z}} = \nabla_{z(N)}L = -\lambda(N) \tag{6.12}$$

$$0_{\mathbb{R}^{n_u}} = \nabla_{u(k)}L = S(k)\hat{u}(k) + B(k)^\top\lambda(k+1) \quad (k = 0,\dots,N-1). \tag{6.13}$$

This proves the assertion. □

The necessary and sufficient conditions in (6.10)–(6.13) together with the linear dynamics are, in fact, a *large-scale, sparse linear system of equations*:

$$\left(\begin{array}{cccc:cccc} \Gamma_0 & & & & E_0^\top & \Omega_0^\top & & \\ & \Gamma_1 & & & & E_1^\top & \ddots & \\ & & \ddots & & & & \ddots & \Omega_{N-1}^\top \\ & & & \Gamma_{N-1} & & & & E_N^\top \\ \hdashline E_0 & & & & & & & \\ \Omega_0 & E_1 & & & & & & \\ & \ddots & \ddots & & & & & \\ & & \Omega_{N-1} & E_N & & & & \end{array}\right) \left(\begin{array}{c} \zeta(0) \\ \zeta(1) \\ \vdots \\ \zeta(N-1) \\ \zeta(N) \\ \hdashline -\sigma \\ \lambda(1) \\ \vdots \\ \lambda(N) \end{array}\right) = \left(\begin{array}{c} 0_{\mathbb{R}^{n_z+n_u}} \\ 0_{\mathbb{R}^{n_z+n_u}} \\ \vdots \\ 0_{\mathbb{R}^{n_z+n_u}} \\ 0_{\mathbb{R}^{n_z}} \\ \hdashline -z_0 \\ 0_{\mathbb{R}^{n_z}} \\ \vdots \\ 0_{\mathbb{R}^{n_z}} \end{array}\right)$$

with $\zeta(k) = (\hat{z}(k), \hat{u}(k))$, $k = 0, \dots, N-1$, $\zeta(N) = \hat{z}(N)$, $E_N = -I_{n_z}$,

$$\Gamma_k = \begin{pmatrix} R(k) & \\ & S(k) \end{pmatrix}, \quad \Omega_k = \begin{pmatrix} A(k) & B(k) \end{pmatrix}, \quad E_k = \begin{pmatrix} -I_{n_z} & 0_{\mathbb{R}^{n_z \times n_u}} \end{pmatrix}$$

for $k = 0, \dots, N-1$. This linear system of equations can be solved numerically by so-called "sparse solvers", which exploit the sparsity structure and use tailored LU-decomposition methods, e. g., MA57 (https://www.hsl.rl.ac.uk/catalogue/ma57.html), PARDISO (https://www.pardiso-project.org/), SuperLU (https://portal.nersc.gov/project/sparse/superlu/).

---

i A re-organization of the matrix using column and row permutations yields an equivalent linear equation system with a banded matrix as in Section 5.3 and Equation (5.44).

There is an alternative way to solve the optimality system, though, which yields a feedback law. We choose the *ansatz*

$$\lambda(k) := P(k)\hat{z}(k) \quad (k = 1, \dots, N).$$

Introducing this ansatz into the equations yields the following:

(i) Since $0_{\mathbb{R}^{n_z}} = \lambda(N) = P(N)\hat{z}(N)$ is supposed to hold for arbitrary $\hat{z}(N)$, we find $P(N) = 0_{\mathbb{R}^{n_z \times n_z}}$.

(ii) The optimal control is given by

$$\hat{u}(k) := -S(k)^{-1}B(k)^\top P(k+1)\hat{z}(k+1) \quad (k = 0, \dots, N-1)$$

(iii) The discrete dynamics yields

$$\hat{z}(k+1) = A(k)\hat{z}(k) + B(k)\hat{u}(k) = A(k)\hat{z}(k) - B(k)S(k)^{-1}B(k)^\top P(k+1)\hat{z}(k+1)$$

Solving for $\hat{z}(k+1)$ yields

$$\hat{z}(k+1) = \left(I_{n_z} + B(k)S(k)^{-1}B(k)^\top P(k+1)\right)^{-1} A(k)\hat{z}(k),$$

provided the inverse exists.

For $k = 1, \dots, N-1$, we thus obtained the following relations:

$$\begin{aligned}
0_{\mathbb{R}^{n_z}} &= \lambda(k) - P(k)\hat{z}(k) \\
&= R(k)\hat{z}(k) + A(k)^\top \lambda(k-1) - P(k)\hat{z}(k) \\
&= R(k)\hat{z}(k) + A(k)^\top P(k+1)\hat{z}(k+1) - P(k)\hat{z}(k) \\
&= \left(R(k) + A(k)^\top P(k+1)W(k)^{-1}A(k) - P(k)\right)\hat{z}(k)
\end{aligned}$$

with

$$W(k) = I_{n_z} + B(k)S(k)^{-1}B(k)^\top P(k+1).$$

These relations are supposed to hold independently of $\hat{z}$ which implies $P(N) = 0_{\mathbb{R}^{n_z \times n_z}}$ and

$$P(k) = R(k) + A(k)^\top P(k+1)W(k)^{-1}A(k)$$

for $k = N-1, \dots, 1$. The latter can be further transformed with the *Sherman–Morrison–Woodbury formula.*

Theorem 6.2.8 (Sherman–Morrison–Woodbury formula). *For matrices X, U, C, V, we have*

$$(X + UCV)^{-1} = X^{-1} - X^{-1}U\left(C^{-1} + VX^{-1}U\right)^{-1}VX^{-1},$$

if the inverses exist. □

Application of the Sherman–Morrison–Woodbury formula with $X = I_{n_z}$, $U = B(k)$, $C = S(k)^{-1}$, $V = B(k)^\top P(k+1)$ yields

$$\begin{aligned}
W(k)^{-1} &= \left(I_{n_z} + B(k)S(k)^{-1}B(k)^\top P(k+1)\right)^{-1} \\
&= I_{n_z} - B(k)\left(S(k) + B(k)^\top P(k+1)B(k)\right)^{-1}B(k)^\top P(k+1)
\end{aligned}$$

and we obtain a discrete counterpart of the Riccati differential equation.

Definition 6.2.9 (Discrete Riccati equation (DRE)). Let $P(N) = 0$ and solve

$$P(k) = R(k) + A(k)^\top \left(P(k+1) - P(k+1)B(k)M(k)^{-1}B(k)^\top P(k+1)\right)A(k)$$

for $k = N-1, \dots, 1$ with

$$M(k) := S(k) + B(k)^\top P(k+1)B(k). \tag{6.14}$$

□

Please note that the matrix $M(k)$ in (6.14) is non-singular, if $S(k)$ is symmetric and positive definite and $P(k+1)$ is symmetric and positive semi-definite.

Another application of the Sherman–Morrison–Woodbury formula finally yields the *linear feedback law*

$$\hat{u}(k) = -M(k)^{-1}B(k)^\top P(k+1)A(k)\hat{z}(k) \quad (k = 0, \dots, N-1)$$

The discrete Riccati equation is the discrete counterpart of the Riccati differential equation.

As in the case with continuous time, it is possible to derive an algebraic Riccati equation in the time-discrete case. To this end, we assume the matrices $A = A(\cdot)$, $B = B(\cdot)$, $R = R(\cdot)$, $S = S(\cdot)$ are time-invariant, that is, independent of k. Taking the limit $N \to \infty$ and assuming that $P(1)$ in DRE converges to a matrix P yields the following:

Definition 6.2.10 (Discrete algebraic Riccati equation (DARE)).

$$P = R + A^\top (P - PB(S + B^\top PB)^{-1} B^\top P)A$$ □

DARE can be solved numerically in MATLAB using the solver `idare`, and we obtain the linear feedback law

$$u(k) = -Cz(k) \quad \text{with } C = (S + B^\top PB)^{-1} B^\top PA$$

The corresponding closed-loop system is

$$\begin{aligned} z(k+1) &= (A - BC)z(k) \quad && (k = 0, 1, 2, \dots) \\ z(0) &= z_0 && (z_0 \text{ given}). \end{aligned}$$

The system is asymptotically stable, if $|\lambda| < 1$ for all eigenvalues λ of $A - BC$.

6.3 Model-predictive control

The approach in Section 6.2 works well as long as the control problem 6.2.1 does not contain control or state constraints. If control or state constraints are present, then the corresponding control and state-constrained linear-quadratic optimal control problem 6.2.3 cannot be solved anymore by the Riccati approach. More elaborate techniques need to be constructed that are able to take into account control or state constraints. In Chapter 5, such methods were discussed already. Unfortunately, the discretization methods in Chapter 5 only provide an open-loop control, that is, the control explicitly depends on time. A feedback dealing with potential deviations from the reference state trajectory is not included so far. However, as we shall see, a (nonlinear) feedback control can be obtained by *model-predictive control (MPC)*, which is based on the repeated solution of optimal control problems on shifting time windows. Figure 6.15 illustrates the operation of MPC in the discrete-time case.

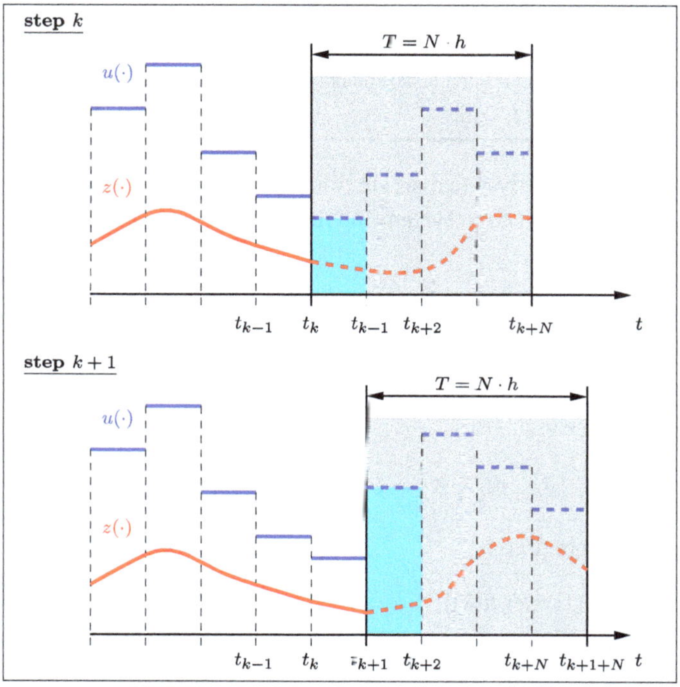

Figure 6.15: MPC scheme: Repeated solution of optimal control problems (in discrete time) on a prediction horizon of length $T = N \cdot h$ (grey box). The first control on the prediction horizon will be implemented (blue box).

MPC exists in several variants, such as continuous-time, discrete-time, multi-step, linear, nonlinear, tracking, and economic MPC. A comprehensive discussion of MPC can be found in [279] and [150]. MPC is characterized, in particular, by the fact that constraints in the form of control constraints and/or state constraints can be taken into account during the control procedure. Conceptually, MPC is therefore very powerful, but the realization of MPC requires solving optimization problems in real-time, which poses a great challenge to the efficiency and robustness of numerical solution methods.

MPC uses a model of the process in terms of differential equations and the following parameters:

(i) Discrete *sampling times* $t_k = k \cdot h$ for $k = 0, 1, 2, \ldots$.
At these time points, the measurement of the actual state and the adjustment of the control takes place.

(ii) *Sampling step-size* $h > 0$.
The MPC controller runs with the frequency $1/h$, e. g., with 100 Hz at $h = 10$ ms.

(iii) *Prediction horizon* $T = N \cdot h$.
MPC uses a preview of length $T = N \cdot h$ for a prediction of the future behavior of the control system based on a model of the dynamic system.

Throughout, we identify the discrete time index k with the time point $t_k = k \cdot h$ and denote by $z(k)$ and $u(k)$ the state $z(t_k)$ and the control $u(t_k)$, respectively.

In the following, we restrict ourselves to discrete-time problems, which often arise from a discretization of a continuous-time system in ODE or DAE formulation as described in Chapters 4 and 5.

6.3.1 Nonlinear MPC (NMPC)

The aim is to construct a *feedback law*

$$\mu_N : \mathbb{N}_0 \times Z \longrightarrow U$$

for the dynamic system in discrete time

$$\begin{aligned} z(k+1) &= f(k, z(k), u(k)), && k = 0, 1, 2, \ldots, \\ z(k) &\in Z, && k = 0, 1, 2, \ldots, \\ u(k) &\in U, && k = 0, 1, 2, \ldots, \\ z(0) &= z_0. \end{aligned}$$

Herein, $z_0 \in \mathbb{R}^{n_z}$ is a given initial state, and $Z \subset \mathbb{R}^{n_z}$ and $U \subset \mathbb{R}^{n_u}$ are given sets. Nonlinear model-predictive control requires solving optimal control problems in discrete time of the following type:

Problem 6.3.1 (D-OCP(k, z_k, N)).
Minimize

$$\sum_{\ell=k}^{k+N-1} f_0(\ell, z(\ell), u(\ell))$$

subject to

$$\begin{aligned} z(\ell+1) &= f(\ell, z(\ell), u(\ell)), && \ell = k, \ldots, k+N-1, \\ z(\ell) &\in Z, && \ell = k, \ldots, k+N, \\ u(\ell) &\in U, && \ell = k, \ldots, k+N-1, \\ z(k) &= z_k. \end{aligned}$$

The optimal control problem in discrete time is at the core of the following feedback law:

Definition 6.3.2 (Nonlinear model-predictive control (NMPC) feedback law). For time $k \in \mathbb{N}_0$, state $z_k \in Z$, and prediction horizon $N \in \mathbb{N}$, let $\hat{u}(\ell)$, $\ell = k, \dots, k+N-1$, be an optimal control sequence of D-OCP(k, z_k, N). The *NMPC feedback law* is defined by

$$\mu_N(k, z_k) := \hat{u}(k).$$

□

The working steps of NMPC are given by the following algorithm.

Algorithm 6.3.3 (NMPC (basic version)).

(0) *Input:* Prediction horizon $N \in \mathbb{N}$. Set $k := 0$.

(1) Measure (or predict) the state z_k at time k.

(2) Solve D-OCP(k, z_k, N) and let $\hat{u}(\ell)$, $\ell = k, \dots, k+N-1$, be an optimal control sequence. Set $\mu_N(k, z_k) = \hat{u}(k)$.

(3) Apply $\mu_N(k, z_k)$ to the control system for the step from k to $k+1$, i. e., in the time interval $[t_k, t_{k+1}]$.

(4) Set $k \leftarrow k+1$ and go to (1). □

Application of the NMPC controller to the nonlinear dynamic system yields the closed loop system in discrete time

$$\begin{aligned} z(k+1) &= f(k, z(k), \mu_N(k, z(k))), \quad k = 0, 1, 2, \dots, \\ z(0) &= z_0. \end{aligned}$$

---

i Please note that the dynamics, the constraints, and the objective function in D-OCP(k, z_k, N) are nonlinear and non-convex in general. Thus, *nonlinear* and in general *non-convex* optimal control problems in discrete time need to be solved in each step of the NMPC algorithm using appropriate numerical methods like, e. g., SQP methods or interior-point methods. Solving the problems fastly and robustly is usually a challenge in online applications.

---

i D-OCP(k, z_k, N) often arises from a discretization of an optimal control problem in continuous time. In practice, the grid, which is used in the discretization, does not have to coincide with the grid determined by the sampling times $t_k = k \cdot h$. For notational convenience, however, we assume both grids coincide.

Definition 6.3.4 (Tracking type NMPC, economic NMPC). We call the optimal control problem D-OCP(k, z_k, N) and the corresponding NMPC algorithm 6.3.3 of *tracking type*, if $f_0 : \mathbb{N}_0 \times \mathbb{R}^{n_z} \times \mathbb{R}^{n_u}$ is of the form

$$f_0(k, z, u) = \frac{1}{2}(z - z_{\text{ref}}(k))^\top R(k)(z - z_{\text{ref}}(k)) + \frac{1}{2}(u - u_{\text{ref}}(k))^\top S(k)(u - u_{\text{ref}}(k))$$

for a given reference trajectory $(z_{\text{ref}}(k), u_{\text{ref}}(k))$ and symmetric matrices $R(k) \in \mathbb{R}^{n_z \times n_z}$ and $S(k) \in \mathbb{R}^{n_u \times n_u}$, $k \in \mathbb{N}_0$.

In all other cases, we call D-OCP(k, z_k, N) and the corresponding NMPC algorithm of *economic type*. □

The following example illustrates the NMPC algorithm of tracking type.

Example 6.3.5 (Pendulum with control bounds). Consider the pendulum differential equations with damping,

$$\dot{z}_1(t) = z_2(t)$$
$$\dot{z}_2(t) = \frac{1}{\ell}(-cz_2(t) - g\sin(z_1(t)) - u(t))$$

with $g = 9.81$, $\ell = 1$, $c = 0.4$, and control bounds $u(t) \in [-7, 7]$. Herein, z_1 denotes the angle between the pendulum and the vertical axis. Let the reference solution be the upper equilibrium position with $z_{\text{ref}}(t) := (\pi, 0)^\top$ and $u_{\text{ref}}(t) := 0$ for all t. The classic Runge–Kutta method of order four is used for the discretization of the differential equations in D-OCP(k, z_k, N).

Figure 6.16 shows the solution of the tracking type NMPC algorithm 6.3.3 for the initial value $z_0 = (-25°, 0)^\top$, $h = 0.1$, $N = 10$, and

$$R(k) = \begin{pmatrix} 12 & 0 \\ 0 & 1 \end{pmatrix}, \quad S(k) = (1) \quad (k \in \mathbb{N}_0).$$

The NMPC algorithm is able to control the pendulum back to the reference solution obeying the control bounds. □

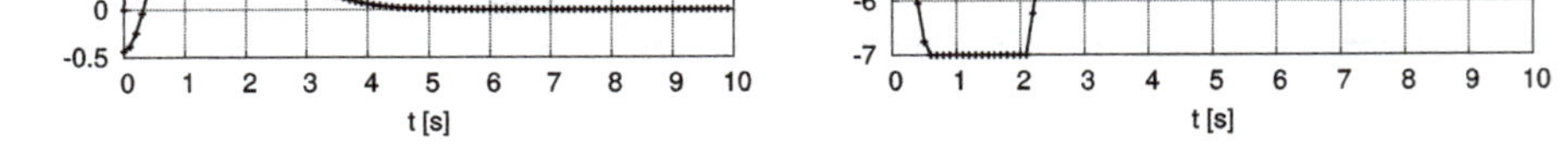

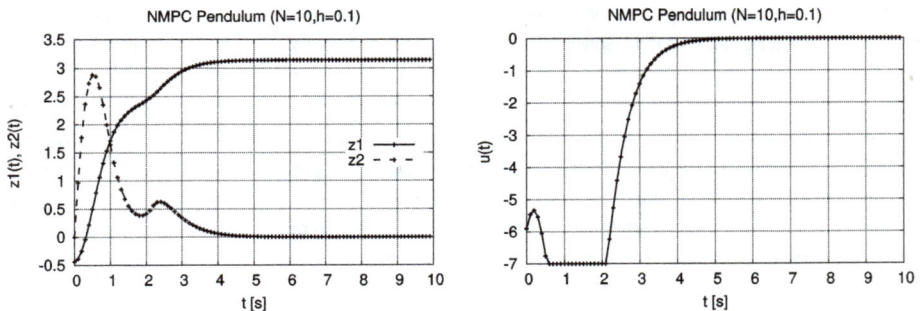

Figure 6.16: Nonlinear model predictive control of a pendulum with $N = 10$, $h = 0.1$, $z_{\text{ref}} \equiv (\pi, 0)^\top$, $u_{\text{ref}} \equiv 0$.

NMPC is frequently used for online and offline path planning tasks for mobile robots. The following example provides details of a path planning problem for an automated vehicle. Herein, the NMPC algorithm is not of tracking type, but of economic type.

Example 6.3.6 (Geometric path planning in curvilinear coordinates). Path planning is a core task in automated driving and NMPC is well-suited for this task. For driving along roads, it is usually more convenient to use so-called curvilinear coordinates instead of Cartesian coordinates (x,y) in the plane. Curvilinear coordinates require a twice continuously differentiable reference curve

$$\gamma_{\mathrm{ref}}(s) = \begin{pmatrix} x_{\mathrm{ref}}(s) \\ y_{\mathrm{ref}}(s) \end{pmatrix},$$

which is parameterized with respect to its arclength $s \in [0,L]$. Herein, L is the length of the curve. The reference curve typically coincides, e. g., with the midline, the left boundary, or the right boundary of the road. The curvature of the reference curve is given by

$$\kappa_{\mathrm{ref}}(s) = x'_{\mathrm{ref}}(s)y''_{\mathrm{ref}}(s) - y'_{\mathrm{ref}}(s)x''_{\mathrm{ref}}(s)$$

and the path angle is

$$\psi_{\mathrm{ref}}(s) = \arctan\left(\frac{y'_{\mathrm{ref}}(s)}{x'_{\mathrm{ref}}(s)}\right).$$

We have the relation

$$\psi'_{\mathrm{ref}}(s) = \kappa_{\mathrm{ref}}(s).$$

The motion of the vehicle is now represented relative to the reference curve in curvilinear coordinates (s,r), where s is the arclength of the projected vehicle position on the reference curve, and r is the lateral distance of the vehicle to the reference curve, compare Figure 6.17 and the derivation in [43]:

$$\dot{s}(t) = \frac{v(t)\cos(\psi(t) - \psi_{\mathrm{ref}}(t))}{1 - r(t)\kappa_{\mathrm{ref}}(s(t))}, \tag{6.15}$$

$$\dot{r}(t) = v(t)\sin(\psi(t) - \psi_{\mathrm{ref}}(t)), \tag{6.16}$$

$$\dot{\psi}(t) = v(t) \cdot \kappa(t), \tag{6.17}$$

$$\dot{\psi}_{\mathrm{ref}}(t) = \kappa_{\mathrm{ref}}(s(t)) \cdot \dot{s}(t). \tag{6.18}$$

Herein, v denotes the velocity of the vehicle, κ is the curvature of the vehicle's path, which serves as a control in this model. Curvature κ and steering angle δ of the automated vehicle are related approximately by the relation $\delta = \arctan(\ell \cdot \kappa)$, where ℓ is the distance of the front and rear axle. The (x,y) position of the vehicle in Cartesian coordinates is given by $x(t) = x_{\mathrm{ref}}(s(t)) - r(t)\sin\psi_{\mathrm{ref}}(s(t))$ and $y(t) = y_{\mathrm{ref}}(s(t)) + r(t)\cos\psi_{\mathrm{ref}}(s(t))$.

In order to simplify the path planning task, the problem is often decoupled into two simpler problems. First, a geometric path in arclength parameterization is sought, i. e., a

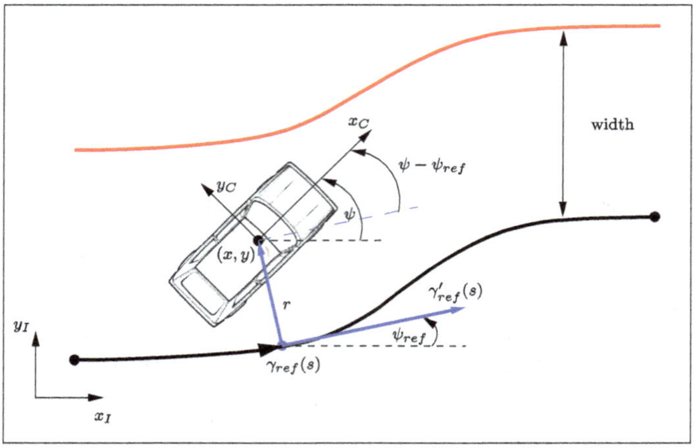

Figure 6.17: Motion in curvilinear coordinates.

path with $v(t) = 1$. Second, a suitable velocity profile on this geometric path is computed. For the latter problem, we refer the reader to [24], where an optimal velocity profile is obtained very efficiently through a suitably defined optimal control problem.

As for the geometric path planning, in each step of the NMPC algorithm, the following optimal control problem is solved using a direct discretization technique of Chapter 5. The objective function is a linear combination of total curvature (to be minimized) and progress relative to the reference curve (to be maximized).

Minimize

$$-\alpha_1 s(t_i + T) + \alpha_2 \int_{t_i}^{t_i+T} \kappa(t)^2 dt$$

subject to the differential equations (6.15)–(6.18) *with* $v(t) = 1$, *initial values* $(s(t_i), r(t_i), \psi(t_i), \psi_{\text{ref}}(t_i)) = (s_i, r_i, \psi_i, \psi_{\text{ref},i})$, *the state constraints*

$$r(t) \in [r_{\min}, r_{\max}],$$

and the control constraints

$$\kappa(t) \in [\kappa_{\min}, \kappa_{\max}].$$

Figure 6.18 shows the results of the NMPC strategy for $\alpha_1 = 6.5 \cdot 10^{-4}$, $\alpha_2 = 1$, $r_{\max} = -r_{\min} = 3$, $\kappa_{\max} = -\kappa_{\min} = 0.1$ on a grid with $N = 50$ intervals, and prediction horizon $T = 100$.

Figure 6.19 shows the intermediate solutions of the NMPC steps. □

A similar task for a flight system is investigated in the following example.

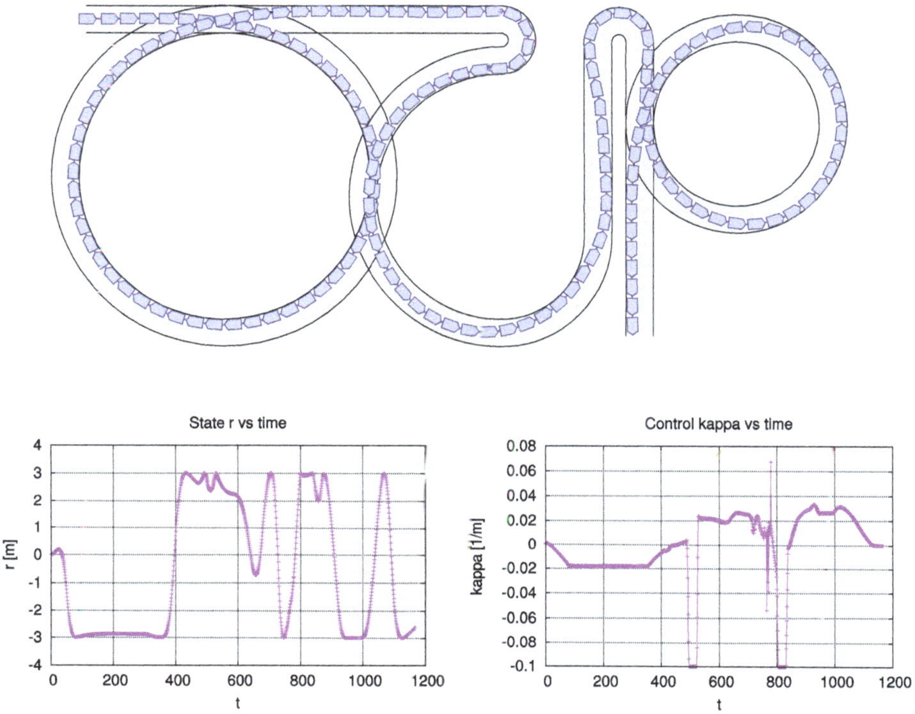

Figure 6.18: Resulting geometric path of the NMPC path planning technique (top), lateral offset r (bottom left), and curvature κ of driven path (bottom right).

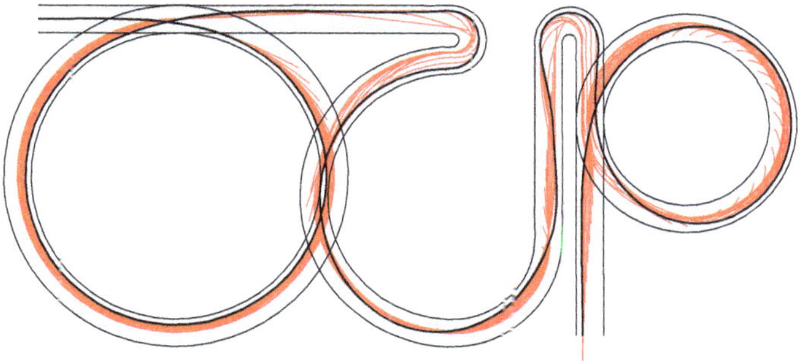

Figure 6.19: Intermediate solutions (in red) of the NMPC steps (every 7th step).

Example 6.3.7 (Path planning for a drone in a flight corridor). The task is to control the flight of a drone in a given flight corridor along a track on the testing area of the Universität der Bundeswehr München using NMPC, compare Figure 6.20.

Figure 6.20: Control of a drone along a flight corridor.

Let a twice continuously differentiable reference curve

$$y_{\text{ref}}(s) = \begin{pmatrix} x_{\text{ref}}(s) \\ y_{\text{ref}}(s) \end{pmatrix}$$

with curvature κ_{ref}, path angle ψ_{ref}, and arclength s be given. This curve describes the midline (in the (x, y)-plane) of the flight corridor. In addition, the z-direction of the flight corridor is subject to altitude constraints

$$z_{\min}(s) \leq z(s) \leq z_{\max}(s).$$

The width of the flight corridor is bounded by

$$r_{\min}(s) \leq r(s) \leq r_{\max}(s).$$

This describes a rectangular flight corridor of variable size. The motion of the drone relative to the planar reference curve is described by the following system of differential equations, which are parameterized with respect to the arclength ℓ of the drone's flight path:

$$\begin{aligned}
\dot{s}(\ell) &= \frac{v_{xy}(\ell) \cdot \cos(\psi(\ell) - \psi_{\text{ref}}(s(\ell)))}{1 - r(\ell) \cdot \kappa_{\text{ref}}(s(\ell))} \\
\dot{r}(\ell) &= v_{xy}(\ell) \cdot \sin(\psi(\ell) - \psi_{\text{ref}}(s(\ell))) \\
\dot{z}(\ell) &= v_z(\ell) \\
m \cdot \dot{v}_x(\ell) &= u_1(\ell) \cdot \cos\phi(\ell) \cdot \sin\theta(\ell) - D_x(v_x(\ell)) \\
m \cdot \dot{v}_y(\ell) &= -u_1(\ell) \cdot \sin\phi(\ell) - D_y(v_y(\ell)) \\
m \cdot \dot{v}_z(\ell) &= u_1(\ell) \cdot \cos\phi(\ell) \cdot \cos\theta(\ell) - m \cdot g - D_z(v_z(\ell)) \\
\dot{\phi}(\ell) &= \frac{u_2(\ell) - \phi(\ell)}{\delta} \\
\dot{\theta}(\ell) &= \frac{u_3(\ell) - \theta(\ell)}{\delta}
\end{aligned}$$

Herein, (s,r,z) denotes the position of the drone in curvilinear coordinates, ϕ is the roll angle, and θ the pitch angle. u_1 denotes the thrust (control), u_2 the commanded roll angle (control), u_3 the commanded pitch angle (control), $v_{xy} = \sqrt{v_x^2 + v_y^2}$ is the projection of the velocity in the (x,y)-plane, v_z the velocity component in z-direction. The angle $\psi = \arctan(v_y/v_x)$ denotes the yaw angle, $\delta > 0$ is a delay constant, m the mass, $g = 9.81$ the acceleration due to gravity, and (D_x, D_y, D_z) denotes the drag vector with $D_* = \frac{c_w F \rho_0}{2} v_*^2$ for $* \in \{x,y,z\}$ with $\rho_0 = 1.249512$, $F = 0.04$, $c_w = 0.9$.

The objective function to be minimized

$$\underbrace{\int_0^L \frac{1}{v(\ell)}\, d\ell}_{\text{flight time}} + \underbrace{\int_0^L u_1(\ell)^2 + u_2(\ell)^2 + u_3(\ell)^2\, d\ell}_{\text{control effort}}$$

is used in the NMPC algorithm.

The following control and state constraints are imposed:

$$\begin{aligned} &z_{\min}(s(\ell)) \le z(\ell) \le z_{\max}(s(\ell)) && \text{(altitude)}\\ &r_{\min}(s(\ell)) \le r(\ell) \le r_{\max}(s(\ell)) && \text{(distance)}\\ &v_{\min} \le v \le v_{\max} && \text{(velocity)}\\ &|\phi| \le \phi_{\max}, \quad |\theta| \le \theta_{\max} && \text{(roll angle)}\\ &u_i \in [u_{i,\min}, u_{i,\max}], \quad i = 1,2,3 && \text{(controls)} \end{aligned}$$

Figure 6.21 shows the results of the NMPC technique for a prediction horizon of 20 [m], $N = 20$, $[u_{1,\min}, u_{1,\max}] = [0,50]$ [N], $[u_{2,\min}, u_{2,\max}] = [u_{3,\min}, u_{3,\max}] = [-45,45]$ [deg], $m = 3$ [kg], $\delta = 0.1$, $[v_{\min}, v_{\max}] = [0,15]$ [m/s], $\phi_{\max} = \theta_{\max} = 45$ [deg] on a laptop with AMD®Ryzen 5 4600H processor and 16 GB RAM. □

We close this section with a "real-world example" and apply the above NMPC technique to plan a path for a LEGO®MINDSTORMS robot.

Example 6.3.8 (Test-drive of a LEGO®MINDSTORMS robot). We consider the test-drive of a LEGO®MINDSTORMS robot around a given track.

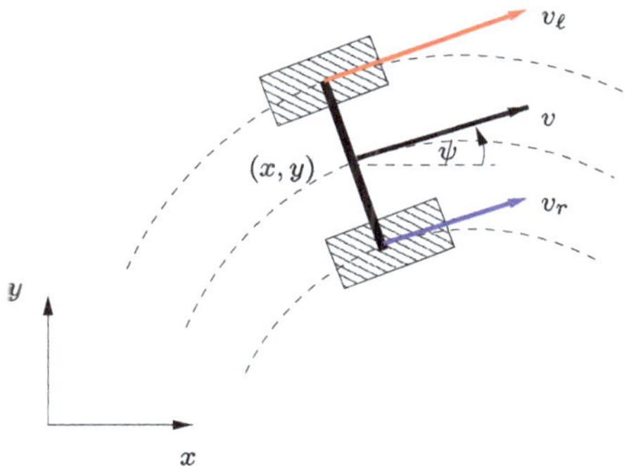

States r, z, v:

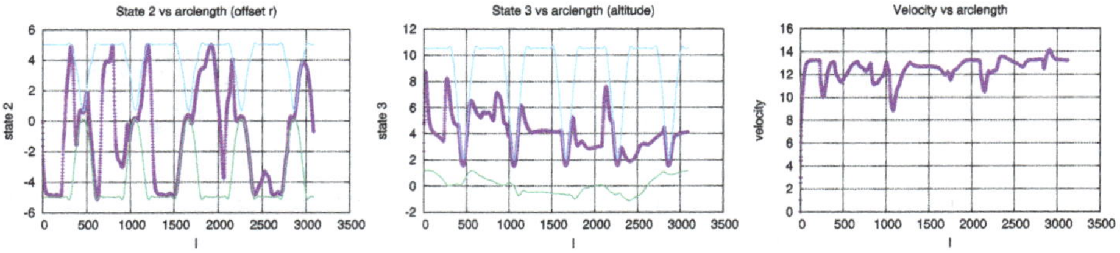

States (x,y), roll and pitch angles:

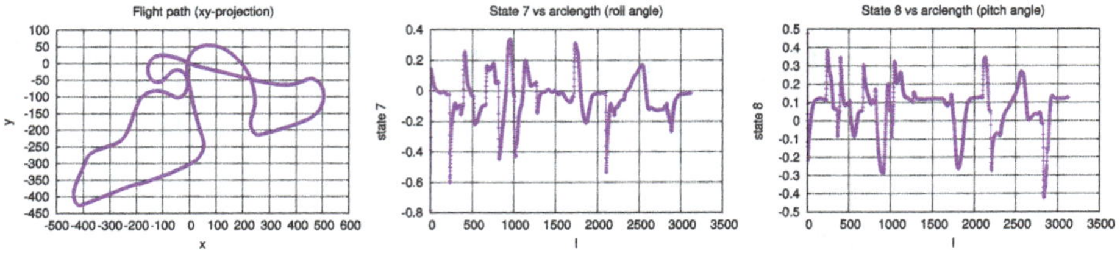

Controls thrust, commanded roll and pitch angles:

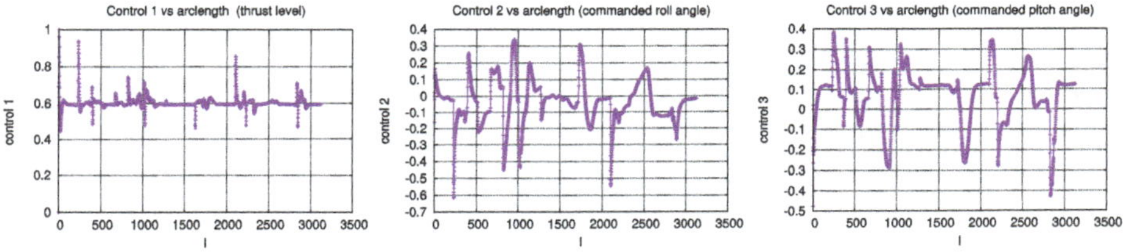

Figure 6.21: Results of the NMPC feedback law for the flight of a drone in a flight corridor. The flight time is 253.87 [s] and the computation time amounts to 152.56 [s] for 3128 NMPC steps (48.77 [ms] per NMPC step).

Herein, (x, y) denotes the center of gravity of the robot, ψ the yaw angle, v_r the velocity of the right wheel, v_ℓ the velocity of the left wheel, and v the velocity of the center of gravity. A simple model for the motion of the robot is given by the following differential equations:

$$
\begin{aligned}
\dot{x}(t) &= \frac{v_\ell(t) + v_r(t)}{2} \cos(\psi(t)), \\
\dot{y}(t) &= \frac{v_\ell(t) + v_r(t)}{2} \sin(\psi(t)), \\
\dot{\psi}(t) &= \frac{v_r(t) - v_\ell(t)}{B}, \\
\dot{v}_\ell(t) &= u_1(t), \\
\dot{v}_r(t) &= u_2(t).
\end{aligned}
$$

The controls u_1 and u_2 control the acceleration of the left and right wheel, respectively, and are subject to the control constraints

$$u_i(t) \in [-0.5, 0.5], \quad i = 1, 2.$$

In addition, state constraints apply that keep the robot on the track. As it is difficult to compute an optimal solution for the whole track, the nonlinear model-predictive control algorithm is applied. The objective function in each optimal control problem is a linear combination of minimum control effort and maximal distance driven on the track. For details, we refer to [114, 128]. The result of the nonlinear model predictive control algorithm is depicted in Figure 6.22. Herein, 113 optimal control problems, each with $N = 30$ grid points, have been solved within 35.78 CPU seconds on a PC with 2.3 GHz. □

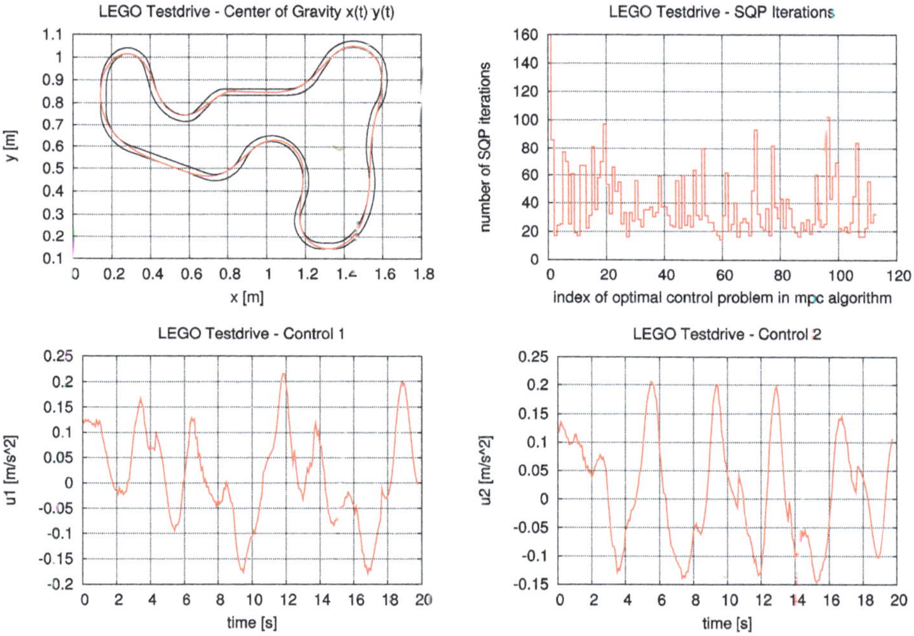

Figure 6.22: Results of the nonlinear model predictive control algorithm for a test-drive along a test-course: center of gravity (top left), SQP iterations for subproblems (top right), controls u_1 and u_2 (bottom).

The existence of a solution of D-OCP(k, z_k, N) is in general not guaranteed, and it is a non-trivial task to guarantee feasibility, especially if state constraints are present.

The success of NMPC (and MPC in general) depends largely on how quickly D-OCP(k, z_k, N) can be solved. In the basic NMPC algorithm 6.3.3, it is assumed that the solution is available instantaneously. However, this is an idealization because the solving time of D-OCP(k, z_k, N) causes a delay in the application of $\hat{u}(k)$. If the solving time of D-OCP(k, z_k, N) is δ, then $\hat{u}(k)$ can be applied only at time $t_k + \delta$. This delay can be taken into account at time t_k by using the model to predict the state at time $t_k + \delta$ and then using $t_k + \delta$ and the predicted state as the initial time and the initial state, respectively, in D-OCP.

With regard to the solution of D-OCP, an approximate optimal solution is sufficient, which is often obtained by performing only one or a few steps of the underlying NLP method, see [76]. In addition, for small deviations between the predicted and measured states, a parametric sensitivity analysis with respect to perturbations in the initial value can be used to update a nominal NMPC solution in real-time using the techniques in Section 6.1, compare also [329].

Furthermore, time can be gained by applying not only $\hat{u}(k)$ but $\hat{u}(k), \dots, \hat{u}(k+M-1)$ with $M \geq 2$. This variant is called M-multistep MPC, compare [261] for a detailed study of stability properties of M-multistep MPC algorithms with and without re-optimization on receding horizons.

If online optimization is too costly, it is still possible to pre-compute a database of optimal solutions offline and to apply sensitivity updates online, see, e. g., [308]. Herein, it is important to cover the region of interest in the state space properly and estimate neighborhoods in which sensitivity updates and the involved linearizations are valid.

6.3.2 Linear-quadratic MPC (LMPC)

As outlined before, solving the fully nonlinear Problem 6.3.1 in real-time is often not possible owing to high computational costs or owing to a lack of robustness caused by nonconvexities. Especially for tracking tasks, it is common practice to rely on a linearization of the problem around the reference solution, assuming small deviations. This yields a linear model-predictive control algorithm, which is typically less demanding than the nonlinear version. As in the nonlinear setting, the aim is again to construct a feedback law

$$\mu_N : \mathbb{N}_0 \times Z \longrightarrow U$$

for the linear time discrete dynamic system

$$\begin{aligned} z(k+1) &= A(k)z(k) + B(k)u(k) + d(k), && k = 0, 1, 2, \dots, \\ z(k) &\in Z := [z_{\min}, z_{\max}], && k = 0, 1, 2, \dots, \\ u(k) &\in U := [u_{\min}, u_{\max}], && k = 0, 1, 2, \dots, \\ z(0) &= z_0. \end{aligned}$$

Herein, $z_0 \in \mathbb{R}^{n_z}$ is a given initial value, and $Z = [z_{\min}, z_{\max}] \subset R^{n_z}$ and $U = [u_{\min}, u_{\max}] \subset \mathbb{R}^{n_u}$ are given vector-valued box constraints for the state and the control, respectively. The vector $d(k) \in \mathbb{R}^{n_z}$ can be viewed as a perturbation.

Similarly, as for the LQ controller in Section 6.2, we consider the following optimal control problem in discrete time with state z_k at time k as input parameters. The matrices R and S are supposed to be symmetric and positive semidefinite, S is even supposed to be positive definite. $N \in \mathbb{N}$ denotes the length of the prediction horizon.

Problem 6.3.9 (D-LQOCP(k, z_k, N)).
Minimize

$$\frac{1}{2}\sum_{\ell=k}^{k+N-1} z(\ell)^\top R(\ell)z(\ell) + u(\ell)^\top S(\ell)u(\ell)$$

subject to

$$\begin{aligned} z(\ell+1) &= A(\ell)z(\ell) + B(\ell)u(\ell) + d(\ell), && \ell = k,\dots,k+N-1,\\ z(\ell) &\in [z_{\min}, z_{\max}], && \ell = k,\dots,k+N,\\ u(\ell) &\in [u_{\min}, u_{\max}], && \ell = k,\dots,k+N-1,\\ z(k) &= z_k. \end{aligned}$$

With the help of this linear-quadratic optimal control problem in discrete time, we are in the position to define a feedback law:

Definition 6.3.10 (Linear model-predictive control (LMPC) feedback law). For time $k \in \mathbb{N}_0$, state $z_k \in Z$, and prediction horizon $N \in \mathbb{N}$, let $\hat{u}(\ell)$, $\ell = k,\dots,k+N-1$, be an optimal control sequence of D-LQOCP(k, z_k, N). The *LMPC feedback law* is defined by

$$\mu_N(k, z_k) := \hat{u}(k).$$ □

Application of the LMPC feedback law to the dynamic system yields the closed-loop system

$$\begin{aligned} z(k+1) &= A(k)z(k) + B(k)\mu_N(k, z(k)) + d(k), \quad k = 0,1,2,\dots,\\ z(0) &= z_0. \end{aligned}$$

As it was mentioned before, the particular strength of the LMPC feedback law is the ability to handle control and state constraints. The following algorithm specifies the steps of the LMPC framework.

Algorithm 6.3.11 (LMPC (basic version)).
(0) *Input:* Prediction horizon $N \in \mathbb{N}$, weight matrices R, S. Set $k := 0$.
(1) Measure (or predict) the state z_k at time k.
(2) Solve D-LQOCP(k, z_k, N) and denote an optimal solution by $\hat{u}(\ell)$, $\ell = k,\dots,k+N-1$. Set $\mu_N(k, z_k) = \hat{u}(k)$.
(3) Apply $\mu_N(k, z_k)$ to the control system for the step from k to $k+1$, i. e., within the time interval $[t_k, t_{k+1}]$.
(4) Set $k \leftarrow k+1$ and go to (1).

For symmetric, positive semidefinite matrices R and S, the discrete-time control problems D-LQOCP(k, z_k, N) to be solved in each step are *convex*. For the numerical solution of linear-quadratic optimization problems, there are many methods, such as primal or dual active set methods, interior-point methods, semi-smooth Newton methods, Lemke's method, and many more, compare [68, Chapter 3] for an overview. Details of the semi-smooth Newton method are provided in Section 5.2.3.

Example 6.3.12. Consider again the pendulum equations with damping,

$$\dot{x}_1(t) = x_2(t),$$
$$\dot{x}_2(t) = \frac{1}{\ell}(-cx_2(t) - g\sin(x_1(t)) - v(t))$$

with $g = 9.81$, $\ell = 1$, $c = 0.4$, and control bounds $v(t) \in [-7, 7]$.

Linearization at the upper equilibrium $x_{\text{ref}} = (\pi, 0)^\top$, $v_{\text{ref}} = 0$ yields the linear differential equation for $z(t) := x(t) - x_{\text{ref}}$ and $u(t) := v(t) - v_{\text{ref}}$:

$$\begin{pmatrix} \dot{z}_1(t) \\ \dot{z}_2(t) \end{pmatrix} = \underbrace{\begin{pmatrix} 0 & 1 \\ \frac{g}{\ell} & -\frac{c}{\ell} \end{pmatrix}}_{=:A} \begin{pmatrix} z_1(t) \\ z_2(t) \end{pmatrix} + \underbrace{\begin{pmatrix} 0 \\ -\frac{1}{\ell} \end{pmatrix}}_{=:B} u(t).$$

Figure 6.23 shows the solution of the LMPC algorithm 6.3.11 for the initial value $z(0) = (-40^\circ, 0)^\top$, the parameters $h = 0.1$, $N = 10$, and the weighting matrix

$$R(k) = \begin{pmatrix} 12 & 0 \\ 0 & 1 \end{pmatrix}, \quad S(k) = (1).$$

The linear MPC algorithm is able to control the pendulum back to the reference solution obeying the control bounds. □

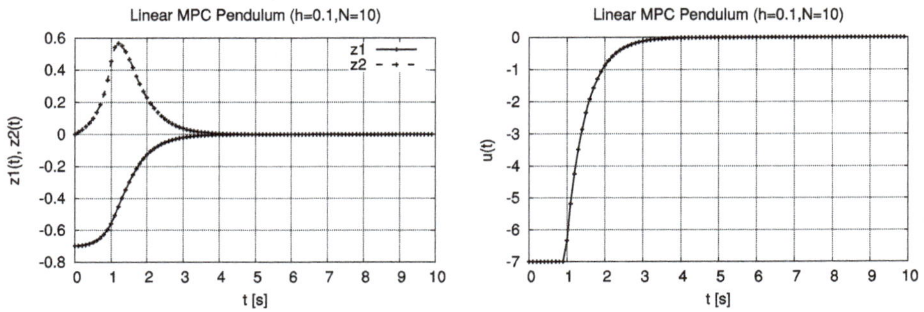

Figure 6.23: Linear model predictive control of a pendulum with $h = 0.1$, $N = 10$ and linearization in $x_{\text{ref}} \equiv (\pi, 0)^\top$, $u_{\text{ref}} \equiv 0$: States z_1, z_2 and control u.

Example 6.3.13 (LMPC tracking controller for a vehicle). An important task in automated driving is the tracking of a geometric reference curve. Similarly, as in Example 6.3.6, let the reference curve be given in terms of a two-dimensional curve

$$y_{\text{ref}}(s) = \begin{pmatrix} x_{\text{ref}}(s) \\ y_{\text{ref}}(s) \end{pmatrix}$$

with $s \in [0, L]$. The variable s denotes the arclength, and L is the length of the curve. Figure 6.24 depicts the reference curve.

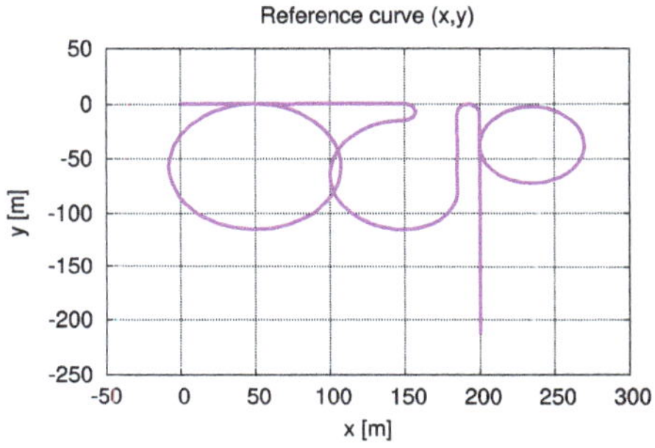

Figure 6.24: Reference curve.

We intend to design an LMPC tracking controller and use the dynamics (6.15)–(6.18) in curvilinear coordinates. In contrast to Example 6.3.6, we do not control the curvature κ directly but its derivative $\dot{\kappa}(t) = u(t)$.

Since we like to track the reference curve, it is reasonable to assume small deviations from the reference curve. For $r \approx r_{\text{ref}} = 0$, $\psi \approx \psi_{\text{ref}}$, $\kappa \approx \kappa_{\text{ref}}$, and $u \approx u_{\text{ref}} = \dot{\kappa}_{\text{ref}}$, we linearize the dynamics and obtain

$$\begin{aligned}
\dot{r}(t) &= v \cdot (\psi(t) - \psi_{\text{ref}}(t)), \\
\dot{\psi}(t) &= v \cdot \kappa(t), \\
\dot{\kappa}(t) &= u(t), \\
\dot{\psi}_{\text{ref}}(t) &= v \cdot \kappa_{\text{ref}}(s(t)), \\
s(t) &= \int_0^t v \, d\tau = v \cdot t.
\end{aligned}$$

and in matrix form

$$\underbrace{\begin{pmatrix} \dot{r}(t) \\ \dot{\psi}(t) \\ \dot{\kappa}(t) \\ \dot{\psi}_{\text{ref}}(t) \end{pmatrix}}_{=\dot{z}(t)} = \underbrace{\begin{pmatrix} 0 & v & 0 & -v \\ 0 & 0 & v & 0 \\ 0 & 0 & 0 & 0 \\ 0 & 0 & 0 & 0 \end{pmatrix}}_{=\tilde{A}} \underbrace{\begin{pmatrix} r(t) \\ \psi(t) \\ \kappa(t) \\ \psi_{\text{ref}}(t) \end{pmatrix}}_{=z(t)} + \underbrace{\begin{pmatrix} 0 \\ 0 \\ 1 \\ 0 \end{pmatrix}}_{=\tilde{B}} u(t) + \underbrace{\begin{pmatrix} 0 \\ 0 \\ 0 \\ v \cdot \kappa_{\text{ref}}(s(t)) \end{pmatrix}}_{=\tilde{c}(t)}.$$

Herein, we consider the velocity v on the prediction horizon $N \cdot h$ as constant. Discretization of the linear system of differential equations with the explicit Euler method and step-size $h > 0$ yields the system in discrete time

$$z(k+1) = \underbrace{(I + h\tilde{A})}_{=:A} z(k) + \underbrace{h\tilde{B}}_{=:B} u(k) + \underbrace{h\tilde{d}(t_k)}_{=:d(k)},$$

where again the convention $k \hat{=} t_k$ with $t_k = kh$, $k = 0, 1, 2, \ldots$, is used.

We apply the LMPC algorithm 6.3.11 to the track in Figure 6.24 with initial state $z(0) = z_0$ and

$$R = \begin{pmatrix} \alpha_1 & 0 & 0 & 0 \\ 0 & \alpha_2 & 0 & -\alpha_2 \\ 0 & 0 & 0 & 0 \\ 0 & -\alpha_2 & 0 & \alpha_2 \end{pmatrix}, \quad S = (\alpha_3)$$

with weights $\alpha_1, \alpha_2, \alpha_3 > 0$. It assumed that r and $\psi - \psi_{\text{ref}}$ can be measured by suitable sensors (e. g., by D-GPS).

Figure 6.25 shows the results of the LMPC algorithm with a sampling time of 20 [ms] (50 Hz) and a prediction horizon of 5 [s] with $N = 20$ grid points in the discretization. Please note that the sampling points and the grid points used in the discretization do not coincide in contrast to the basic LMPC algorithm. Consequently, the subproblems D-LQOCP(k, z_k, N) in the LMPC algorithm need to be solved in less than 20 [ms]. In addition, a velocity controller was used to control the velocity such that a maximal lateral acceleration of 5 [m/s] and a maximal velocity of 10 [m/s] are obeyed. The used weights are $\alpha_1 = \alpha_2 = 1$, $\alpha_3 = 0.1$. In addition, there are control bounds $u \in [-0.5, 0.5]$. □

We have presented here only the basics of model-predictive control. This will be followed by further questions, such as:

- Under which conditions are the LMPC or NMPC controllers (asymptotically) stable?
- Under which conditions are the LMPC or NMPC controllers robust with respect to disturbances?
- How can we guarantee the reliability of the controllers when state constraints are present?
- What are efficient methods to achieve real-time capability? Here implementation details and structure exploitation play a major role.
- Are there optimality estimates for $N \to \infty$ in the feedback law μ_N?

Many of these questions are answered in the special literature [279] and [150] and the references therein.

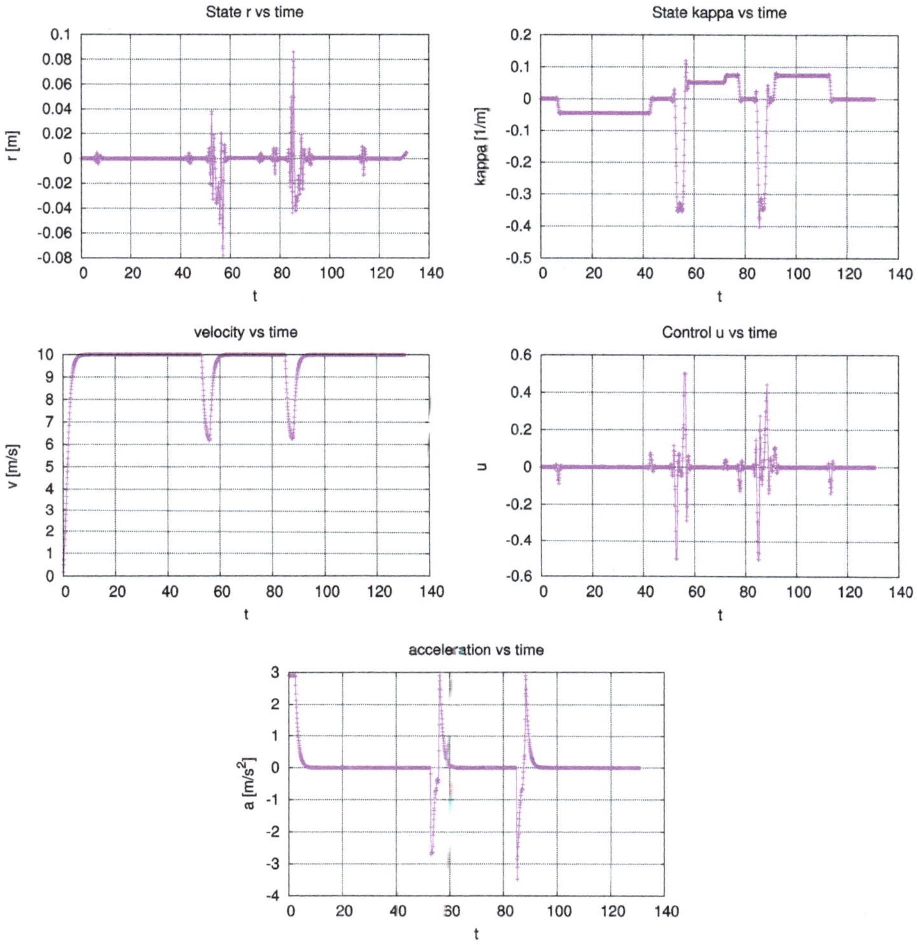

Figure 6.25: Results of the LMPC controller.

6.4 Exercises

Consider the following optimization problem:

$$\begin{aligned} &\textit{Minimize} \quad J(z_1, z_2) := -0.1(z_1 - 4)^2 + z_2^2 \\ &\textit{subject to} \quad G(z_1, z_2) := 1 - z_1^2 - z_2^2 \leq 0. \end{aligned}$$

Check the KKT conditions and the second-order sufficient conditions.

For $y \leq \sqrt{2}$, consider the parametric optimization problem:

$$\begin{aligned} \textit{Minimize} \quad & J(z_1, z_2) := -(z_1+1)^2 - (z_2+1)^2 \\ \textit{subject to} \quad & z_1^2 + z_2^2 - 2 \leq 0, \\ & z_1 - y \leq 0. \end{aligned}$$

Compute the solutions $\hat{z}(y)$ and verify second-order sufficient conditions.

For the parameters $x = (x_1, \ldots, x_m)^\top$ and $y = (y_1, \ldots, y_p)^\top$, let the following parametric nonlinear optimization problem be given:

$$\begin{aligned} \textit{Minimize} \quad & J(z) \\ \textit{subject to} \quad & G_i(z) - x_i \leq 0, \quad i = 1, \ldots, m, \\ & H_j(z) - y_j = 0, \quad j = 1, \ldots, p. \end{aligned}$$

Let $(\hat{z}, \hat{\mu}, \hat{\lambda})$ be a KKT point of the nominal problem for the nominal parameters $\hat{x} = (\hat{x}_1, \ldots, \hat{x}_m)^\top = 0_{\mathbb{R}^m}$ and $\hat{y} = (\hat{y}_1, \ldots, \hat{y}_p)^\top = 0_{\mathbb{R}^p}$.

Use the sensitivity theorem (assuming that the assumptions are satisfied) to show that the following relations hold:

$$\begin{aligned} \left. \frac{\partial J(z(x,y))}{\partial x} \right|_{(x,y)=(\hat{x},\hat{y})} &= -\hat{\mu}^\top, \\ \left. \frac{\partial J(z(x,y))}{\partial y} \right|_{(x,y)=(\hat{x},\hat{y})} &= -\hat{\lambda}^\top. \end{aligned}$$

The Lagrange multipliers indicate the sensitivity of the objective function with respect to perturbations in the constraints!

Consider the following parametric nonlinear optimization problem:

P(ω) *Minimize*

$$-\left(\frac{1}{2} + \omega_4\right)\sqrt{z_1} - \left(\frac{1}{2} - \omega_4\right) z_2$$

subject to the constraints $z \in \mathbb{R}^2$ *and*

$$\begin{aligned} -z_1 + 0.1 &\leq \omega_1, \\ -z_2 &\leq \omega_2, \\ z_1 + z_2 - 1 &\leq \omega_3. \end{aligned}$$

Herein, $\omega \in \mathbb{R}^4$ is a parameter vector with $\|\omega\|$ small.

Solve the nominal problem $P(0_{\mathbb{R}^4})$ using the KKT conditions and compute the derivatives

$$\frac{\partial J(z(\omega),\omega)}{\partial \omega_1}(0_{\mathbb{R}^4}), \quad \frac{\partial J(z(\omega),\omega)}{\partial \omega_2}(0_{\mathbb{R}^4}), \quad \frac{\partial J(z(\omega),\omega)}{\partial \omega_3}(0_{\mathbb{R}^4}), \quad \frac{\partial J(z(\omega),\omega)}{\partial \omega_4}(0_{\mathbb{R}^4})$$

at the nominal parameter $\omega = 0_{\mathbb{R}^4}$, where $z(\omega)$ denotes the solution of $P(\omega)$ with Lagrange multiplier $\mu(\omega)$ for $\|\omega\|$ small.

Compute the sensitivities

$$\frac{\partial z}{\partial \omega}(0_{\mathbb{R}^4}), \quad \frac{\partial \mu}{\partial \omega}(0_{\mathbb{R}^4}).$$

Consider the following perturbed quadratic optimization problem with $Q \in \mathbb{R}^{n_z \times n_z}$ symmetric and positive definite, $A \in \mathbb{R}^{m \times n_z}$ of rank m, and parameter $p \in \mathbb{R}^m$:

$$\mathrm{QP}(p) \quad \textit{Minimize } \frac{1}{2} z^\top Q z \textit{ with respect to } z \in \mathbb{R}^{n_z} \textit{ subject to } Az = b + p.$$

Show that the *value function*

$$\Phi(p) := \inf\left\{ \frac{1}{2} z^\top Q z \mid Az = b + p,\ z \in \mathbb{R}^{n_z} \right\}$$

satisfies

$$\Phi(p) = \Phi(0) - \lambda(0)^\top p + \frac{1}{2} p^\top \left(AQ^{-1}A^\top\right)^{-1} p,$$

where $\lambda(0)$ denotes the Lagrange multiplier of $\mathrm{QP}(0_{\mathbb{R}^m})$.

Consider a car with steering. Let $v(t)$ denote the velocity of the car, $\psi(t)$ the yaw angle, and $\ell > 0$ the length. The differential equations for v and ψ are

$$\begin{aligned} \dot v(t) &= a(t), \\ \dot\psi(t) &= \frac{v(t)}{\ell} \tan \delta(t), \end{aligned}$$

where δ (steering angle) and a (acceleration) are the controls. Find the linearization at $(v_{\mathrm{ref}}(t), \psi_{\mathrm{ref}}(t))$ and $(a_{\mathrm{ref}}(t), \delta_{\mathrm{ref}}(t))$! Find the specific linearization at $v_{\mathrm{ref}} = 10$, $\psi_{\mathrm{ref}} = 0$, $a_{\mathrm{ref}} = 0$, $\delta_{\mathrm{ref}} = 0$!

Consider the equations of motion of a mobile robot,

$$\begin{aligned} \dot x(t) &= \frac{v_\ell(t) + v_r(t)}{2} \cos(\psi(t)), \\ \dot y(t) &= \frac{v_\ell(t) + v_r(t)}{2} \sin(\psi(t)), \\ \dot\psi(t) &= \frac{v_r(t) - v_\ell(t)}{B}, \\ \dot v_\ell(t) &= u_\ell(t), \\ \dot v_r(t) &= u_r(t). \end{aligned}$$

Find the linearization for driving straight ahead, i. e., at constant yaw angle ψ_{ref} and constant velocity $v_{\ell,\text{ref}} = v_{r,\text{ref}} = v_{\text{ref}}$ with given velocity v_{ref}. Find the linearization for the specific values $\psi_{\text{ref}} = 0\,[\text{rad}]$ and $v_{\text{ref}} = 1\,[\text{m/s}]$.

Write down the necessary optimality conditions for the following optimal control problem, compare Example 6.2.5, and solve the resulting linear boundary value problem for the data $g = 9.81$, $\ell = g$, $c = \ell$, $r_1 = 10$, $r_2 = 0$, $s = 1$, $t_f = 10$, $z_{1,0} = 0.1$, $z_{2,0} = 0$.

Minimize

$$\frac{1}{2}\int_{t_0}^{t_f} r_1 \cdot z_1(t)^2 + r_2 \cdot z_2(t)^2 + s \cdot v(t)^2\,dt$$

subject to

$$\begin{pmatrix} \dot z_1(t) \\ \dot z_2(t) \end{pmatrix} = \underbrace{\begin{pmatrix} 0 & 1 \\ \frac{g}{\ell} & -\frac{c}{\ell} \end{pmatrix}}_{=A} \begin{pmatrix} z_1(t) \\ z_2(t) \end{pmatrix} + \underbrace{\begin{pmatrix} 0 \\ -\frac{1}{\ell} \end{pmatrix}}_{=B} v(t)$$

with initial values $z_1(t_0) = z_{1,0}$ *and* $z_2(t_0) = z_{2,0}$.

Let a controlled process in $[\hat t, t_f]$ with t_f fixed be defined by

$$\dot x(t) = u(t), \quad x(\hat t) = \hat x.$$

The task is to control the state x as close as possible to the target state $x(t_f) = x_f$ using a moderate control effort. This can be achieved by minimizing the objective function

$$J(x,u) = \frac{c}{2}\left(x(t_f) - x_f\right)^2 + \frac{1}{2}\int_{t_0}^{t_f} u(t)^2 dt$$

with a weight factor $c \geq 0$.

Compute the optimal control and express it in feedback form

$$u(t) = -K(\hat t)(\hat x - x_f).$$

Discuss the cases $c \to \infty$ and $c \to 0$.

Consider the following linear-quadratic optimal control problem, where only the vector $y(t) := C(t)z(t) + D(t)u(t)$ can be observed:

Minimize

$$\frac{1}{2}\int_{0}^{t_f} y(t)^\top R(t) y(t) + u(t)^\top S(t) u(t) dt$$

subject to the constraints

$$\begin{aligned} \dot{z}(t) &= A(t)z(t) + B(t)u(t), \\ y(t) &= C(t)z(t) + D(t)u(t), \\ z(0) &= z_0. \end{aligned}$$

Exploit the local minimum principle to find a feedback control law for the control. Formulate a suitable Riccati differential equation.

Consider the following linear-quadratic optimal control problem:

Minimize

$$J(x,u) = \frac{1}{2}\int_0^1 x^\top Qx + u^\top Ru - 2x^\top Su + 2z^\top x + 2w^\top u dt$$

subject to

$$\begin{aligned} \dot{x} &= Ax + Bu + b, \\ C_0x(0) + C_1x(1) &= c, \\ Ex + Fu &\leq e. \end{aligned}$$

Herein, $Q, R, S, z, w, A, B, b, E, F, e$ are time-dependent matrices and vectors of appropriate dimension. Q, R, S are supposed to be symmetric, and

$$\begin{pmatrix} Q & S \\ S^\top & R \end{pmatrix}$$

is supposed to be positive semidefinite. C_0, C_1, c are constant matrices and vectors.

Prove: If $(\hat{x}, \hat{u})$ is feasible and satisfies the local minimum principle with $\ell_0 = 1$, then $(\hat{x}, \hat{u})$ is a global minimum of the optimal control problem.

We consider a car with steering and would like to control the yaw angle ψ such that it follows a reference yaw angle ψ_{ref}. Let $v > 0$ denote the given velocity of the car and $\ell > 0$ its length. The differential equation for ψ reads

$$\dot{\psi}(t) = \frac{v(t)}{\ell}\tan\delta(t),$$

where δ denotes the control (steering angle). Design a linear-quadratic regulator for this task!

(Autopilot)
Consider the flight of an aircraft in the 2D-plane given by the equations of motion

$$\dot{h}(t) = v(t)\sin\gamma(t), \qquad h(0) = 3000,$$

$$\dot{v}(t) = \frac{1}{m}\Big(T(t) - D\big(v(t), h(t), C_L(t)\big)\Big) - g\sin\gamma(t), \quad v(0) = 300,$$
$$\dot{\gamma}(t) = \frac{1}{mv(t)} L\big(v(t), h(t), C_L(t)\big) - \frac{g}{v(t)}\cos\gamma(t), \quad \gamma(0) = 0,$$

where h denotes the altitude, v the velocity, and γ the pitch angle. Moreover,

$$L(v, h, C_L) = q(v, h)\cdot F\cdot C_L, \quad D(v, h, C_L) = q(v, h)\cdot F\cdot \big(C_{D,0} + kC_L^2\big),$$
$$q(v, h) = \frac{1}{2}\cdot \rho(h)\cdot v^2, \quad \rho(h) = \rho_0\cdot \exp(-\beta h),$$

with constants

$$F = 26, \quad C_{D,0} = 0.0165, \quad k = 1, \quad \rho_0 = 1.225, \quad \beta = 1/6900, \quad m = 7500, \quad g = 9.81.$$

The aircraft can be controlled by the lift coefficient $C_L \in [0.01, 0.18326]$ and the thrust $T \in [0, 59300]$.

The task is to track the equilibrium solution

$$h_{\text{ref}} := 3000, \quad v_{\text{ref}} := 300, \quad \gamma_{\text{ref}} := 0,$$

and

$$C_{L,\text{ref}} := \frac{m\cdot g}{q(v_{\text{ref}}, h_{\text{ref}})\cdot F}, \quad T_{\text{ref}} := D(v_{\text{ref}}, h_{\text{ref}}, C_{L,\text{ref}}).$$

(a) Find the linearization in the cruise mode, i. e., at constant velocity v_{ref}, constant altitude h_{ref}, and flight path angle $\gamma_{\text{ref}} = 0$.

(b) Develop a linear-quadratic regulator taking into account (6.9) to track the reference solution and test the controller by perturbing the equilibrium solution.

(c) Use the nonlinear model-predictive control algorithm 6.3.3 to track the reference solution and compare it to the controller in (b).

(Collision Avoidance Project)

This collision avoidance project can be either simulated on a computer or actually realized using, e. g., LEGO®MINDSTORMS robots. The task is to avoid a collision with a fixed obstacle, see figure. The obstacle is supposed to be stationary, i. e., non-moving, and its measurements and position are supposed to be known to the avoiding vehicle.

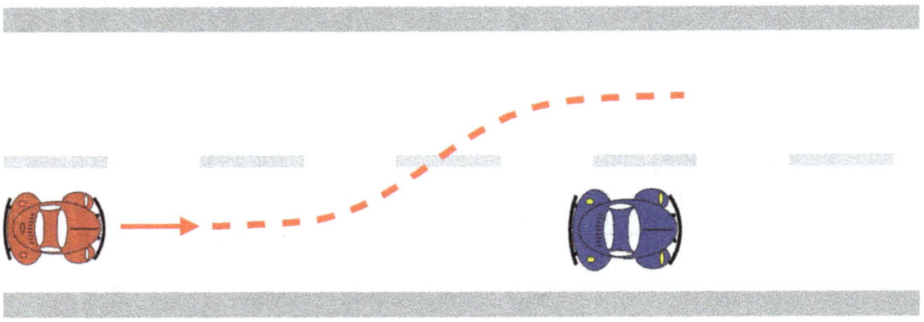

Project steps:

(a) Formulate a suitable optimal control problem to avoid a collision and compute an optimal collision avoidance trajectory using the direct discretization methods in Chapter 5 for a given initial distance to the obstacle and a given initial velocity of the avoiding car. Use the model in Example 6.3.8 for the avoiding car.

(b) Track the computed optimal collision avoidance trajectory in (a) as follows:
   (i) Drive in a straight line with top speed towards the obstacle. In a real vehicle, a controller might be necessary to drive a straight line!
   (ii) Measure repeatedly the distance to the obstacle using, e. g., an ultrasonic sensor.
   (iii) If the distance is reached for which the optimal collision avoidance trajectory in (a) has been computed, then track the collision avoidance trajectory using a suitable controller (P-, I-, D-, PID-controller, Riccati, MPC).

7 Mixed-integer optimal control

Technical or economical processes often involve *discrete* control variables, which are used to model finitely many decisions, discrete resources, or switching structures, like gear shifts in a car or operating modes of a device. This leads to optimal control problems with non-convex and partly discrete control set $\mathcal{U}$. More specifically, some of the control variables may still assume any real value within a given convex set with a non-empty interior; those are called *continuous-valued control variables* in the sequel, while other control variables are restricted to a finite set of values, those are called *discrete control variables* in the sequel.

An optimal control problem involving continuous-valued and discrete control variables is called *mixed-integer optimal control problem*. Mixed-integer optimal control is a field of increasing importance, and practical applications can be found in [167, 32, 120, 122, 304, 291, 178, 292]. For a website of further benchmark problems, please refer to [289].

One approach to solving mixed-integer optimal control problems is by exploiting necessary optimality conditions. Unfortunately, the assumptions imposed in Chapter 3 on the set $\mathcal{U}$, i. e., convexity and non-empty interior, do not hold for mixed-integer optimal control problems. Hence, the techniques used to prove the local minimum principles in Chapter 3 based on the Fritz John conditions in Chapter 2 cannot be applied to mixed-integer optimal control problems. New techniques are required. In Section 7.1, we will exploit the idea of Dubovitskii and Milyutin, see [83, 82], [139, p. 95], [169, p. 148], who used a time transformation to transform the mixed-integer optimal control problem into an equivalent optimal control problem without discrete control variables. Necessary conditions are then obtained by applying the local minimum principles of Chapter 3 to the transformed problem. The result is necessary conditions in terms of *global minimum principles*. The global minimum principle is valid even for discrete control sets. Global minimum principles for DAE optimal control problems can be found in [287] for Hessenberg DAE optimal control problems, in [66] for semi-explicit index-one DAEs, in [75] for implicit control systems, in [17] for quasilinear DAEs, and in [194] for nonlinear DAEs of arbitrary index. The approach in [287] exploits known results for the ODE case and applies them to the DAE setting by first solving an algebraic constraint for the algebraic variable, expressing it as a function of the control and the differential state using the implicit function theorem. Then, the algebraic variable is replaced by this implicitly defined function, and a standard ODE optimal control problem is obtained. Finally, the Pontryagin minimum principle is applied, and the resulting minimum principle is translated to the DAE formulation. Although this approach is convenient, we do not follow this approach as we intend to work with the original DAE optimal control formulation.

The global minimum principle can be exploited numerically using an indirect approach, but a very good initial guess of the problem's switching structure is needed. Such an initial guess is often unavailable for practical applications. Based on the minimum principle, a graph-based solution method was developed in [176], which, however, is limited to single-state problems.

https://doi.org/10.1515/9783110797893-007

The time transformation method of Dubovitskii and Milyutin can be exploited numerically and leads to the variable time transformation method in Section 7.2. This method couples the direct discretization method of Chapter 5 and the variable time transformation in Section 7.1, see [204, 203, 202, 317, 304, 120, 122].

Direct discretization methods based on relaxations and sum-up-rounding strategies are investigated in [293, 290, 294]. These methods have shown their ability to solve difficult real-world examples.

Section 7.3 particularly addresses the problem of including switching costs in mixed-integer optimal control problems. Switching costs apply each time the discrete control switches. From a mathematical point of view, switching costs can be assigned to avoid so-called chattering discrete controls, which cannot be realized in practice. The discussion is restricted to discretized optimal control problems in combination with a dynamic programming approach, which exploits Bellman's optimality principle.

7.1 Global minimum principle

We consider the following autonomous optimal control problem on a fixed time interval $[t_0, t_f]$:

Problem 7.1.1 (Mixed-integer optimal control problem). Let $\mathcal{I} := [t_0, t_f] \subset \mathbb{R}$ be a non-empty compact time interval with $t_0 < t_f$ fixed. Let

$$\begin{aligned} \varphi &: \mathbb{R}^{n_x} \times \mathbb{R}^{n_x} \longrightarrow \mathbb{R}, \\ f_0 &: \mathbb{R}^{n_x} \times \mathbb{R}^{n_y} \times \mathbb{R}^{n_u} \longrightarrow \mathbb{R}, \\ f &: \mathbb{R}^{n_x} \times \mathbb{R}^{n_y} \times \mathbb{R}^{n_u} \longrightarrow \mathbb{R}^{n_x}, \\ g &: \mathbb{R}^{n_x} \longrightarrow \mathbb{R}^{n_y}, \\ \psi &: \mathbb{R}^{n_x} \times \mathbb{R}^{n_x} \longrightarrow \mathbb{R}^{n_\psi} \end{aligned}$$

be sufficiently smooth functions and $\mathcal{U} \subseteq \mathbb{R}^{n_u}$ a non-empty set. Let the index-two Assumption 3.1.3 hold.

Minimize the objective function

$$\varphi(x(t_0), x(t_f)) + \int_{t_0}^{t_f} f_0(x(t), y(t), u(t))dt$$

with respect to

$$x \in W_{1,\infty}^{n_x}(\mathcal{I}), \quad y \in L_\infty^{n_y}(\mathcal{I}), \quad u \in L_\infty^{n_u}(\mathcal{I}),$$

subject to the semi-explicit DAE

$$\begin{aligned} \dot{x}(t) &= f(x(t), y(t), u(t)) && \text{a. e. in } \mathcal{I}, \\ 0_{\mathbb{R}^{n_y}} &= g(x(t)) && \text{in } \mathcal{I}, \end{aligned}$$

the boundary condition

$$\psi(x(t_0), x(t_f)) = 0_{\mathbb{R}^{n_\psi}},$$

and the set constraint

$$u(t) \in \mathcal{U} \quad a.e.\ in\ \mathcal{I}.$$

The set $\mathcal{U}$ in Problem 7.1.1 is supposed to be an arbitrary set. We particularly allow that $\mathcal{U}$ may only contain finitely many vectors so that Problem 7.1.1 contains problems with discrete controls.

Example 7.1.2 ([304, Section 4.4.3]). The following optimal control problem has a discrete control set:

Minimize

$$x(2)^2 + \int_0^2 \left(\sin\left(\frac{\pi}{2}t\right) - x(t)\right)^2 dt$$

with respect to $x \in W_{1,\infty}([0,2])$ *and* $u \in L_\infty^3([0,2])$ *subject to the constraints*

$$\dot{x}(t) = u_1(t) - u_2(t) + 2tu_3(t), \quad x(0) = 0,$$

and

$$\begin{pmatrix} u_1(t) \\ u_2(t) \\ u_3(t) \end{pmatrix} \in \mathcal{U} := \left\{ \begin{pmatrix} 1 \\ 0 \\ 0 \end{pmatrix}, \begin{pmatrix} 0 \\ 1 \\ 0 \end{pmatrix}, \begin{pmatrix} 0 \\ 0 \\ 1 \end{pmatrix} \right\}.$$

□

When dealing with discrete controls, it is important to relax the usual terminology of a local minimum.

Definition 7.1.3 (Strong and weak local minima). Let $(\hat{x}, \hat{y}, \hat{u})$ be feasible for Problem 7.1.1.

(a) $(\hat{x}, \hat{y}, \hat{u})$ is called a *(weak) local minimum of Problem* 7.1.1, if $(\hat{x}, \hat{y}, \hat{u})$ minimizes the objective function among all feasible (x, y, u) with $\|x - \hat{x}\|_{1,\infty} < \varepsilon$, $\|y - \hat{y}\|_\infty < \varepsilon$, and $\|u - \hat{u}\|_\infty < \varepsilon$ for some $\varepsilon > 0$.

(b) $(\hat{x}, \hat{y}, \hat{u})$ is called a *strong local minimum of Problem* 7.1.1, if $(\hat{x}, \hat{y}, \hat{u})$ minimizes the objective function among all feasible (x, y, u) with $\|x - \hat{x}\|_\infty < \varepsilon$. □

Notice that strong local minima are also weak local minima. The converse is not true. Strong local minima are minimal with respect to a larger class of algebraic variables and controls. Especially if the control or algebraic variable is discontinuous, the difference between strong and weak local minima becomes apparent, since an ε-neighborhood of a discontinuous function with regard to the L_∞-norm basically consists of functions with the same location of discontinuities and slightly varied function values in the continuous parts. Even worse, weak neighborhoods of feasible discrete controls typically only contain the discrete control.

In proving necessary conditions, we cannot exploit a special structure of $\mathcal{U}$ as it was done in Chapter 3 by assuming that the set was convex with non-empty interior. Hence, the necessary optimality conditions in Chapter 3 do not hold for Problem 7.1.1. However, the necessary optimality conditions in Chapter 3 are not worthless. As we shall see, it is possible to use a special time transformation to transform Problem 7.1.1 into an equivalent problem with a nice convex control set with non-empty interior for which the necessary optimality conditions of Chapter 3 are valid. This time transformation is due to Dubovitskii and Milyutin, and the following proof techniques can be found in [169, p. 148] and [139, p. 95] in the case of ODEs. The results are extended to the DAE setting in Problem 7.1.1. To this end, let

$$\mathcal{H}(x,y,u,\lambda_f,\lambda_g,\ell_0) := \ell_0 f_0(x,y,u) + (\lambda_f^\top + \lambda_g^\top g_x'(x))f(x,y,u)$$

denote the Hamilton function for Problem 7.1.1, and let $(\hat{x},\hat{y},\hat{u})$ be a strong local minimum of Problem 7.1.1.

For $\tau \in [0,1]$, we introduce the *time transformation*

$$t(\tau) := t_0 + \int_0^\tau w(s)ds, \quad t(0) = t_0, \quad t(1) = t_f, \quad w(s) \geq 0, \tag{7.1}$$

compare Figure 7.1. The inverse mapping is defined by

$$\tau(s) := \inf\{\tau \in [0,1] \mid t(\tau) = s\}. \tag{7.2}$$

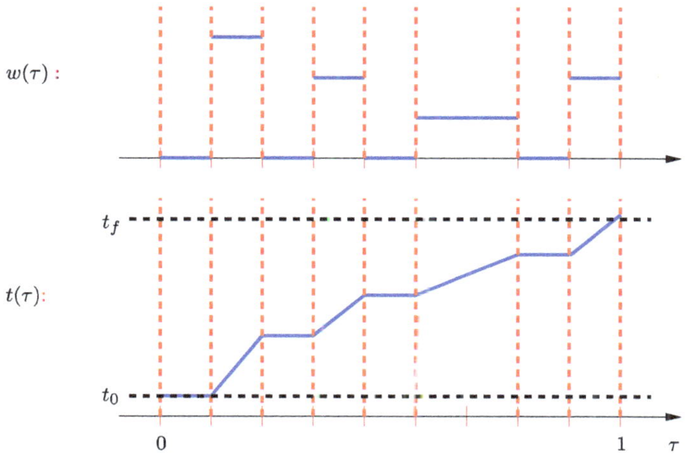

Figure 7.1: Time transformation $t(\tau)$ for a given function $w(\tau) \geq 0$: If $w(\tau) = 0$, then the time t stands still; otherwise, the time t proceeds.

For any function $w \in L_\infty([0,1],\mathbb{R})$ satisfying (7.1), define

$$\tilde{u}(\tau) := \begin{cases} \hat{u}(t(\tau)), & \text{for } \tau \in \Delta_w, \\ \text{arbitrary}, & \text{for } \tau \in [0,1] \setminus \Delta_w, \end{cases}$$
$$\tilde{x}(\tau) := \hat{x}(t(\tau)),$$
$$\tilde{y}(\tau) := \begin{cases} \hat{y}(t(\tau)), & \text{for } \tau \in \Delta_w, \\ \text{suitable}, & \text{for } \tau \in [0,1] \setminus \Delta_w, \end{cases}$$

where

$$\Delta_w := \{\tau \in [0,1] \mid w(\tau) > 0\}.$$

Suitable values for $\tilde{y}$ on $[0,1] \setminus \Delta_w$ will be provided later.

The functions $\tilde{x}$, $\tilde{y}$, $\tilde{u}$ are feasible for the following auxiliary DAE optimal control problem in which w is considered a *control* and $\tilde{u}$ a *fixed* function. In addition, it is exploited that the original index-two DAE in Problem 7.1.1 is mathematically equivalent to the index-one DAE

$$\dot{x}(t) = f(x(t),y(t),u(t)),$$
$$0_{\mathbb{R}^{n_y}} = g_x'(x(t))f(x(t),y(t),u(t))$$

with the additional boundary condition $g(x(t_0)) = 0_{\mathbb{R}^{n_y}}$.

Problem 7.1.4 (Auxiliary DAE optimal control problem).

Minimize

$$\varphi(x(0),x(1)) + \int_0^1 w(\tau)f_0(x(\tau),y(\tau),\tilde{u}(\tau))d\tau$$

with respect to

$$x \in W_{1,\infty}^{n_x}([0,1]), \quad y \in L_\infty^{n_y}([0,1]), \quad t \in W_{1,\infty}([0,1]), \quad w \in L_\infty([0,1])$$

subject to the constraints

$$\begin{aligned} \dot{x}(\tau) &= w(\tau)f(x(\tau),y(\tau),\tilde{u}(\tau)) && \text{a.e. in } [0,1], \\ 0_{\mathbb{R}^{n_y}} &= g_x'(x(\tau))f(x(\tau),y(\tau),\tilde{u}(\tau)) && \text{a.e. in } [0,1], \\ \dot{t}(\tau) &= w(\tau) && \text{a.e. in } [0,1], \\ \psi(x(0),x(1)) &= 0_{\mathbb{R}^{n_\psi}}, \\ g(x(0)) &= 0_{\mathbb{R}^{n_y}}, \\ t(0) &= t_0, \\ t(1) &= t_f, \\ w(\tau) &\geq 0 && \text{a.e. in } [0,1]. \end{aligned}$$

Note that the control w in Problem 7.1.4 is only restricted by the control constraint $w(\tau) \geq 0$, and hence it is not a discrete control! Note further that $(\tilde{x}, \tilde{y}, w)$ is actually a weak local minimum of Problem 7.1.4 for any feasible control w. This is because solutions of the transformed DAE subject to the time transformation, and solutions of the original DAE coincide almost everywhere. A formal proof can be conducted similarly to [169, pp. 149–156].

Note: If $w \equiv 0$ on some interval, then on this interval

$$\dot{x}(\tau) \equiv 0 \quad \Longrightarrow \quad x(\tau) \equiv \text{const},$$
$$\dot{t}(\tau) \equiv 0 \quad \Longrightarrow \quad t(\tau) \equiv \text{const}.$$

Note further: If instead of the index-reduced DAE the original index-two DAE was used in Problem 7.1.4, then the algebraic variable y would not be defined on regions with $w \equiv 0$, since differentiation of the algebraic constraint would lead to

$$0_{\mathbb{R}^{n_y}} = w(\tau)g_x'(x(\tau))f(x(\tau), y(\tau), \tilde{u}(\tau))$$

and the index would not be defined in case $w \equiv 0$. This is the reason, why we prefer to use the index-reduced formulation in Problem 7.1.4.

We apply the first-order necessary optimality conditions in Theorem 3.4.4 to Problem 7.1.4 and note that the assumptions (ii)–(v) and 3.4.2 are automatically satisfied for Problem 7.1.4 for $(\tilde{x}, \tilde{y}, w)$ under Assumption 3.1.3. Assuming smoothness as in Assumption 2.2.5, we may apply Theorem 3.4.4 to Problem 7.1.4. To this end, let the *reduced Hamilton function* of Problem 7.1.1 be defined by

$$\mathcal{R}(x, y, u, \lambda_f, \ell_0) := \ell_0 f_0(x, y, u) + \lambda_f^\top f(x, y, u),$$

and the augmented Hamilton function of Problem 7.1.4 by

$$\begin{aligned}
&\bar{\mathcal{H}}(\tau, x, y, t, w, \lambda_f, \lambda_g, \lambda_t, \eta, \ell_0) \\
&\quad := w(\ell_0 f_0(x, y, \tilde{u}(\tau)) + \lambda_f^\top f(x, y, \tilde{u}(\tau)) + \lambda_t - \eta) + \lambda_g^\top g_x'(x) f(x, y, \tilde{u}(\tau)) \\
&\quad = w(\mathcal{R}(x, y, \tilde{u}(\tau), \lambda_f, \ell_0) + \lambda_t - \eta) + \lambda_g^\top r(x, y, \tilde{u}(\tau)),
\end{aligned}$$

where $r : \mathbb{R}^{n_x} \times \mathbb{R}^{n_y} \times \mathbb{R}^{n_u} \longrightarrow \mathbb{R}^{n_y}$ is defined by

$$r(x, y, u) := g_x'(x) f(x, y, u).$$

The first-order necessary optimality conditions in Theorem 3.4.4 read as follows: There exist multipliers $\tilde{\ell}_0 \in \mathbb{R}$, $\tilde{\lambda}_f \in W_{1,\infty}^{n_x}([0,1])$, $\tilde{\lambda}_g \in L_\infty^{n_y}([0,1])$, $\tilde{\lambda}_t \in W_{1,\infty}([0,1])$, $\tilde{\eta} \in L_\infty([0,1])$, $\tilde{\sigma} \in \mathbb{R}^{n_\psi}$, and $\tilde{\zeta} \in \mathbb{R}^{n_y}$ not all zero with
(a) $\tilde{\ell}_0 \geq 0$.

(b) In $[0,1]$, we have the adjoint equation

$$\begin{aligned}\frac{d}{d\tau}\tilde{\lambda}_f(\tau) &= -w(\tau)\mathcal{R}'_x(\tilde{x}(\tau),\tilde{y}(\tau),\tilde{u}(\tau),\tilde{\lambda}_f(\tau),\tilde{\ell}_0)^\top - r'_x(\tilde{x}(\tau),\tilde{y}(\tau),\tilde{u}(\tau))^\top\tilde{\lambda}_g(\tau),\\ 0_{\mathbb{R}^{n_y}} &= \ w(\tau)\mathcal{R}'_y(\tilde{x}(\tau),\tilde{y}(\tau),\tilde{u}(\tau),\tilde{\lambda}_f(\tau),\tilde{\ell}_0)^\top + r'_y(\tilde{x}(\tau),\tilde{y}(\tau),\tilde{u}(\tau))^\top\tilde{\lambda}_g(\tau),\\ \frac{d}{d\tau}\tilde{\lambda}_t(\tau) &= 0.\end{aligned}$$

In particular, $\tilde{\lambda}_t$ is constant.

(c) Transversality conditions:

$$\tilde{\lambda}_f(0)^\top = -(\tilde{\ell}_0\varphi'_{x_0} + \tilde{\sigma}^\top\psi'_{x_0} + \tilde{\zeta}^\top g'_x[0]), \quad \tilde{\lambda}_f(1)^\top = \tilde{\ell}_0\varphi'_{x_f} + \tilde{\sigma}^\top\psi'_{x_f}.$$

(d) Almost everywhere in $[0,1]$, it holds

$$0 = \mathcal{R}(\tilde{x}(\tau),\tilde{y}(\tau),\tilde{u}(\tau),\tilde{\lambda}_f(\tau),\tilde{\ell}_0) + \tilde{\lambda}_t(\tau) - \tilde{\eta}(\tau).$$

Owing to the complementarity condition in (e), we thus have

$$\mathcal{R}(\tilde{x}(\tau),\tilde{y}(\tau),\tilde{u}(\tau),\tilde{\lambda}_f(\tau),\tilde{\ell}_0) + \tilde{\lambda}_t(\tau) \begin{cases} = 0, & \text{if } \tau \in \Delta_w,\\ \geq 0, & \text{if } \tau \notin \Delta_w.\end{cases}$$

(e) Almost everywhere in $[0,1]$, it holds

$$\tilde{\eta}(\tau)w(\tau) = 0 \quad \text{and} \quad \tilde{\eta}(t) \geq 0.$$

Using the inverse time transformation defined in (7.2), we may define

$$\ell_0 := \tilde{\ell}_0, \quad \sigma := \tilde{\sigma}, \quad \zeta := \tilde{\zeta}, \quad \lambda_f(t) := \tilde{\lambda}_f(\tau(t)), \quad \lambda_t(t) := \tilde{\lambda}_t(\tau(t)),$$

and

$$\lambda_g(t)^\top := -\mathcal{R}'_y(\hat{x}(t),\hat{y}(t),\hat{u}(t),\lambda_f(t),\ell_0)(g'_x(\hat{x}(t))f'_y(\hat{x}(t),\hat{y}(t),\hat{u}(t)))^{-1}.$$

Then, for almost every $\tau \in \Delta_w$, it holds

$$\lambda_f(t(\tau)) = \tilde{\lambda}_f(\tau), \quad w(\tau)\lambda_g(t(\tau)) = \tilde{\lambda}_g(\tau), \quad \lambda_t(t(\tau)) = \tilde{\lambda}_t(\tau)$$

and a short calculation shows that λ_f and λ_g satisfy the adjoint equation

$$\begin{aligned}\dot{\lambda}_f(t) &= -\mathcal{H}'_x(\hat{x}(t),\hat{y}(t),\hat{u}(t),\lambda_f(t),\lambda_g(t),\ell_0)^\top,\\ 0_{\mathbb{R}^{n_y}} &= \ \mathcal{H}'_y(\hat{x}(t),\hat{y}(t),\hat{u}(t),\lambda_f(t),\lambda_g(t),\ell_0)^\top,\end{aligned}$$

and the transversality conditions

$$\lambda_f(t_0)^\top = -(\ell_0 \varphi'_{x_0} + \sigma^\top \psi'_{x_0} + \zeta^\top g'_x[t_0]), \quad \lambda_f(t_f)^\top = \ell_0 \varphi'_{x_f} + \sigma^\top \psi'_{x_f}.$$

The above conditions hold for every w. Now we will choose w in a special way to exploit this degree of freedom. The following construction follows [169, p. 157], compare Figure 7.2. This construction principle will be exploited numerically in Section 7.2.

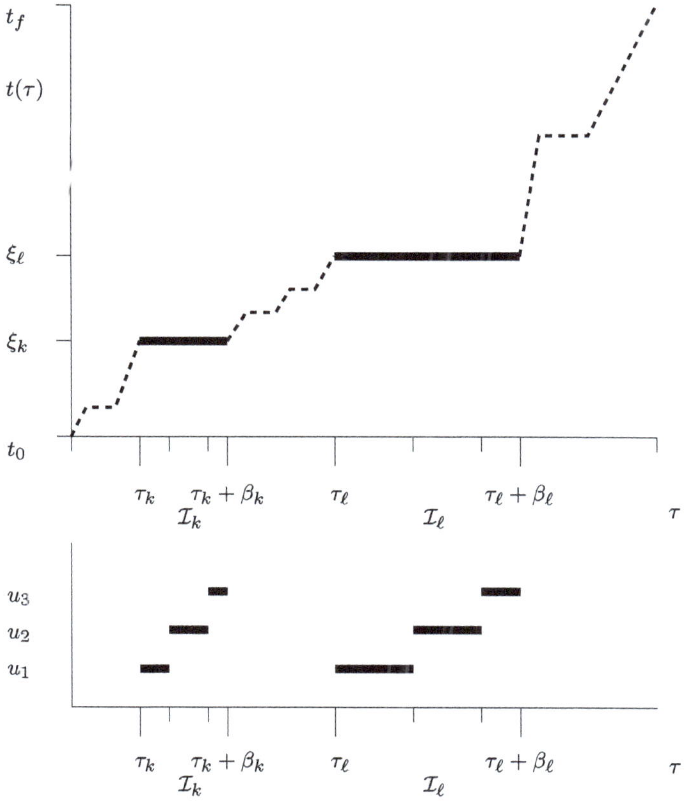

Figure 7.2: Construction principle of w.

Let $w(\tau)$ vanish on the intervals $\mathcal{I}_k := (\tau_k, \tau_k + \beta_k]$, $k = 1, 2, \ldots$, which are to be constructed such that the image of $\bigcup_k \mathcal{I}_k$ under the mapping $\tau \mapsto t(\tau)$ is dense in $[t_0, t_f]$.

Therefore, let $\{\xi_1, \xi_2, \ldots\}$ be a countable dense subset of $[t_0, t_f]$. Choose $\beta_k > 0$ with $\sum_k \beta_k = \frac{1}{2}$, and let

$$\tau_k := \frac{\xi_k - t_0}{2(t_f - t_0)} + \sum_{j:\xi_j < \xi_k} \beta_j.$$

Then the intervals $\mathcal{I}_k = (\tau_k, \tau_k + \beta_k]$ are pairwise disjoint. Define

$$w(\tau) := \begin{cases} 0, & \text{if } \tau \in \bigcup_k \mathcal{I}_k, \\ 2(t_f - t_0), & \text{if } \tau \notin \bigcup_k \mathcal{I}_k. \end{cases}$$

We will show that $t(\tau) = \xi_k$ for any $\tau \in \mathcal{I}_k$. As $\{\xi_k\}$ was chosen to be dense in $[t_0, t_f]$, so is the image of $\bigcup_k \mathcal{I}_k$ under the mapping $\tau \mapsto t(\tau)$.

We note that $\tau_j < \tau_k$, if and only if $\xi_j < \xi_k$ and $t(\tau) = t(\tau_k)$ for all $\tau \in \mathcal{I}_k$. For $\tau \in \mathcal{I}_k$, we find

$$\begin{aligned} t(\tau) &= t_0 + \int_0^{\tau} w(\xi)d\xi \\ &= t_0 + 2(t_f - t_0)\Big(\tau_k - \sum_{j:\tau_j<\tau_k} \beta_j\Big) \\ &= t_0 + 2(t_f - t_0)\Big(\tau_k - \sum_{j:\xi_j<\xi_k} \beta_j\Big) \\ &= t_0 + (t_f - t_0)\frac{\xi_k - t_0}{t_f - t_0} = \xi_k. \end{aligned}$$

Now let

(i) $\mathcal{I}_k := \bigcup_j \mathcal{I}_{kj}$, where $\mathcal{I}_{kj}$ are non-empty closed from the right intervals.
(ii) $\{u_1, u_2, \ldots\}$ be a countable dense subset of $\mathcal{U}$.
(iii) $\tilde{u}(\tau) := u_j$ if $\tau \in \mathcal{I}_{kj}$.
(iv) $\tilde{y}(\tau) := y_j$ with

$$0_{\mathbb{R}^{n_y}} = r(\tilde{x}(\tau), y_j, u_j) = g_x'(\tilde{x}(\tau))f(\tilde{x}(\tau), y_j, u_j),$$

if $\tau \in \mathcal{I}_{kj}$.

According to (d), we have for almost every $\tau \in \bigcup_k \mathcal{I}_k$ the inequality

$$\mathcal{R}(\tilde{x}(\tau), \tilde{y}(\tau), \tilde{u}(\tau), \tilde{\lambda}_f(\tau), \tilde{\ell}_0) + \tilde{\lambda}_t(\tau) \geq 0.$$

As every interval $\mathcal{I}_{kj}$ has a positive measure, there exists $\tau \in \mathcal{I}_{kj}$ with $t(\tau) = \xi_k$ such that

$$\begin{aligned} &\mathcal{R}(\tilde{x}(\tau), \tilde{y}(\tau), \tilde{u}(\tau), \tilde{\lambda}_f(\tau), \tilde{\ell}_0) + \tilde{\lambda}_t(\tau) \\ &\quad = \mathcal{R}(\hat{x}(\xi_k), y_j, u_j, \lambda_f(\xi_k), \ell_0) + \lambda_t(\xi_k) \geq 0. \end{aligned}$$

Since the set $\{\xi_1, \xi_2, \ldots\}$ is dense in $[t_0, t_f]$, and since $\{u_1, u_2, \ldots\}$ is dense in $\mathcal{U}$, and since

$$h(t, y, u) := \mathcal{R}(\hat{x}(t), y, u, \lambda_f(t), \ell_0)$$

is continuous, it follows for almost all $t \in [t_0, t_f]$ that

$$\mathcal{R}(\hat{x}(t), y, u, \lambda_f(t), \ell_0) + \lambda_t(t) \geq 0$$

for all

$$(u, y) \in \Omega(\hat{x}(t)) := \{(u, y) \in \mathcal{U} \times \mathbb{R}^{n_y} \mid 0_{\mathbb{R}^{n_y}} = g_x'(\hat{x}(t)) f(\hat{x}(t), y, u)\}.$$

Note that $\mathcal{R} \equiv \mathcal{H}$, whenever $(u, y) \in \Omega(\hat{x}(t))$.

On the other hand, according to (d), for almost every $\tau \in \Delta_w$, and thus for almost every $t \in \mathcal{I}$, it holds

$$\mathcal{R}(\hat{x}(t), \hat{y}(t), \hat{u}(t), \lambda_f(t), \ell_0) + \lambda_t(t) = 0.$$

Putting both relations together yields the minimality of the reduced Hamilton function for almost every $t \in [t_0, t_f]$:

$$\begin{aligned} &\mathcal{R}(\hat{x}(t), \hat{y}(t), \hat{u}(t), \lambda_f(t), \ell_0) \\ &\quad \leq \mathcal{R}(\hat{x}(t), y, u, \lambda_f(t), \ell_0) \quad \text{for all } (u, y) \in \Omega(\hat{x}(t)). \end{aligned}$$

Moreover, since $\hat{u}$ is essentially bounded and the function $h(t, y, u)$ is continuous, it follows that

$$\mathcal{R}(\hat{x}(t), \hat{y}(t), \hat{u}(t), \lambda_f(t), \ell_0) + \lambda_t(t) \equiv 0$$

almost everywhere. Since λ_t is continuous and constant according to (b), so is $\mathcal{R}$ as a function of time. Exploiting $\mathcal{R} \equiv \mathcal{H}$ whenever $(u, y) \in \Omega(\hat{x}(t))$, we have thus proved the following global minimum principle:

Theorem 7.1.5 (Global minimum principle). *Let the following assumptions be fulfilled for the optimal control problem* 7.1.1.

(i) *Let Assumption* 2.2.5 *hold for* φ, ψ, *and the functions*

$$\begin{aligned} \tilde{f}_0(t, x, y) &:= f_0(x, y, \hat{u}(t)), \\ \tilde{f}(t, x, y) &:= f(x, y, \hat{u}(t)). \end{aligned}$$

Let g *be twice continuously differentiable.*

(ii) *Let* $(\hat{x}, \hat{y}, \hat{u})$ *be a strong local minimum of the optimal control problem* 7.1.1.

(iii) *Let Assumption* 3.1.3 *be valid.*

Then, there exist multipliers

$$\ell_0 \in \mathbb{R}, \quad \lambda_f \in W_{1,\infty}^{n_x}(\mathcal{I}), \quad \lambda_g \in L_\infty^{n_y}(\mathcal{I}), \quad \zeta \in \mathbb{R}^{n_y}, \quad \sigma \in \mathbb{R}^{n_\psi}$$

such that the following conditions are satisfied:

(a) $\ell_0 \geq 0$, $(\ell_0, \zeta, \sigma, \lambda_f, \lambda_g) \neq \Theta$.

(b) Adjoint equations: *Almost everywhere in $\mathcal{I}$, it holds*

$$\begin{aligned}\dot{\lambda}_f(t) &= -\mathcal{H}'_x(\hat{x}(t),\hat{y}(t),\hat{u}(t),\lambda_f(t),\lambda_g(t),\ell_0)^\top,\\ 0_{\mathbb{R}^{n_y}} &= \mathcal{H}'_y(\hat{x}(t),\hat{y}(t),\hat{u}(t),\lambda_f(t),\lambda_g(t),\ell_0)^\top.\end{aligned}$$

(c) Transversality conditions:

$$\begin{aligned}\lambda_f(t_0)^\top &= -(\ell_0\varphi'_{x_0}(\hat{x}(t_0),\hat{x}(t_f)) + \sigma^\top\psi'_{x_0}(\hat{x}(t_0),\hat{x}(t_f)) + \zeta^\top g'_x(\hat{x}(t_0))),\\ \lambda_f(t_f)^\top &= \ell_0\varphi'_{x_f}(\hat{x}(t_0),\hat{x}(t_f)) + \sigma^\top\psi'_{x_f}(\hat{x}(t_0),\hat{x}(t_f)).\end{aligned}$$

(d) Optimality condition: *Almost everywhere in $\mathcal{I}$, it holds*

$$\mathcal{H}(\hat{x}(t),\hat{y}(t),\hat{u}(t),\lambda_f(t),\lambda_g(t),\ell_0) \le \mathcal{H}(\hat{x}(t),y,u,\lambda_f(t),\lambda_g(t),\ell_0)$$

for all $(u,y) \in \Omega(\hat{x}(t))$, where

$$\Omega(x) = \{(u,y) \in \mathcal{U}\times\mathbb{R}^{n_y} \mid g'_x(x)f(x,y,u) = 0_{\mathbb{R}^{n_y}}\}.$$

(e) *The Hamilton function is constant with respect to time:*

$$\mathcal{H}(\hat{x}(t),\hat{y}(t),\hat{u}(t),\lambda_f(t),\lambda_g(t),\ell_0) \equiv \mathit{const.}$$

□

Example 7.1.6. Consider the following optimal control problem:

Minimize

$$\int_0^1 x(t)^2 + \alpha\big(y(t) - u(t)\big)^2 dt$$

subject to the constraints

$$\begin{aligned}\dot{x}(t) &= y(t) - u(t), \quad x(0) = 0,\\ 0 &= x(t),\\ u(t) &\in \mathcal{U} := [-1,1].\end{aligned}$$

Apparently, *every* feasible control is optimal!

The Hamilton function:

$$\mathcal{H}(x,y,u,\lambda_f,\lambda_g,\ell_0) = \ell_0 x^2 + \alpha(y-u)^2 + \lambda_f(y-u) + \lambda_g(y-u).$$

Minimization of $\mathcal{H}$ with respect to

$$(u,y) \in \Omega = \{(u,y) \in \mathcal{U}\times\mathbb{R} \mid y - u = 0\}$$

yields that every u satisfies the global minimum principle.

Please note that the simultaneous coupling between u and y by means of the set Ω is important in the global minimum principle. A wrong condition would be obtained, if the Hamilton function was first minimized with respect to u (assuming y to be fixed), and the consistent algebraic variable y corresponding to the minimizing u was determined afterwards. For instance, consider the case $\alpha = -1$ and $\hat{u} \equiv 0$. The algebraic equation, the adjoint equation, and the transversality condition yield

$$\dot{\lambda}_f(t) = -\ell_0 \hat{x}(t) = 0, \quad \lambda_f(1) = 0 \quad \Longrightarrow \quad \lambda_f(t) \equiv 0.$$

Moreover, $g_x'(\hat{x}(t)) f(\hat{x}(t), \hat{y}(t), \hat{u}(t)) \equiv 0$ and hence

$$\mathcal{H}(\hat{x}, \hat{y}, \hat{u}, \lambda_f, \lambda_g, \ell_0) = \alpha(\hat{y} - \hat{u})^2 = -(\hat{y} - \hat{u})^2.$$

Minimizing $\mathcal{H}$ with respect to $u \in [-1, 1]$ yields either $\hat{u} = +1$ or $\hat{u} = -1$ depending on $\hat{y} \in [-1, 1]$. This contradicts $\hat{u} \equiv 0$. □

The global minimum principle allows proving additional properties of the Hamilton function in the case of a free final time. In this case, the Hamilton function vanishes almost everywhere.

Theorem 7.1.7. *Let the assumptions of Theorem 7.1.5 hold, and let the final time in Problem 7.1.1 be free. Then, $\mathcal{H}$ vanishes almost everywhere and*

$$\mathcal{H}(\hat{x}(t_f), \hat{y}(t_f), \hat{u}(t_f), \lambda_f(t_f), \lambda_g(t_f), \ell_0) = 0.$$

Proof. We use the transformation technique in Section 1.2 to transform the problem into an equivalent problem on a fixed time interval:

Minimize

$$\varphi(\bar{x}(0), \bar{x}(1)) + \int_0^1 (t_f(\tau) - t_0) f_0(\bar{x}(\tau), \bar{y}(\tau), \bar{u}(\tau)) d\tau$$

subject to the constraints

$$\begin{aligned}
\frac{d}{d\tau}\bar{x}(\tau) &= (t_f(\tau) - t_0) f(\bar{x}(\tau), \bar{y}(\tau), \bar{u}(\tau)) && \text{a. e. in } [0,1], \\
0_{\mathbb{R}^{n_y}} &= g(\bar{x}(\tau)) && \text{in } [0,1], \\
\frac{d}{d\tau} t_f(\tau) &= 0 && \text{in } [0,1], \\
\psi(\bar{x}(0), \bar{x}(1)) &= 0_{n_\psi}, && \\
\bar{u}(\tau) &\in \mathcal{U} && \text{a. e. in } [0,1].
\end{aligned}$$

The Hamilton function for the transformed problem reads as

$$\bar{\mathcal{H}}(\bar{x},\bar{y},t_f,\bar{u},\bar{\lambda}_f,\bar{\lambda}_g,\bar{\lambda}_t,\bar{\ell}_0) = (t_f - t_0)\mathcal{H}(\bar{x},\bar{y},\bar{u},\bar{\lambda}_f,\bar{\lambda}_g,\bar{\ell}_0),$$

where $\mathcal{H}$ denotes the Hamilton function of the original problem.

The necessary optimality conditions in Theorem 7.1.5 yield the adjoint equation for the adjoint variable $\bar{\lambda}_t$

$$\frac{d}{d\tau}\bar{\lambda}_t(\tau) = -\bar{\mathcal{H}}'_{t_f}[\tau] = -\mathcal{H}[\tau]$$

and the transversality conditions

$$\bar{\lambda}_t(0) = \bar{\lambda}_t(1) = 0.$$

According to part (e) of Theorem 7.1.5, the Hamilton function $\bar{\mathcal{H}}$ is constant almost everywhere, and thus $\mathcal{H}$ is constant almost everywhere as well, since t_f is constant.

From

$$\bar{\lambda}'_t(\tau)^\top = -\mathcal{H}[\tau] = \text{const a.e. in } [0,1],\ \bar{\lambda}_t(0) = \bar{\lambda}_t(1) = 0,$$

it follows $\mathcal{H}[\tau] = 0$ almost everywhere and particularly after back-transformation

$$\mathcal{H}(\hat{x}(t_f),\hat{y}(t_f),\hat{u}(t_f),\lambda_f(t_f),\lambda_g(t_f),\ell_0) = 0. \qquad \square$$

7.1.1 Singular controls

Singular controls may occur, if the control u and the algebraic variable y enter linearly into the optimal control problem. For simplicity, the discussion is restricted to problems with only one control and one algebraic variable.

Problem 7.1.8 (DAE optimal control problem with control appearing linearly). Let $\mathcal{I} := [t_0, t_f] \subset \mathbb{R}$ be a non-empty compact time interval with $t_0 < t_f$ fixed. Let

$$\begin{aligned}
\varphi &: \mathbb{R}^{n_x} \times \mathbb{R}^{n_x} \longrightarrow \mathbb{R},\\
\alpha_0, \beta_0, \gamma_0 &: \mathbb{R}^{n_x} \longrightarrow \mathbb{R},\\
\alpha, \beta, \gamma &: \mathbb{R}^{n_x} \longrightarrow \mathbb{R}^{n_x},\\
g &: \mathbb{R}^{n_x} \longrightarrow \mathbb{R},\\
\psi &: \mathbb{R}^{n_x} \times \mathbb{R}^{n_x} \longrightarrow \mathbb{R}^{n_\psi}
\end{aligned}$$

be sufficiently smooth functions and $\mathcal{U} := [u_{\min}, u_{\max}]$ with $-\infty < u_{\min} < u_{\max} < \infty$.

Minimize

$$\varphi(x(t_0), x(t_f)) + \int_{t_0}^{t_f} \alpha_0(x(t)) + \beta_0(x(t))y(t) + \gamma_0(x(t))u(t)dt$$

with respect to $x \in W_{1,\infty}^{n_x}(\mathcal{I})$, $y \in L_\infty(\mathcal{I})$, $u \in L_\infty(\mathcal{I})$ *subject to the constraints*

$$\begin{aligned} \dot{x}(t) &= \alpha(x(t)) + \beta(x(t))y(t) + \gamma(x(t))u(t) && \text{a. e. in } \mathcal{I}, \\ 0_{\mathbb{R}^{n_y}} &= g(x(t)) && \text{in } \mathcal{I}, \\ 0_{\mathbb{R}^{n_\psi}} &= \psi(x(t_0), x(t_f)), \\ u(t) &\in \mathcal{U} = [u_{\min}, u_{\max}] && \text{a. e. in } \mathcal{I}. \end{aligned}$$

Let $(\hat{x}, \hat{y}, \hat{u})$ be a minimum of Problem 7.1.8.

Assumption 7.1.9 (Index-two assumption). Let a constant $c > 0$ exist with

$$|g_x'(\hat{x}(t))\beta(\hat{x}(t))| \geq c \quad \text{in } \mathcal{I}.$$ □

The Hamilton function for Problem 7.1.8 reads as

$$\begin{aligned} \mathcal{H}(x, y, u, \lambda_f, \lambda_g, \ell_0) = {} & \ell_0(\alpha_0(x) + \beta_0(x)y + \gamma_0(x)u) \\ & + (\lambda_f^\top + \lambda_g^\top g_x'(x))(\alpha(x) + \beta(x)y + \gamma(x)u). \end{aligned}$$

Evaluation of the optimality condition (d) in the global minimum principle in Theorem 7.1.5 yields

$$(\hat{u}(t), \hat{y}(t)) = \arg \min_{(u,y) \in \Omega(\hat{x}(t))} \mathcal{H}(\hat{x}(t), y, u, \lambda_f(t), \lambda_g(t), \ell_0)$$

with

$$\Omega(x) = \{(u, y) \in [u_{\min}, u_{\max}] \times \mathbb{R} \mid g_x'(x)(\alpha(x) + \beta(x)y + \gamma(x)u) = 0\}.$$

By Assumption 7.1.9, it holds $(\hat{u}(t), \hat{y}(t)) \in \Omega(\hat{x}(t))$ if and only if

$$\hat{y}(t) = -(g_x'(\hat{x}(t))\beta(\hat{x}(t)))^{-1} g_x'(\hat{x}(t))(\alpha(\hat{x}(t)) + \gamma(\hat{x}(t))\hat{u}(t))$$

and hence the optimality condition reduces to

$$\hat{u}(t) = \arg \min_{u \in [u_{\min}, u_{\max}]} \Gamma(\hat{x}(t), \lambda_f(t), \ell_0)u, \tag{7.3}$$

where Γ denotes the switching function, which is defined in

Definition 7.1.10 (Switching function). The function $\Gamma : \mathbb{R}^{n_x} \times \mathbb{R}^{n_x} \times \mathbb{R} \longrightarrow \mathbb{R}$ with

$$\Gamma(x, \lambda_f, \ell_0) := \ell_0(\gamma_0(x) - \beta_0(x)(g_x'(x)\beta(x))^{-1} g_x'(x)\gamma(x))$$

$$+\lambda_f^\top\left(\gamma(x)-\beta(x)\left(g_x'(x)\beta(x)\right)^{-1}g_x'(x)\gamma(x)\right)$$

is called *switching function* of Problem 7.1.8. □

Since u appears linearly in (7.3), it follows

$$\hat{u}(t)=\begin{cases}u_{\min}, & \text{if } \Gamma(\hat{x}(t),\lambda_f(t),\ell_0)>0,\\ u_{\max}, & \text{if } \Gamma(\hat{x}(t),\lambda_f(t),\ell_0)<0,\\ \text{undefined}, & \text{if } \Gamma(\hat{x}(t),\lambda_f(t),\ell_0)\equiv 0 \text{ on some interval.}\end{cases} \tag{7.4}$$

If the switching function Γ vanishes on some interval $[t_1,t_2]\subseteq\mathcal{I}$, $t_1<t_2$, then the minimum principle does not explicitly define the optimal control.

Definition 7.1.11 (Bang-bang control, singular control). Let $[t_1,t_2]\subseteq\mathcal{I}$ with $t_1<t_2$.

(a) u is called *bang-bang control* in $[t_1,t_2]$, if Γ has only isolated zeros in $[t_1,t_2]$. The isolated zeros of Γ are called *switching points*.

(b) u is called *singular* in $[t_1,t_2]$, if $\Gamma\equiv 0$ in $[t_1,t_2]$. The time points t_1,t_2 are called *junction points*, if u in $[t_1-\varepsilon,t_1]$ and in $[t_2,t_2+\varepsilon]$ for some sufficiently small $\varepsilon>0$ is a bang-bang control. □

Figure 7.3 illustrates the terms bang-bang control and singular control.

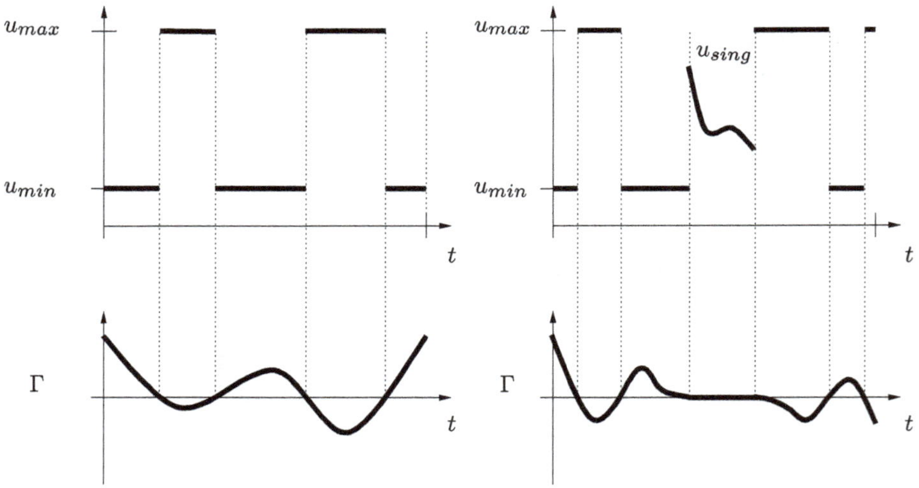

Figure 7.3: Bang-bang control for isolated zeros of the switching function (left) and bang-bang control with singular subarc (right).

If Γ has only isolated zeros t_i, $i=1,\dots,q$, then the optimal control $\hat{u}$ is a bang-bang control, which is defined by (7.4). The switching points t_i, $i=1,\dots,q$, are implicitly defined by the conditions

$$\Gamma(\hat{x}(t_i), \lambda_f(t_i), \ell_0) = 0, \quad i = 1, \dots, q.$$

Computing a singular control is more involved. To this end, let $[t_1, t_2] \subseteq \mathcal{I}$ be an interval with $\Gamma \equiv 0$. The idea to find the singular control on $[t_1, t_2]$ is closely related to the determination of the differentiation index for DAEs. In fact, the original DAE on a singular subarc can be interpreted again as a DAE (on $[t_1, t_2]$) with the additional algebraic constraint $\Gamma \equiv 0$. To find the singular control, the identity $\Gamma \equiv 0$ is being differentiated in $[t_1, t_2]$ with respect to t until the control u appears for the first time. Herein, the DAE for x and y and the adjoint DAE are exploited. In particular, we use the relations

$$\begin{aligned} y &= -(g_x'(x)\beta(x))^{-1} g_x'(x)(\alpha(x) + \gamma(x)u) =: Y(x,u), \\ \lambda_g^\top &= -(\ell_0 \beta_0(x) + \lambda_f^\top \beta(x))(g_x'(x)\beta(x))^{-1} =: \Lambda_g(x, \lambda_f, \ell_0), \end{aligned}$$

which result from the algebraic constraints of the respective DAEs. Notice that Y depends linearly on u.

Let the functions $\Gamma^{(k)}(x, \lambda_f, u, \ell_0)$, $0 \le k \le \infty$, be defined recursively by

$$\Gamma^{(0)}(x, \lambda_f, u, \ell_0) := \Gamma(x, \lambda_f, \ell_0)$$

and for $k = 0, 1, 2, \dots$ by

$$\begin{aligned} &\Gamma^{(k+1)}(x, \lambda_f, u, \ell_0) \\ &\quad := \nabla_x \Gamma^{(k)}(x, \lambda_f, u, \ell_0)^\top (\alpha(x) + \beta(x) Y(x,u) + \gamma(x)u) \\ &\quad + \nabla_\lambda \Gamma^{(k)}(x, \lambda_f, u, \ell_0)^\top (-\mathcal{H}_x'(x, Y(x,u), u, \lambda_f, \Lambda_g(x, \lambda_f, \ell_0), \ell_0)^\top). \end{aligned}$$

$\Gamma^{(k)}$, $k = 0, 1, 2, \dots$, can be expressed inductively as

$$\Gamma^{(k)}(x, \lambda_f, u, \ell_0) =: A_k(x, \lambda_f, \ell_0) + B_k(x, \lambda_f, \ell_0)u$$

with suitable matrices A_k and B_k, since u always appears linearly.

Until u appears explicitly for the first time in $\Gamma^{(k)}$ and if $\Gamma^{(k)}$ is evaluated at a solution of the DAE and the adjoint DAE, then $\Gamma^{(k)}$ is just the kth derivative of Γ with respect to t and it holds

$$\Gamma^{(k)}(\hat{x}(t), \lambda_f(t), \hat{u}(t), \ell_0) = 0 \quad \text{for all } t \in [t_1, t_2], \ k = 0, 1, 2, \dots.$$

Two cases may occur:

(a) There exists $k < \infty$ with

$$\frac{\partial}{\partial u} \Gamma^{(j)}(x, \lambda_f, u, \ell_0) = B_j(x, \lambda_f, \ell_0) = 0 \quad \text{for all } 0 \le j \le k-1, \tag{7.5}$$

and

$$\frac{\partial}{\partial u}\Gamma^{(k)}(x,\lambda_f,u,\ell_0) = B_k(x,\lambda_f,\ell_0) \neq 0. \tag{7.6}$$

In this case, the equation

$$\Gamma^{(k)}(\hat{x}(t),\lambda_f(t),\hat{u}(t),\ell_0) = 0 \quad \text{for } t \in [t_1,t_2],$$

can be solved for $\hat{u}$ and the singular optimal control is given by

$$\hat{u}(t) = \hat{u}_{\text{sing}}(t) = -\frac{A_k(\hat{x}(t),\lambda_f(t),\ell_0)}{B_k(\hat{x}(t),\lambda_f(t),\ell_0)}, \quad t \in [t_1,t_2]. \tag{7.7}$$

(b) For all $0 \leq k \leq \infty$, it holds

$$\frac{\partial}{\partial u}\Gamma^{(k)}(x,\lambda_f,u,\ell_0) = B_k(x,\lambda_f,\ell_0) = 0.$$

In this case, $\hat{u}$ cannot be determined.

This proves

Corollary 7.1.12. *Let there exist $k < \infty$ with (7.5) and (7.6). Then the singular control $[t_1,t_2]$ is given by (7.7).* □

Unfortunately, there is no rule that indicates whether a singular subarc occurs or not. Indirect methods therefore rely on hypotheses regarding existence and location of singular subarcs.

Example 7.1.13. Consider the following DAE optimal control problem on $\mathcal{I} := [0,3]$:

Minimize

$$\frac{1}{2}\int_0^3 x_2(t)^2\,dt$$

subject to the constraints

$$\begin{aligned} \dot{x}_1(t) &= u(t), \quad x_1(0) = x_1(3) = 1,\\ \dot{x}_2(t) &= y(t), \quad x_2(0) = 1,\\ 0 &= x_1(t) - x_2(t),\\ u(t) &\in [-1,1]. \end{aligned}$$

Differentiation of the algebraic constraint yields $0 = u(t) - y(t)$, and hence the DAE has index two. The Hamilton function with $x := (x_1,x_2)^\top$ and $\lambda_f := (\lambda_{f,1},\lambda_{f,2})^\top$ is given by

$$\mathcal{H}(t,x,y,u,\lambda_f,\lambda_g,\ell_0) = \frac{\ell_0}{2}x_2^2 + \lambda_{f,1}u + \lambda_{f,2}y + \lambda_g(u-y).$$

The adjoint DAE reads as

$$\dot{\lambda}_{f,1}(t) = 0,$$
$$\dot{\lambda}_{f,2}(t) = -\ell_0 \hat{x}_2(t), \quad \lambda_{f,2}(3) = 0,$$

and hence $\lambda_{f,1}$ is a constant. The switching function computes to

$$\Gamma(\lambda_f) = \lambda_{f,1} + \lambda_{f,2}$$

and the optimality condition yields

$$\hat{u}(t) = \begin{cases} -1, & \text{if } \Gamma(\lambda_f(t)) > 0, \\ 1, & \text{if } \Gamma(\lambda_f(t)) < 0, \\ \text{undefined}, & \text{if } \Gamma(\lambda_f(t)) \equiv 0 \text{ in some interval.} \end{cases}$$

The case $\ell_0 = 0$ can be excluded by contradiction. Hence, we may set $\ell_0 = 1$.

The objective function suggests that it is optimal to control x_2 (respectively x_1) as quickly as possible to zero and to stay there as long as possible. Hence, it is conjectured that the optimal switching structure is *bang–singular–bang*.

On the singular subarc, it holds

$$\begin{aligned} & 0 = \lambda_{f,1}(t) + \lambda_{f,2}(t) \\ \Longrightarrow \quad & 0 = \dot{\lambda}_{f,2}(t) = -\hat{x}_2(t) \\ \Longrightarrow \quad & 0 = \ddot{\lambda}_{f,2}(t) = -\hat{y}(t) = -\hat{u}(t). \end{aligned}$$

Hence, $u_{\text{sing}} = 0$.

Owing to the symmetry of the problem, we then find that the following functions satisfy the global minimum principle:

$$\hat{u}(t) = \hat{y}(t) = \begin{cases} -1, & \text{if } t \in [0,1), \\ 0, & \text{if } t \in [1,2), \\ 1, & \text{if } t \in [2,3], \end{cases} \qquad \hat{x}_1(t) = \hat{x}_2(t) = \begin{cases} 1-t, & \text{if } t \in [0,1), \\ 0, & \text{if } t \in [1,2), \\ t-2, & \text{if } t \in [2,3], \end{cases}$$

and

$$\lambda_{f,1}(t) = -\frac{1}{2}, \quad \lambda_{f,2}(t) = \begin{cases} \frac{1}{2} + \frac{1}{2}(1-t)^2, & \text{if } t \in [0,1), \\ \frac{1}{2}, & \text{if } t \in [1,2), \\ \frac{1}{2} - \frac{1}{2}(t-2)^2, & \text{if } t \in [2,3]. \end{cases}$$

More generally: For $\mathcal{I} = [0, t_f]$ with $t_f > 2$, a singular subarc occurs in the interval $[1, t_f - 1]$ with singular control $\hat{u}(t) = u_{\text{sing}}(t) = 0$. The control structure is *lower bound–singular–upper bound*. For $t_f \leq 2$, no singular subarc appears. The control structure is *lower bound–upper bound*. □

Example 7.1.14 (Simplified robot model). A simplified robot model in three space dimensions is given by the task to control the acceleration of the joint angles by the control $u_i(t) \in [-u_{\max}, u_{\max}]$, $i = 1, 2, 3$, $u_{\max} > 0$, such that the robot moves in minimum time t_f from a given initial position $q_0 \in \mathbb{R}^3$ with velocity $v(0) = 0_{\mathbb{R}^3}$ to a given final position $q_f \in \mathbb{R}^3$ with velocity $v(t_f) = 0_{\mathbb{R}^3}$. This leads to the following optimal control problem:

Minimize

$$t_f = \int_0^{t_f} 1dt$$

subject to the constraints

$$\begin{aligned}
\dot{q}(t) &= v(t), \quad q(0) = q_0, \; q(t_f) = q_f, \\
\dot{v}(t) &= u(t), \quad v(0) = v(t_f) = 0_{\mathbb{R}^3}, \\
u_i(t) &\in [-u_{\max}, u_{\max}], \quad i = 1, 2, 3.
\end{aligned}$$

This problem admits an analytical solution. To this end, the Hamilton function (with $\ell_0 = 1$) reads as

$$\mathcal{H}(q, v, u, \lambda_q, \lambda_v) = 1 + \lambda_q^\top v + \lambda_v^\top u.$$

The first-order necessary optimality conditions for a minimum $(\hat{q}, \hat{v}, \hat{u})$ are given by the minimum principle, that is, there exist λ_q, λ_v with

$$\begin{aligned}
\dot{\lambda}_q(t) &= -H_q'[t]^\top = 0_{\mathbb{R}^3}, \\
\dot{\lambda}_v(t) &= -H_v'[t]^\top = -\lambda_q(t), \\
\hat{u}_i(t) &= \arg \min_{u_i \in [-u_{\max}, u_{\max}]} \mathcal{H}(\hat{q}(t), \hat{v}(t), u, \lambda_q(t), \lambda_v(t)) \\
&= \arg \min_{u_i \in [-u_{\max}, u_{\max}]} \lambda_{v,i}(t) u_i, \quad i \in \{1, 2, 3\}, \\
0 \equiv H[t] &= 1 + \lambda_q(t)^\top \hat{v}(t) + \lambda_v(t)^\top \hat{u}(t).
\end{aligned}$$

The adjoint equations yield

$$\lambda_q(t) = c_q \in \mathbb{R}^3, \quad \lambda_v(t) = -c_q t + c_v, \quad c_v \in \mathbb{R}^3.$$

Hence, the optimal control is given by

$$\hat{u}_i(t) = \begin{cases} u_{\max}, & \text{if } -c_{q,i}t + c_{v,i} < 0, \\ -u_{\max}, & \text{if } -c_{q,i}t + c_{v,i} > 0, \\ \text{undefined}, & \text{if } -c_{q,i}t + c_{v,i} \equiv 0, \end{cases} \quad i = 1, 2, 3.$$

The latter singular case for $i \in \{1,2,3\}$ with $-c_{q,i}t + c_{v,i} \equiv 0$ can only occur, if $c_{q,i} = c_{v,i} = 0$ and thus $\lambda_{q,i} \equiv 0 \equiv \lambda_{v,i}$. The minimum principle does not provide any information about $\hat{u}_i$ in this case, except that it has to be feasible.

If $c_{v,i}^2 + c_{q,i}^2 > 0$, then there will be at most one switching point $0 \le t_{s,i} \le t_f$ of the control $\hat{u}_i$. The switching point is determined by

$$t_{s,i} = \frac{c_{v,i}}{c_{q,i}} \quad \text{if } c_{q,i} \neq 0.$$

Let us assume that the control $\hat{u}_i$ does not switch, i. e. either $\hat{u}_i = u_{\max}$ or $\hat{u}_i = -u_{\max}$. Then the corresponding velocity satisfies $\hat{v}_i(t) = \pm u_{\max} t$ and, in particular, $\hat{v}_i(t_f) = \pm u_{\max} t_f \neq 0$ provided $t_f > 0$. Hence, the boundary conditions cannot be satisfied using $\hat{u}_i = \pm u_{\max}$, and this control structure does not occur.

Notice furthermore that not all $c_{q,i}, c_{v,i}$, $i = 1,2,3$, vanish, because this would lead to the contradiction $0 \equiv H[t] = 1$. Hence, there is at least one index i for which $c_{q,i}$ and $c_{v,i}$ do not vanish simultaneously. For this index, the optimal control will be of bang-bang type.

Case I. Let $i \in \{1,2,3\}$ and

$$\hat{u}_i(t) = \begin{cases} u_{\max}, & \text{if } 0 \le t < t_{s,i}, \\ -u_{\max}, & \text{otherwise.} \end{cases}$$

Then,

$$\hat{v}_i(t) = \begin{cases} u_{\max} t, & \text{if } 0 \le t < t_{s,i}, \\ u_{\max}(2t_{s,i} - t), & \text{otherwise,} \end{cases}$$

$$\hat{q}_i(t) = \begin{cases} q_{0,i} + \frac{1}{2} u_{\max} t^2, & \text{if } 0 \le t < t_{s,i}, \\ q_{0,i} + u_{\max}(t_{s,i}^2 - \frac{1}{2}(2t_{s,i} - t)^2), & \text{otherwise.} \end{cases}$$

The boundary conditions yield

$$u_{\max}(2t_{s,i} - t_{f,i}) = 0, \quad q_{0,i} + u_{\max}\left(t_{s,i}^2 - \frac{1}{2}(2t_{s,i} - t_{f,i})^2 \right) = q_{f,i},$$

where $t_{f,i}$ is the individual final time for the ith component that is necessary to reach the final position $q_i(t_{f,i}) = q_{f,i}$ and $v_i(t_{f,i}) = 0$.

The first equation yields $2t_{s,i} - t_{f,i} = 0$, and thus the second equation yields

$$t_{s,i} = \sqrt{\frac{q_{f,i} - q_{0,i}}{u_{\max}}}$$

and hence

$$t_{f,i} = 2\sqrt{\frac{q_{f,i} - q_{0,i}}{u_{\max}}}.$$

These expressions are valid for $q_{f,i} - q_{0,i} \geq 0$.

Note that $t_{f,i}$ is not necessarily the final time for all three components $i = 1, 2, 3$. The overall final time is $t_f = \max\{t_{f,1}, t_{f,2}, t_{f,3}\}$. If the ith component needs less time to reach the terminal condition, then one can set $\hat{u}_i(t) = 0$ for $t_{f,i} \leq t \leq t_f$. This is possible because the individual motions are decoupled in this model. The corresponding adjoints $\lambda_{q,i}$ and $\lambda_{v,i}$ are set to zero.

Case II. Let $i \in \{1, 2, 3\}$ and

$$\hat{u}_i(t) = \begin{cases} -u_{\max}, & \text{if } 0 \leq t < t_{s,i}, \\ u_{\max}, & \text{otherwise.} \end{cases}$$

Then

$$v_i(t) = \begin{cases} -u_{\max} t, & \text{if } 0 \leq t < t_{s,i}, \\ -u_{\max}(2t_{s,i} - t), & \text{otherwise,} \end{cases}$$

$$q_i(t) = \begin{cases} q_{0,i} - \frac{1}{2} u_{\max} t^2, & \text{if } 0 \leq t < t_{s,i}, \\ q_{0,i} - u_{\max}(t_{s,i}^2 - \frac{1}{2}(2t_{s,i} - t)^2), & \text{otherwise.} \end{cases}$$

The boundary conditions yield

$$-u_{\max}(2t_{s,i} - t_{f,i}) = 0, \quad q_{0,i} - u_{\max}\left(t_{s,i}^2 - \frac{1}{2}(2t_{s,i} - t_{f,i})^2 \right) = q_{f,i}.$$

The first equation yields $2t_{s,i} - t_{f,i} = 0$, and thus the second equation yields

$$t_{s,i} = \sqrt{\frac{q_{0,i} - q_{f,i}}{u_{\max,i}}}$$

and hence

$$t_{f,i} = 2\sqrt{\frac{q_{0,i} - q_{f,i}}{u_{\max,i}}}.$$

These expressions are valid for $q_{0,i} - q_{f,i} \geq 0$.

Summarizing, an optimal solution can be obtained as follows:

(a) For $i = 1, 2, 3$ set

$$t_{s,i} = \begin{cases} \sqrt{\frac{q_{0,i} - q_{f,i}}{u_{\max}}}, & \text{if } q_{f,i} < q_{0,i}, \\ \sqrt{\frac{q_{f,i} - q_{0,i}}{u_{\max}}}, & \text{otherwise.} \end{cases}$$

Set $t_{f,i} = 2t_{s,i}$ and $t_f = \max\{t_{f,1}, t_{f,2}, t_{f,3}\}$.

(b) For $i = 1, 2, 3$ set

$$\hat{u}_i(t) = \begin{cases} -u_{\max}, & \text{if } q_{f,i} < q_{0,i}, 0 \le t \le t_{s,i}, \\ u_{\max}, & \text{if } q_{f,i} < q_{0,i}, t_{s,i} < t \le t_{f,i}, \\ u_{\max}, & \text{if } q_{f,i} \ge q_{0,i}, 0 \le t \le t_{s,i}, \\ -u_{\max}, & \text{if } q_{f,i} \ge q_{0,i}, t_{s,i} < t \le t_{f,i}, \\ 0, & \text{if } t_{f,i} < t \le t_f. \end{cases}$$

(c) Integrate $\dot{q}(t) = v(t)$ and $\dot{v}(t) = \hat{u}(t)$.

Note that this solution is not necessarily unique. □

7.2 Variable time transformation method

The time transformation (7.1) was used in Section 7.1 as a theoretical tool to prove the global minimum principle. Interestingly, the same variable time transformation can be used numerically to solve mixed-integer optimal control problems, see [203, 204, 317, 304, 122], time optimal control problems, see [203], and singular optimal control problems, see [305]. A method for solving nonlinear mixed-integer programming problems based on a suitable formulation of an equivalent optimal control problem was introduced in [202].

Next, we distinguish between continuous-valued controls u with values in the closed convex set $\mathcal{U} \subseteq \mathbb{R}^{n_u}$ with $\text{int}(\mathcal{U}) \neq \emptyset$ and discrete controls v with values in the discrete finite set

$$\mathcal{V} := \{v_1, \dots, v_M\}, \quad v_i \in \mathbb{R}^{n_v}, \; M \in \mathbb{N}. \tag{7.8}$$

We investigate

Problem 7.2.1 (Mixed-integer optimal control problem (MIOCP)). Let $\mathcal{I} := [t_0, t_f] \subset \mathbb{R}$ be a non-empty compact time interval with $t_0 < t_f$ fixed. Let

$$\begin{aligned} \varphi &: \mathbb{R}^{n_x} \times \mathbb{R}^{n_x} \longrightarrow \mathbb{R}, \\ f &: \mathbb{R}^{n_x} \times \mathbb{R}^{n_y} \times \mathbb{R}^{n_u} \times \mathbb{R}^{n_v} \longrightarrow \mathbb{R}^{n_x}, \\ g &: \mathbb{R}^{n_x} \times \mathbb{R}^{n_y} \times \mathbb{R}^{n_u} \times \mathbb{R}^{n_v} \longrightarrow \mathbb{R}^{n_y}, \\ s &: \mathbb{R}^{n_x} \longrightarrow \mathbb{R}^{n_s}, \\ \psi &: \mathbb{R}^{n_x} \times \mathbb{R}^{n_x} \longrightarrow \mathbb{R}^{n_\psi} \end{aligned}$$

be sufficiently smooth functions, $\mathcal{U} \subseteq \mathbb{R}^{n_u}$ closed and convex with non-empty interior, and $\mathcal{V}$ as in (7.8).

Minimize the objective function

$$\varphi(x(t_0), x(t_f))$$

with respect to

$$x \in W_{1,\infty}^{n_x}(\mathcal{I}), \quad y \in L_{\infty}^{n_y}(\mathcal{I}), \quad u \in L_{\infty}^{n_u}(\mathcal{I}), \quad v \in L_{\infty}^{n_v}(\mathcal{I})$$

subject to the semi-explicit DAE

$$\begin{aligned} \dot{x}(t) &= f(x(t), y(t), u(t), v(t)) \quad \text{a. e. in } \mathcal{I}, \\ 0_{\mathbb{R}^{n_y}} &= g(x(t), y(t), u(t), v(t)) \quad \text{a. e. in } \mathcal{I}, \end{aligned}$$

the state constraint

$$s(x(t)) \leq 0_{\mathbb{R}^{n_s}},$$

the boundary condition

$$\psi(x(t_0), x(t_f)) = 0_{\mathbb{R}^{n_\psi}},$$

and the set constraints

$$\begin{aligned} u(t) &\in \mathcal{U} \quad \text{a. e. in } \mathcal{I}, \\ v(t) &\in \mathcal{V} \quad \text{a. e. in } \mathcal{I}. \end{aligned}$$

The semi-explicit DAE has to been seen as a generic DAE model for which additional structural properties have to be imposed whenever necessary, e. g., uniform non-singularity of g_y' (index-one case) or uniform non-singularity of $g_x' f_y'$ while y, u, and v do not appear in g (index-two case). As we intend to present the basic idea of the time transformation method and not the DAE specific theory, we prefer to use a generic DAE model.

(Branch&Bound) Straightforward application of a direct discretization approach of Section 5 to Problem 7.2.1 leads to a finite dimensional mixed-integer nonlinear and non-convex optimization problem, which can be solved by a Branch&Bound method, see [120]. In each node of the Branch&Bound tree, a relaxed discretized optimal control problem has to be solved to global optimality. Numerical tests, however, show that the Branch&Bound method is computationally extremely expensive for Problem 7.2.1.

The variable time transformation method is based on a discretization. For simplicity, only equally spaced grids are discussed. A generalization towards non-equidistant grids is straightforward. Let the *major grid*

$$\mathbb{G}_N := \{t_i = t_0 + ih \mid i = 0, \dots, N\}, \quad h = \frac{t_f - t_0}{N}$$

with $N \in \mathbb{N}$ intervals be given. Each major grid interval is subdivided into M equally spaced subintervals, where M denotes the number of values in the discrete control set $\mathcal{V}$ in (7.8). This leads to the *minor grid*

$$\mathbb{G}_{N,M} := \left\{\tau_{i,j} = t_i + j\frac{h}{M} \mid j = 0, \dots, M,\ i = 0, \dots, N-1\right\}.$$

Similarly as in Figure 7.2, we define the fixed and piecewise constant function

$$v_{\mathbb{G}_{N,M}}(\tau) := v_j \quad \text{for } \tau \in [\tau_{i,j-1}, \tau_{i,j}),\ i = 0, \ldots, N-1,\ j = 1, \ldots, M, \tag{7.9}$$

see Figure 7.4.

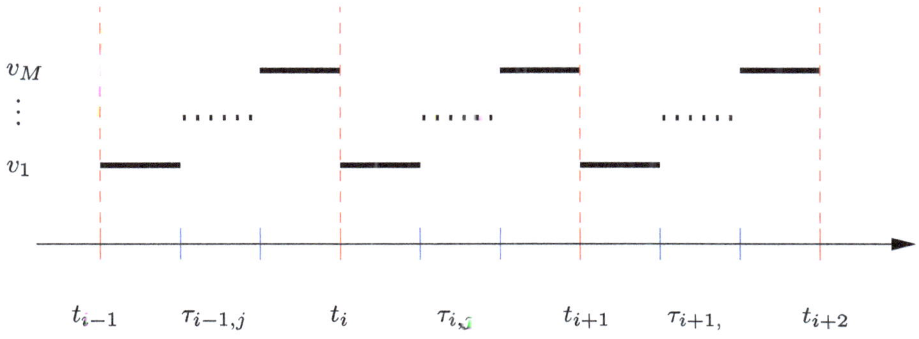

Figure 7.4: Piecewise constant staircase function $v_{\mathbb{G}_{N,M}}$ used for re-parameterization of the discrete controls.

Now, the time transformation

$$t(\tau) = t_0 + \int_{t_0}^{\tau} w(s)ds, \quad \tau \in [t_0, t_f],$$

with

$$\int_{t_0}^{t_f} w(s)ds = t_f - t_0$$

is invoked to control the length of the intervals $[t(\tau_{i,j}), t(\tau_{i,j+1})]$ by proper choice of w, compare (7.1). Herein, w is considered a control. Note that this transformation maps $[t_0, t_f]$ onto itself but changes the speed of running through this interval. In particular, it holds

$$\frac{dt}{d\tau}(\tau) = w(\tau) \quad \text{for } \tau \in [t_0, t_f].$$

It is convenient to consider monotone time transformations by imposing the constraint

$$w(\tau) \geq 0 \quad \text{for almost every } \tau \in [t_0, t_f]. \tag{7.10}$$

The function values $w(\tau)$ with $\tau \in [\tau_{i,j}, \tau_{i,j+1})$ are related to the length of the interval $[t(\tau_{i,j}), t(\tau_{i,j+1})]$ according to

$$\int_{\tau_{i,j}}^{\tau_{i,j+1}} w(\tau)d\tau = t(\tau_{i,j+1}) - t(\tau_{i,j}).$$

The interval $[t(\tau_{i,j}), t(\tau_{i,j+1})]$ shrinks to the point $\{t(\tau_{i,j})\}$, if

$$w(\tau) = 0 \quad \text{in } [\tau_{i,j}, \tau_{i,j+1}).$$

Figure 7.5 illustrates the variable time transformation.

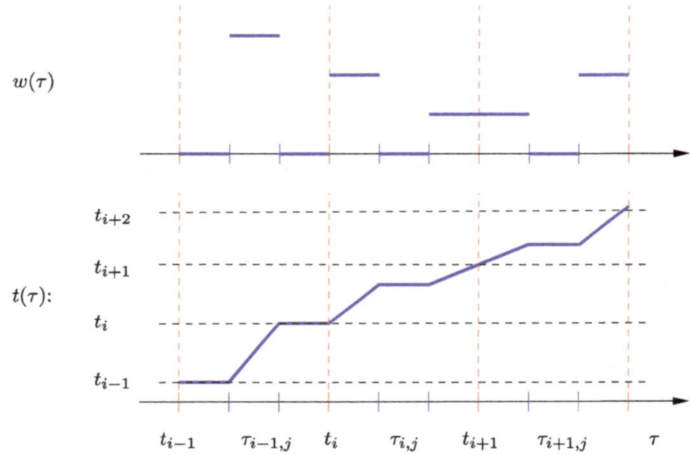

Figure 7.5: Illustration of variable time transformation: Parameterization w (top) and corresponding time $t = t(\tau)$ (bottom).

Later on, the state constraint $s(x(t)) \leq 0_{\mathbb{R}^{n_s}}$ will be discretized at the major grid points t_i, $i = 0, \ldots, N$, subject to the time transformation. If the time transformation was applied without further restrictions within an optimization procedure, then it can be observed that the optimal time transformation mainly 'optimizes' the position of the time points $t(t_i)$ (in original time t) such that state constraints $s(x(t(t_i))) \leq 0_{\mathbb{R}^{n_s}}$ can be satisfied conveniently. In extreme cases, almost all time points $t(t_i)$, $i = 0, 1, \ldots, N$, are moved into regions, where the state constraints are inactive. To avoid this undesired behavior, we impose the additional constraints

$$\int_{t_i}^{t_{i+1}} w(\tau)d\tau = t_{i+1} - t_i = h, \quad i = 0, \ldots, N-1, \tag{7.11}$$

which ensure that the major grid points t_i are fixed points subject to the time transformation. Moreover, the parameterized time actually passes through the entire major grid interval $[t_i, t_{i+1}]$ with a non-zero w.

Joining the function $v_{\mathbb{G}_{N,M}}$ from (7.9) and any w satisfying the conditions (7.10) and (7.11) yields a feasible discrete control $v(t) \in \mathcal{V}$ defined by

$$v(s) := v_{\mathbb{G}_{N,M}}(t^{-1}(s)), \quad s \in [t_0, t_f]$$

see Figure 7.6. Notice that minor intervals with $w(\tau) = 0$ do not contribute to $v(s)$.

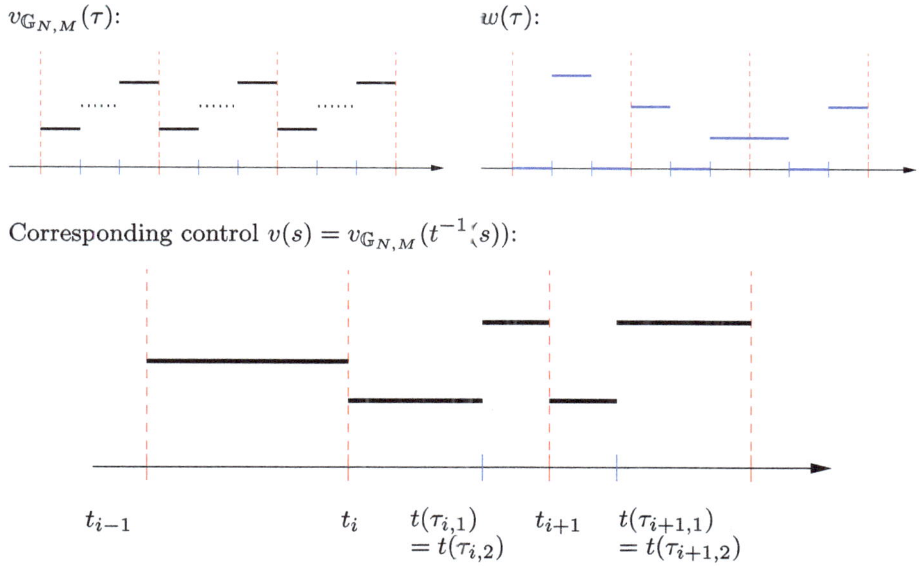

Figure 7.6: Back-transformation v (bottom) of variable time transformation for given w and fixed $v_{\mathbb{G}_{N,M}}$ (top).

Vice versa, every piecewise constant discrete control v on the major grid $\mathbb{G}_N$ can be described by $v_{\mathbb{G}_{N,M}}$ and some feasible w. Owing to the preference of values given by the definition of the fixed function $v_{\mathbb{G}_{N,M}}$ on the minor grid $\mathbb{G}_{N,M}$, it is not possible to realize any discrete control with values in $\mathcal{V}$ on the minor grid. For instance, a discrete control switching from v_M to v_1 within a major grid interval cannot be represented. Admittedly, the order in which $v_{\mathbb{G}_{N,M}}$ is defined is somehow arbitrary and any other order of the values v_1 to v_M would be feasible as well. However, one should keep in mind that already the major grid introduces a discretization with step-size $h > 0$. Moreover, any discrete control with finitely many jumps can be approximated arbitrarily close on the major grid, if h tends to zero. Hence, one can expect sufficiently good approximations for mixed-integer optimal control problems with finitely many jumps in v, if h is sufficiently small.

Summarizing, the time transformation leads to the following partly discretized optimal control problem:

Problem 7.2.2.
Minimize

$$\varphi\big(x(t_0), x(t_f)\big)$$

with respect to

$$x \in W_{1,\infty}^{n_x}(\mathcal{I}), \quad y \in L_\infty^{n_y}(\mathcal{I}), \quad u \in L_\infty^{n_u}(\mathcal{I}), \quad w \in L_\infty(\mathcal{I})$$

subject to the constraints

$$\begin{aligned}
\dot{x}(\tau) &= w(\tau) f\big(x(\tau), y(\tau), u(\tau), v_{\mathbb{G}_{N,M}}(\tau)\big) && \text{a. e. in } \mathcal{I},\\
0_{\mathbb{R}^{n_y}} &= g\big(x(\tau), y(\tau), u(\tau), v_{\mathbb{G}_{N,M}}(\tau)\big) && \text{a. e. in } \mathcal{I},\\
s\big(x(\tau)\big) &\le 0_{\mathbb{R}^{n_s}} && \text{in } \mathcal{I},\\
\psi\big(x(t_0), x(t_f)\big) &= 0_{\mathbb{R}^{n_\psi}},\\
u(\tau) &\in \mathcal{U} && \text{a. e. in } \mathcal{I},\\
w &\in \mathcal{W}.
\end{aligned}$$

Herein, the control set $\mathcal{W}$ is defined by

$$\mathcal{W} := \left\{ w \in L_\infty(\mathcal{I}) \;\middle|\; \begin{array}{l} w(\tau) \ge 0,\\ w \text{ piecewise constant on } \mathbb{G}_{N,M},\\ \displaystyle\int_{t_i}^{t_{i+1}} w(\tau)d\tau = t_{i+1} - t_i,\ i = 0, \dots, N \end{array} \right\}.$$

Problem 7.2.2 has only continuous-valued controls and can be solved by the direct discretization methods of Chapter 5 with minor adaptions. Let $(\hat{x}, \hat{y}, \hat{u}, \hat{w})$ be an optimal solution of Problem 7.2.2. The inverse time transformation

$$x(s) := \hat{x}(t^{-1}(s)), \quad y(s) := \hat{y}(t^{-1}(s)), \quad u(s) := \hat{u}(t^{-1}(s)), \quad v(s) := v_{G_{N,M}}(t^{-1}(s))$$

with

$$t(\tau) = t_0 + \int_{t_0}^{\tau} \hat{w}(s)ds$$

and $t^{-1}(s)$ according to (7.2) yields an approximate solution of Problem 7.2.1.

Notice that this approach is not limited to v being a scalar function. For instance, consider the case of $n_v = 2$ discrete controls each assuming values in $\{0, 1\}$. Then the control set $\mathcal{V}$ is given by all possible combinations of values:

$$\mathcal{V} = \left\{ \begin{pmatrix} 0\\0 \end{pmatrix}, \begin{pmatrix} 1\\0 \end{pmatrix}, \begin{pmatrix} 0\\1 \end{pmatrix}, \begin{pmatrix} 1\\1 \end{pmatrix} \right\}.$$

Example 7.2.3 (See [289]). Consider the following nonlinear optimal control problem with a discrete control v for an F8 aircraft:

Minimize t_f subject to the constraints

$$\begin{aligned}
\dot{x}_1(t) &= -0.877x_1(t) + x_3(t) - 0.088x_1(t)x_3(t) + 0.47x_1(t)^2 \\
&\quad - 0.019x_2(t)^2 - x_1(t)^2x_3(t) + 3.846x_1(t)^3 - 0.215v(t) \\
&\quad + 0.28x_1(t)^2v + 0.47x_1(t)v(t)^2 + 0.63v(t)^3, \\
\dot{x}_2(t) &= x_3(t), \\
\dot{x}_3(t) &= -4.208x_1(t) - 0.396x_3(t) - 0.47x_1(t)^2 - 3.564x_1(t)^3 \\
&\quad - 20.967v(t) + 6.265x_1(t)^2v(t) + 46x_1(t)v(t)^2 + 61.4v(t)^3, \\
v(t) &\in \{-0.05236, 0.05236\},
\end{aligned}$$

and

$$x(0) = (0.4655, 0, 0)^\top, \quad x(t_f) = (0, 0, 0)^\top.$$

Figure 7.7 shows the solution of the variable time transformation method with $N = 500$ and $t_f = 5.728674$. Notice that this solution is just a local minimum of the transformed optimal control Problem 7.2.2. Better solutions can be found by techniques from global optimization or by choosing more appropriate initial guesses. □

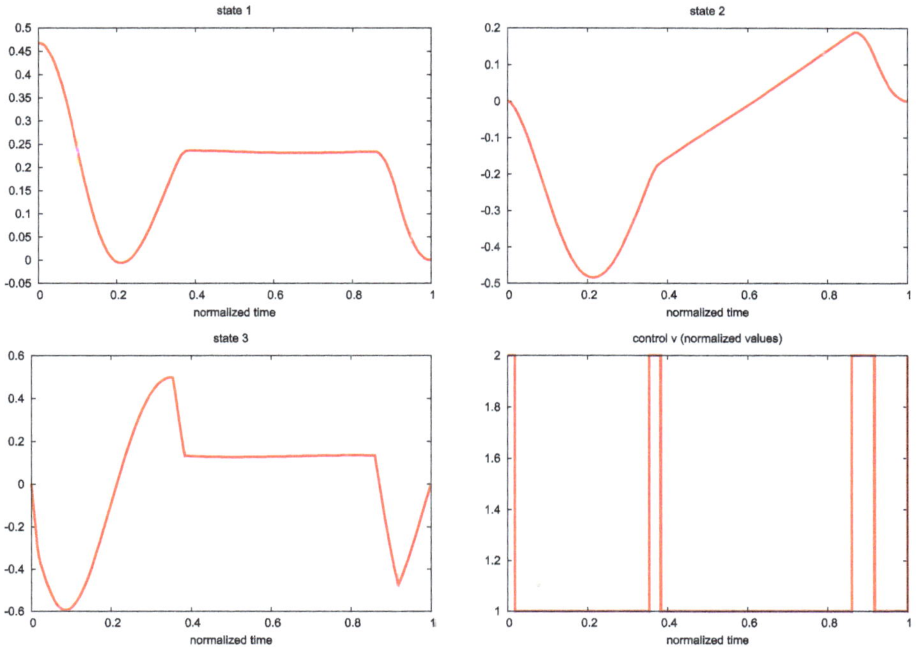

Figure 7.7: Numerical solution of the F8 aircraft problem.

We will consider a particular optimal control problem arising from automobile test-driving with gear shifts. Herein, one component of the control is discrete.

Example 7.2.4 (Test-drive of a car, see [120]). For modeling the test-drive of a car, we need three ingredients: a mathematical model of the car, a test-course, and a driver.

Model of a car

We use the single-track model, which is a simplified car model that neglects the rolling and pitching behavior of the car, see [245, 282, 254, 250]. These simplifying assumptions allow combining the wheels of each axle in one virtual wheel in the middle. In addition, the center of gravity is assumed to be on the ground level; hence, it suffices to consider the motion in the plane.

The following car model has four controls: The steering angle velocity $|w_\delta| \leq 0.5$ [rad/s], the braking force $0 \leq F_B \leq 15000$ [N], the gear $\mu \in \{1, 2, 3, 4, 5\}$, and the gas pedal position $\phi \in [0, 1]$. The configuration is depicted in Figure 7.8:

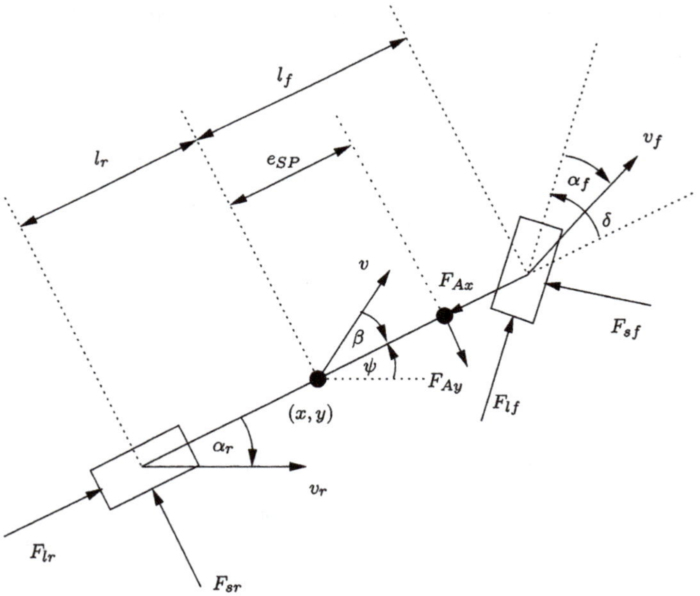

Figure 7.8: Geometric description of the single-track car model.

The following notion is used:

- (x, y): center of gravity of car.
- v, v_f, v_r: velocity of car, front wheel, and rear wheel, respectively.
- δ, β, ψ: steering angle, side slip angle, yaw angle.
- α_f, α_r: side slip angle at front and rear wheel, respectively.

- F_{sf}, F_{sr}: lateral tire forces at front and rear wheel, respectively.
- F_{lf}, F_{lr}: longitudinal tire forces at front and rear wheel, respectively.
- l_f, l_r, e_{SP}: measurements.
- F_{Ax}, F_{Ay}: air resistance forces in x- and y-direction, respectively.
- m: mass of car.

The equations of motion are given by the differential equations

$$\dot{x} = v\cos(\psi - \beta), \tag{7.12}$$

$$\dot{y} = v\sin(\psi - \beta), \tag{7.13}$$

$$\dot{v} = \frac{1}{m}\left[(F_{lr} - F_{Ax})\cos\beta + F_{lf}\cos(\delta + \beta) - (F_{sr} - F_{Ay})\sin\beta - F_{sf}\sin(\delta + \beta)\right], \tag{7.14}$$

$$\dot{\beta} = w_z - \frac{1}{m \cdot v}\left[(F_{lr} - F_{Ax})\sin\beta + F_{lf}\sin(\delta + \beta) + (F_{sr} - F_{Ay})\cos\beta + F_{sf}\cos(\delta + \beta)\right], \tag{7.15}$$

$$\dot{\psi} = w_z, \tag{7.16}$$

$$\dot{w}_z = \frac{1}{I_{zz}}\left[F_{sf} \cdot l_f \cdot \cos\delta - F_{sr} \cdot l_r - F_{Ay} \cdot e_{SP} - F_{lf} \cdot l_f \cdot \sin\delta\right], \tag{7.17}$$

$$\dot{\delta} = w_\delta. \tag{7.18}$$

The lateral tire forces depend on the side slip angles. A popular model is the *magic formula* by Pacejka [260]:

$$F_{sf}(\alpha_f) = D_f \sin(C_f \arctan(B_f\alpha_f - E_f(B_f\alpha_f - \arctan(B_f\alpha_f)))),$$
$$F_{sr}(\alpha_r) = D_r \sin(C_r \arctan(B_r\alpha_r - E_r(B_r\alpha_r - \arctan(B_r\alpha_r)))),$$

see Figure 7.9. $B_f, B_r, C_f, C_r, D_f, D_r, E_f, E_r$ are constants depending on the tire.

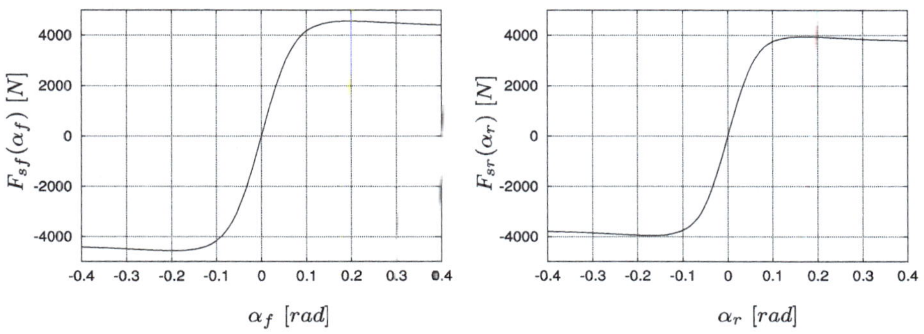

Figure 7.9: Lateral tire forces at front (left) and rear (right) wheel as functions of the side slip angle.

The side slip angles compute to

$$\alpha_f = \delta - \arctan\left(\frac{l_f\dot{\psi} - v\sin\beta}{v\cos\beta}\right), \quad \alpha_r = \arctan\left(\frac{l_r\dot{\psi} + v\sin\beta}{v\cos\beta}\right),$$

compare [249, p. 23]. The air drag is given by

$$F_{Ax} = \frac{1}{2}\cdot c_w \cdot \rho \cdot A \cdot v^2,$$

where c_w denotes the cw-value, ρ is the air density, and A is the effective surface. For simplicity, side wind is neglected, that is $F_{Ay} = 0$.

For a car with rear wheel drive, the longitudinal tire force at the front wheel is given by

$$F_{lf} = -F_{Bf} - F_{Rf},$$

where F_{Bf} denotes the braking force, and F_{Rf} is the rolling resistance at the front wheel. Accordingly, the longitudinal tire force at the rear wheel is given by

$$F_{lr} = \frac{M_{\text{wheel}}(\phi,\mu)}{R} - F_{Br} - F_{Rr},$$

where $M_{\text{wheel}}(\phi,\mu)$ is the motor torque at the rear wheel.

According to [254], it holds

$$M_{\text{wheel}}(\phi,\mu) = i_g(\mu)\cdot i_t \cdot M_{\text{mot}}(\phi,\mu), \tag{7.19}$$

where

$$M_{\text{mot}}(\phi,\mu) = f_1(\phi)\cdot f_2(w_{\text{mot}}(\mu)) + (1 - f_1(\phi))f_3(w_{\text{mot}}(\mu))$$

denotes the motor torque and

$$w_{\text{mot}}(\mu) = \frac{v\cdot i_g(\mu)\cdot i_t}{R} \tag{7.20}$$

denotes the rotary frequency of the motor depending on the gear μ. Notice that this relation is based on the assumption that the longitudinal slip can be neglected. The functions f_1, f_2, and f_3 are given by

$$\begin{aligned}
f_1(\phi) &= 1 - \exp(-3\phi),\\
f_2(w_{\text{mot}}) &= -37.8 + 1.54\cdot w_{\text{mot}} - 0.0019\cdot w_{\text{mot}}^2,\\
f_3(w_{\text{mot}}) &= -34.9 - 0.04775\cdot w_{\text{mot}}.
\end{aligned}$$

It is assumed that the braking force F_B, which is controlled by the driver, is distributed as follows:

$$F_{Bf} = \frac{2}{3}F_B, \quad F_{Br} = \frac{1}{3}F_B.$$

Finally, the rolling resistance forces are given by

$$F_{Rf} = f_R(v) \cdot F_{zf}, \quad F_{Rr} = f_R(v) \cdot F_{zr},$$

where

$$f_R(v) = f_{R0} + f_{R1}\frac{v}{100} + f_{R4}\left(\frac{v}{100}\right)^4 \quad (v \text{ in [km/h]}),$$

is the friction coefficient and

$$F_{zf} = \frac{m \cdot l_r \cdot g}{l_f + l_r}, \quad F_{zr} = \frac{m \cdot l_f \cdot g}{l_f + l_r}$$

denote the static tire loads at the front and rear wheel, respectively, see [282].

For the upcoming numerical computations, we used the parameters summarized in Table 7.1.

Table 7.1: Parameters for the single-track car model (partly taken from [282, 254]).

Parameter	Value	Description
m	1239 [kg]	mass of car
g	9.81 [m/s^2]	acceleration due to gravity
l_f/l_r	1.19016/1.37484 [m]	dist. center of gravity to front/rear wheel
e_{SP}	0.5 [m]	dist. center of gravity to drag mount point
R	0.302 [m]	wheel radius
I_{zz}	1752 [kg m^2]	moment of inertia
c_w	0.3	air drag coefficient
ρ	1.249512 [N/m^2]	air density
A	1.4378946874 [m^2]	effective flow surface
$i_g(1)$	3.91	first gear
$i_g(2)$	2.002	second gear
$i_g(3)$	1.33	third gear
$i_g(4)$	1.0	fourth gear
$i_g(5)$	0.805	fifth gear
i_t	3.91	motor torque transmission
B_f/B_r	10.96/12.67	Pacejka-model (stiffness factor)
$C_f = C_r$	1.3	Pacejka-model (shape factor)
D_f/D_r	4560.40/3947.81	Pacejka-model (peak value)
$E_f = E_r$	−0.5	Pacejka-model (curvature factor)
$f_{R0}/f_{R1}/f_{R4}$	0.009/0.002/0.0003	coefficients

Test-course

We model a standard maneuver in automotive industry – the double-lane-change maneuver. The driver has to complete the test-course in Figure 7.10.

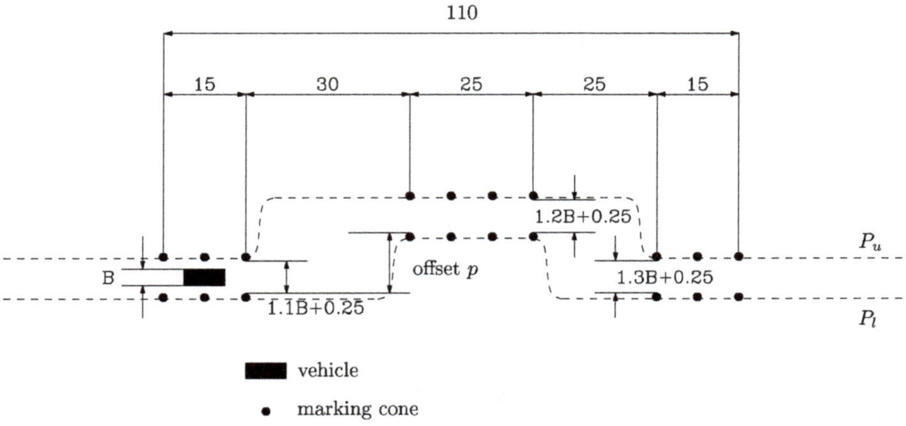

Figure 7.10: Measurements of the double-lane-change maneuver with boundaries P_l and P_u (dashed), compare [331].

The boundaries of the test-course impose state constraints $P_l(x)$ (lower boundary) and $P_u(x)$ (upper boundary) for a car of width $B = 1.5$ [m] by:

$$P_l(x) = \begin{cases} 0, & \text{if } x \le 44\,, \\ 4 \cdot h_2 \cdot (x-44)^3, & \text{if } 44 < x \le 44.5, \\ 4 \cdot h_2 \cdot (x-45)^3 + h_2, & \text{if } 44.5 < x \le 45, \\ h_2, & \text{if } 45 < x \le 70, \\ 4 \cdot h_2 \cdot (70-x)^3 + h_2, & \text{if } 70 < x \le 70.5, \\ 4 \cdot h_2 \cdot (71-x)^3, & \text{if } 70.5 < x \le 71, \\ 0, & \text{if } x > 71, \end{cases}$$

$$P_u(x) = \begin{cases} h_1, & \text{if } x \le 15, \\ 4 \cdot (h_3-h_1) \cdot (x-15)^3 + h_1, & \text{if } 15 < x \le 15.5, \\ 4 \cdot (h_3-h_1) \cdot (x-16)^3 + h_3, & \text{if } 15.5 < x \le 16, \\ h_3, & \text{if } 16 < x \le 94, \\ 4 \cdot (h_3-h_4) \cdot (94-x)^3 + h_3, & \text{if } 94 < x \le 94.5, \\ 4 \cdot (h_3-h_4) \cdot (95-x)^3 + h_4, & \text{if } 94.5 < x \le 95, \\ h_4, & \text{if } x > 95, \end{cases}$$

with $h_1 = 1.1 \cdot B + 0.25$, $h_2 = 3.5$, $h_3 = 1.2 \cdot B + 3.75$, $h_4 = 1.3 \cdot B + 0.25$.

Model of the driver

The driver is modeled by formulating a suitable optimal control problem with free final time t_f. The boundaries of the test-course are obeyed by the state constraints

$$y(t) \le P_u(x(t)) - B/2, \quad y(t) \ge P_l(x(t)) + B/2. \tag{7.21}$$

Let the initial position of the car at $t = 0$ be

$$(x(0), y(0), v(0), \beta(0), \psi(0), w_z(0), \delta(0)) = (-30, \mathit{free}, 10, 0, 0, 0, 0). \tag{7.22}$$

To force the driver to complete the test-course, the terminal conditions

$$x(t_f) = 140, \quad \psi(t_f) = 0 \tag{7.23}$$

are imposed. The latter guarantees that the longitudinal axis of the car at final time is in parallel with the track. This constraint ensures that the car can continue its drive after time t_f without leaving the track immediately.

Finally, the (optimal) driver is modeled by minimizing a linear combination of steering effort and final time. This driver is a compromise between a race driver, who aims at minimizing time, and a secure driver, who aims at minimizing steering effort.

Summarizing, the double-lane-change maneuver is modeled by the following optimal control problem:

Minimize

$$t_f + \int_0^{t_f} w_\delta(t)^2 dt$$

subject to the differential equations (7.12)–(7.18), the boundary conditions (7.22) and (7.23), the state constraints (7.21), and the control constraints

$$w_\delta(t) \in [-0.5, 0.5], \quad F_B(t) \in [0, 15000], \quad \phi \in [0, 1], \quad \mu(t) \in \{1, 2, 3, 4, 5\}.$$

Note that this is a mixed-integer optimal control problem as the gear $\mu(t)$ is restricted to a discrete set of available gears.

Results

Figure 7.11 shows the numerical solution for the variable time transformation method for $N = 40$. A direct discretization method for Problem 7.2.2 was used in combination with an SQP method. It took 9 minutes and 39.664 seconds to solve the problem on a CPU with 1.6 GHz with objective function value 6.787982 and final time $t_f = 6.783380$ [s]. Herein, the derivatives were computed simply by finite differences. More efficient techniques would allow for a further reduction of CPU times.

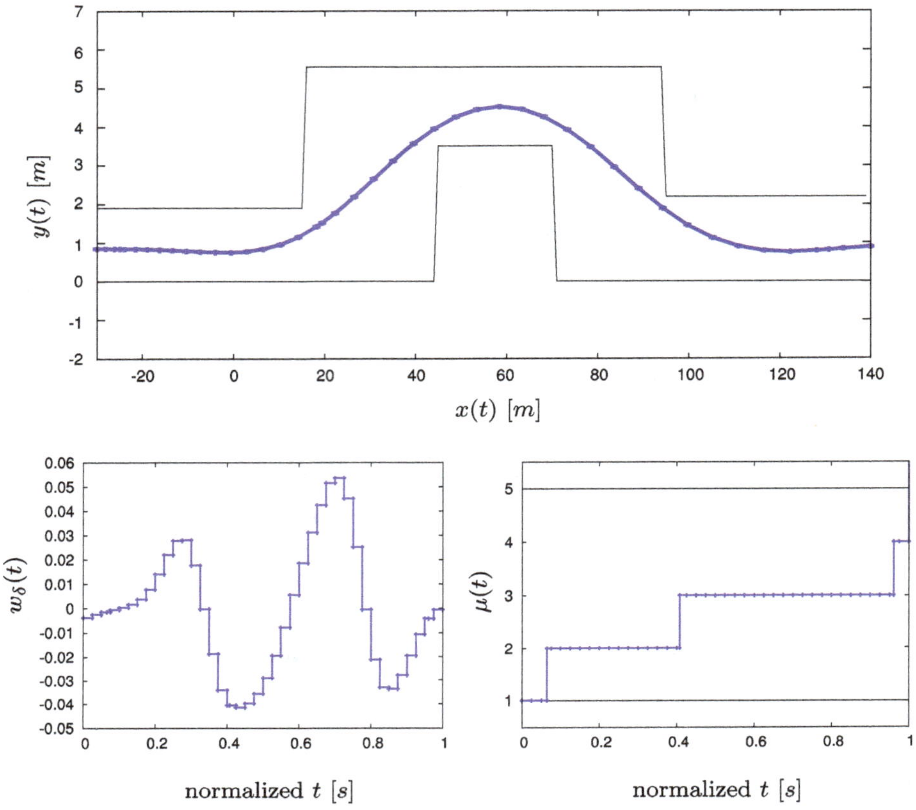

Figure 7.11: Numerical result for $N = 40$: Path $(x(t), y(t))$ of the car's center of gravity (top), steering angle velocity $w_\delta(t)$ (bottom, left), and gear shift $\mu(t)$ (bottom, right).

In comparison, the Branch&Bound method in [120] needed 232 hours, 25 minutes and 31 seconds of CPU time to solve the problem on a CPU with 750 MHz and yields the optimal objective function value 6.791374 and final time $t_f = 6.786781\,[\mathrm{s}]$. The Branch&Bound tree consists of 146941 nodes where each node requires solving a relaxed mixed-integer optimal control problem.

In all cases, the optimal braking forces and the optimal gas pedal positions compute to $F_B \equiv 0$ and $\varphi \equiv 1$, respectively. □

7.3 Switching costs, dynamic programming, Bellman's optimality principle

One of the first methods for solving (discretized) optimal control problems is the so-called *dynamic programming method*, which is based on *Bellman's optimality principle*. The method is applicable to very general problem settings, but it suffers from the *curse of dimensionality* as the resulting equations tend to require a large amount of memory and computing power. For a more detailed discussion, we refer to the monographs [22, 23, 33, 255]. Many applications and examples can be found in [328].

Throughout, we focus on discrete time optimal control problems, which exist in their own rights but also can be viewed as discretizations of continuous time optimal control problems. We particularly distinguish continuous-valued controls and discrete controls as we intend to include *switching costs* into the problem formulation. Switching costs only apply, if a discrete control switches from a discrete value to another value on two consecutive time points. The resulting problems will be solved by a dynamic programming method that is based on Bellman's famous optimality principle.

7.3.1 Dynamic optimization model with switching costs

For $N \in \mathbb{N}$, let a fixed grid

$$\mathbb{G}_N = \{t_0 < t_1 < \cdots < t_N\}$$

be given.

The grid function $z : \mathbb{G}_N \longrightarrow \mathbb{R}^{n_z}$, $t_j \mapsto z(t_j)$, is called *state variable*, and it is restricted by the *state constraints*

$$z(t_i) \in \mathcal{Z}(t_i), \quad i = 0, \dots, N,$$

where

$$\mathcal{Z}(t_i) \subseteq \mathbb{R}^{n_z}, \quad \mathcal{Z}(t_i) \neq \emptyset, \quad i = i = 0, \dots, N,$$

are closed connected sets.

The grid function $u : \mathbb{G}_N \longrightarrow \mathbb{R}^{n_u}$, $t_j \mapsto u(t_j)$, is called *continuous-valued control*, and it is restricted by the *control constraints*

$$u(t_i) \in \mathcal{U}(t_i, z(t_i)), \quad i = 0, \dots, N,$$

where

$$\mathcal{U}(t_i, z) \subseteq \mathbb{R}^{n_u}, \quad \mathcal{U}(t_i, z) \neq \emptyset, \quad i = 0, \dots, N,$$

are closed connected sets for $z \in \mathbb{R}^{n_z}$.

The grid function $v : \mathbb{G}_N \longrightarrow \mathbb{R}^{n_v}$, $t_j \mapsto v(t_j)$, is called *discrete control*, and it is restricted by the constraints

$$v(t_i) \in \mathcal{V}(t_i), \quad i = 0, \dots, N,$$

where

$$\mathcal{V}(t_i) := \{v^1(t_i), \ldots, v^{M_i}(t_i)\}, \quad v^j(t_i) \in \mathbb{R}^{n_v},\ j = 1, \ldots, M_i,$$

are discrete sets with $M_i \in \mathbb{N}$, $i = 0, \ldots, N$, elements at the respective time points.

Next, we aim at assigning costs at time t_i, if and only if the discrete control v switches, that is, if $v(t_i) \neq v(t_{i+1})$ holds for subsequent time points t_i and t_{i+1}. Hence, jumps in the grid function v can be measured with the help of the *discrete variation* $dv : \mathbb{G}_N \longrightarrow \mathbb{R}^{n_v}$ defined by

$$dv(t_i) = \begin{cases} v(t_{i+1}) - v(t_i), & \text{for } i = 0, \ldots, N-1, \\ v(t_N) - v(t_{N-1}), & \text{for } i = N. \end{cases}$$

A jump of v at t_i occurs, if and only if $dv(t_i) \neq 0$. To this end, dv detects switchings and measures even the variation itself. Switching costs, however, can have different nature, for instance, it may merely be important that a switch does occur, while the variation itself is not important and does not influence the switching costs. Contrarily, it may not only be important that a switch occurs, but the switching costs may also depend on the variation and/or direction of the switch.

In order to model different kinds of switching costs in the discrete control v, we will make use of a function $s = (s_1, \ldots, s_{n_v})^\top : \mathbb{G}_N \times \mathbb{R}^{n_v} \longrightarrow \mathbb{R}^{n_v}$ with the property

$$w_j = 0 \quad \Longrightarrow \quad s_j(t, w) = 0 \quad \text{for all } t \in \mathbb{G}_N,\ w = (w_1, \ldots, w_{n_v})^\top \in \mathbb{R}^{n_v}.$$

Example 7.3.1 (Switching functions). Typical examples of s for a scalar w are as follows:

(a) The function

$$s(t, w) = \begin{cases} 0, & \text{if } w = 0, \\ 1, & \text{otherwise} \end{cases}$$

can be used to detect switches in v when applied to the grid function dv. The height or direction of the switch is not measured by s.

(b) The function

$$s(t, w) = w$$

can be used to measure height and direction of a switch in v, when applied to the grid function dv.

(c) The function

$$s(t, w) = |w|$$

can be used to measure the height of a switch in v, when applied to the grid function dv. The direction of a switch is not measured.

(d) The function

$$s(t,w) = \operatorname{sign}(w) = \begin{cases} 0, & \text{if } w = 0, \\ 1, & \text{if } w > 0, \\ -1, & \text{if } w < 0, \end{cases}$$

can be used to detect the direction of a switch in v, when applied to the grid function dv. The height of the switch is not measured. □

With the above functions a dynamic optimization problem in discrete time that takes into account switching costs caused by the discrete control v reads as follows:

Problem 7.3.2 (DMIOCP).

Minimize

$$\begin{aligned} J(z,u,v) &= \varphi(z(t_N)) + \sum_{i=0}^{N-1} f_0(t_i, z(t_i), u(t_i), v(t_i)) + \sum_{i=0}^{N-2} g(t_i, z(t_i), u(t_i), v(t_i))^\top s(t_i, dv(t_i)) \\ &= \varphi(z(t_N)) + \sum_{i=0}^{N-1} f_0(t_i, z(t_i), u(t_i), v(t_i)) + \sum_{i=0}^{N-2} g(t_i, z(t_i), u(t_i), v(t_i))^\top s(t_i, v(t_{i+1}) - v(t_i)) \end{aligned}$$

with respect to grid functions $z : \mathbb{G}_N \longrightarrow \mathbb{R}^{n_z}$, $u : \mathbb{G}_N \longrightarrow \mathbb{R}^{n_u}$, and $v : \mathbb{G}_N \longrightarrow \mathbb{R}^{n_v}$ subject to the constraints

$$\begin{aligned} z(t_{i+1}) &= f(t_i, z(t_i), u(t_i), v(t_i)), && i = 0,1,\ldots,N-1, \\ z(t_0) &= z_0, \\ z(t_i) &\in \mathcal{Z}(t_i), && i = 0,1,\ldots,N, \\ u(t_i) &\in \mathcal{U}(t_i, z(t_i)), && i = 0,1,\ldots,N-1, \\ v(t_i) &\in \mathcal{V}(t_i), && i = 0,1,\ldots,N-1. \end{aligned}$$

The function $g : \mathbb{G}_N \times \mathbb{R}^{n_z} \times \mathbb{R}^{n_u} \times \mathbb{R}^{n_v} \longrightarrow \mathbb{R}^{n_v}$ is used to model the switching costs. Note that v is interpreted as a piecewise constant control, and hence, the control $v(t_N)$ does not have any influence in the problem. Consequently, only switches at the time points $t_1, \ldots, t_{N-1}$ are taken into account. Next, we will make use of the convention $v_N = v_{N-1}$ whenever useful.

7.3.2 A dynamic programming approach

Let $t_k \in \mathbb{G}_N$ with $k \in \{0,\ldots,N\}$ be a fixed time point and

$$\mathbb{G}_N^k := \{t_j \mid j = k, k+1, \ldots, N\}.$$

DMIOCP is embedded into a family of perturbed problems:

Problem 7.3.3 (DMIOCP(t_k, z_k, v_k)). Let $k \in \mathbb{N}$, $t_k \in \mathbb{G}_N$, $z_k \in \mathbb{R}^{n_z}$, and $v_k \in \mathcal{V}(t_k)$ be given.

Minimize

$$J_k(z,u,v) = \varphi(z(t_N)) + \sum_{i=k}^{N-1} f_0(t_i, z(t_i), u(t_i), v(t_i)) + \sum_{i=k}^{N-2} g(t_i, z(t_i), u(t_i), v(t_i))^\top s(t_i, dv(t_i))$$

with respect to grid functions $z : \mathbb{G}_N^k \longrightarrow \mathbb{R}^{n_z}$, $u : \mathbb{G}_N^k \longrightarrow \mathbb{R}^{n_u}$, $v : \mathbb{G}_N^k \longrightarrow \mathbb{R}^{n_v}$ *subject to the constraints*

$$\left.\begin{aligned} z(t_{i+1}) &= f(t_i, z(t_i), u(t_i), v(t_i)), & i &= k,\dots,N-1,\\ z(t_k) &= z_k, & &\\ z(t_i) &\in \mathcal{Z}(t_i), & i &= k,\dots,N,\\ u(t_i) &\in \mathcal{U}(t_i, z(t_i)), & i &= k,\dots,N-1,\\ v(t_i) &\in \mathcal{V}(t_i), & i &= k,\dots,N-1,\\ v(t_k) &= v_k. & & \end{aligned}\right\} \tag{7.24}$$

Definition 7.3.4. The function $W : \mathbb{G}_N \times \mathbb{R}^{n_z} \times \mathbb{R}^{n_v} \longrightarrow \mathbb{R}$ that assigns to $(t_k, z_k, v_k) \in \mathbb{G}_N \times \mathbb{R}^{n_z} \times \mathcal{V}(t_k)$ the optimal objective function value of DMIOCP(t_k, z_k, v_k) is called *optimal value function of DMIOCP*:

$$W(t_k, z_k, v_k) := \begin{cases} \inf\limits_{z,u,v \text{ with } (7.24)} J_k(z,u,v), & \text{if DMIOCP}(t_k, z_k, v_k) \text{ is feasible},\\ \infty, & \text{otherwise}. \end{cases}$$ □

Bellman's famous optimality principle holds for DMIOCP. In essence, the optimality principle states: The decisions in the periods $k, k+1, \dots, N$ of Problem DMIOCP for a given state z_k and a given discrete control v_k are independent of the decisions in the periods $t_0, t_1, \dots, t_{k-1}$, compare Figure 7.12.

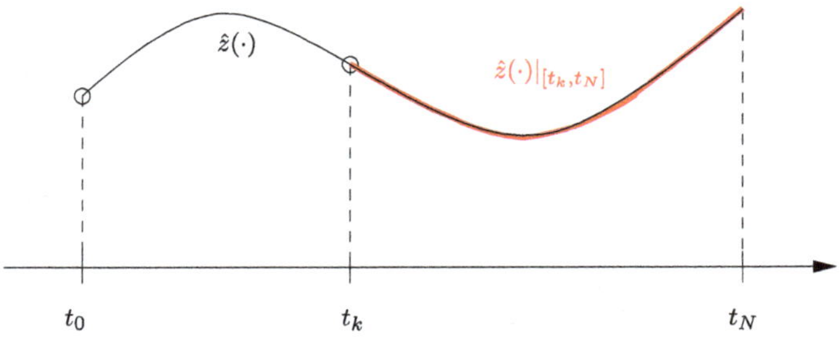

Figure 7.12: Bellman's optimality principle: Remaining trajectories of optimal trajectories remain optimal.

Theorem 7.3.5 (Bellman optimality principle). *Let* $(\hat{z}, \hat{u}, \hat{v})$ *be optimal grid functions for DMIOCP and let* $\mathbb{G}_N^k := \mathbb{G}_N \setminus \{t_0, \dots, t_{k-1}\}$. *Then the restrictions on* $\mathbb{G}_N^k$ *given by* $\hat{z}|_{\mathbb{G}_N^k}$, $\hat{u}|_{\mathbb{G}_N^k}$, *and* $\hat{v}|_{\mathbb{G}_N^k}$ *are optimal for DMIOCP*($t_k, \hat{z}(t_k), \hat{v}(t_k)$) *for all* k.

Proof. Assume that $\hat{z}|_{\mathbb{G}_N^k}$, $\hat{u}|_{\mathbb{G}_N^k}$, and $\hat{v}|_{\mathbb{G}_N^k}$ for some $k \in \{0,\ldots,N-1\}$ are not optimal for DMIOCP$(t_k, \hat{z}(t_k), \hat{v}(t_k))$. Then there exist feasible grid functions

$$\tilde{z} : \mathbb{G}_N^k \longrightarrow \mathbb{R}^{n_z}, \quad \tilde{u} : \mathbb{G}_N^k \longrightarrow \mathbb{R}^{n_u}, \quad \tilde{v} : \mathbb{G}_N^k \longrightarrow \mathbb{R}^{n_v}$$

for DMIOCP$(t_k, \hat{z}(t_k), \hat{v}(t_k))$ with

$$J_k(\tilde{z}, \tilde{u}, \tilde{v}) < J_k(\hat{z}, \hat{u}, \hat{v}), \tag{7.25}$$

$\tilde{z}(t_k) = \hat{z}(t_k)$, and $\tilde{v}(t_k) = \hat{v}(t_k)$. Hence, the trajectories

$$z : \mathbb{G}_N \longrightarrow \mathbb{R}^{n_z}, \; u : \mathbb{G}_N \longrightarrow \mathbb{R}^{n_u}, \; v : \mathbb{G}_N \longrightarrow \mathbb{R}^{n_v}$$

with

$$z(t_i) := \begin{cases} \hat{z}(t_i), & \text{for } i = 0,1,\ldots,k-1, \\ \tilde{z}(t_i), & \text{for } i = k,k+1,\ldots,N, \end{cases}$$

$$u(t_i) := \begin{cases} \hat{u}(t_i), & \text{for } i = 0,1,\ldots,k-1, \\ \tilde{u}(t_i), & \text{for } i = k,k+1,\ldots,N, \end{cases}$$

$$v(t_i) := \begin{cases} \hat{v}(t_i), & \text{for } i = 0,1,\ldots,k-1, \\ \tilde{v}(t_i), & \text{for } i = k,k+1,\ldots,N, \end{cases}$$

are feasible for DMIOCP and satisfy

$$\begin{aligned} J(z,u,v) &= \varphi(z(t_N)) + \sum_{i=0}^{N-1} f_0(t_i, z(t_i), u(t_i), v(t_i)) \\ &\quad + \sum_{i=0}^{N-2} g(t_i, z(t_i), u(t_i), v(t_i))^\top s(t_i, dv(t_i)) \\ &= \varphi(\tilde{z}(t_N)) + \sum_{i=0}^{k-1} f_0(t_i, \hat{z}(t_i), \hat{u}(t_i), \hat{v}(t_i)) \\ &\quad + \sum_{i=0}^{k-1} g(t_i, \hat{z}(t_i), \hat{u}(t_i), \hat{v}(t_i))^\top s(t_i, \hat{v}(t_{i+1}) - \hat{v}(t_i)) \\ &\quad + \sum_{i=k}^{N-1} f_0(t_i, \tilde{z}(t_i), \tilde{u}(t_i), \tilde{v}(t_i)) \\ &\quad + \sum_{i=k}^{N-2} g(t_i, \tilde{z}(t_i), \tilde{u}(t_i), \tilde{v}(t_i))^\top s(t_i, \tilde{v}(t_{i+1}) - \tilde{v}(t_i)) \\ &< \varphi(\hat{z}(t_N)) + \sum_{i=0}^{k-1} f_0(t_i, \hat{z}(t_i), \hat{u}(t_i), \hat{v}(t_i)) \end{aligned}$$

$$\begin{aligned}
&+\sum_{i=0}^{k-1} g(t_i,\hat{z}(t_i),\hat{u}(t_i),\hat{v}(t_i))^\top s(t_i,\hat{v}(t_{i+1})-\hat{v}(t_i))\\
&+\sum_{i=k}^{N-1} f_0(t_i,\hat{z}(t_i),\hat{u}(t_i),\hat{v}(t_i))\\
&+\sum_{i=k}^{N-2} g(t_i,\hat{z}(t_i),\hat{u}(t_i),\hat{v}(t_i))^\top s(t_i,\hat{v}(t_{i+1})-\hat{v}(t_i))\\
=\,&\varphi(\hat{z}(t_N))+\sum_{i=0}^{N-1} f_0(t_i,\hat{z}(t_i),\hat{u}(t_i),\hat{v}(t_i))\\
&+\sum_{i=0}^{N-2} g(t_i,\hat{z}(t_i),\hat{u}(t_i),\hat{v}(t_i))^\top s(t_i,d\hat{v}(t_i))\\
=\,&J(\hat{z},\hat{u},\hat{v}),
\end{aligned}$$

where (7.25) and $\tilde{v}(t_k)=\hat{v}(t_k)$ are exploited. This contradicts the optimality of $\hat{z}(\cdot)$, $\hat{u}(\cdot)$, and $\hat{v}(\cdot)$. □

For the validity of the optimality principle, it is essential that the discrete dynamic optimization problem can be divided into stages, i. e. the state z at t_{j+1} only depends on the values of z, u, and v at the previous stage t_j and not on the respective values at, e. g., t_0 and t_N. Likewise, the objective function is defined stage-wise, and the constraints only restrict z, u, and v at t_j. This allows for application of a stepwise optimization procedure. From the definition of the optimal value function, one immediately obtains

$$W(t_N,z_N,v_N)=\begin{cases}\varphi(z_N), & \text{if } z_N\in\mathcal{Z}(t_N),\\ \infty, & \text{otherwise.}\end{cases}\tag{7.26}$$

Bellman's optimality principle yields the following result:

Theorem 7.3.6. The optimal value function in Definition 7.3.4 satisfies the recursion

$$\begin{aligned}W(t_k,z_k,v_k)=\inf_{u\in\mathcal{U}(t_k,z_k),v\in\mathcal{V}(t_{k+1})}&\{f_0(t_k,z_k,u,v_k)\\&+g(t_k,z_k,u,v_k)^\top s(t_k,v-v_k)+W(t_{k+1},f(t_k,z_k,u,v_k),v)\},\end{aligned}$$

which has to hold for $(t_k,z_k,v_k)\in\mathbb{G}_N\times\mathcal{Z}(t_k)\times\mathcal{V}(t_k)$ and $k=N-1,\dots,0$. Herein, W at $t=t_N$ is given by (7.26) and the convention $W(t_k,z_k,v_k)=\infty$ is used whenever $(z_k,v_k)\notin\mathcal{Z}(t_k)\times\mathcal{V}(t_k)$.

Proof. Let $(t_k,z_k,v_k)\in\mathbb{G}_N\times\mathcal{Z}(t_k)\times\mathcal{V}(t_k)$ and $k\in\{0,\dots,N-1\}$ be given. If $(z_k,v_k)\notin\mathcal{Z}(t_k)\times\mathcal{V}(t_k)$, then $W(t_k,z_k,v_k)=\infty$ by definition.

If $f(t_k,z_k,u,v_k)\notin\mathcal{Z}(t_{k+1})$ for all $u\in\mathcal{U}(t_k,z_k)$, then DMIOCP(t_k,z_k,v_k) and DMIOCP$(t_{k+1},f(t_k,z_k,u,v_k),v)$ are infeasible for every $(u,v)\in\mathcal{U}(t_k,z_k)\times\mathcal{V}(t_{k+1})$ and hence $W(t_k,z_k,v_k)=\infty$.

For arbitrary $u \in \mathcal{U}(t_k, z_k)$ and $v \in \mathcal{V}(t_{k+1})$ with $f(t_k, z_k, u, v_k) \in \mathcal{Z}(t_{k+1})$, the definition of the optimal value function yields

$$W(t_k, z_k, v_k) \leq f_0(t_k, z_k, u, v_k) + g(t_k, z_k, u, v_k)^\top s(t_k, v - v_k) + W(t_{k+1}, f(t_k, z_k, u, v_k), v).$$

Taking the infimum over all $(u, v) \in \mathcal{U}(t_k, z_k) \times \mathcal{V}(t_{k+1})$ yields the first part of the assertion.

Now, let $\varepsilon > 0$ and feasible $\tilde{z}, \tilde{u}, \tilde{v}$ with $\tilde{z}(t_k) = z_k$, $\tilde{v}(t_k) = v_k$ and

$$J_k(\tilde{z}, \tilde{u}, \tilde{v}) \leq W(t_k, z_k, v_k) + \varepsilon$$

be given. Then,

$$\begin{aligned} W(t_k, z_k, v_k) &\geq J_k(\tilde{z}, \tilde{u}, \tilde{v}) - \varepsilon \\ &\geq f_0(t_k, z_k, \tilde{u}(t_k), v_k) \\ &\quad + g(t_k, z_k, \tilde{u}(t_k), v_k)^\top s(t_k, \tilde{v}(t_{k+1}) - v_k) \\ &\quad + W(t_{k+1}, f(t_k, z_k, \tilde{u}(t_k), v_k), \tilde{v}(t_{k+1})) - \varepsilon \\ &\geq \inf_{u \in \mathcal{U}(t_k, z_k), v \in \mathcal{V}(t_{k+1})} \{f_0(t_k, z_k, u, v_k) \\ &\quad + g(t_k, z_k, u, v_k)^\top s(t_k, v - v_k) \\ &\quad + W(t_{k+1}, f(t_k, z_k, u, v_k), v)\} - \varepsilon. \end{aligned}$$

As $\varepsilon > 0$ was arbitrary, the assertion follows. □

Theorem 7.3.6 allows deducing the following generic dynamic programming algorithm.

Algorithm 7.3.7 (Dynamic programming). **Init:** Let $\mathbb{G}_N = \{t_0 < t_1 < \cdots < t_N\}$ be given. Set

$$W(t_N, z_N, v_N) = \begin{cases} \varphi(z_N), & \text{if } z_N \in \mathcal{Z}(t_N), \\ \infty, & \text{otherwise} \end{cases}$$

for all $z_N \in \mathbb{R}^{n_x}$ and all $v_N \in \mathcal{V}(t_N)$.

Phase 1: Backward solution

For $k = N - 1, \ldots, 0$ compute

$$\begin{aligned} W(t_k, z_k, v_k) = \inf_{u \in \mathcal{U}(t_k, z_k), v \in \mathcal{V}(t_{k+1})} &\{f_0(t_k, z_k, u, v_k) \\ &+ g(t_k, z_k, u, v_k)^\top s(t_k, v - v_k) \\ &+ W(t_{k+1}, f(t_k, z_k, u, v_k), v)\} \end{aligned} \tag{7.27}$$

for all $z_k \in \mathbb{R}^{n_x}$ and all $v_k \in \mathcal{V}(t_k)$.

Phase 2: Forward solution

(i) Find

$$\hat{v}(t_0) = \arg \min_{v \in \mathcal{V}(t_0)} W(t_0, z_0, v)$$

and set $\hat{z}(t_0) = z_0$.

(ii) For $k = 0, \ldots, N-1$, find

$$\begin{aligned}(\hat{u}(t_k), \hat{v}(t_{k+1})) = \arg \min_{(u,v) \in \mathcal{U}(t_k, \hat{z}(t_k)) \times \mathcal{V}(t_{k+1})} & \{f_0(t_k, \hat{z}(t_k), u, \hat{v}(t_k)) \\ & + g(t_k, \hat{z}(t_k), u, \hat{v}(t_k))^\top s(t_k, v - \hat{v}(t_k)) \\ & + W(t_{k+1}, f(t_k, \hat{z}(t_k), u, \hat{v}(t_k)), v)\}\end{aligned} \tag{7.28}$$

and set $\hat{z}(t_{k+1}) = f(t_k, \hat{z}(t_k), \hat{u}(t_k), \hat{v}(t_k))$. □

Note that due to the presence of $v(t_{i+1}) - v(t_i)$ in DMIOCP, the recursive formula for the optimal value function is not of standard type.

Practical issues in the dynamic programming algorithm

In a numerical implementation, it is impossible to evaluate the optimal value function for every $z \in \mathbb{R}^{n_z}$. First, it is necessary to restrict the z-range to a compact set

$$\Omega = \{z \in \mathbb{R}^{n_z} \mid z_\ell \le z \le z_u\} \subseteq \mathbb{R}^{n_z}$$

with lower and upper bounds $z_\ell, z_u \in \mathbb{R}^{n_z}$ and $-\infty < z_\ell < z_u < \infty$, say. Ω should contain $\mathcal{Z}(t_k)$ for all $k = 0, \ldots, N$. Second, the set Ω needs to be discretized. For simplicity, we choose an equidistant partition

$$\Omega_{N_z} = \left\{ (z_1, \ldots, z_{n_z})^\top \in \Omega \;\middle|\; \begin{array}{l} z_j = z_{\ell,j} + i \frac{z_{u,j} - z_{\ell,j}}{N_z}, \\ j = 1, \ldots, n_z, \\ i \in \{0, \ldots, N_z\} \end{array} \right\}$$

with $N_z \in \mathbb{N}$. More sophisticated partitions using finite element meshes and adaptive schemes based on error estimates have been investigated in [149].

Third, during the backward phase and the forward phase the optimal value function, W needs to be evaluated at points of type $(t_{k+1}, \tilde{z}, v)$, where $\tilde{z} = f(t_k, z_k, u, v_k)$. One cannot always expect $\tilde{z}$ to be an element of Ω_{N_z}, and as W is only defined on $\mathbb{G}_N \times \Omega_{N_z} \times \mathcal{V}(\cdot)$, the value of the optimal function needs to be approximated in a suitable way. We will use an interpolating polynomial as follows:

(a) Assume that $\tilde{z} \notin \Omega_{N_z}$. If $\tilde{z} \notin \Omega$, then use ∞ as an approximation of $W(t_{k+1}, \tilde{z}, v)$. Otherwise, find a point $z^1 \in \Omega_{N_z}$ with $z^1 \le \tilde{z} < z^1 + H$, where $H = (H_1, \ldots, H_{n_z})^\top$, $H_j = \frac{z_{u,j} - z_{\ell,j}}{N_z}, j = 1, \ldots, n_z$.

(b) Determine the interpolating polynomial function

$$\tilde{W}(z) := \sum_{i_1=0}^{1} \cdots \sum_{i_{n_z}=0}^{1} a_{i_1,\ldots,i_{n_z}} z_1^{i_1} \cdots z_{n_z}^{i_{n_z}}$$

that interpolates $W(t_{k+1}, \cdot, v)$ at the points

$$\{(z_1, \ldots, z_{n_z})^\top \mid z_j = z_j^1 + \delta H_j,\ j = 1, \ldots, n_z,\ \delta \in \{0,1\}\}.$$

(c) Use $\tilde{W}(\tilde{z})$ as an approximation of $W(t_{k+1}, \tilde{z}, v)$.

For simplicity, we will assume that the minimization problems (7.27) and (7.28) can be solved by appropriate methods from nonlinear programming. For instance, an SQP method might be used to solve the resulting minimization problem with respect to u for each $v \in \mathcal{V}(t_{k+1})$ fixed. Note that this procedure is appropriate, if the cardinalities M_i of the discrete sets $\mathcal{V}(t_i)$, $i = 0, \ldots, N$, are low.

The main drawback of Bellman's dynamic programming method is the so-called *curse of dimensionality*. As it can be seen from the recursive formula in Theorem 7.3.6, the method requires computing and storing the values $W(t_k, z, v)$ for each $k = N, N-1, \ldots, 0$, each $z \in \mathcal{Z}$, and each $v \in \mathcal{V}$. Depending on the value N, this can become a really huge number. In the worst case, each discrete trajectory emanating from z_0 has to be considered.

7.3.3 Examples

Three illustrative examples are discussed. The first only involves continuous-valued controls and no discrete controls. It serves to illustrate the dynamic programming method. The second and third examples contain discrete controls.

Example 7.3.8. Consider the following discrete dynamic optimization problem with constants $k > 0$, $c > 0$, $b = 0.6$ and $N = 5$:

Minimize

$$-\sum_{j=0}^{N-1} c(1 - u(j))z(j)$$

subject to the constraints

$$z(j+1) = z(j)\big(0.9 + 0.6u(j)\big), \quad j = 0, 1, \ldots, N-1,$$

$$z(0) = k > 0,$$

$$0 \le u(j) \le 1, \quad j = 0, 1, \ldots, 4.$$

Since $k > 0$ and $u(j) \geq 0$, it holds $z(j) > 0$ for all j. Next, we use the abbreviation $z_j := z(j)$. A discrete control does not appear. The recursion for the optimal value function for $N = 5$ are given by

$$\begin{aligned}
&W(5, z_5) = 0,\\
&W(j, z_j) = \min_{0\leq u_j\leq 1} \{-cz_j(1-u_j) + W(j+1, z_j(0.9+0.6u_j))\},\\
&\quad 0 \leq j \leq N-1.
\end{aligned}$$

Evaluation of the recursion and observing $c > 0, y_j > 0$ yields

$$\begin{aligned}
W(4, z_4) &= \min_{0\leq u_4\leq 1} \{-cz_4(1-u_4) + \underbrace{W(5, z_4(0.9+0.6u_4))}_{=0}\} = -cz_4, \quad \hat{u}_4 = 0,\\
W(3, z_3) &= \min_{0\leq u_3\leq 1} \{-cz_3(1-u_3) + W(4, z_3(0.9+0.6u_3))\}\\
&= \min_{0\leq u_3\leq 1} \{-cz_3(1-u_3) - cz_3(0.9+0.6u_3)\}\\
&= cz_3 \min_{0\leq u_3\leq 1} \{-1.9 + 0.4u_3\} = -1.9cz_3, \quad \hat{u}_3 = 0,\\
W(2, z_2) &= \min_{0\leq u_2\leq 1} \{-cz_2(1-u_2) + W(3, z_2(0.9+0.6u_2))\}\\
&= \min_{0\leq u_2\leq 1} \{-cz_2(1-u_2) - 1.9cz_2(0.9+0.6u_2)\}\\
&= cz_2 \min_{0\leq u_2\leq 1} \{-2.71 - 0.14u_2\} = -2.85cz_2, \quad \hat{u}_2 = 1,\\
W(1, z_1) &= \min_{0\leq u_1\leq 1} \{-cz_1(1-u_1) + W(2, z_1(0.9+0.6u_1))\}\\
&= \min_{0\leq u_1\leq 1} \{-cz_1(1-u_1) - 2.85cz_1(0.9+0.6u_1)\}\\
&= cz_1 \min_{0\leq u_1\leq 1} \{-3.565 - 0.71u_1\} = -4.275cz_1, \quad \hat{u}_1 = 1,\\
W(0, z_0) &= \min_{0\leq u_0\leq 1} \{-cz_0(1-u_0) + W(1, z_0(0.9+0.6u_0))\}\\
&= \min_{0\leq u_0\leq 1} \{-cz_0(1-u_0) - 4.275cz_0(0.9+0.6u_0)\}\\
&= cz_0 \min_{0\leq u_0\leq 1} \{-4.8475 - 1.565u_0\} = -6.4125cz_0, \quad \hat{u}_0 = 1.
\end{aligned}$$

Hence, the optimal control is $\hat{u}_0 = \hat{u}_1 = \hat{u}_2 = 1$, $\hat{u}_3 = \hat{u}_4 = 0$. Forward evaluation leads to $\hat{z}_0 = k$, $\hat{z}_1 = 1.5 \cdot k$, $\hat{z}_2 = 2.25 \cdot k$, $\hat{z}_3 = 3.375 \cdot k$, $\hat{z}_4 = 3.0375 \cdot k$, $\hat{z}_5 = 2.73375 \cdot k$. The optimal objective value is $-c\hat{z}_3 - c\hat{z}_4 = -c(\hat{z}_3 + \hat{z}_4) = -6.4125 \cdot c \cdot k$. □

Example 7.3.9 (Compare [304]). Let

$$\mathbb{G}_N := \{t_i = ih \mid i = 0, \ldots, N\}, \quad h = \frac{2}{N}, \; N \in \mathbb{N},$$

and $\alpha \geq 0$ be given. Let $s : \mathbb{R} \longrightarrow \mathbb{R}$ be defined by

$$s(w) = \begin{cases} 1, & \text{if } w \neq 0,\\ 0, & \text{if } w = 0. \end{cases}$$

Minimize

$$J(z,v) := z(t_N)^2 + h\sum_{i=0}^{N-1}\left(\sin\left(\frac{\pi}{2}t_i\right) - z(t_i)\right)^2 + \alpha\sum_{i=0}^{N-2}\left(s\left(dv_1(t_i)\right) + s\left(dv_2(t_i)\right) + s\left(dv_3(t_i)\right)\right)$$

subject to the constraints

$$z(t_{i+1}) = z(t_i) + h\left(v_1(t_i) - v_2(t_i) + 2t_i v_3(t_i)\right), \quad i = 0,1,\ldots,N-1,$$

$$z(t_0) = 0,$$

$$\begin{pmatrix} v_1(t_i) \\ v_2(t_i) \\ v_3(t_i) \end{pmatrix} \in \mathcal{V} := \left\{\begin{pmatrix} 1 \\ 0 \\ 0 \end{pmatrix}, \begin{pmatrix} 0 \\ 1 \\ 0 \end{pmatrix}, \begin{pmatrix} 0 \\ 0 \\ 1 \end{pmatrix}\right\}, \quad i = 0,1,\ldots,N-1.$$

The following numerical experiments use $N = 1000$ and $N_z = 18000$. Figure 7.13 shows the solution of the dynamic program without switching costs ($\alpha = 0$). If switching costs are neglected, a high number of switchings can be observed, which appears to be similar to a chattering control on singular sub-arcs in continuous time optimal control problems. In a practical applications, this chattering is not desired as it cannot be implemented and has to be avoided. Figures 7.14, 7.15, 7.16 illustrate solutions, if switching costs are considered in the objective function by choosing different values $\alpha > 0$.

Finally, Figure 7.17 depicts the optimal value function. □

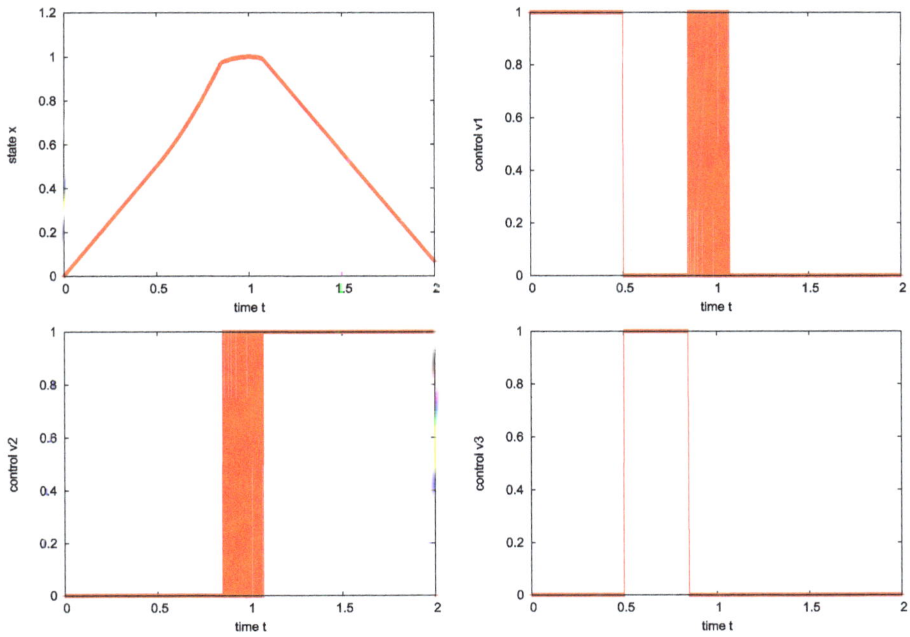

Figure 7.13: Optimal solution for $\alpha = 0$ (no switching costs) with objective function value 0.03164375965004069: Chattering controls with a high number of switches appear.

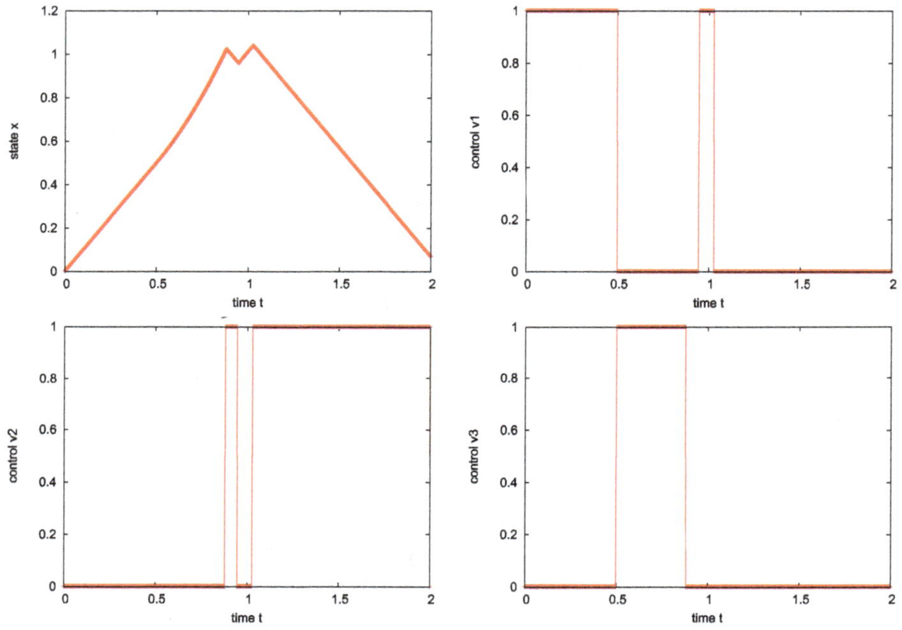

Figure 7.14: Optimal solution for $\alpha = 0.0001$ (switching costs with small weighting) with objective function value 0.03257074843026389: The number of switches has reduces drastically.

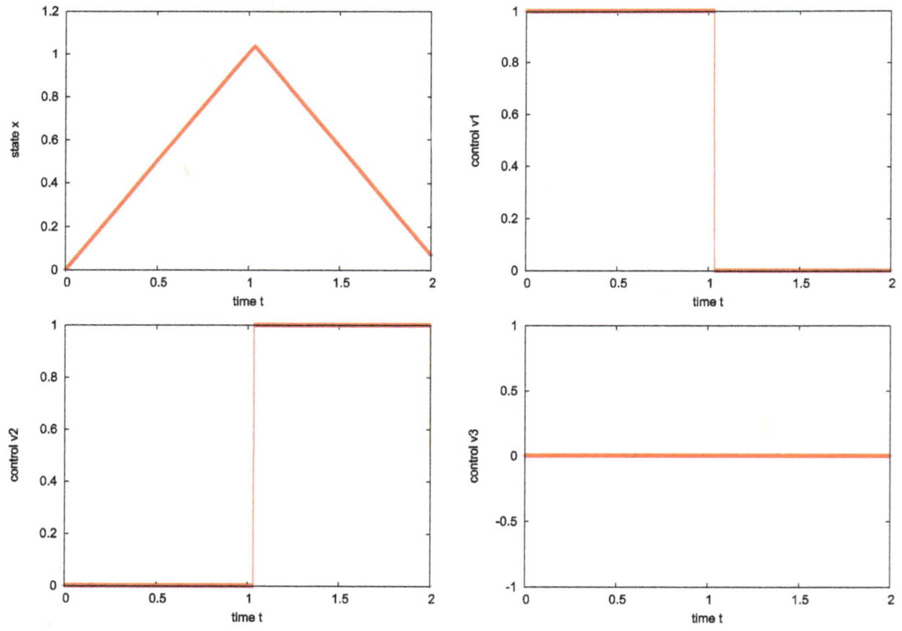

Figure 7.15: Optimal solution for $\alpha = 0.01$ (switching costs with moderate weighting) with objective function value 0.05611173096324493: The number of switches has reduced to one.

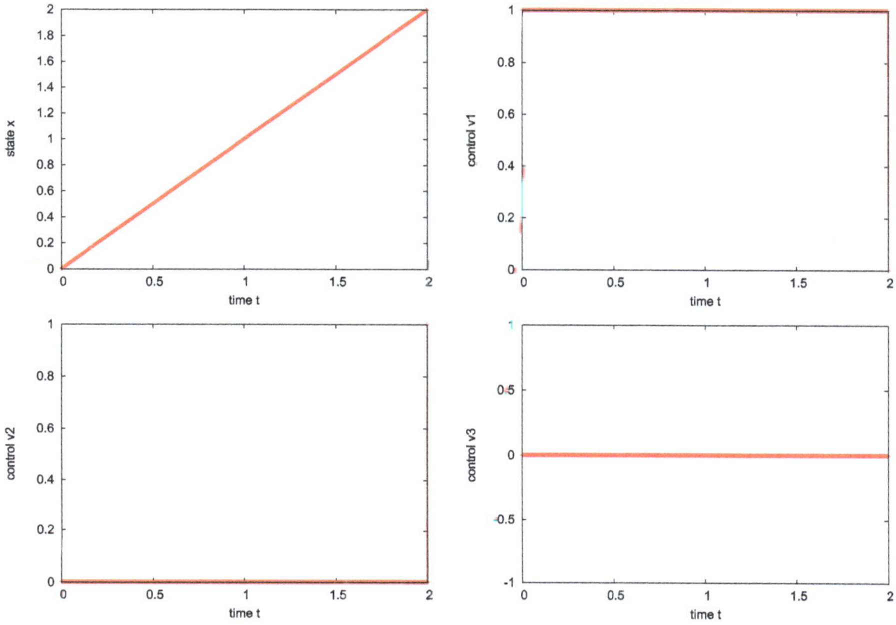

Figure 7.16: Optimal solution for $\alpha = 10$ (switching costs with high weighting) with objective function value 5.10020700492513: No switching occurs. The switching costs became dominant.

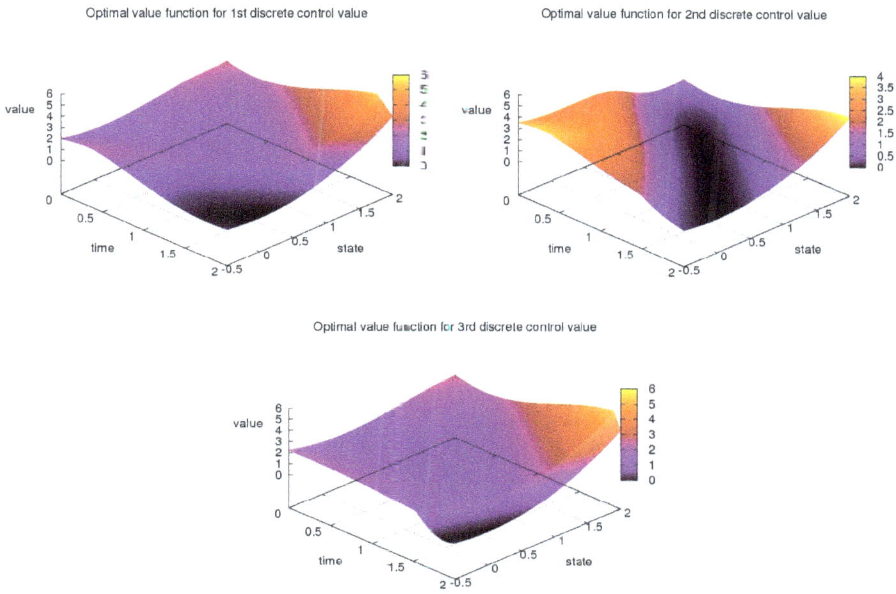

Figure 7.17: Optimal value function $W(\cdot, \cdot, v)$ for $\alpha = 0.75$, $N = N_z = 1000$ and $v = v^1$ (top left), $v = v^2$ (top right), $v = v^3$ (bottom).

Example 7.3.10 (Lotka–Volterra fishing problem, see [289]). Consider the following optimal control problem:

Minimize

$$\int_0^{12} (x_1(t)-1)^2 + (x_2(t)-1)^2 dt$$

subject to the constraints

$$\begin{aligned}
\dot{x}_1(t) &= x_1(t) - x_1(t)x_2(t) - 0.4x_1(t)v(t),\\
\dot{x}_2(t) &= -x_2(t) + x_1(t)x_2(t) - 0.2x_2(t)v(t),\\
v(t) &\in \{0,1\},\\
x(0) &= (0.5, 0.7)^\top.
\end{aligned}$$

Figure 7.18 shows the solution of the relaxed problem with control $v(t) \in [0,1]$. The control exhibits a bang-bang arc until time $t \approx 4$ and a singular sub-arc thereafter.

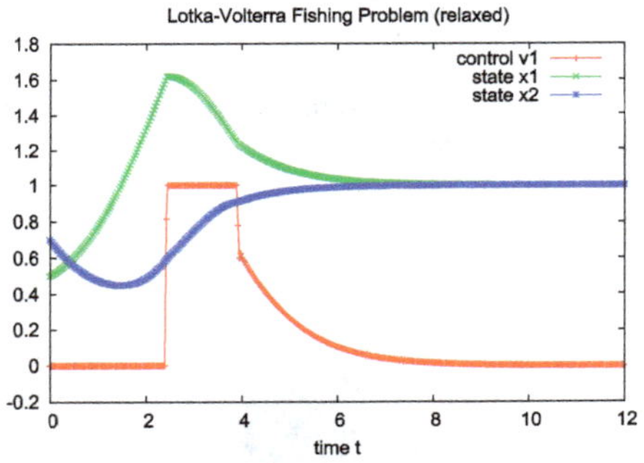

Figure 7.18: Solution of the relaxed problem.

The solution of the binary problem with $v(t) \in \{0,1\}$ in Figure 7.19 shows the same bang-bang arc until $t \approx 4$ and a *chattering control* thereafter, which switches between $v = 0$ and $v = 1$ back and forth.

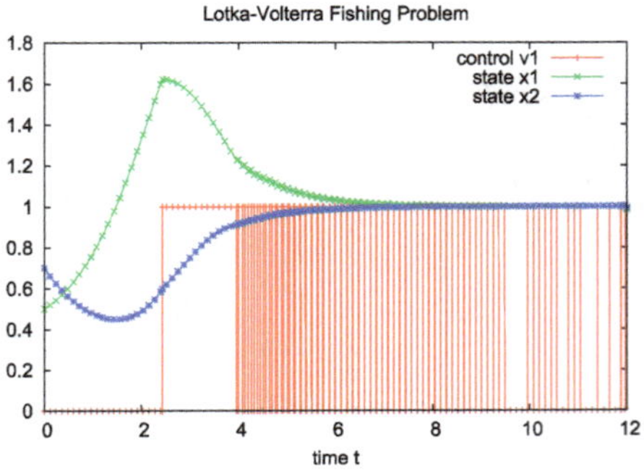

Figure 7.19: Solution of the binary problem without switching costs.

Finally, the solution of the binary problem taking into account switching costs is depicted in Figure 7.20. Only two switches occur. □

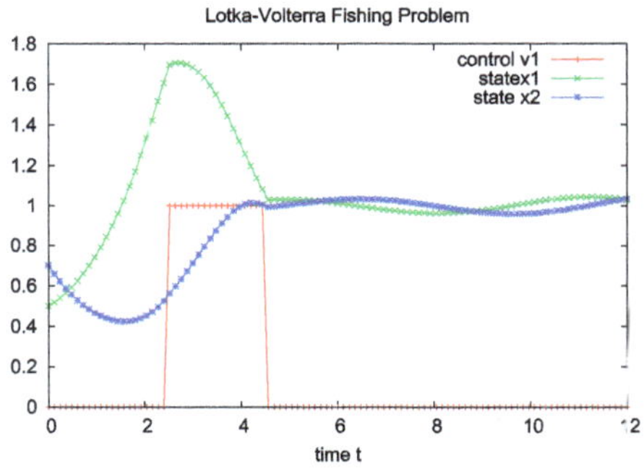

Figure 7.20: Solution of the relaxed problem with switching costs weighted by $\alpha = 0.75$.

(Extension to Optimal Control Problems)
A natural extension to optimal control problems would be to measure switching costs by an integral

$$\int_0^1 g\big(t, x(t), u(t), v(t)\big) s\big(t, dv(t)\big)$$

of the Riemann–Stieltjes type in the objective function. For $s(t, w) = w$, the integral becomes a standard Riemann–Stieltjes integral, assuming that the discrete control v is of bounded variation. For any other function s, the integral has to be defined appropriately, for instance, according to

$$\int_0^1 g\big(t, x(t), u(t), v(t)\big) s\big(t, dv(t)\big) := \lim_{\mathbb{G}_N\,:\,\mathbb{G}_N=\{0=t_0<t_1<\cdots<t_N=1\}} \sum_{i=0}^{N} g\big(\xi_i, x(\xi_i), u(\xi_i), v(\xi_i)\big) s\big(t_i, v(t_{i+1}) - v(t_i)\big),$$

where $\xi_i \in [t_i, t_{i+1})$ provided that u is piecewise continuous.

(Hamilton–Jacobi–Bellman Equation)
The dynamic programming method can be extended to the optimal control setting. As a result, the famous *Hamilton–Jacobi–Bellman equation* has to be solved. This equation is a first-order partial differential equation whose space dimension equals the state dimension of the optimal control problem. In the simplest case, the equation reads as

$$\frac{\partial V}{\partial t}(t, x) + \mathcal{H}\left(t, x, \hat{u}(t), \frac{\partial V}{\partial x}(t, x)\right) = 0,$$

where $\mathcal{H}$ denotes the Hamilton function. There exists a vast literature on solution methods and properties of this type of equation, see, for instance, [149, 20].

7.4 Exercises

Formulate the necessary optimality conditions for the optimal control problem in Example 7.1.13 as a boundary value problem and solve it using a shooting technique.

Herein, the correct switching structure 'lower bound–singular–upper bound' can be assumed to be known, but the switching points and initial values of the adjoints are supposed to be unknown.

Prove an analog global minimum principle as in Theorem 7.1.5 for the following index-one DAE optimal control problem with Assumption 3.4.2:

Minimize the objective function

$$\varphi(x(t_0), x(t_f)) + \int_{t_0}^{t_f} f_0(x(t), y(t), u(t))dt$$

with respect to $x \in W_{1,\infty}^{n_x}(\mathcal{I})$, $y \in L_{\infty}^{n_y}(\mathcal{I})$, $u \in L_{\infty}^{n_u}(\mathcal{I})$ *subject to*

$$\begin{aligned}
\dot{x}(t) &= f(x(t), y(t), u(t)),\\
0_{\mathbb{R}^{n_y}} &= g(x(t), y(t), u(t)),\\
0_{\mathbb{R}^{n_\psi}} &= \psi(x(t_0), x(t_f)),\\
u(t) &\in \mathcal{U}.
\end{aligned}$$

Use the variable time transformation method to solve Example 7.1.2 with $N = 4$ and $N = 80$.

Use the dynamic programming method to solve the following problem for $N = 5$:

Minimize

$$\sum_{j=0}^{N-1} u(j)^2$$

subject to

$$\begin{aligned}
x_1(j+1) &= x_1(j) + 2x_2(j), \quad & j = 0, 1, \ldots, N-1,\\
x_2(j+1) &= 2u(j) - x_2(j), \quad & j = 0, 1, \ldots, N-1,\\
x_1(0) &= 0,\\
x_1(N) &= 4,\\
x_2(0) &= 0,\\
x_2(N) &= 0.
\end{aligned}$$

Consider the following optimization problem:

Minimize

$$\sum_{j=1}^{n} u_j^\top A_j u_j + 2x_j^\top B_j u_j + x_j^\top C_j x_j$$

subject to

$$\begin{aligned} x_{j+1} &= D_j x_j + E_j u_j, \quad j = 1, \ldots, n-1, \\ x_{j+1} &\in \mathbb{R}^m, \\ u_j &\in \mathbb{R}^r, \\ x_1 &= x_a. \end{aligned}$$

The matrices C_j and A_j are supposed to be symmetric. Moreover, A_j is supposed to be positive definite, and $\begin{pmatrix} C_j & B_j \\ B_j^\top & A_j \end{pmatrix}$ is supposed to be positive semi-definite.

(a) Write down the Bellman equation for the optimal value function.

(b) Prove: If $W(j+1, x_{j+1})$ is of type

$$W(j+1, x_{j+1}) = x_{j+1}^\top Q_{j+1} x_{j+1}$$

with a symmetric and positive semi-definite matrix Q_{j+1}, then $W(j, x_j)$ is also of type

$$W(j, x_j) = x_j^\top Q_j x_j$$

with a symmetric and positive semi-definite matrix Q_j.

(c) Find feedback controls u_j in terms of x_j using the matrices Q_{j+1} and the Bellman equation.

(Compare [33, Exercise 8.8])
Three different methods are available in each time period of a production process at different costs, which are given by the following table:

method	I	II	III
volume of production	0	15	30
costs	0	500	800

The fixed costs for a positive production volume amount to 300 Euros per period. The inventory costs amount to 15 Euros per part and period. The demand per period amounts to 25 units of the product. Find a production plan with minimal costs for three time periods, if the stock level at the beginning amounts to 35 parts and if the stock level after three time periods is supposed to be 20 parts. Assume that inventory costs apply at the beginning of each time period.

(Compare [328])
Let us assume that a computer works, if and only if three components A, B, and C work properly. In order to increase the reliability of the computer system, it is possible to add certain emergency systems to each

component. It costs 100 dollars to add such an emergency system to the first component, 300 dollars for the second component, and 200 dollars for the third component. Furthermore, it is assumed that at most two emergency systems for each component can be added. The probability that a component works properly depends on the number of emergency systems added to the component according to the following table:

number/system	**A**	**B**	**C**
0	0.85	0.60	0.70
1	0.90	0.85	0.90
2	0.95	0.95	0.98

We are looking for a configuration that maximizes the reliability of the computer subject to the additional restriction that at most 600 dollars can be spent for additional emergency systems. Use dynamic programming to solve this problem.

(Compare [33, Exercise 8.9])
A manager has to choose co-workers for a project. There are four eligible co-workers to choose from. Each of them has a number assigned that indicates the capability of the respective co-worker. The respective numbers are 3, 5, 2, and 4. The costs for the co-workers are 30, 50, 20, and 40 thousand dollars, respectively. The manager has a maximal budget of 90 thousand dollars available for the personal expenditure. Use dynamic programming to decide which co-workers the manager should choose for the project.

8 Function space methods

This chapter addresses solution methods for optimal control problems that are formulated in a function space setting. The basic idea is to consider the optimal control problem as an infinite-dimensional optimization problem in a suitable Banach space, see Chapter 3. Then well-known algorithms for finite-dimensional nonlinear optimization problems like gradient methods, sequential quadratic programming (SQP) methods, or penalty methods are extended to the Banach space setting and are applied directly to the infinite-dimensional optimization problem. An early overview of this function space approach can be found in the paper [273], which provides an extensive list of references. The Lagrange–Newton method and the SQP method are extended to Banach spaces in [3, 4] and are applied to optimal control problems in [7, 8] and [216]. The function space approach is very famous in PDE-constrained optimal control, see the recent monographs [318, 170, 166] for the state-of-the-art in this field.

The function space approach is appealing because the algorithms typically work in the same spaces where the optimal control problem is stated. Hence, no discretization error is introduced at this stage as it was done in the direct discretization method. Moreover, the structure of the optimal control problem can be exploited very well. In order to realize the function space approach on a computer, discretizations become necessary at a lower level of the optimization algorithms. As we shall see later, the search direction in a function space optimization algorithm like the gradient method or the Lagrange–Newton method is given by initial value problems or boundary value problems, which in general can only be solved numerically and, hence, suitable discretizations need to be introduced. It has to be mentioned, though, that the direct discretization method and the function space approach with suitable discretizations at lower level may coincide to some degree. However, there may be subtle deviations as we shall demonstrate for the gradient method.

In Section 8.1, the well-known gradient method for finite-dimensional unconstrained optimization problems is extended to optimal control problems in a function space setting. In order to define the method, one needs to identify a suitable gradient for the function space problem. This task is not as straightforward as it might appear, see [141].

In Section 8.2, the Lagrange–Newton method is applied to first-order necessary optimality conditions for optimal control problems without control and state constraints. The Lagrange–Newton method exhibits a locally quadratic convergence rate under suitable assumptions, and thus it is very efficient. Extensions towards problems with control and/or state constraints can be found in [4] and [216].

https://doi.org/10.1515/9783110797893-008

8.1 Gradient method

One of the most basic methods for solving (unconstrained) optimization problems is the *gradient method* or *method of steepest descent*. The idea of this method is to iteratively follow the direction of the steepest descent of a given objective function at a current iterate to converge to a stationary point. In order to achieve convergence to a stationary point from arbitrary starting points, the gradient method is usually combined with a line-search strategy like Armijo's rule. We summarize the gradient method for minimizing the continuously differentiable function $J : \mathbb{R}^n \longrightarrow \mathbb{R}$, $u \mapsto J(u)$, in

Algorithm 8.1.1 (Gradient method for finite dimensional problems).

(0) Let $J : \mathbb{R}^n \longrightarrow \mathbb{R}$ be continuously differentiable, $u^{(0)} \in \mathbb{R}^n$, $\beta \in (0,1)$, $\sigma \in (0,1)$, and $k := 0$.

(1) Compute $d^{(k)} := -\nabla J(u^{(k)})$.

(2) If $\|d^{(k)}\| \approx 0$, STOP.

(3) Perform line-search: Find smallest $j \in \{0,1,2,\ldots\}$ with

$$J(u^{(k)} + \beta^j d^{(k)}) \leq J(u^{(k)}) - \sigma\beta^j \|\nabla J(u^{(k)})\|_2^2$$

and set $\alpha_k := \beta^j$.

(4) Set $u^{(k+1)} := u^{(k)} + \alpha_k d^{(k)}$, $k \leftarrow k+1$, and go to (1). □

Extensions of this approach towards constrained optimization problems with simple constraints like box constraints exist and use projections onto the feasible region. A convergence proof for a projected gradient method for finite-dimensional optimization problems can be found in, e. g., [111, pp. 300–302].

In the following, we derive a gradient method for an optimal control problem. Herein, we restrict ourselves to optimal control problems subject to index-one DAEs without state and control constraints:

Problem 8.1.2. Let $\mathcal{I} := [0,1]$, $\bar{x} \in \mathbb{R}^{n_x}$ a given vector, and

$$\begin{aligned} \varphi &: \mathbb{R}^{n_x} \longrightarrow \mathbb{R}, \\ f_0 &: \mathbb{R}^{n_x} \times \mathbb{R}^{n_y} \times \mathbb{R}^{n_u} \longrightarrow \mathbb{R}, \\ f &: \mathbb{R}^{n_x} \times \mathbb{R}^{n_y} \times \mathbb{R}^{n_u} \longrightarrow \mathbb{R}^{n_x}, \\ g &: \mathbb{R}^{n_x} \times \mathbb{R}^{n_y} \times \mathbb{R}^{n_u} \longrightarrow \mathbb{R}^{n_y} \end{aligned}$$

continuously differentiable functions.

Minimize

$$\Gamma(x,y,u) := \varphi(x(1)) + \int_0^1 f_0(x(t),y(t),u(t))dt$$

with respect to $x \in W_{1,\infty}^{n_x}(\mathcal{I})$, $y \in L_\infty^{n_y}(\mathcal{I})$, $u \in L_\infty^{n_u}(\mathcal{I})$ subject to the constraints

$$\dot{x}(t) = f(x(t), y(t), u(t)) \quad a.e.\ in\ \mathcal{I}, \tag{8.1}$$

$$0_{\mathbb{R}^{n_y}} = g(x(t), y(t), u(t)) \quad a.e.\ in\ \mathcal{I}, \tag{8.2}$$

$$x(0) = \bar{x}. \tag{8.3}$$

A reduction approach is applied to Problem 8.1.2 in order to eliminate the constraints. The idea is to solve the constraints (8.1)–(8.3) for x and y depending on u. This gives rise to the *control-state mappings*

$$u \mapsto x(u), \qquad u \mapsto y(u),$$

and we impose the following assumptions:

Assumption 8.1.3.

(a) The inverse matrix $(g_y'(x, y, u))^{-1}$ exists and is bounded for every (x, y, u).

(b) The initial value problem (8.1)–(8.3) possesses a unique solution $(x(u), y(u)) \in W_{1,\infty}^{n_x}(\mathcal{I}) \times L_\infty^{n_y}(\mathcal{I})$ for every control $u \in L_\infty^{n_u}(\mathcal{I})$.

(c) The mapping $L_\infty^{n_u}(\mathcal{I}) \ni u \mapsto (x(u), y(u)) \in W_{1,\infty}^{n_x}(\mathcal{I}) \times L_\infty^{n_y}(\mathcal{I})$ is continuously Fréchet-differentiable. □

Some remarks are in order.

Assumption 8.1.3 (a) ensures that the DAE has index one.

Assumption 8.1.3 (b) and (c) hold *locally* according to the implicit function theorem in some neighborhood of an optimal solution $(\hat{x}, \hat{y}, \hat{u})$, if the linear operator $H_{(x,y)}'(\hat{x}, \hat{y}, \hat{u}) : W_{1,\infty}^{n_x}(\mathcal{I}) \times L_\infty^{n_y}(\mathcal{I}) \longrightarrow L_\infty^{n_x}(\mathcal{I}) \times L_\infty^{n_y}(\mathcal{I}) \times \mathbb{R}^{n_x}$ defined by

$$H_{(x,y)}'(\hat{x}, \hat{y}, \hat{u})(x, y) := \begin{pmatrix} \dot{x}(\cdot) - f_x'[\cdot]x(\cdot) - f_y'[\cdot]y(\cdot) \\ g_x'[\cdot]x(\cdot) + g_y'[\cdot]y(\cdot) \\ x(0) \end{pmatrix}$$

is continuous and bijective.

By Assumption 8.1.3 (b), the constrained Problem 8.1.2 is reduced to an equivalent unconstrained minimization problem by solving the constraints for x and y depending on u and minimizing the reduced functional

$$J(u) := \Gamma(x(u), y(u), u) \tag{8.4}$$

with respect to $u \in L_\infty^{n_u}(\mathcal{I})$:

Problem 8.1.4 (Reduced problem).

Minimize $J(u)$ with respect to $u \in L_\infty^{n_u}(\mathcal{I})$.

In order to construct the gradient method for Problem 8.1.4, we need to define what the gradient of J with respect to u is supposed to be. Note that the gradient for a Fréchet-differentiable mapping $J : \mathbb{R}^n \longrightarrow \mathbb{R}$, $u \mapsto J(u)$, is well-defined as the vector of partial derivatives of J with respect to the components of u, that is

$$\nabla J(\hat{u}) := J'(\hat{u})^\top = \begin{pmatrix} \frac{\partial J}{\partial u_1}(\hat{u}) \\ \vdots \\ \frac{\partial J}{\partial u_n}(\hat{u}) \end{pmatrix}.$$

In particular, the directional derivative of J at $\hat{u}$ in direction u satisfies

$$J'(\hat{u})(u) = \nabla J(\hat{u})^\top u = \langle J(\hat{u}), u\rangle_{\mathbb{R}^n},$$

where $\langle \cdot, \cdot \rangle_{\mathbb{R}^n}$ denotes the standard inner product in $\mathbb{R}^n$.

More generally, if J was considered as a Fréchet-differentiable mapping from some Hilbert space U into $\mathbb{R}$, that is, $J : U \longrightarrow \mathbb{R}$, then the derivative $J'(\hat{u})(\cdot)$ at $\hat{u} \in U$ is an element of the dual space U^* of U, that is $J'(\hat{u})(\cdot) : U \longrightarrow R$ is linear and continuous. According to Riesz' Theorem 2.1.10, for every $\hat{u}$, there exists a unique element $\eta(\hat{u}) \in U$ such that

$$J'(\hat{u})(u) = \langle \eta(\hat{u}), u \rangle_U \tag{8.5}$$

for all $u \in U$, where $\langle \cdot, \cdot \rangle_U$ denotes the inner product of the Hilbert space U. In accordance with the finite-dimensional case, we call $\eta(\hat{u}) \in U$ the *gradient of J at $\hat{u}$* and use the notion $\nabla J(\hat{u})$ for $\eta(\hat{u})$.

For the functional $J : L_\infty^{n_u}(\mathcal{I}) \longrightarrow \mathbb{R}$ in (8.4), it is less obvious how the gradient should be defined, since $L_\infty^{n_u}(\mathcal{I})$ is not a Hilbert space and Theorem 2.1.10 does not apply. Golomb and Tapia [141] address this situation and define the so-called *metric gradient* in normed linear spaces.

Next, we aim at deriving an expression similar to (8.5) for our optimal control problem setting. To this end, let the Hamilton function be defined by

$$\mathcal{H}(x, y, u, \lambda_f, \lambda_g) := f_0(x, y, u) + \lambda_f^\top f(x, y, u) + \lambda_g^\top g(x, y, u).$$

Let $\hat{u} \in L_\infty^{n_u}(\mathcal{I})$ be fixed and denote by

$$(\hat{x}, \hat{u}) := (x(\hat{u}), y(\hat{u})) \in W_{1,\infty}^{n_x}(\mathcal{I}) \times L_\infty^{n_y}(\mathcal{I})$$

the corresponding solution of (8.1)–(8.3) and by

$$(S^x, S^y) := (x'(\hat{u})(u), y'(\hat{u})(u)) \in W_{1,\infty}^{n_x}(\mathcal{I}) \times L_\infty^{n_y}(\mathcal{I})$$

the so-called *sensitivity functions*, which denote the Fréchet-derivative of $(x(\cdot), y(\cdot))$ at $\hat{u}$ in the direction $u \in L_{\infty}^{n_u}(\mathcal{I})$. As the initial value $x(0) = \bar{x}$ is fixed, it follows $S^x(0) = 0_{\mathbb{R}^{n_x}}$.

Define the auxiliary functional

$$\begin{aligned}\tilde{J}(\hat{u}) &:= J(\hat{u}) + \left\langle \lambda_f(\cdot), f(\hat{x}(\cdot), \hat{y}(\cdot), \hat{u}(\cdot)) - \frac{d}{dt}\hat{x}(\cdot) \right\rangle_{L_2} + \langle \lambda_g(\cdot), g(\hat{x}(\cdot), \hat{y}(\cdot), \hat{u}(\cdot)) \rangle_{L_2} \\ &= \varphi(\hat{x}(1)) + \int_0^1 \mathcal{H}(\hat{x}(t), \hat{y}(t), \hat{u}(t), \lambda_f(t), \lambda_g(t)) - \lambda_f(t)^\top \frac{d}{dt}\hat{x}(t)dt,\end{aligned}$$

where $\lambda_f \in W_{1,\infty}^{n_x}(\mathcal{I})$ and $\lambda_g \in L_{\infty}^{n_y}(\mathcal{I})$ are functions that will be specified later. Partial integration of the last term yields

$$\begin{aligned}\tilde{J}(\hat{u}) &= \varphi(\hat{x}(1)) - [\lambda_f(t)^\top \hat{x}(t)]_0^1 \\ &\quad + \int_0^1 \mathcal{H}(\hat{x}(t), \hat{y}(t), \hat{u}(t), \lambda_f(t), \lambda_g(t)) + \left(\frac{d}{dt}\lambda_f(t)\right)^\top \hat{x}(t)dt.\end{aligned}$$

Formal differentiation of $\tilde{J}$ at $\hat{u}$ in the direction u and exploitation of $S^x(0) = 0_{\mathbb{R}^{n_x}}$ yields

$$\begin{aligned}\tilde{J}'(\hat{u})(u) &= (\varphi'(\hat{x}(1)) - \lambda_f(1)^\top)S^x(1) \\ &\quad + \int_0^1 (\mathcal{H}_x'[t] + (\dot{\lambda}_f(t))^\top)S^x(t) + \mathcal{H}_y'[t]S^y(t) + \mathcal{H}_u'[t]u(t)dt.\end{aligned}$$

As the sensitivity functions S^x and S^y are expensive to compute, λ_f and λ_g are chosen in such a way that the terms involving S^x and S^y are eliminated. This yields the index-one adjoint DAE

$$\dot{\lambda}_f(t) = -\mathcal{H}_x'(\hat{x}(t), \hat{y}(t), \hat{u}(t), \lambda_f(t), \lambda_g(t))^\top, \tag{8.6}$$

$$0_{\mathbb{R}^{n_y}} = \mathcal{H}_y'(\hat{x}(t), \hat{y}(t), \hat{u}(t), \lambda_f(t), \lambda_g(t))^\top, \tag{8.7}$$

$$\lambda_f(1) = \varphi'(\hat{x}(1))^\top, \tag{8.8}$$

and $\tilde{J}'(\hat{u})(u)$ reduces to

$$\tilde{J}'(\hat{u})(u) = \int_0^1 \mathcal{H}_u'[t]u(t)dt.$$

The adjoint DAE (8.6)–(8.7) is well-defined according to Assumption 8.1.3, (a).

Recall that we are looking for a direction u of steepest descent. Minimizing $\tilde{J}'(\hat{u})(u)$ with respect to u subject to $\|u\|_2 = 1$ yields

$$\tilde{u}(t) = -\frac{1}{\|\mathcal{H}_u'\|_2}\mathcal{H}_u'[t]^\top,$$

because for every u with $\|u\|_2 = 1$, it holds

$$|\tilde{J}'(\hat{u})(u)| \leq \|\mathcal{H}_u'\|_2 \cdot \|u\|_2 = \|\mathcal{H}_u'\|_2$$

by the Schwarz inequality and

$$\tilde{J}'(\hat{u})(\tilde{u}) = -\|\mathcal{H}_u'\|_2.$$

Hence, we may choose

$$d(t) = -\mathcal{H}_u'[t]^\top$$

as search direction at $\hat{u}$. The following result links $\tilde{J}'(\hat{u})$ to $J'(\hat{u})$.

Theorem 8.1.5. *Let $\hat{u} \in L_\infty^{n_u}(\mathcal{I})$ be given, let Assumption 8.1.3 hold, and let λ_f and λ_g satisfy the adjoint DAE (8.6)–(8.8). Then it holds $\tilde{J}'(\hat{u})(u) = J'(\hat{u})(u)$ for all $u \in L_\infty^{n_u}(\mathcal{I})$.*

Proof. For arbitrary $u \in L_\infty^{n_u}(\mathcal{I})$, it holds

$$\begin{aligned} x(u)(t) &= \bar{x} + \int_0^t f(x(u)(\tau), y(u)(\tau), u(\tau))d\tau \quad && \text{for all } t \in [0,1], \\ 0_{\mathbb{R}^{n_y}} &= g(x(u)(t), y(u)(t), u(t)) && \text{a. e. in } [0,1]. \end{aligned}$$

Differentiation of these identities in u with respect to u at $\hat{u} \in L_\infty^{n_u}(\mathcal{I})$ yields

$$\begin{aligned} S^x(t) &= \int_0^t f_x'[\tau]S^x(\tau) + f_y'[\tau]S^y(\tau) + f_u'[\tau]u(\tau)d\tau \quad && \text{for all } t \in [0,1], \\ 0_{\mathbb{R}^{n_y}} &= g_x'[t]S^x(t) + g_y'[t]S^y(t) + g_u'[t]u(t) && \text{a. e. in } [0,1], \end{aligned} \tag{8.9}$$

where all derivatives of f and g are evaluated at $\hat{x} = x(\hat{u}), \hat{y} = y(\hat{u})$, and $\hat{u}$. Differentiation of the absolutely continuous function $S^x(\cdot)$ and multiplication of (8.9) by $\lambda_g(\cdot)^\top$ from the left yields

$$\begin{aligned} \dot{S}^x(t) &= f_x'[t]S^x(t) + f_y'[t]S^y(t) + f_u'[t]u(t), \\ 0 &= \lambda_g(t)^\top(g_x'[t]S^x(t) + g_y'[t]S^y(t) + g_u'[t]u(t)). \end{aligned} \tag{8.10}$$

By exploitation of $0_{\mathbb{R}^{n_y}} = \mathcal{H}_y'[t]^\top$ and (8.10), we find

$$\begin{aligned} \frac{d}{dt}(\lambda_f(t)^\top S^x(t)) &= \dot{\lambda}_f(t)^\top S^x(t) + \lambda_f(t)^\top \dot{S}^x(t) \\ &= -\mathcal{H}_x'[t]S^x(t) \end{aligned}$$

$$
\begin{aligned}
&+ \lambda_f(t)^\top (f_x'[t]S^x(t) + f_y'[t]S^y(t) + f_u'[t]u(t)) \\
&= -\mathcal{H}_x'[t]S^x(t) - \mathcal{H}_y'[t]S^y(t) \\
&\quad + \lambda_f(t)^\top (f_x'[t]S^x(t) + f_y'[t]S^y(t) + f_u'[t]u(t)) \\
&\quad + \lambda_g(t)^\top (g_x'[t]S^x(t) + g_y'[t]S^y(t) + g_u'[t]u(t)) \\
&= -f_{0x}'[t]S^x(t) - f_{0y}'[t]S^y(t) - f_{0u}'[t]u(t) + \mathcal{H}_u'[t]u(t).
\end{aligned}
$$

Hence,

$$
\begin{aligned}
\tilde{J}'(\hat{u})(u) &= \int_0^1 \mathcal{H}_u'[t]u(t)dt \\
&= [\lambda_f(t)^\top S^x(t)]_0^1 + \int_0^1 f_{0x}'[t]S^x(t) + f_{0y}'[t]S^y(t) + f_{0u}'[t]u(t)dt \\
&= \varphi'(\hat{x}(1))S^x(1) + \int_0^1 f_{0x}'[t]S^x(t) + f_{0y}'[t]S^y(t) + f_{0u}'[t]u(t)dt \\
&= J'(\hat{u})(u).
\end{aligned}
$$

□

Definition 8.1.6 (Gradient of reduced functional). Let $\hat{u} \in L_\infty^{n_u}(\mathcal{I})$ be given, let Assumption 8.1.3 hold, and let λ_f and λ_g satisfy the adjoint DAE (8.6)–(8.8).

The *gradient* $\nabla J(\hat{u}) \in L_\infty^{n_u}(\mathcal{I})$ *of* J *at* $\hat{u}$ is defined by

$$
\nabla J(\hat{u})(\cdot) := \mathcal{H}_u'(\hat{x}(\cdot), \hat{y}(\cdot), \hat{u}(\cdot), \lambda_f(\cdot), \lambda_g(\cdot))^\top .
$$

□

Exploitation of this result allows formulating

Algorithm 8.1.7 (Gradient method).

(0) Choose $u^{(0)} \in L_\infty^{n_u}(\mathcal{I})$, $\beta \in (0,1)$, $\sigma \in (0,1)$, and set $k := 0$.

(1) Solve (8.1)–(8.3) with $u = u^{(k)}$. Let $x^{(k)} = x(u^{(k)})$ and $y^{(k)} = y(u^{(k)})$ denote the corresponding solution.

Solve the adjoint DAE

$$
\begin{aligned}
\dot{\lambda}_f(t) &= -\mathcal{H}_x'(x^{(k)}(t), y^{(k)}(t), u^{(k)}(t), \lambda_f(t), \lambda_g(t))^\top, \\
0_{\mathbb{R}^{n_y}} &= \mathcal{H}_y'(x^{(k)}(t), y^{(k)}(t), u^{(k)}(t), \lambda_f(t), \lambda_g(t))^\top, \\
\lambda_f(1) &= \varphi'(x^{(k)}(1))^\top .
\end{aligned}
$$

Let $\lambda_f^{(k)}$ and $\lambda_g^{(k)}$ denote the corresponding solution.

(2) If $\|\mathcal{H}_u'\|_2 \approx 0$, STOP.

(3) Set

$$
d^{(k)}(t) := -\mathcal{H}_u'(x^{(k)}(t), y^{(k)}(t), u^{(k)}(t), \lambda_f^{(k)}(t), \lambda_g^{(k)}(t))^\top .
$$

(4) Perform an Armijo line-search: Find smallest $j \in \{0, 1, 2, \ldots\}$ with

$$J(u^{(k)} + \beta^j d^{(k)}) \leq J(u^{(k)}) - \sigma\beta^j \|\mathcal{H}'_u(x^{(k)}(\cdot), y^{(k)}(\cdot), u^{(k)}(\cdot), \lambda_f^{(k)}(\cdot), \lambda_g^{(k)}(\cdot))\|_2^2$$

and set $\alpha_k := \beta^j$.

(5) Set $u^{(k+1)} := u^{(k)} + \alpha_k d^{(k)}$, $k \leftarrow k + 1$, and go to (1). □

Theorem 8.1.8. *Let Assumption 8.1.3 hold. Suppose that Algorithm 8.1.7 does not terminate. Let u_* be an accumulation point of the sequence $\{u^{(k)}\}_{k\in\mathbb{N}}$ generated by Algorithm 8.1.7 and $x_* := x(u_*)$, $y_* := y(u_*)$.*

Then it holds

$$\|\nabla J(u_*)\|_2 = 0.$$

Proof. Assume that $\{u^{(k_j)}\}_{j\in\mathbb{N}}$ is a subsequence with $u^{(k_j)} \longrightarrow u_*$ and $\|\nabla J(u_*)\|_2 \neq 0$.

By construction, the sequence $\{J(u^{(k_j)})\}_{j\in\mathbb{N}}$ is monotonically non-increasing, and by continuity of J and ∇J with regard to the argument u, it holds

$$J(u^{(k_j)}) \longrightarrow J(u_*), \tag{8.11}$$

$$d^{(k_j)} = -\nabla J(u^{(k_j)}) \longrightarrow -\nabla J(u_*) =: d_* \tag{8.12}$$

as $j \longrightarrow \infty$. In addition, the line-search in step (4) of Algorithm 8.1.7 yields

$$J(u^{(k_{j+1})}) \leq J(u^{(k_j+1)}) \leq J(u^{(k_j)}) - \sigma\alpha_{k_j} \|\nabla J(u^{(k_j)})\|_2^2$$

for $j \in \mathbb{N}$. Rearranging terms yields

$$0 \leq \sigma\alpha_{k_j} \|\nabla J(u^{(k_j)})\|_2^2 \leq J(u^{(k_j)}) - J(u^{(k_{j+1})}).$$

Owing to $\|\nabla J(u_*)\|_2 \neq 0$ and (8.11)–(8.12), it follows necessarily

$$\alpha_{k_j} \longrightarrow 0 \quad \text{as } j \longrightarrow \infty.$$

Hence, there is some index $j_0 \in \mathbb{N}$ such that $\alpha_{k_j} < 1$ for every $j \geq j_0$. Moreover, for every $j \geq j_0$, there exists some index $\ell_j \in \{0, 1, 2, \ldots\}$ with $\alpha_{k_j} = \beta^{\ell_j+1} \longrightarrow 0$,

$$J(u^{(k_j)} + \beta^{\ell_j} d^{(k_j)}) > J(u^{(k_j)}) - \sigma\beta^{\ell_j} \|\nabla J(u^{(k_j)})\|_2^2$$

and

$$\frac{J(u^{(k_j)} + \beta^{\ell_j} d^{(k_j)}) - J(u^{(k_j)})}{\beta^{\ell_j}} > -\sigma\|\nabla J(u^{(k_j)})\|_2^2,$$

respectively. Taking the limit $j \longrightarrow \infty$ on both sides yields

$$J'(u_*)(d_*) = J'(u_*)(-\nabla J(u_*)) = -\|\nabla J(u_*)\|_2^2 > -\sigma\|\nabla J(u_*)\|_2^2.$$

Herein, (8.12) and Lemma 8.1.9 below are exploited.

This inequality leads to a contradiction since $\sigma \in (0,1)$. □

Lemma 8.1.9. *Let X be a Banach space and $f : X \longrightarrow \mathbb{R}$ continuously Fréchet-differentiable. Let $\{x^{(k)}\}_{k\in\mathbb{N}} \subseteq X$, $\{d^{(k)}\}_{k\in\mathbb{N}} \subseteq X$, and $\{\alpha_k\}_{k\in\mathbb{N}} \subseteq \mathbb{R}_+$ be sequences with*

$$\lim_{k\longrightarrow\infty} x^{(k)} = x_* \in X, \quad \lim_{k\longrightarrow\infty} d^{(k)} = d_* \in X, \quad \lim_{k\longrightarrow\infty} \alpha_k = 0.$$

Then it holds

$$\lim_{k\longrightarrow\infty} \frac{f(x^{(k)} + \alpha_k d^{(k)}) - f(x^{(k)})}{\alpha_k} = f'(x_*)(d_*).$$

Proof. (i) Owing to the continuity of $f'(\cdot)$ in X, for every $\varepsilon > 0$, there exists some $\delta > 0$ with

$$\begin{aligned}
&|f'(x^{(k)})(d^{(k)}) - f'(x_*)(d^{(k)})| \\
&\quad = \|d^{(k)}\|_X \cdot \left|f'(x^{(k)})\left(\frac{d^{(k)}}{\|d^{(k)}\|_X}\right) - f'(x_*)\left(\frac{d^{(k)}}{\|d^{(k)}\|_X}\right)\right| \\
&\quad \le \|d^{(k)}\|_X \cdot \sup_{\|d\|_X=1} |f'(x^{(k)})(d) - f'(x_*)(d)| \\
&\quad = \|d^{(k)}\|_X \cdot \|f'(x^{(k)}) - f'(x_*)\|_{\mathcal{L}(X,\mathbb{R})} \\
&\quad \le \varepsilon\|d^{(k)}\|_X
\end{aligned}$$

for all $\|x^{(k)} - x_*\|_X \le \delta$. Hence,

$$\lim_{k\longrightarrow\infty} |f'(x^{(k)})(d^{(k)}) - f'(x_*)(d^{(k)})| = 0.$$

(ii) As $f'(x_*)(\cdot)$ is a linear and continuous operator, it follows

$$|f'(x_*)(d^{(k)}) - f'(x_*)(d_*)| \le C\|d^{(k)} - d_*\|_X$$

for some constant $C > 0$. Hence,

$$\lim_{k\longrightarrow\infty} |f'(x_*)(d^{(k)}) - f'(x_*)(d_*)| = 0.$$

(iii) The mean value theorem [169, p. 27] yields

$$\begin{aligned}
&\left|\frac{f(x^{(k)} + \alpha_k d^{(k)}) - f(x^{(k)})}{\alpha_k} - f'(x^{(k)})(d^{(k)})\right| \\
&\quad = \frac{1}{\alpha_k}|f(x^{(k)} + \alpha_k d^{(k)}) - f(x^{(k)}) - f'(x^{(k)})(\alpha_k d^{(k)})| \\
&\quad = \frac{1}{\alpha_k}\left|\int_0^1 f'(x^{(k)} + t\alpha_k d^{(k)})(\alpha_k d^{(k)})dt - f'(x^{(k)})(\alpha_k d^{(k)})\right|
\end{aligned}$$

$$\leq \int_0^1 \|f'(x^{(k)} + t\alpha_k d^{(k)}) - f'(x^{(k)})\|_{\mathcal{L}(X,\mathbb{R})} dt \cdot \|d^{(k)}\|_X.$$

By continuity of $f'(\cdot)$, it follows

$$\lim_{k\to\infty} \left| \frac{f(x^{(k)} + \alpha_k d^{(k)}) - f(x^{(k)})}{\alpha_k} - f'(x^{(k)})(d^{(k)}) \right| = 0.$$

(iv) The assertion follows from

$$\begin{aligned}
&\left| \frac{f(x^{(k)} + \alpha_k d^{(k)}) - f(x^{(k)})}{\alpha_k} - f'(x_*)(d_*) \right| \\
&\quad \leq \left| \frac{f(x^{(k)} + \alpha_k d^{(k)}) - f(x^{(k)})}{\alpha_k} - f'(x^{(k)})(d^{(k)}) \right| \\
&\qquad + |f'(x^{(k)})(d^{(k)}) - f'(x_*)(d^{(k)})| + |f'(x_*)(d^{(k)}) - f'(x_*)(d_*)|
\end{aligned}$$

and (i), (ii), and (iii). □

Example 8.1.10. Consider the following optimal control problem subject to an ordinary differential equation:

Minimize $x_2(1)$ subject to the constraints

$$\begin{aligned}
\dot{x}_1(t) &= -x_1(t) + \sqrt{3}u(t), \\
\dot{x}_2(t) &= \frac{1}{2}\left(x_1(t)^2 + u(t)^2\right), \\
x_1(0) &= 2, \\
x_2(0) &= 0.
\end{aligned}$$

The gradient method with initial guess $u^{(0)}(t) \equiv 0$, $\beta = 0.9$, $\sigma = 0.1$, and a symplectic discretization scheme based on the explicit Euler method for the dynamics and the adjoint dynamics with $N = 100$ equally spaced time intervals yields the result in Table 8.1 and Figure 8.1. □

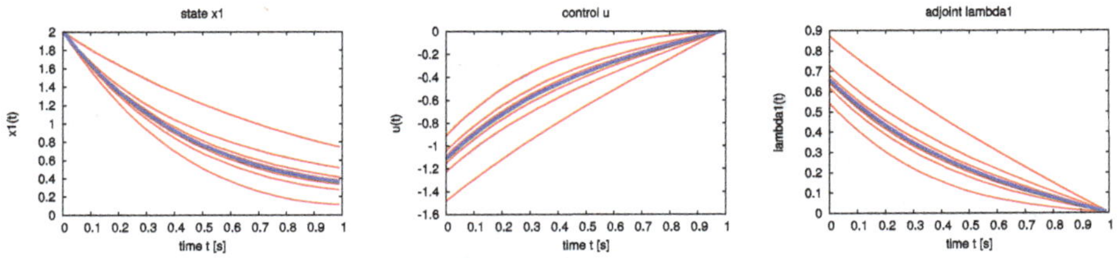

Figure 8.1: Some iterates $(x^{(k)}, u^{(k)}, \lambda^{(k)})$ (thin lines) and converged solution (thick line).

Table 8.1: Summary of iterations of gradient method. The iteration was stopped as soon as $\|\mathcal{H}'_u\|_\infty < 10^{-6}$.

k	α_k	$J(u^{(k)})$	$\|\mathcal{H}'_u\|_\infty$	$\|\mathcal{H}'_u\|_2^2$
0	0.00000000e+00	0.87037219e+00	0.14877655e+01	0.65717322e+00
1	0.10000000e+01	0.72406641e+00	0.61765831e+00	0.21168343e+00
2	0.10000000e+01	0.68017548e+00	0.35175249e+00	0.72633493e-01
3	0.10000000e+01	0.66515486e+00	0.20519966e+00	0.24977643e-01
4	0.10000000e+01	0.65999003e+00	0.12021286e+00	0.85902348e-02
⋮	⋮	⋮	⋮	⋮
23	0.10000000e+01	0.65728265e+00	0.47781223e-05	0.13406009e-10
24	0.10000000e+01	0.65728265e+00	0.28158031e-05	0.46203574e-11
25	0.10000000e+01	0.65728265e+00	0.16459056e-05	0.15877048e-11
26	0.10000000e+01	0.65728265e+00	0.97422070e-06	0.54326016e-12

Example 8.1.11. Consider the following optimal control problem subject to an ordinary differential equation:

Minimize $x_2(1) + 2.5(x_1(1) - 1)^2$ *subject to the constraints*

$$\begin{aligned} \dot{x}_1(t) &= u(t) - 15\exp(-2t), \\ \dot{x}_2(t) &= \frac{1}{2}\left(u(t)^2 + x_1(t)^3\right), \\ x_1(0) &= 4, \\ x_2(0) &= 0. \end{aligned}$$

The gradient method with initial guess $u^{(0)}(t) \equiv 0$, $\beta = 0.9$, $\sigma = 0.1$, and a symplectic discretization scheme based on the explicit Euler method for the dynamics and the adjoint dynamics with $N = 100$ equally spaced time intervals yields the result in Table 8.2 and Figure 8.2. □

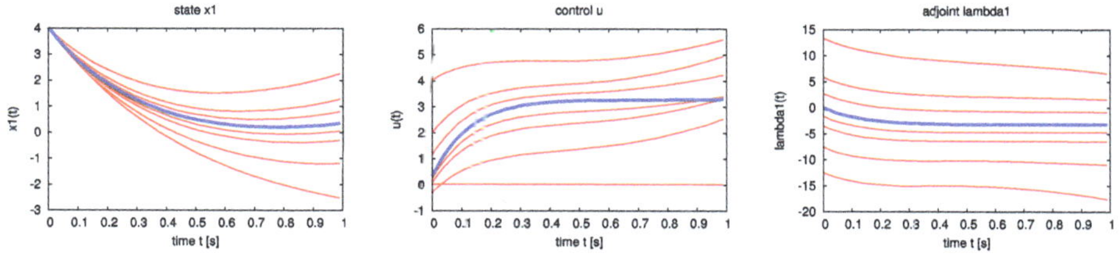

Figure 8.2: Some iterates $(x^{(k)}, u^{(k)}, \lambda^{(k)})$ (thin lines) and converged solution (thick line).

Now, let us compare the function space approach with the direct discretization method. To this end, we consider the following discretization of Problem 8.1.2 and fol-

Table 8.2: Summary of iterations of gradient method. The iteration was stopped as soon as $\|\mathcal{H}'_u\|_\infty < 10^{-6}$.

k	α_k	$J(u^{(k)})$	$\|\mathcal{H}'_u\|_\infty$	$\|\mathcal{H}'_u\|_2^2$
0	0.00000000e+00	0.32263509e+02	0.17750257e+02	0.23786744e+03
1	0.31381060e+00	0.21450571e+02	0.16994586e+02	0.18987670e+03
2	0.25418658e+00	0.15604667e+02	0.91485607e+01	0.78563930e+02
3	0.28242954e+00	0.11414005e+02	0.75417222e+01	0.39835695e+02
4	0.25418658e+00	0.97695774e+01	0.42526367e+01	0.15677990e+02
⋮	⋮	⋮	⋮	⋮
61	0.28242954e+00	0.84261386e+01	0.22532128e-05	0.28170368e-11
62	0.28242954e+00	0.84261386e+01	0.16626581e-05	0.17267323e-11
63	0.28242954e+00	0.84261386e+01	0.16639237e-05	0.13398784e-11
64	0.22876792e+00	0.84261386e+01	0.83196204e-06	0.21444710e-12

low the same steps as in the function space setting to derive the gradient method for the discretized problem:

Problem 8.1.12. Let $\bar{x} \in \mathbb{R}^{n_x}$ be a given vector, and let

$$\begin{aligned} \varphi &: \mathbb{R}^{n_x} \longrightarrow \mathbb{R}, \\ f_0 &: \mathbb{R}^{n_x} \times \mathbb{R}^{n_y} \times \mathbb{R}^{n_u} \longrightarrow \mathbb{R}, \\ f &: \mathbb{R}^{n_x} \times \mathbb{R}^{n_y} \times \mathbb{R}^{n_u} \longrightarrow \mathbb{R}^{n_x}, \\ g &: \mathbb{R}^{n_x} \times \mathbb{R}^{n_y} \times \mathbb{R}^{n_u} \longrightarrow \mathbb{R}^{n_y} \end{aligned}$$

be continuously differentiable functions. Let

$$\mathbb{G}_N := \{t_i \mid t_i = ih,\ i = 0,1,\dots,N\}, \quad h = \frac{1}{N},\ N \in \mathbb{N},$$

be a grid.

Minimize

$$\Gamma_N(x,y,u) := \varphi(x_N) + h\sum_{i=1}^{N} f_0(x_i,y_i,u_i)$$

with respect to $x = (x_0,x_1,\dots,x_N)^\top \in \mathbb{R}^{(N+1)n_x}$, $y = (y_1,\dots,y_N)^\top \in \mathbb{R}^{Nn_y}$, $u = (u_1,\dots,u_N)^\top \in \mathbb{R}^{Nn_u}$ *subject to the constraints*

$$\frac{x_i - x_{i-1}}{h} = f(x_i,y_i,u_i) \quad i = 1,\dots,N, \tag{8.13}$$

$$0_{\mathbb{R}^{n_y}} = g(x_i,y_i,u_i) \quad i = 1,\dots,N, \tag{8.14}$$

$$x_0 = \bar{x}. \tag{8.15}$$

Similarly as in Assumption 8.1.3, we assume

Assumption 8.1.13.

(a) The inverse matrix $(g'_y(x,y,u))^{-1}$ exists and is bounded for every (x,y,u).

(b) The system (8.13)–(8.15) possesses a unique solution $x(u) \in \mathbb{R}^{(N+1)n_x}$ and $y(u) \in \mathbb{R}^{Nn_y}$ for every $u \in \mathbb{R}^{Nn_u}$.
(c) The mapping $u \mapsto (x(u), y(u))$ is continuously Fréchet-differentiable. □

Solving the constraints (8.13)–(8.15) with respect to x and y yields

Problem 8.1.14 (Reduced discretized problem).

$$\textit{Minimize } J_N(u) := \Gamma_N(x(u), y(u), u) \textit{ with respect to } u \in \mathbb{R}^{Nn_u}.$$

Let $\hat{u} \in \mathbb{R}^{Nn_u}$ be given, $\hat{x} := x(\hat{u})$ and $\hat{y} := y(\hat{u})$. Define the auxiliary functional

$$\tilde{J}_N(\hat{u}) := \varphi(\hat{x}_N) + h\sum_{i=1}^{N}\left(\mathcal{H}(\hat{x}_i, \hat{y}_i, \hat{u}_i, \lambda_{f,i-1}, \lambda_{g,i-1}) - \lambda_{f,i-1}^\top\left(\frac{\hat{x}_i - \hat{x}_{i-1}}{h}\right)\right).$$

The Fréchet-derivative of $\tilde{J}_N$ with respect to $\hat{u}$ computes to

$$\begin{aligned}
\tilde{J}_N'(\hat{u}) &= \varphi'(\hat{x}_N)S_N^x + h\sum_{i=1}^{N}\left(\mathcal{H}_x'[t_i]S_i^x + \mathcal{H}_y'[t_i]S_i^y + \mathcal{H}_u'[t_i]\frac{\partial \hat{u}_i}{\partial \hat{u}} - \lambda_{f,i-1}^\top\left(\frac{S_i^x - S_{i-1}^x}{h}\right)\right)\\
&= (\varphi'(\hat{x}_N) + h\mathcal{H}_x'[t_N] - \lambda_{f,N-1}^\top)S_N^x + \lambda_{f,0}^\top S_0^x\\
&\quad + h\sum_{i=1}^{N-1}\left(\mathcal{H}_x'[t_i] + \frac{1}{h}(\lambda_{f,i}^\top - \lambda_{f,i-1}^\top)\right)S_i^x\\
&\quad + h\sum_{i=1}^{N}\left(\mathcal{H}_y'[t_i]S_i^y + \mathcal{H}_u'[t_i]\frac{\partial \hat{u}_i}{\partial \hat{u}}\right),
\end{aligned}$$

where the abbreviations

$$\begin{aligned}
S_i^x &:= \frac{\partial x_i}{\partial u}(\hat{u}) \in \mathbb{R}^{n_x \times Nn_u}, \quad & i = 0, \ldots, N,\\
S_i^y &:= \frac{\partial y_i}{\partial u}(\hat{u}) \in \mathbb{R}^{n_y \times Nn_u}, \quad & i = 1, \ldots, N,
\end{aligned}$$

are used. These *sensitivity matrices* are defined recursively by differentiating (8.13)–(8.15) with respect to u at $\hat{u}$:

$$\frac{S_i^x - S_{i-1}^x}{h} = f_x'[t_i]S_i^x + f_y'[t_i]S_i^y + f_u'[t_i]\frac{\partial \hat{u}_i}{\partial \hat{u}}, \quad i = 1, \ldots, N, \tag{8.16}$$

$$0_{\mathbb{R}^{n_y \times Nn_u}} = g_x'[t_i]S_i^x + g_y'[t_i]S_i^y + g_u'[t_i]\frac{\partial \hat{u}_i}{\partial \hat{u}}, \quad i = 1, \ldots, N, \tag{8.17}$$

$$S_0^x = 0_{\mathbb{R}^{n_x \times Nn_u}}. \tag{8.18}$$

In order to eliminate terms involving S_i^x and S_i^y, λ_f and λ_g are chosen such that the *discretized adjoint DAE*

$$\frac{\lambda_{f,i} - \lambda_{f,i-1}}{h} = -\mathcal{H}'_x(\hat{x}_i, \hat{y}_i, \hat{u}_i, \lambda_{f,i-1}, \lambda_{g,i-1})^\top, \quad i = 1, \dots, N, \tag{8.19}$$

$$0_{\mathbb{R}^{n_y}} = \mathcal{H}'_y(\hat{x}_i, \hat{y}_i, \hat{u}_i, \lambda_{f,i-1}, \lambda_{g,i-1})^\top, \quad i = 1, \dots, N, \tag{8.20}$$

$$\lambda_{f,N} = \varphi'(\hat{x}_N)^\top \tag{8.21}$$

holds. Then,

$$\tilde{J}'_N(\hat{u}) = h \sum_{i=1}^{N} \mathcal{H}'_u[t_i] \frac{\partial \hat{u}_i}{\partial \hat{u}} = h \left(\mathcal{H}'_u[t_1] \quad \mathcal{H}'_u[t_2] \quad \cdots \quad \mathcal{H}'_u[t_N] \right).$$

The following discrete counterpart of Theorem 8.1.5 links $\tilde{J}'_N(\hat{u}_N)$ to the gradient $\nabla J_N(\hat{u})$.

Theorem 8.1.15. *Let $\hat{u} \in \mathbb{R}^{Nn_u}$ be given, let Assumption 8.1.13 hold, and let $\lambda_f \in \mathbb{R}^{(N+1)n_x}$ and $\lambda_g \in \mathbb{R}^{Nn_y}$ satisfy the discretized adjoint DAE (8.19)–(8.21). Then it holds $\nabla J_N(\hat{u}) = \tilde{J}'_N(\hat{u})^\top$.*

Proof. Multiplication of (8.16)–(8.17) with $h\lambda_{f,i-1}^\top$ and $h\lambda_{g,i-1}^\top$, respectively, yields

$$h\lambda_{f,i-1}^\top f'_u[t_i] \frac{\partial \hat{u}_i}{\partial \hat{u}} = \lambda_{f,i-1}^\top (S_i^x - S_{i-1}^x) - h\lambda_{f,i-1}^\top f'_x[t_i] S_i^x - h\lambda_{f,i-1}^\top f'_y[t_i] S_i^y,$$

$$h\lambda_{g,i-1}^\top g'_u[t_i] \frac{\partial \hat{u}_i}{\partial \hat{u}} = -h\lambda_{g,i-1}^\top g'_x[t_i] S_i^x - h\lambda_{g,i-1}^\top g'_y[t_i] S_i^y$$

for $i = 1, \dots, N$. By these expressions, it follows

$$\begin{aligned}
h\mathcal{H}'_u[t_i] \frac{\partial \hat{u}_i}{\partial \hat{u}} &= hf'_{0,u}[t_i] \frac{\partial \hat{u}_i}{\partial \hat{u}} \\
&\quad + \lambda_{f,i-1}^\top (S_i^x - S_{i-1}^x) - h\lambda_{f,i-1}^\top f'_x[t_i] S_i^x - h\lambda_{f,i-1}^\top f'_y[t_i] S_i^y \\
&\quad - h\lambda_{g,i-1}^\top g'_x[t_i] S_i^x - h\lambda_{g,i-1}^\top g'_y[t_i] S_i^y \\
&= hf'_{0,u}[t_i] \frac{\partial \hat{u}_i}{\partial \hat{u}} + hf'_{0,x}[t_i] S_i^x + hf'_{0,y}[t_i] S_i^y \\
&\quad + \lambda_{f,i-1}^\top (S_i^x - S_{i-1}^x) - h\mathcal{H}'_x[t_i] S_i^x - h\mathcal{H}'_y[t_i] S_i^y
\end{aligned}$$

for $i = 1, \dots, N$.

Summation and exploitation of (8.19)–(8.21) and (8.18) yields

$$\begin{aligned}
\tilde{J}'_N(\hat{u}) &= h \sum_{i=1}^{N} \mathcal{H}'_u[t_i] \frac{\partial \hat{u}_i}{\partial \hat{u}} \\
&= h \sum_{i=1}^{N} \left(f'_{0,u}[t_i] \frac{\partial \hat{u}_i}{\partial \hat{u}} + f'_{0,x}[t_i] S_i^x + f'_{0,y}[t_i] S_i^y \right) \\
&\quad + h \sum_{i=1}^{N-1} \left(\frac{\lambda_{f,i-1}^\top - \lambda_{f,i}^\top}{h} - \mathcal{H}'_x[t_i] \right) S_i^x \\
&\quad - \lambda_{f,0}^\top S_0^x + (\lambda_{f,N-1}^\top - h\mathcal{H}'_x[t_N]) S_N^x
\end{aligned}$$

$$= \varphi'(\hat{x}_N)S_N^x + h\sum_{i=1}^{N}\left(f'_{0,u}[t_i]\frac{\partial u_i}{\partial \hat{u}} + f'_{0,x}[t_i]S_i^x + f'_{0,y}[t_i]S_i^y\right)$$
$$= J_N'(\hat{u}) = \nabla J_N(\hat{u})^\top.$$
□

In summary, the gradient method 8.1.1 applied to the reduced discretized problem 8.1.14 uses the search direction

$$d^{(k)} = -\nabla J_N(u^{(k)}) = -h\begin{pmatrix}\mathcal{H}_u'[t_1]^\top \\ \vdots \\ \mathcal{H}_u'[t_N]^\top\end{pmatrix}$$

in iteration k.

Algorithm 8.1.1 for Problem 8.1.14 reads as follows:

Algorithm 8.1.16 (Gradient method).
(0) Choose $u^{(0)} \in \mathbb{R}^{Nn_u}$, $\beta \in (0,1)$, $\sigma \in (0,1)$, and set $k := 0$.
(1) Solve (8.13)–(8.15) with $u = u^{(k)}$. Let $x^{(k)} = x(u^{(k)})$ and $y^{(k)} = y(u^{(k)})$ denote the corresponding solution.
Solve the adjoint DAE (8.19)–(8.21). Let $\lambda_f^{(k)}$ and $\lambda_g^{(k)}$ denote the corresponding solution.
(2) Set

$$d^{(k)} := -\nabla J_N(u^{(k)}) = -h\begin{pmatrix}\mathcal{H}_u'[t_1]^\top \\ \vdots \\ \mathcal{H}_u'[t_N]^\top\end{pmatrix}$$

(3) If $\|d^{(k)}\|_2 \approx 0$, STOP.
(4) Perform an Armijo line-search: Find smallest $j \in \{0,1,2,\dots\}$ with

$$J_N(u^{(k)} + \beta^j d^{(k)}) \le J_N(u^{(k)}) - \sigma\beta^j \|d^{(k)}\|_2^2$$

and set $\alpha_k := \beta^j$.
(5) Set $u^{(k+1)} := u^{(k)} + \alpha_k d^{(k)}$, $k \leftarrow k+1$, and go to (1). □

Clearly, (8.13)–(8.15) is a discretization of the DAE (8.1)–(8.3) by means of the implicit Euler method. Accordingly, (8.19)–(8.21) is a discretization of the adjoint DAE (8.6)–(8.8). The combined discretization scheme is a symplectic integration scheme, compare Subsection 5.4.2.1 and Theorem 5.4.2. The search direction in step (2) of Algorithm 8.1.16 is a discretization of the search direction in step (3) of Algorithm 8.1.7, except that the former is scaled by a factor of h. As a consequence, the scaling of the search direction depending on the discretization parameter h may lead to smaller steps in the gradient method when applied to the discretized problem compared to the function space gradient method, if an Armijo-type line-search with initial step-size $\beta^0 = 1$ is used. Hence, one has to expect a difference in computational performance between the function space gradient method with suitable discretizations of the DAE and adjoint DAE and the direct discretization method. For the above examples, the function space approach needs significantly fewer iterations than the direct discretization method with a scaled search direction. Note that both methods would match up to the error introduced by discretization if β^j/h was used as a step-size in Armijo's line-search procedure in Algorithm 8.1.1. We illustrate this effect by revisiting Example 8.1.10.

Example 8.1.17 (Compare Example 8.1.10). Consider the following optimal control problem subject to an ordinary differential equation:

Minimize $x_2(1)$ *subject to the constraints*

$$\begin{aligned}\dot{x}_1(t) &= -x_1(t) + \sqrt{3}u(t),\\ \dot{x}_2(t) &= \frac{1}{2}\left(x_1(t)^2 + u(t)^2\right),\\ x_1(0) &= 2,\\ x_2(0) &= 0.\end{aligned}$$

We follow the *first discretize – then optimize approach* and discretize the problem for simplicity by the explicit Euler method, which was also used in Example 8.1.10. We then apply Algorithm 8.1.1 directly with initial guess $u^{(0)} = 0_{\mathbb{R}^N}$, $N = 100$, $\beta = 0.9$, and $\sigma = 0.1$. The gradient $\nabla J(u^{(k)})$ is approximated by a finite difference scheme. According to the previous analysis, we have to expect that the search-direction $d^{(k)} = -\nabla J(u^{(k)})$ in Algorithm 8.1.1 is approximately the search direction in Algorithm 8.1.7 scaled by a factor of $h = 1/N = 10^{-2}$. Likewise, we have to expect that Algorithm 8.1.1 makes slower progress than Algorithm 8.1.7. This is actually the case as the results of the gradient method in Algorithm 8.1.1 for the reduced discretized problem 8.1.14 in Table 8.3 show. □

8.2 Lagrange–Newton method

The Lagrange–Newton method is based on the numerical solution of first-order necessary optimality conditions for an optimal control problem without control and state constraints using Newton's method. The necessary conditions for such problems lead to a nonlinear operator equation of type

$$T(z) = \Theta_Y \tag{8.22}$$

for the operator $T : Z \longrightarrow Y$, where Z and Y denote suitable Banach spaces endowed with norms $\|\cdot\|_Z$ and $\|\cdot\|_Y$. Application of Newton's method to (8.22) results in the well-known *Lagrange–Newton method* provided that T is continuously Fréchet-differentiable. More specifically, the Lagrange–Newton method generates sequences

Table 8.3: Summary of iterations of gradient method for the reduced discretized problem. The iteration was stopped as soon as $\|d^{(k)}\|_\infty < 10^{-6}$.

k	α_k	$J(u^{(k)})$	$\|\mathcal{H}_x'\|_\infty$	$\|\mathcal{H}_u'\|_2^2$
0	0.00000000e+00	0.87037219e+00	0.14877677e-01	0.65717185e-02
1	0.10000000e+01	0.86385155e+00	0.14671320e-01	0.63689360e-02
2	0.10000000e+01	0.85753209e+00	0.14468015e-01	0.61724878e-02
3	0.10000000e+01	0.85140755e+00	0.14267676e-01	0.59821683e-02
4	0.10000000e+01	0.84547186e+00	0.14070156e-01	0.57976919e-02
5	0.10000000e+01	0.83971919e+00	0.13875455e-01	0.56189844e-02
6	0.10000000e+01	0.83414385e+00	0.13683565e-01	0.54457966e-02
7	0.10000000e+01	0.82874034e+00	0.13494428e-01	0.52779702e-02
8	0.10000000e+01	0.82350335e+00	0.13307910e-01	0.51153368e-02
9	0.10000000e+01	0.81842770e+00	0.13124130e-01	0.49577892e-02
⋮	⋮	⋮	⋮	⋮
812	0.10000000e+01	0.65728265e+00	0.11053883e-05	0.79804802e-11
813	0.10000000e+01	0.65728265e+00	0.10356781e-05	0.76292003e-11
814	0.10000000e+01	0.65728265e+00	0.10655524e-05	0.77345706e-11
815	0.10000000e+01	0.65728265e+00	0.98588398e-06	0.74390524e-11

$\{z^{(k)}\}$, $\{d^{(k)}\}$, and – if a globalization procedure is added – $\{\alpha_k\}$ related by the iteration

$$z^{(k+1)} = z^{(k)} + \alpha_k d^{(k)}, \quad k = 0, 1, 2, \ldots .$$

Herein, the search direction $d^{(k)}$ is the solution of the linear operator equation

$$T'(z^{(k)})(d^{(k)}) = -T(z^{(k)}),$$

and the step-length $\alpha_k > 0$ is determined by a line-search procedure of the Armijo type for a suitably defined merit function. We investigate the details of the outlined Lagrange–Newton method for the following problem setting:

Problem 8.2.1. Let $\mathcal{I} := [0, 1]$, and let

$$\begin{aligned} \varphi &: \mathbb{R}^{n_x} \times \mathbb{R}^{n_x} \longrightarrow \mathbb{R}, \\ f_0 &: \mathbb{R}^{n_x} \times \mathbb{R}^{n_y} \times \mathbb{R}^{n_u} \longrightarrow \mathbb{R}, \\ f &: \mathbb{R}^{n_x} \times \mathbb{R}^{n_y} \times \mathbb{R}^{n_u} \longrightarrow \mathbb{R}^{n_x}, \\ g &: \mathbb{R}^{n_x} \times \mathbb{R}^{n_y} \times \mathbb{R}^{n_u} \longrightarrow \mathbb{R}^{n_y}, \\ \psi &: \mathbb{R}^{n_x} \times \mathbb{R}^{n_x} \longrightarrow \mathbb{R}^{n_\psi} \end{aligned}$$

be twice continuously differentiable functions.

Minimize

$$\Gamma(x,y,u) := \varphi\big(x(0),x(1)\big) + \int_0^1 f_0\big(x(t),y(t),u(t)\big)dt$$

with respect to $x \in W^{n_x}_{1,\infty}(\mathcal{I})$, $y \in L^{n_y}_{\infty}(\mathcal{I})$, $u \in L^{n_u}_{\infty}(\mathcal{I})$ *subject to the constraints*

$$\dot{x}(t) = f\big(x(t),y(t),u(t)\big) \quad \text{a. e. in } [0,1], \tag{8.23}$$

$$0_{\mathbb{R}^{n_y}} = g\big(x(t),y(t),u(t)\big) \quad \text{a. e. in } [0,1], \tag{8.24}$$

$$0_{\mathbb{R}^{n_\psi}} = \psi\big(x(0),x(1)\big). \tag{8.25}$$

Note that Problem 8.2.1 is more general than Problem 8.1.2 as it contains general boundary conditions of type (8.25). As a consequence, a complete reduction approach, as it became possible owing to Assumption 8.1.3, (b) in Section 8.1, cannot be applied anymore as the boundary conditions (8.25), in general, will not be feasible for an arbitrary control u. However, It is still possible to apply a *partially* reduced approach where (8.23)–(8.24) are solved with respect to $x = x(u, x_0)$ and $y = y(u, x_0)$ for a given control u and a given initial value $x_0 = x(0)$. The reduced problem then leads to a constrained optimization problem with the aim to minimize the reduced functional $J(u, x_0) := \Gamma(x(u, x_0), y(u, x_0), u)$ with respect to u and x_0 subject to the constraint $\psi(x_0, x(u, x_0)(1)) = 0_{\mathbb{R}^{n_\psi}}$. This approach is the function-space-equivalent to the direct multiple shooting method in Subsection 5.1.2 in Chapter 5. However, this partially reduced approach is not followed here. Problem 8.2.1 is considered instead as it is without a further reduction.

Although the application of Newton's method to the optimality system seems to be straightforward, several issues have to be addressed, for instance, the computation of search directions, local speed of convergence, and globalization. For simplicity, we restrict ourselves to semi-explicit index-one DAEs again.

Assumption 8.2.2. The inverse matrix $(g'_y(x,y,u))^{-1}$ exists and is bounded for every (x,y,u). □

To this end, first-order necessary optimality conditions for a weak local minimizer (x_*, y_*, u_*) of Problem 8.2.1 under the Mangasarian–Fromowitz constraint qualification in Theorem 3.4.5 read as follows, compare Theorem 3.4.3:

There exist Lagrange multipliers $\lambda_{f,*} \in W^{n_x}_{1,\infty}(\mathcal{I})$, $\lambda_{g,*} \in L^{n_y}_{\infty}(\mathcal{I})$, and $\sigma_* \in \mathbb{R}^{n_\psi}$ with

$$\dot{x}_*(t) - f(x_*(t), y_*(t), u_*(t)) = 0_{\mathbb{R}^{n_x}}, \tag{8.26}$$

$$g(x_*(t), y_*(t), u_*(t)) = 0_{\mathbb{R}^{n_y}}, \tag{8.27}$$

$$\dot{\lambda}_{f,*}(t) + \mathcal{H}'_x(x_*(t), y_*(t), u_*(t), \lambda_{f,*}(t), \lambda_{g,*}(t))^\top = 0_{\mathbb{R}^{n_x}}, \tag{8.28}$$

$$\mathcal{H}'_y(x_*(t), y_*(t), u_*(t), \lambda_{f,*}(t), \lambda_{g,*}(t))^\top = 0_{\mathbb{R}^{n_y}}, \tag{8.29}$$

$$\psi(x_*(0), x_*(1)) = 0_{\mathbb{R}^{n_\psi}}, \tag{8.30}$$

$$\lambda_{f,*}(0) + \kappa'_{x_0}(x_*(0), x_*(1), \sigma_*)^\top = 0_{\mathbb{R}^{n_x}}, \tag{8.31}$$

$$\lambda_{f,*}(1) - \kappa'_{x_1}(x_*(0), x_*(1), \sigma_*)^\top = 0_{\mathbb{R}^{n_x}}, \tag{8.32}$$

$$\mathcal{H}'_u(x_*(t), y_*(t), u_*(t), \lambda_{f,*}(t), \lambda_{g,*}(t))^\top = 0_{\mathbb{R}^{n_u}}. \tag{8.33}$$

Herein, we used the abbreviation $\kappa := \varphi + \sigma^\top \psi$ and the Hamilton function

$$\mathcal{H}(x, y, u, \lambda_f, \lambda_g) := f_0(x, y, u) + \lambda_f^\top f(x, y, u) + \lambda_g^\top g(x, y, u).$$

Let $z_* := (x_*, y_*, u_*, \lambda_{f,*}, \lambda_{g,*}, \sigma_*)$, and let the Banach spaces

$$Z = W^{n_x}_{1,\infty}(\mathcal{I}) \times L^{n_y}_\infty(\mathcal{I}) \times L^{n_u}_\infty(\mathcal{I}) \times W^{n_x}_{1,\infty}(\mathcal{I}) \times L^{n_y}_\infty(\mathcal{I}) \times \mathbb{R}^{n_\psi},$$

$$Y = L^{n_x}_\infty(\mathcal{I}) \times L^{n_y}_\infty(\mathcal{I}) \times L^{n_x}_\infty(\mathcal{I}) \times L^{n_y}_\infty(\mathcal{I}) \times \mathbb{R}^{n_\psi} \times \mathbb{R}^{n_x} \times \mathbb{R}^{n_x} \times L^{n_u}_\infty(\mathcal{I})$$

be equipped with the maximum norm for product spaces. Then the necessary conditions (8.26)–(8.33) are equivalent with the nonlinear equation

$$T(z_*) = \Theta_Y, \tag{8.34}$$

where $T : Z \longrightarrow Y$ is given by

$$T(z)(\cdot) := \begin{pmatrix} \dot{x}(\cdot) - f(x(\cdot), y(\cdot), u(\cdot)) \\ g(x(\cdot), y(\cdot), u(\cdot)) \\ \dot{\lambda}_f(\cdot) + H'_x(x(\cdot), y(\cdot), u(\cdot), \lambda_f(\cdot), \lambda_g(\cdot))^\top \\ H'_y(x(\cdot), y(\cdot), u(\cdot), \lambda_f(\cdot), \lambda_g(\cdot))^\top \\ \psi(x(0), x(1)) \\ \lambda_f(0) + \kappa'_{x_0}(x(0), x(1), \sigma)^\top \\ \lambda_f(1) - \kappa'_{x_1}(x(0), x(1), \sigma)^\top \\ H'_u(x(\cdot), y(\cdot), u(\cdot), \lambda_f(\cdot), \lambda_g(\cdot))^\top \end{pmatrix}. \tag{8.35}$$

The standard approach to solve (8.35) numerically is to apply Newton's method.

Algorithm 8.2.3 (Lagrange–Newton method).
(0) Choose $z^{(0)} \in Z$ and set $k := 0$.
(1) If $\|T(z^{(k)})\|_Y \approx 0$, STOP.
(2) Compute the search direction $d^{(k)}$ from the linear equation

$$T'(z^{(k)})(d^{(k)}) = -T(z^{(k)}). \tag{8.36}$$

(3) Set $z^{(k+1)} := z^{(k)} + d^{(k)}$, $k \leftarrow k + 1$, and go to (1). □

Theorem 8.2.4. *Let z_* be a zero of T. Suppose that there exist constants $\Delta > 0$ and $C > 0$ such that for every $z \in B_\Delta(z_*)$ the derivative $T'(z)$ is non-singular and*

$$\|T'(z)^{-1}\|_{\mathcal{L}(Y,Z)} \leq C.$$

(a) *If φ, f_0, f, g, ψ are twice continuously differentiable, then there exists $\delta > 0$ such that Algorithm* 8.2.3 *is well-defined for every $z^{(0)} \in B_\delta(z_*)$, and the sequence $\{z^{(k)}\}_{k\in\mathbb{N}}$ converges superlinearly to z_* for every $z^{(0)} \in B_\delta(z_*)$.*

(b) *If the second derivatives of φ, f_0, f, g, ψ are locally Lipschitz continuous, then the convergence in (a) is quadratic.*

(c) *If in addition to the assumption in (a) $T(z^{(k)}) \neq \Theta_Y$ for all k, then the residual values converge superlinearly:*

$$\lim_{k\longrightarrow\infty} \frac{\|T(z^{(k+1)})\|_Y}{\|T(z^{(k)})\|_Y} = 0.$$

Proof. Owing to the non-singularity of $T'(z)$ and the uniform boundedness of $T'(z)^{-1}$ in $B_\Delta(z_*)$, the algorithm is well-defined in $B_\Delta(z_*)$. For $z^{(k)} \in B_\Delta(z_*)$, it holds

$$\begin{aligned} T'(z^{(k)})(z^{(k+1)} - z_*) &= T'(z^{(k)})(z^{(k)} + d^{(k)} - z_*) \\ &= T'(z^{(k)})(z^{(k)} - z_*) + T'(z^{(k)})(d^{(k)}) \\ &= T'(z^{(k)})(z^{(k)} - z_*) - T(z^{(k)}) + T(z_*), \end{aligned}$$

and

$$\begin{aligned} \|z^{(k+1)} - z_*\|_Z &= \|T'(z^{(k)})^{-1}(T'(z^{(k)})(z^{(k)} - z_*) - T(z^{(k)}) + T(z_*))\|_Z \\ &\leq \|T'(z^{(k)})^{-1}\|_{\mathcal{L}(Y,Z)} \cdot \|T(z^{(k)}) - T(z_*) - T'(z^{(k)})(z^{(k)} - z_*)\|_Y \\ &\leq C \cdot \|T(z^{(k)}) - T(z_*) - T'(z^{(k)})(z^{(k)} - z_*)\|_Y. \end{aligned} \tag{8.37}$$

(i) If T is continuously Fréchet-differentiable in some neighborhood of z_*, then by the mean value theorem, [169, p. 28],

$$\begin{aligned} &\|T(z^{(k)}) - T(z_*) - T'(z^{(k)})(z^{(k)} - z_*)\|_Y \\ &\quad \leq \sup_{0\leq t\leq 1} \|T'(z_* + t(z^{(k)} - z_*)) - T'(z_*)\|_{\mathcal{L}(Z,Y)} \cdot \|z^{(k)} - z_*\|_Z. \end{aligned} \tag{8.38}$$

Owing to the differentiability assumptions in (a), $T'(\cdot)$ is continuous in some neighborhood of z_*. Hence, for every $\varepsilon > 0$, there exists some $0 < \delta \leq \Delta$ such that

$$\|T'(z_* + t(z^{(k)} - z_*)) - T'(z_*)\|_{\mathcal{L}(Z,Y)} \leq \varepsilon \tag{8.39}$$

for every $\|z^{(k)} - z_*\|_Z \leq \delta$ and every $t \in [0, 1]$. Hence,

$$\|T(z^{(k)}) - T(z_*) - T'(z^{(k)})(z^{(k)} - z_*)\|_Y \leq \varepsilon\|z^{(k)} - z_*\|_Z$$

for every $\|z^{(k)} - z_*\|_Z \leq \delta$.
Particularly, for $\varepsilon < \frac{1}{C}$, it holds $\|z^{(k+1)} - z_*\|_Z \leq \delta \leq \Delta$, and the sequence $\{z^{(k)}\}_{k\in\mathbb{N}}$ stays in $B_\delta(z_*)$.

As $\varepsilon > 0$ was arbitrary, this shows the locally superlinear convergence of $\{z^{(k)}\}_{k\in\mathbb{N}}$ to z_* in (a).

(ii) If the second derivatives of φ, f_0, f, g, ψ are locally Lipschitz continuous, then T' satisfies locally the Lipschitz condition

$$\|T'(z_* + t(z^{(k)} - z_*)) - T'(z_*)\|_{\mathcal{L}(Z,Y)} \leq L(z_*)\|z^{(k)} - z_*\|_Z$$

with Lipschitz constant $L(z_*)$ for every $z^{(k)}$ in that neighborhood. With (8.37) and (8.38), this shows the locally quadratic convergence in (b).

(iii) Let $\varepsilon > 0$ be arbitrary. Because of (8.37), (8.38), and (8.39), there exists $0 < \delta \leq \Delta$ with

$$\|z^{(k+1)} - z_*\|_Z \leq \varepsilon\|z^{(k)} - z_*\|_Z \quad \text{whenever } \|z^{(k)} - z_*\|_Z \leq \delta.$$

Notice that for any $\delta > 0$, there exists some $k_0(\delta)$ such that $\|z^{(k)} - z_*\| \leq \delta$ for every $k \geq k_0(\delta)$ since $z^{(k)}$ converges to z_*. As T is continuously Fréchet-differentiable, it satisfies the Lipschitz condition

$$\begin{aligned}\|T(z^{(k+1)})\|_Y &= \|T(z^{(k+1)}) - T(z_*)\|_Y \\ &\leq \tilde{L}(z_*)\|z^{(k+1)} - z_*\|_Z \\ &\leq \tilde{L}(z_*)\varepsilon\|z^{(k)} - z_*\|_Z\end{aligned}$$

locally around z_*, and the Newton iteration implies

$$\|z^{(k+1)} - z^{(k)}\|_Z \leq \|T'(z^{(k)})^{-1}\|_{\mathcal{L}(Y,Z)} \cdot \|T(z^{(k)})\|_Y \leq C\|T(z^{(k)})\|_Y.$$

Thus,

$$\begin{aligned}\|z^{(k)} - z_*\|_Z &\leq \|z^{(k+1)} - z^{(k)}\|_Z + \|z^{(k+1)} - z_*\|_Z \\ &\leq C\|T(z^{(k)})\|_Y + \|z^{(k+1)} - z_*\|_Z \\ &\leq C\|T(z^{(k)})\|_Y + \varepsilon\|z^{(k)} - z_*\|_Z\end{aligned}$$

and

$$\|z^{(k)} - z_*\|_Z \leq \frac{C}{1-\varepsilon}\|T(z^{(k)})\|_Y.$$

Finally,

$$\|T(z^{(k+1)})\|_Y \leq \tilde{L}(z_*)\varepsilon\|z^{(k)} - z_*\|_Z \leq \frac{\tilde{L}(z_*)\varepsilon C}{1-\varepsilon}\|T(z^{(k)})\|_Y.$$

Since $T(z^{(k)}) \neq \Theta_Y$ and ε was arbitrary, this shows the assertion in (c). □

Similarly as in the reduced gradient method in Algorithm 8.1.7, a globalization technique can be employed to enlarge the convergence region of the local Lagrange–Newton method in Algorithm 8.2.3. To this end, the squared L_2-norm of T, i. e.

$$\gamma(z) := \frac{1}{2}\|T(z)\|_2^2,$$

is used as a merit function in combination with a line-search procedure of the Armijo type leading to

Algorithm 8.2.5 (Globalized Lagrange–Newton method).

(0) Choose $z^{(0)} \in Z$, $\beta \in (0,1)$, $\sigma \in (0,1/2)$.

(1) If $\gamma(z^{(k)}) \approx 0$, STOP.

(2) Compute the search direction $d^{(k)}$ from (8.36).

(3) Find smallest $j \in \{0,1,2,\dots\}$ with

$$\gamma(z^{(k)} + \beta^j d^{(k)}) \le \gamma(z^{(k)}) + \sigma\beta^j \gamma'(z^{(k)})(d^{(k)})$$

and set $\alpha_k := \beta^j$.

(4) Set $z^{(k+1)} := z^{(k)} + \alpha_k d^{(k)}$, $k \leftarrow k+1$, and go to (1). □

The function $\gamma : Z \longrightarrow \mathbb{R}$ is Fréchet-differentiable, if φ, f_0, f, g, ψ are twice continuously differentiable, and it holds

$$\gamma'(z^{(k)})(d^{(k)}) = -2\gamma(z^{(k)}) = -\|T(z^{(k)})\|_2^2, \tag{8.40}$$

where $d^{(k)}$ satisfies $T'(z^{(k)})(d^{(k)}) = -T(z^{(k)})$. As a consequence, $d^{(k)}$ is a direction of descent of γ at $z^{(k)}$, and the line-search in Algorithm 8.2.5 is well-defined unless $z^{(k)}$ is a zero of T.

8.2.1 Computation of the search direction

The main effort in Algorithms 8.2.3 and 8.2.5 is caused by the computation of the search direction $d^{(k)}$ in step (2) for which the operator equation (8.36) needs to be solved. The Fréchet-derivative of T at $z^{(k)} = (x^{(k)}, y^{(k)}, u^{(k)}, \lambda_f^{(k)}, \lambda_g^{(k)}, \sigma^{(k)}) \in Z$ in the direction $z = (x, y, u, \lambda_f, \lambda_g, \sigma) \in Z$ is given by

$$T'(z^{(k)})(z) = \begin{pmatrix} \dot{x} - f_x' x - f_y' y - f_u' u \\ g_x' x + g_y' y + g_u' u \\ \dot{\lambda}_f + \mathcal{H}_{xx}'' x + \mathcal{H}_{xy}'' y + \mathcal{H}_{xu}'' u + \mathcal{H}_{x\lambda_f}'' \lambda_f + \mathcal{H}_{x\lambda_g}'' \lambda_g \\ \mathcal{H}_{yx}'' x + \mathcal{H}_{yy}'' y + \mathcal{H}_{yu}'' u + \mathcal{H}_{y\lambda_f}'' \lambda_f + \mathcal{H}_{y\lambda_g}'' \lambda_g \\ \psi_{x_0}' x(0) + \psi_{x_1}' x(1) \\ \lambda_f(0) + \kappa_{x_0 x_0}'' x(0) + \kappa_{x_0 x_1}'' x(1) + \kappa_{x_0 \sigma}'' \sigma \\ \lambda_f(1) - \kappa_{x_1 x_0}'' x(0) - \kappa_{x_1 x_1}'' x(1) - \kappa_{x_1 \sigma}'' \sigma \\ \mathcal{H}_{ux}'' x + \mathcal{H}_{uy}'' y + \mathcal{H}_{uu}'' u + \mathcal{H}_{u\lambda_f}'' \lambda_f + \mathcal{H}_{u\lambda_g}'' \lambda_g \end{pmatrix}.$$

Herein, all partial derivatives of the functions φ, f_0, f, g, and ψ are evaluated at the current iterate $z^{(k)}$. The linear operator equation (8.36) with $d^{(k)} = (x, y, u, \lambda_f, \lambda_g, \sigma)$ in step (2) of Algorithms 8.2.3 and 8.2.5 reads as

$$\begin{pmatrix} \dot{x} \\ \dot{\lambda}_f \\ 0_{\mathbb{R}^{n_y}} \\ 0_{\mathbb{R}^{n_y}} \\ 0_{\mathbb{R}^{n_u}} \end{pmatrix} - \begin{pmatrix} f_x' & \Theta & f_y' & \Theta & f_u' \\ -\mathcal{H}_{xx}'' & -\mathcal{H}_{x\lambda_f}'' & -\mathcal{H}_{xy}'' & -\mathcal{H}_{x\lambda_g}'' & -\mathcal{H}_{xu}'' \\ -g_x' & \Theta & -g_y' & \Theta & -g_u' \\ -\mathcal{H}_{yx}'' & -\mathcal{H}_{y\lambda_f}'' & -\mathcal{H}_{yy}'' & -\mathcal{H}_{y\lambda_g}'' & -\mathcal{H}_{yu}'' \\ -\mathcal{H}_{ux}'' & -\mathcal{H}_{u\lambda_f}'' & -\mathcal{H}_{uy}'' & -\mathcal{H}_{u\lambda_g}'' & -\mathcal{H}_{uu}'' \end{pmatrix} \begin{pmatrix} x \\ \lambda_f \\ y \\ \lambda_g \\ u \end{pmatrix}$$

$$= - \begin{pmatrix} \dot{x}^{(k)} - f \\ \dot{\lambda}_f^{(k)} + (\mathcal{H}_x')^\top \\ g \\ (\mathcal{H}_y')^\top \\ (\mathcal{H}_u')^\top \end{pmatrix} \tag{8.41}$$

and

$$\begin{pmatrix} \psi_{x_0}' & \Theta & \Theta \\ \kappa_{x_0x_0}'' & I_{n_x} & \kappa_{x_0\sigma}'' \\ -\kappa_{x_1x_0}'' & \Theta & -\kappa_{x_1\sigma}'' \end{pmatrix} \begin{pmatrix} x(0) \\ \lambda_f(0) \\ \sigma \end{pmatrix} + \begin{pmatrix} \psi_{x_1}' & \Theta & \Theta \\ \kappa_{x_0x_1}'' & \Theta & \Theta \\ -\kappa_{x_1x_1}'' & I_{n_x} & \Theta \end{pmatrix} \begin{pmatrix} x(1) \\ \lambda_f(1) \\ \sigma \end{pmatrix}$$

$$= - \begin{pmatrix} \psi \\ \lambda_f^{(k)}(0) + (\kappa_{x_0}')^\top \\ \lambda_f^{(k)}(1) - (\kappa_{x_1}')^\top \end{pmatrix}. \tag{8.42}$$

Equations (8.41) and (8.42) define a *linear DAE boundary value problem* with differential variables x and λ_f and algebraic variables y, λ_g, and u.

Such linear DAE boundary value problems can be solved numerically by the collocation method in [195] or the shooting methods in Section 4.6. This requires introducing a grid on which the solution of the DAE boundary value problem is approximated. Of course, this introduces a discretization error, but only at this stage, a discretization method enters the scene.

The following result provides a sufficient condition for the DAE being index-one. It follows immediately from (8.41).

Theorem 8.2.6. *The DAE (8.41) has index one, if the matrix function*

$$M(t) := \begin{pmatrix} g_y'[t] & \Theta & g_u'[t] \\ \mathcal{H}_{yy}''[t] & \mathcal{H}_{y\lambda_g}''[t] & \mathcal{H}_{yu}''[t] \\ \mathcal{H}_{uy}''[t] & \mathcal{H}_{u\lambda_g}''[t] & \mathcal{H}_{uu}''[t] \end{pmatrix} = \begin{pmatrix} g_y'[t] & \Theta & g_u'[t] \\ \mathcal{H}_{yy}''[t] & (g_y'[t])^\top & \mathcal{H}_{yu}''[t] \\ \mathcal{H}_{uy}''[t] & (g_u'[t])^\top & \mathcal{H}_{uu}''[t] \end{pmatrix}$$

is non-singular for almost every $t \in [0,1]$ and $\|M(t)^{-1}\| \le C$ for some constant C and almost every $t \in [0,1]$. □

If the operator $M(\cdot)$ is not invertible, the situation becomes more involved as (8.41) becomes a higher-index DAE in this case. Especially, the computation of consistent initial values imposes additional difficulties as it may happen that, owing to the linearization at some iterate $z^{(k)}$, algebraic constraints and boundary conditions contradict themselves and eventually lead to an infeasible boundary value problem.

In order to demonstrate the performance of the Lagrange–Newton method, we first consider a slight modification of the trolley example in [126].

Example 8.2.7. We consider an optimal control problem for a trolley of mass m_1 moving in a high rack storage area. A load of mass m_2 is attached to the trolley by a rigid cable of length ℓ, see Figure 8.3. Herein, x_1 and x_3 denote the x-coordinate of the trolley and its velocity, respectively, and x_2 and x_4 refer to the angle between the vertical axis and cable and its velocity, respectively.

The acceleration of the trolley can be controlled by the control u. The equations of motion of the trolley are given by the following differential equations for the state $x = (x_1, x_2, x_3, x_4)^\top$:

$$\dot{x}_1 = x_3, \tag{8.43}$$

$$\dot{x}_2 = x_4, \tag{8.44}$$

$$\begin{aligned}\dot{x}_3 = {} & \frac{m_2^2\ell^3 \sin(x_2)x_4^2 - m_2\ell^2 u + m_2 I_y \ell x_4^2 \sin(x_2) - I_y u}{-m_1 m_2 \ell^2 - m_1 I_y - m_2^2\ell^2 - m_2 I_y + m_2^2\ell^2 \cos(x_2)^2} \\ & + \frac{m_2^2\ell^2 g \cos(x_2)\sin(x_2)}{-m_1 m_2 \ell^2 - m_1 I_y - m_2^2\ell^2 - m_2 I_y + m_2^2\ell^2 \cos(x_2)^2},\end{aligned} \tag{8.45}$$

$$\dot{x}_4 = \frac{m_2\ell(m_2\ell\cos(x_2)x_4^2\sin(x_2) - \cos(x_2)u + g\sin(x_2)(m_1 + m_2))}{-m_1 m_2 \ell^2 - m_1 I_y - m_2^2\ell^2 - m_2 I_y + m_2^2\ell^2 \cos(x_2)^2}. \tag{8.46}$$

Figure 8.3: Configuration of the trolley and the load.

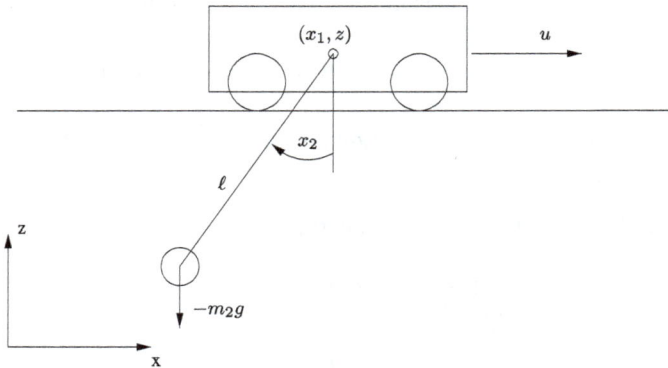

The optimal control problem is defined as follows:

Minimize

$$\frac{1}{2}\int_0^{t_f} u(t)^2 + 5x_4(t)^2 \, dt$$

subject to (8.43)–(8.46), the initial conditions

$$x_1(0) = x_2(0) = x_3(0) = x_4(0) = 0,$$

and the terminal conditions

$$x_1(t_f) = 1, \ x_2(t_f) = x_3(t_f) = x_4(t_f) = 0$$

within the fixed time $t_f = 2.7$.

The objective function aims at minimizing the steering effort and the angle velocity of the cable to avoid extensive swinging of the load.

The following parameters were used for the numerical computations:

$$g = 9.81, \quad m_1 = 0.3, \quad m_2 = 0.5, \quad \ell = 0.75, \quad r = 0.1, \quad I_y = 0.002.$$

Table 8.4 summarizes CPU times for the Lagrange–Newton method depending on the number N of equidistant intervals used in the linear boundary value problems. Table 8.4 shows the output of the Lagrange–Newton method with line-search, i. e., step-size α_k, residual norm $\|T(z^{(k)})\|_2^2$, and search direction $\|d^{(k)}\|_\infty$. The iterations show the rapid quadratic convergence.

Table 8.4: Output of globalized Lagrange–Newton method for the trolley example for $N = 1000$ subintervals and the Euler discretization: local quadratic convergence.

k	α_k	$\|T(z^{(k)})\|_2^2$	$\|d^{(k)}\|_\infty$
0	0.000000e+00	0.100000e+01	0.451981e-01
1	0.100000e+01	0.688773e-03	0.473501e-02
2	0.100000e+01	0.809983e-12	0.118366e-06
3	0.100000e+01	0.160897e-24	0.141058e-11

Table 8.5 summarizes results for different step-sizes N. The number of iterations is nearly constant, which indicates – at least numerically – the mesh independence of the method. Furthermore, the CPU time grows at a linear rate with N. Mesh independence of the Lagrange–Newton method has been established in [5] and for a certain class of semi-smooth Newton methods in [165]. Finally, Figure 8.4 illustrates the iterates of the Lagrange–Newton method. □

Table 8.5: Results for the trolley example for different step-sizes N.

N	CPU time [s]	Iterations
100	0.022	3
200	0.050	3
400	0.093	3
800	0.174	3
1600	0.622	3
3200	0.822	3
6400	1.900	4
12800	3.771	4
25600	7.939	4

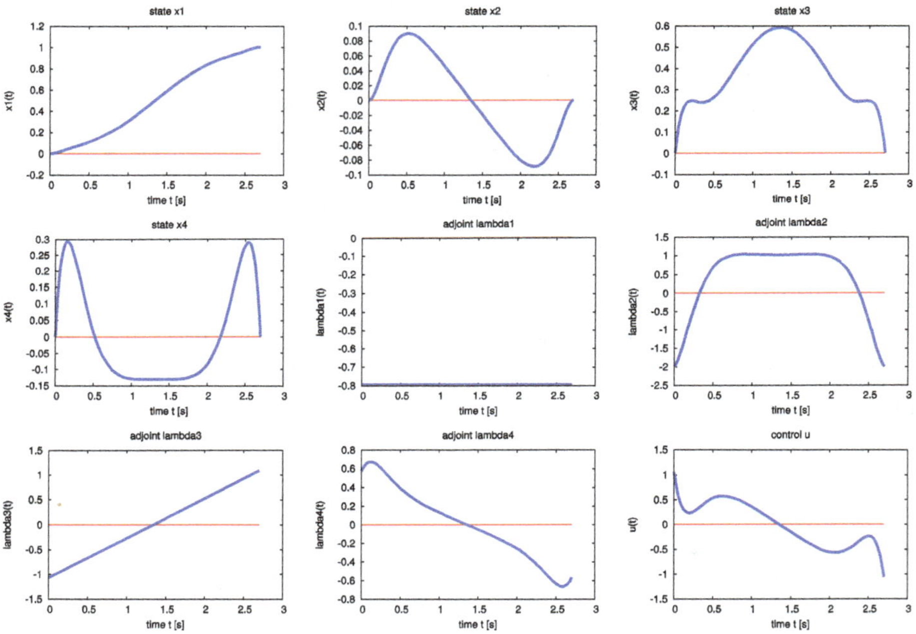

Figure 8.4: Numerical solution of the trolley example for $N = 1000$ Euler steps: States, adjoints, and control at initial guess (thin lines) and converged solution (thick lines).

Example 8.2.8 (Compare [322, 164]). A model for a specific chemical reaction

$$A + B \longrightarrow C \quad \text{and} \quad C + D \longrightarrow D$$

of substances A, B, C, and D is modeled by the following index-one DAE:

$$\dot{M}_A = -V \cdot A_1 \cdot e^{-E_1/T_R} \cdot C_A \cdot C_B, \tag{8.47}$$

$$\dot{M}_B = F_B - V(A_1 e^{-E_1/T_R} \cdot C_A \cdot C_B + A_2 \cdot e^{-E_2/T_R} \cdot C_B \cdot C_C), \tag{8.48}$$

$$\dot{M}_C = V(A_1 e^{-E_1/T_R} \cdot C_A \cdot C_B - A_2 \cdot e^{-E_2/T_R} \cdot C_B \cdot C_C), \tag{8.49}$$

$$\dot{M}_D' = V \cdot A_2 \cdot e^{-E_2/T_R} \cdot C_B \cdot C_C, \tag{8.50}$$

$$\dot{H} = 20F_B - Q - V(-A_1 e^{-E_1/T_R} \cdot C_A \cdot C_B - 75A_2 \cdot e^{-E_2/T_R} \cdot C_B \cdot C_C) \tag{8.51}$$

$$0 = H - \sum_{i=A,B,C,D} M_i\left(\alpha_i(T_R - T_{\text{ref}}) + \frac{\beta_i}{2}(T_R^2 - T_{\text{ref}}^2)\right), \tag{8.52}$$

where

$$V = \sum_{i=A,B,C,D} \frac{M_i}{\rho_i}, \quad C_i = M_i/V, \quad i = A, B, C, D.$$

Herein, F_B is the feed rate of substance B and Q the cooling power. M_i denotes the mass of substance $i = A, B, C, D$. H is the total energy. $T_{\text{ref}} = 298$ is a given target temperature, and the algebraic variable T_R denotes the temperature.

The parameters $\alpha_i, \beta_i, \rho_i, i = A, B, C, D, A_1, A_2, E_1, E_2$, are given by

$$\begin{aligned}
\alpha_A &= 0.1723, & \beta_A &= 0.000473, & \rho_A &= 11250, & A_1 &= 0.8, \\
\alpha_B &= 0.2, & \beta_B &= 0.0005, & \rho_B &= 16000, & A_2 &= 0.002, \\
\alpha_C &= 0.16, & \beta_C &= 0.00055, & \rho_C &= 10400, & E_1 &= 3000, \\
\alpha_D &= 0.155, & \beta_D &= 0.000323, & \rho_D &= 10000, & E_2 &= 2400.
\end{aligned}$$

Initial conditions are given by

$$M_A(0) = 9000, \quad M_B(0) = 0, \quad M_C(0) = 0, \quad M_D(0) = 0, \quad H(0) = 152509.97. \tag{8.53}$$

The optimal control problem reads as follows:

Minimize

$$-M_C(t_f) + 10^{-2} \int_0^{t_f} F_B(t)^2 + Q(t)^2 dt$$

with $t_f = 20$ subject to the DAE (8.47)–(8.52) and the initial conditions (8.53).

Table 8.6 summarizes the output of the globalized Lagrange–Newton method in Algorithm 8.2.5 for the parameters $\beta = 0.9$ and $\sigma = 0.1$.

The initial vector $z^{(0)}$ was chosen to be constant with respect to time. For the state vector, the respective constants were given by the initial values in (8.53), for the algebraic variable T_R, the initial guess was chosen to be T_{ref}, all remaining components of $z^{(0)}$ were set to zero.

Figures 8.5, 8.6, 8.7 illustrate the states, adjoints, and controls obtained in the Lagrange–Newton iterations. □

Table 8.6: Output of globalized Lagrange–Newton method for the chemical reaction example for $N = 20000$ subintervals and the Euler discretization.

k	α_k	$\|T(z^{(k)})\|_2^2$	$\|d^{(k)}\|_\infty$
0	0.000000e+00	0.465186e+12	0.599335e+05
1	0.100000e+01	0.759821e+10	0.523150e+07
2	0.147809e-01	0.755228e+10	0.127262e+07
3	0.423912e-01	0.745716e+10	0.835212e+05
4	0.100000e+01	0.351002e+09	0.344908e+06
5	0.121577e+00	0.340325e+09	0.667305e+05
6	0.100000e+01	0.131555e+08	0.370395e+05
7	0.100000e+01	0.295389e+07	0.169245e+04
8	0.100000e+01	0.114958e+01	0.336606e+00
9	0.100000e+01	0.852576e-11	0.104227e-02
10	0.100000e+01	0.125527e-11	0.658317e-03
11	0.100000e+01	0.283055e-14	0.605453e-05

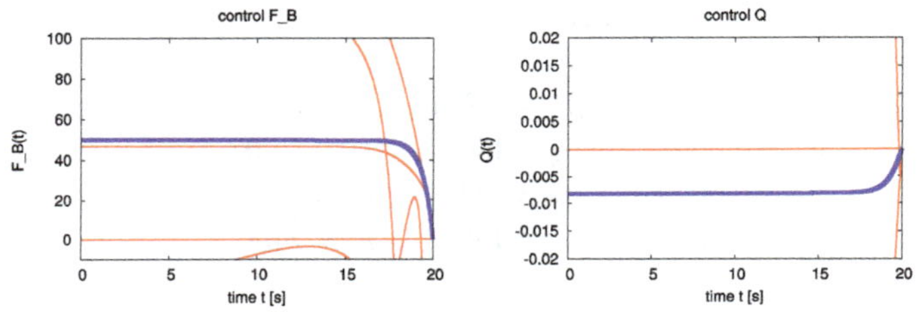

Figure 8.5: Numerical solution of the chemical reaction example for $N = 20000$ Euler steps: Controls at intermediate iterates (thin lines) and converged solution (thick lines).

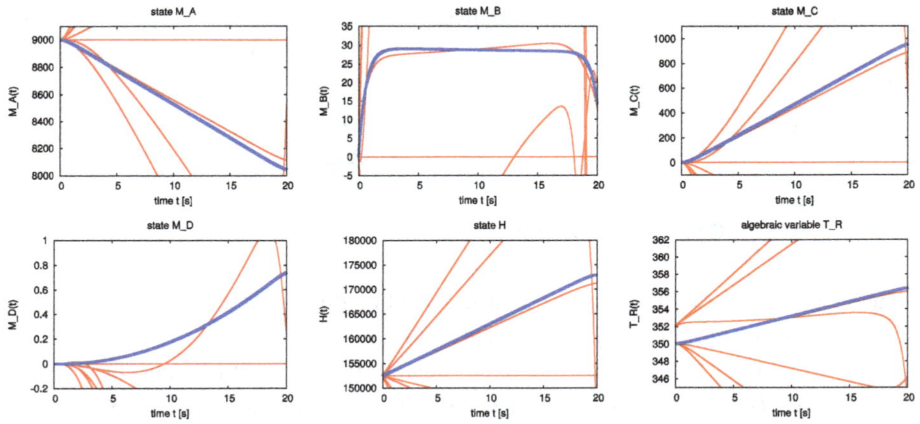

Figure 8.6: Numerical solution of the chemical reaction example for $N = 20000$ Euler steps: Differential and algebraic states at intermediate iterates (thin lines) and converged solution (thick lines).

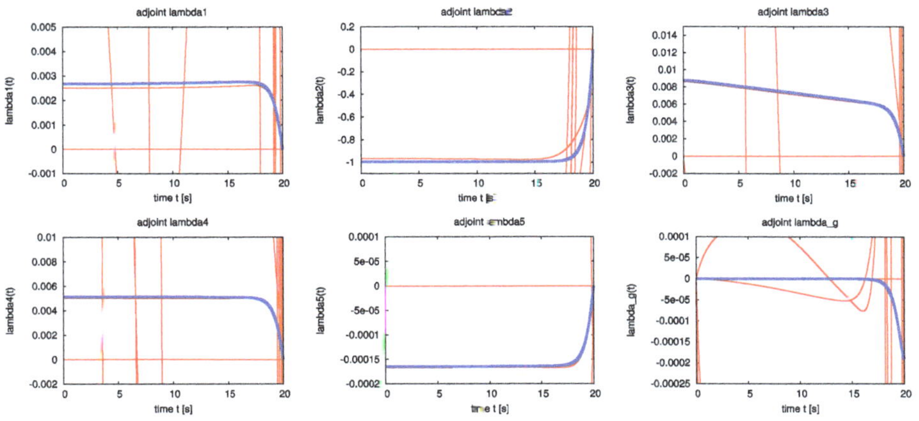

Figure 8.7: Numerical solution of the chemical reaction example for $N = 20000$ Euler steps: Adjoints at intermediate iterates (thin lines) and converged solution (thick lines).

The Lagrange–Newton method can be extended in a similar way to index-two DAEs, if the necessary conditions in Theorem 3.1.9 are used, and some safeguards regarding consistent initial values for the resulting linear index-two DAE boundary value problems are obeyed, see [130] for details. The following example from [130] demonstrates the application of the Lagrange–Newton method to an optimal control problem with the Navier–Stokes equation.

Example 8.2.9 (Unconstrained 2D Navier–Stokes problem, communicated by M. Kunkel). We illustrate the Lagrange–Newton method for the distributed control of the two-dimensional non-stationary incompressible Navier–Stokes equations on $Q := (0, t_f) \times \Omega$ with $\Omega = (0,1) \times (0,1)$.

The task is to minimize the distance to the desired velocity field

$$y_d(t, x_1, x_2) = (-q(t, x_1)q'_{x_2}(t, x_2), q(t, x_2)q'_{x_1}(t, x_1))^\top,$$
$$q(t, z) = (1 - z)^2(1 - \cos(2\pi z t))$$

within a given time $t_f > 0$:

Minimize

$$\frac{1}{2}\int_Q \|y(t,x_1,x_2) - y_d(t,x_1,x_2)\|^2 \, dx_1\, dx_2\, dt + \frac{\delta}{2}\int_Q \|u(t,x_1,x_2)\|^2 \, dx_1\, dx_2\, dt$$

with respect to the velocity vector field $y = (v, w)^\top$, *the pressure* p *and the control vector field* $u = (u_1, u_2)^\top$ *subject to the 2D Navier–Stokes equations*

$$y_t = \frac{1}{Re}\Delta y - (y \cdot \nabla)y - \nabla p + u \quad \text{in } Q,$$

$$
\begin{aligned}
0 &= \operatorname{div}(y) && \text{in } Q,\\
0_{\mathbb{R}^2} &= y(0,x_1,x_2) && \text{for } (x_1,x_2) \in \Omega,\\
0_{\mathbb{R}^2} &= y(t,x_1,x_2) && \text{for } (t,x_1,x_2) \in (0,t_f)\times\partial\Omega.
\end{aligned}
$$

The non-stationary incompressible Navier–Stokes equations can be viewed as a partial differential-algebraic equation with differential variable y and algebraic variable p. We discretize the problem in space on an equally spaced mesh with step-length $h = \frac{1}{N}$, $N \in \mathbb{N}$, while the time stays continuous. Let

$$
\begin{aligned}
y_{ij}(t) = (v_{ij}(t), w_{ij}(t))^\top &\approx y(t, x_{1,i}, x_{2,j}),\\
p_{ij}(t) &\approx p(t, x_{1,i}, x_{2,j}),\\
u_{ij}(t) &\approx u(t, x_{1,i}, x_{2,j}),
\end{aligned}
$$

for $i = 0,\dots,N$ and $j = 0,\dots,N$ denote the approximations at the grid points. The operators in the Navier–Stokes equations are approximated at the points $(t, x_{1,i}, x_{2,j})$, $i,j = 1,\dots,N-1$, by the finite difference schemes

$$
\begin{aligned}
\Delta y &\approx \frac{1}{h^2}(y_{i+1,j}(t) + y_{i-1,j}(t) + y_{i,j+1}(t) + y_{i,j-1}(t) - 4y_{i,j}(t)),\\
(y\cdot\nabla)y &\approx \frac{1}{2h}(v_{ij}(t)(y_{i+1,j}(t) - y_{i-1,j}(t)) + w_{ij}(t)(y_{i,j+1}(t) - y_{i,j-1}(t))),\\
\nabla p &\approx \frac{1}{h}(p_{i+1,j}(t) - p_{i,j}(t), p_{i,j+1}(t) - p_{i,j}(t))^\top,\\
\operatorname{div}(y) &\approx \frac{1}{h}(v_{i,j}(t) - v_{i-1,j}(t) + w_{i,j}(t) - w_{i,j-1}(t)).
\end{aligned}
$$

The undefined pressure components $p_{i,j}$ with $i = N$ or $j = N$ are set to zero.

Introducing these approximations into the Navier–Stokes equations and exploiting the boundary conditions $y_{i,0}(t) = y_{i,N}(t) = 0$ for $i = 1,\dots,N-1$ and $y_{0,j}(t) = y_{N,j}(t) = 0$ for $j = 1,\dots,N-1$ yields the following index-two DAE optimal control problem:

Minimize

$$
\frac{1}{2}\int_0^{t_f} \|y_h(t) - y_{d,h}(t)\|^2\,dt + \frac{\delta}{2}\int_0^{t_f} \|u_h(t)\|^2\,dt
$$

subject to the DAE

$$
\begin{aligned}
\dot{y}_h(t) &= \frac{1}{Re}A_h y_h(t) - \frac{1}{2}\begin{pmatrix} y_h(t)^\top H_{h,1} y_h(t)\\ \vdots \\ y_h(t)^\top H_{h,2(N-1)^2} y_h(t)\end{pmatrix} - B_h p_h(t) + u_h(t),\\
0_{\mathbb{R}^{(N-1)^2}} &= B_h^\top y_h(t),\\
y_h(0) &= 0_{\mathbb{R}^{(N-1)^2}}.
\end{aligned}
$$

Herein, we used the vectors

$$y_h = (y_{1,1}, \dots, y_{N-1,1}, y_{1,2}, \dots, y_{N-1,2}, \dots, y_{1,N-1}, \dots, y_{N-1,N-1})^\top,$$
$$p_h = (p_{1,1}, \dots, p_{N-1,1}, p_{1,2}, \dots, p_{N-1,2}, \dots, p_{1,N-1}, \dots, p_{N-1,N-1})^\top,$$
$$u_h = (u_{1,1}, \dots, u_{N-1,1}, u_{1,2}, \dots, u_{N-1,2}, \dots, u_{1,N-1}, \dots, u_{N-1,N-1})^\top.$$

The matrices $A_h \in \mathbb{R}^{2(N-1)^2 \times 2(N-1)^2}$ and $B_h \in \mathbb{R}^{2(N-1)^2 \times (N-1)^2}$ represent the discretized Laplacian and the discretized gradient, respectively. $H_{h,i} \in \mathbb{R}^{2(N-1)^2 \times 2(N-1)^2}$ is the Hessian of the ith component of the discretized convective term with respect to y_h, that is

$$H_{h,2(i+j(N-1))-1} := \nabla^2_{y_h} q_{ij,1}(t), \quad i,j = 1, \dots, N-1,$$
$$H_{h,2(i+j(N-1))} := \nabla^2_{y_h} q_{ij,2}(t), \quad i,j = 1, \dots, N-1,$$

where

$$q_{ij}(t) := (y \cdot \nabla)y|_{(t,x_{1,i},x_{2,j})} \in \mathbb{R}^2.$$

Note that the matrices $H_{h,k}$, $k = 1, \dots, 2(N-1)^2$, do not depend on $y_h(t)$ since the convective term is quadratic, and thus, $H_{h,k}$ may be calculated once in a preprocessing step.

Figures 8.8 and 8.9 show the results of the Lagrange–Newton method applied to the discretized Navier–Stokes problem for the parameters $t_f = 2$, $\delta = 10^{-5}$, $Re = 1$, $N = 31$, $N_t = 60$. In this case, the DAE optimal control problem has $n_x = 2(N-1)^2 = 1800$ differential variables, $n_y = (N-1)^2 = 900$ algebraic variables, and $n_u = 1800$ controls.

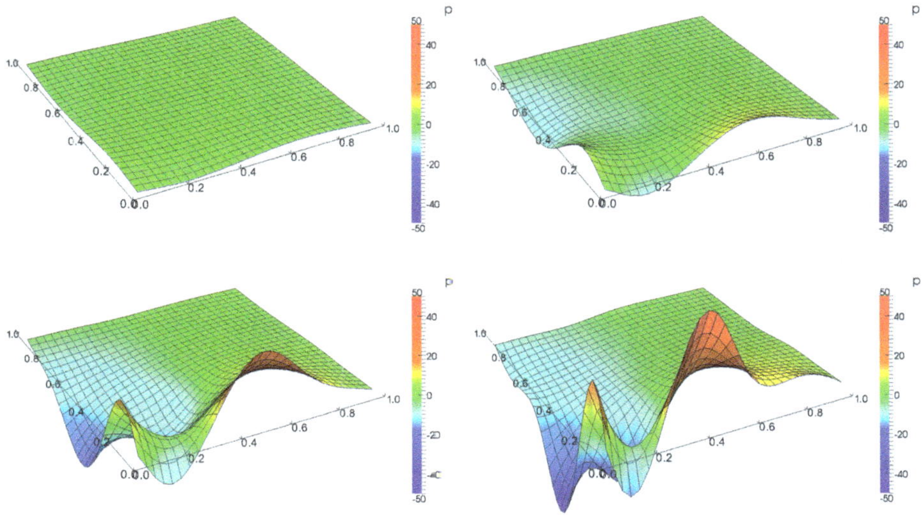

Figure 8.8: Optimal control of Navier–Stokes equations: Pressure p at $t = 0.6$, $t = 1.0$, $t = 1.4$ and $t = 1.967$.

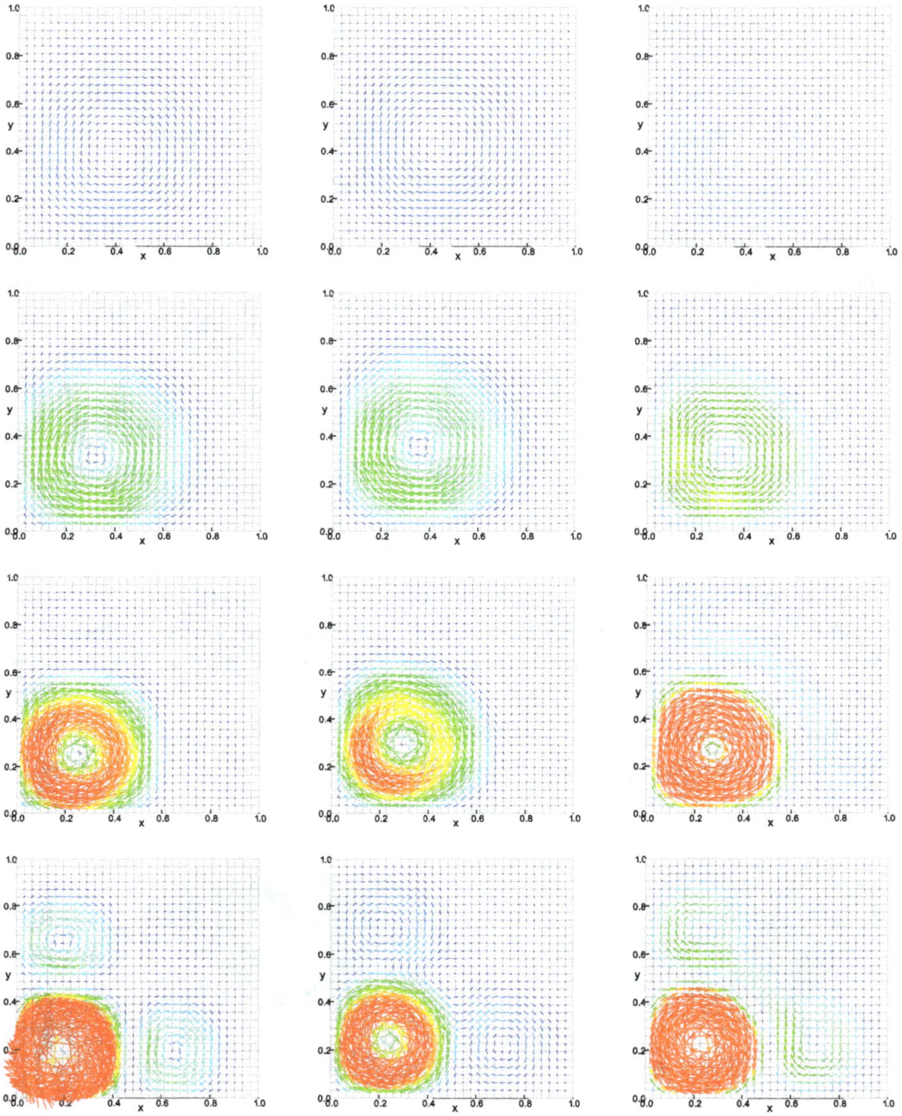

Figure 8.9: Optimal control of Navier–Stokes equations. Desired flow (left), controlled flow (middle), and control (right) at $t = 0.6$, $t = 1.0$, $t = 1.4$ and $t = 1.967$.

Recall that the search direction satisfies a linear DAE boundary value problem that is being solved by a collocation method, see [195]. To this end, in each iteration of the Lagrange–Newton method, a linear equation with almost two million variables has to be solved to find the search direction.

As an initial guess for the Lagrange–Newton method, we use the solution of the Stokes problem that is obtained for $Re = 1$, $Q_{h,i} = 0$, $i = 1, \ldots, n_x$.

The desired flow, controlled flow, and the control are depicted in Figure 8.9. The pressure is depicted in Figure 8.8. The progress of the algorithm is shown in Table 8.7.

Table 8.7: Optimal control of Navier–Stokes equations: Progress of the Lagrange–Newton method.

Solve the Stokes problem:

k	$\int_0^1 f_0[t]dt$	α_{k-1}	$\|T(z^{(k)})\|_2$
0	1.763432802e+03		5.938741958e+01
1	3.986109778e+02	1.0000000000000	7.255927209e-10

Solve the Navier–Stokes problem with solution of the Stokes problem as initial guess:

k	$\int_0^1 f_0[t]dt$	α_{k-1}	$\|T(z^{(k)})\|_2$
0	3.986109778e+02		1.879632471e+04
1	3.988553051e+02	1.0000000000000	1.183777963e+01
2	3.988549264e+02	1.0000000000000	4.521653291e-04
3	3.988549264e+02	1.0000000000000	9.576226032e-10

Instead of finite difference schemes, one often prefers finite element methods (FEM) for the space discretization of the Navier–Stokes equations. FEM leads to a similar DAE where y_h is multiplied from the left by a symmetric and positive definite mass matrix M. By (formally) multiplying with M^{-1}, one can transform the FEM DAE into a semi-explicit DAE of index two which can be treated as above. □

Extensions of the Lagrange–Newton method towards optimal control problems with mixed control-state constraints use semi-smooth Newton methods, which have been investigated in finite dimensions in [277, 278] and in [129] for discretized optimal control problems. Extensions to infinite spaces can be found in [187, 188, 60, 320, 319, 170]. The application of a semi-smooth Newton method to DAE optimal control problems can be found in [130]. SQP methods in function space, see [3, 4], extend the Lagrange–Newton method towards optimal control problems with control and/or state constraints. Details can be found in [216, 7, 8].

8.3 Exercises

(Speed of Convergence of Gradient Method)
Apply the gradient method in Algorithm 8.1.1 to the function $J(u_1, u_2) = u_1^2 + 100\,u_2^2$ with $u^{(0)} = (1, 0.01)^\top$.

Instead of the Armijo line-search in step (3) use exact step-sizes, which are obtained in iteration k by one-dimensional minimization of $J(u^{(k)} + \alpha d^{(k)})$ with respect to α.

Provide explicit representations of the iterates $u^{(k)}$ and show

$$J(u^{(k+1)}) - J(\hat{u}) = \left(\frac{\lambda_{\max} - \lambda_{\min}}{\lambda_{\max} + \lambda_{\min}}\right)^2 \left(J(u^{(k)}) - J(\hat{u})\right),$$

where λ_{max} and λ_{min}, respectively, denote the largest and the smallest eigenvalue of the Hessian matrix $J''(\hat{u})$ with $\hat{u} = (0,0)^\top$.

(Compare [110, Exercise 6.2])
Implement Algorithm 8.1.1 for the minimization of $J : \mathbb{R} \longrightarrow \mathbb{R}$ and test it for the following functions with $u^{(0)} = 0$, $\beta = 0.9$, and $\sigma = 10^{-4}$:

(a) $J(u) = -\frac{u}{u^2+2}$.
(b) $J(u) = (u + 0.004)^5 - 2(u + 0.004)^4$.
(c) $J(u) = w(c_1)\sqrt{(1-u)^2 + c_2^2} + w(c_2)\sqrt{u^2 + c_1^2}$ with $w(c) = \sqrt{1+c^2} - c$ and $c_1 = 0.01$, $c_2 = 0.001$ and $c_1 = 0.001$, $c_2 = 0.01$, respectively.

Compare the results with those for the following modification of the line-search procedure in step (3) of the algorithm:

In step (3) consider the *Armijo line-search with expansion phase*:

(i) Let $\beta \in (0,1)$, and $\sigma \in (0,1)$. Set $\alpha := 1$.
(ii) If

$$J\left(u^{(k)} + \alpha d^{(k)}\right) \le J\left(u^{(k)}\right) - \sigma\alpha\left\|\nabla J\left(u^{(k)}\right)\right\|_2^2$$

is not satisfied, proceed as in step (3) of Algorithm 8.1.1. Otherwise, go to (iii).

(iii) Set $\alpha_k := \alpha$. Then set $\alpha := \frac{\alpha}{\beta}$ (expansion) and go to (iv).
(iv) If

$$J\left(u^{(k)} + \alpha d^{(k)}\right) \le J\left(u^{(k)}\right) - \sigma\alpha\left\|\nabla J\left(u^{(k)}\right)\right\|_2^2$$

is satisfied, go to (iii). Otherwise, stop the line-search with step-size α_k.

Solve the following optimal control problems using the gradient method in Algorithm 8.1.7 and the Lagrange–Newton method in Algorithm 8.2.3:

(a) *Minimize*

$$\frac{1}{2}\int_0^{10} \left(10\left(x(t) - 1\right)^2 + u(t)^2\right) dt$$

subject to

$$\dot{x}(t) = u(t), \quad x(0) = 0.$$

(b) *Minimize*

$$\int_0^1 \frac{1}{2}\left(x(t)^2 + u(t)^2\right) dt$$

subject to

$$\dot{x}(t) = -x(t) + \sqrt{3}u(t), \quad x(0) = 2.$$

Solve the following optimal control problem using the Lagrange–Newton method in Algorithm 8.2.3:

Minimize

$$\int_0^{4.5} u(t)^2 + x_1(t)^2 dt$$

subject to

$$\begin{aligned} \dot{x}_1(t) &= x_2(t), & x_1(0) &= -5,\ x_1(4.5) = 0, \\ \dot{x}_2(t) &= -x_1(t) + x_2(t)y(t) + 4u(t), & x_2(0) &= -5,\ x_2(4.5) = 0, \\ 0 &= 1.4 - 0.14x_2(t)^2 - y(t). \end{aligned}$$

You can use a single shooting method applying the implicit Euler method to solve the linear boundary value problems in each step.

Consider the linear index-two DAE

$$\begin{aligned} \dot{x}(t) &= A(t)x(t) + B(t)y(t) + u(t), \quad x(t_0) = x_0, \\ 0_{\mathbb{R}^{n_y}} &= C(t)x(t) \end{aligned}$$

on the compact interval $\mathcal{I} = [t_0, t_f]$ with consistent initial value x_0 and

$$A(\cdot) \in L_\infty^{n_x \times n_x}(\mathcal{I}), \quad B(\cdot) \in L_\infty^{n_x \times n_y}(\mathcal{I}), \quad C(\cdot) \in W_{1,\infty}^{n_y \times n_x}(\mathcal{I}), \quad u(\cdot) \in L_\infty^{n_u}(\mathcal{I}).$$

Let $M(t) := C(t) \cdot B(t)$ be non-singular almost everywhere in $\mathcal{I}$, and let $M(\cdot)^{-1}$ be essentially bounded on $\mathcal{I}$.

Prove that the control to state mapping

$$L_\infty^{n_u}(\mathcal{I}) \ni u \mapsto \big(x(u), y(u)\big) \in W_{1,\infty}^{n_x}(\mathcal{I}) \times L_\infty^{n_y}(\mathcal{I})$$

is continuously Fréchet-differentiable.

For $t \in \mathcal{I}, \mathcal{I} \subset \mathbb{R}$ compact, consider the matrix function

$$M(t) = \begin{pmatrix} Q(t) & A(t)^\top \\ A(t) & \Theta \end{pmatrix} \quad \text{with } Q \in L_\infty^{n \times n}(\mathcal{I}), A \in L_\infty^{m \times n}(\mathcal{I}).$$

Suppose that

(a) Let there exist $\alpha > 0$ such that almost everywhere in $\mathcal{I}$, it holds

$$d^\top Q(t) d \geq \alpha \|d\|^2 \quad \text{for all } d \in \mathbb{R}^n \text{ with } A(t)d = 0_{\mathbb{R}^m}.$$

(b) Let there exist $\beta > 0$ with

$$\left\|A(t)^\top d\right\| \geq \beta \|d\| \quad \text{for all } d \in \mathbb{R}^m.$$

Prove that $M(t)$ is non-singular for almost every $t \in \mathcal{I}$ and that $M(t)^{-1}$ is essentially bounded.

Bibliography

[1] H. W. Alt. *Lineare Funktionalanalysis*, 4th edition. Springer, Berlin-Heidelberg-New York, 2002.

[2] W. Alt, R. Baier, F. Lempio, and M. Gerdts. Approximation of linear control problems with bang-bang solutions. *Optimization: A Journal of Mathematical Programming and Operations Research*, 2011. https://doi.org/10.1080/02331934.2011.568619.

[3] W. Alt. The Lagrange-Newton method for infinite dimensional optimization problems. *Numerical Functional Analysis and Optimization*, 11:201–224, 1990.

[4] W. Alt. Sequential quadratic programming in Banach spaces. In W. Oettli and D. Pallaschke, editors, *Advances in Optimization*, pages 281–301. Springer, Berlin, 1991.

[5] W. Alt. Mesh-independence of the Lagrange–Newton method for nonlinear optimal control problems and their discretizations. *Annals of Operations Research*, 101:101–117, 2001.

[6] W. Alt. *Nichtlineare Optimierung: Eine Einführung in Theorie, Verfahren und Anwendungen*. Vieweg, Braunschweig/Wiesbaden, 2002.

[7] W. Alt and K. Malanowski. The Lagrange–Newton method for nonlinear optimal control problems. *Computational Optimization and Applications*, 2:77–100, 1993.

[8] W. Alt and K. Malanowski. The Lagrange–Newton method for state constrained optimal control problems. *Computational Optimization and Applications*, 4:217–239, 1995.

[9] P. Amodio and F. Mazzia. Numerical solution of differential algebraic equations and computation of consistent initial/boundary conditions. *Journal of Computational and Applied Mathematics*, 87:135–146, 1997.

[10] M. Arnold. A perturbation analysis for the dynamical simulation of mechnical multibody systems. *Applied Numerical Mathematics*, 18(1):37–56, 1995.

[11] M. Arnold. Half-explicit Runge–Kutta methods with explicit stages for differential-algebraic systems of index 2. *BIT*, 38(3):415–438, 1998.

[12] M. Arnold and A. Murua. Non-stiff integrators for differential-algebraic systems of index 2. *Numerical Algorithms*, 19(1–4):25–41, 1998.

[13] U. M. Ascher, R. M. M. Mattheij, and R. D. Russell. *Numerical Solution of Boundary Value Problems for Ordinary Differential Equations*, volume 13 of Classics in Applied Mathematics. SIAM, Philadelphia, 1995.

[14] U. M. Ascher and L. R. Petzold. Projected implicit Runge–Kutta methods for differential-algebraic equations. *SIAM Journal on Numerical Analysis*, 28(4):1097–1120, 1991.

[15] U. M. Ascher and R. J. Spiteri. Collocation software for boundary value differential-algebraic equations. *SIAM Journal on Scientific Computing*, 15(4):938–952, 1994.

[16] D. Augustin and H. Maurer. Computational sensitivity analysis for state constrained optimal control problems. *Annals of Operations Research*, 101:75–99, 2001.

[17] A. Backes. *Extremalbedingungen für Optimierungs-Probleme mit Algebro-Differentialgleichungen*. PhD thesis, Mathematisch-Naturwissenschaftliche Fakultät, Humboldt-Universität Berlin, Berlin, Germany, 2006.

[18] B. Bank, J. Guddat, D. Klatte, B. Kummer, and K. Tammer. *Non-linear Parametric Optimization*. Birkhäuser, Basel, 1983.

[19] A. Barcley, P. E. Gill, and J. B. Rosen. *SQP methods and their application to numerical optimal control*. Report NA 97-3, Department of Mathematics, University of California, San Diego, 1997.

[20] M. Bardi and I. Capuzzo-Dolcetta. *Optimal Control and Viscosity Solutions of Hamilton–Jacobi–Bellman Equations*. Birkhäuser, Basel, 2008. Reprint of the 1997 original.

[21] M. S. Bazaraa, H. D. Sherali, and C. M. Shetty. *Nonlinear Programming: Theory and Algorithms*, 2nd edition. John Wiley & Sons, 1993.

[22] R. E. Bellman. *Dynamic Programming*. University Press, Princeton, New Jersey, 1957.

[23] R. E. Bellman and S. E. Dreyfus. *Applied Dynamic Programming*. University Press, Princeton, New Jersey, 1971.

https://doi.org/10.1515/9783110797893-009

[24] E. Bertolazzi and M. Frego. Semianalytical minimum-time solution for the optimal control of a vehicle subject to limited acceleration. *Optimal Control Applications & Methods*, 39(2):774–791, 2018.

[25] J. T. Betts. Sparse Jacobian updates in the collocation method for optimal control problems. *Journal of Guidance, Control, and Dynamics*, 13(3):409–415, 1990.

[26] J. T. Betts. *Practical Methods for Optimal Control Using Nonlinear Programming*, volume 3 of Advances in Design and Control. SIAM, Philadelphia, 2001.

[27] J. T. Betts and W. P. Huffman. Application of sparse nonlinear programming to trajectory optimization. *Journal of Guidance, Control, and Dynamics*, 15(1):198–206, 1992.

[28] J. T. Betts and W. P. Huffman. Exploiting sparsity in the direct transcription method for optimal control. *Computational Optimization and Applications*, 14(2):179–201, 1999.

[29] H. G. Bock and K. J. Plitt. A multiple shooting algorithm for direct solution of optimal control problems. In *Proceedings of the 9th IFAC Worldcongress*, Budapest, Hungary, 1984.

[30] H. G. Bock. *Randwertproblemmethoden zur Parameteridentifizierung in Systemen nichtlinearer Differentialgleichungen*, volume 183 of Bonner Mathematische Schriften. Bonn, 1987.

[31] P. T. Boggs, J. W. Tolle, and P. Wang. On the local convergence of quasi-Newton methods for constrained optimization. *SIAM Journal on Control and Optimization*, 20(2):161–171, 1982.

[32] J. Boland, V. Gaitsgory, P. Howlett, and P. Pudney. Stochastic optimal control of a solar car. In *Optimization and Related Topics*, volume 47 of Applied Optimization, pages 71–81. Kluwer Academic Publishers, 2001.

[33] I. M. Bomze and W. Grossmann. *Optimierung – Theorie und Algorithmen*. BI-Wissenschaftsverlag, Mannheim, 1993.

[34] J. F. Bonnans and A. Shapiro. *Perturbation Analysis of Optimization Problems*. Springer Series in Operations Research. Springer, New York, 2000.

[35] K. E. Brenan, S. L. Campbell, and L. R. Petzold. *Numerical Solution of Initial-Value Problems in Differential-Algebraic Equations*, volume 14 of Classics in Applied Mathematics. SIAM, Philadelphia, 1996.

[36] K. E. Brenan. Differential-algebraic equations issues in the direct transcription of path constrained optimal control problems. *Annals of Numerical Mathematics*, 1:247–263, 1994.

[37] K. E. Brenan and B. E. Engquist. Backward differentiation approximations of nonlinear differential/algebraic systems. *Mathematics of Computation*, 51(184):659–676, 1988.

[38] K. E. Brenan and L. R. Petzold. The numerical solution of higher index differential/algebraic equations by implicit methods. *SIAM Journal on Numerical Analysis*, 26(4):976–996, 1989.

[39] A. Britzelmeier and M. Gerdts. A nonsmooth Newton method for linear model-predictive control in tracking tasks for a mobile robot with obstacle avoidance. *IEEE Control Systems Letters*, 4(4):886–891, 2020.

[40] P. N. Brown, A. C. Hindmarsh, and L. R. Petzold. Consistent initial condition calculation for differential-algebraic systems. *SIAM Journal on Scientific Computing*, 19(5):1495–1512, 1998.

[41] A. E. Bryson and Y.-C. Ho. *Applied Optimal Control*. Hemisphere Publishing Corporation, Washington, 1975.

[42] R. Bulirsch. Die Mehrzielmethode zur numerischen Lösung von nichtlinearen Randwertproblemen und Aufgaben der optimalen Steuerung. Report der Carl-Cranz-Gesellschaft, 1971.

[43] M. Burger and M. Gerdts. DAE aspects in vehicle dynamics and mobile robotics. In *Applications of Differential-Algebraic Equations: Examples and Benchmarks, Differential-Algebraic Equations Forum*, pages 37–80. Springer, Cham, 2019.

[44] J. V. Burke and S. P. Han. A robust sequential quadratic programming method. *Mathematical Programming*, 43:277–303, 1989.

[45] C. Büskens. *Optimierungsmethoden und Sensitivitätsanalyse für optimale Steuerprozesse mit Steuer- und Zustandsbeschränkungen*. PhD thesis, Fachbereich Mathematik, Westfälische Wilhems-Universität Münster, Münster, Germany, 1998.

[46] C. Büskens. Real-time solutions for perturbed optimal control problems by a mixed open- and closed-loop strategy. In M. Grötschel, S. O. Krumke and J. Rambau, editors, *Online Optimization of Large Scale Systems*, pages 105–116. Springer, 2001.

[47] C. Büskens and M. Gerdts. Numerical solution of optimal control problems with DAE systems of higher index. In *Optimalsteuerungsprobleme in der Luft- und Raumfahrt, Workshop in Greifswald des Sonderforschungsbereichs 255: Transatmospärische Flugsysteme*, pages 27–38, München, 2000.

[48] C. Büskens and M. Gerdts. Emergency landing of a hypersonic flight system: a corrector iteration method for admissible real–time optimal control approximations. In *Optimalsteuerungsprobleme in der Luft- und Raumfahrt, Workshop in Greifswald des Sonderforschungsbereichs 255: Transatmospärische Flugsysteme*, pages 51–60, München, 2003.

[49] C. Büskens and H. Maurer. Sensitivity analysis and real-time control of nonlinear optimal control systems via nonlinear programming methods. In *Proceedings of the 12th Conference on Calculus of Variations, Optimal Control and Applications*, Trassenheide, 1996.

[50] C. Büskens and H. Maurer. Sensitivity analysis and real-time control of parametric optimal control problems using nonlinear programming methods. In M. Grötschel, S. O. Krumke and J. Rambau, editors, *Online Optimization of Large Scale Systems*, pages 56–68. Springer, 2001.

[51] C. Büskens and H. Maurer. Sensitivity analysis and real-time optimization of parametric nonlinear programming problems. In M. Grötschel, S. O. Krumke and J. Rambau, editors, *Online Optimization of Large Scale Systems*, pages 3–16. Springer, 2001.

[52] S. L. Campbell. Least squares completions for nonlinear differential algebraic equations. *Numerische Mathematik*, 65(1):77–94, 1993.

[53] S. L. Campbell and C. W. Gear. The index of general nonlinear DAEs. *Numerische Mathematik*, 72:173–196, 1995.

[54] S. L. Campbell and E. Griepentrog. Solvability of general differential algebraic equations. *SIAM Journal on Scientific Computing*, 16(2):257–270, 1995.

[55] S. L. Campbell and R. Hollenbeck. Automatic differentiation and implicit differential equations. In *Computational Differentiation: Techniques, Applications and Tools*, pages 215–227. SIAM, Philadelphia, 1996.

[56] S. L. Campbell, C. T. Kelley, and K. D. Yeomans. Consistent initial conditions for unstructured higher index DAEs: a computational study. In *Computational Engineering in Systems Applications*, pages 416–421, France, 1996.

[57] S. L. Campbell, E. Moore, and Y. Zhong. Constraint preserving integrators for unstructured higher index DAEs. *Zeitschrift für Angewandte Mathematik und Mechanik*, 251–254, 1996.

[58] Y. Cao, S. Li, L. R. Petzold, and R. Serban. Adjoint sensitivity analysis for differential-algebraic equations: The adjoint DAE system and its numerical solution. *SIAM Journal on Scientific Computing*, 24(3):1076–1089, 2003.

[59] M. Caracotsios and W. E. Stewart. Sensitivity analysis of initial-boundary-value problems with mixed PDEs and algebraic equations. *Computers & Chemical Engineering*, 19(9):1019–1030, 1985.

[60] X. Chen, Z. Nashed, and L. Qi. Smoothing methods and semismooth methods for nondifferentiable operator equations. *SIAM Journal on Numerical Analysis*, 38(4):1200–1216, 2000.

[61] F. L. Chernousko and A. A. Lyubushin. Method of successive approximations for solution of optimal control problems. *Optimal Control Applications & Methods*, 3:101–114, 1982.

[62] F. H. Clarke. *Optimization and Nonsmooth Analysis*. John Wiley & Sons, New York, 1983.

[63] T. F. Coleman, B. S. Garbow, and J. J. Moré. Software for estimating sparse Hessian matrices. *ACM Transactions on Mathematical Software*, 11:369–377, 1985.

[64] A. R. Conn, N. I. Gould, and P. L. Toint. *Trust-Region Methods*. SIAM, Society for Industrial and Applied Mathematics; MPS, Mathematical Programming Society, Philadelphia, 2000.

[65] C. F. Curtiss and J. O. Hirschfelder. Integration of stiff equations. *Proceedings of the National Academy of Sciences of the United States of America*, 38:235–243, 1952.

[66] M. de Pinho and R. B. Vinter. Necessary conditions for optimal control problems involving nonlinear differential algebraic equations. *Journal of Mathematical Analysis and Applications*, 212:493–516, 1997.

[67] J.-P. Demailly. *Gewöhnliche Differentialgleichungen*. Vieweg, Braunschweig, 1991.

[68] A. De Marchi. *Augmented Lagrangian and Proximal Methods for Constrained Structured Optimization*. PhD thesis, Fakultät für Luft- und Raumfahrttechnik, Universität der Bundeswehr München, Neubiberg, Germany, 2021. https://doi.org/10.5281/zenodo.4972536.

[69] J. W. Demmel, S. C. Eisenstat, J. R. Gilbert, X. S. Li, and J. W. H. Liu. A supernodal approach to sparse partial pivoting. *SIAM Journal on Matrix Analysis and Applications*, 20(3):720–755, 1999.

[70] P. Deuflhard. A modified Newton method for the solution of ill-conditioned systems of nonlinear equations with apllication to multiple shooting. *Numerische Mathematik*, 22:289–315, 1974.

[71] P. Deuflhard. A stepsize control for continuation methods and its special application to multiple shooting techniques. *Numerische Mathematik*, 33:115–146, 1979.

[72] P. Deuflhard and F. Bornemann. *Scientific Computing with Ordinary Differential Equations*, volume 42 of Texts in Applied Mathematics. Springer, New York, 2002.

[73] P. Deuflhard and A. Hohmann. *Numerische Mathematik*. de Gruyter, Berlin, 1991.

[74] P. Deuflhard, H. J. Pesch, and P. Rentrop. A modified continuation method for the numerical solution of nonlinear two-point boundary value problems by shooting techniques. *Numerische Mathematik*, 26:327–343, 1976.

[75] E. N. Devdariani and Y. S. Ledyaev. Maximum principle for implicit control systems. *Applied Mathematics & Optimization*, 40:79–103, 1999.

[76] M. Diehl. *Real-Time Optimization for Large Scale Nonlinear Processes*. PhD thesis, Naturwissenschaftlich-Mathematische Gesamtfakultät, Universität Heidelberg, Heidelberg, Germany, 2001.

[77] M. Diehl, H. G. Bock, and J. P. Schlöder. *Newton-type Methods for the Approximate Solution of Nonlinear Programming Problems in Real-time*. Kluwer Academic Publishers, Boston, MA, 2003.

[78] M. Diehl, H. G. Bock, and J. P. Schlöder. A real-time iteration scheme for nonlinear optimization in optimal feedback control. *SIAM Journal on Control and Optimization*, 43(5):1714–1736, 2005.

[79] H.-J. Diekhoff, P. Lory, H. J. Oberle, H.-J. Pesch, P. Rentrop, and R. Seydel. Comparing routines for the numerical solution of initial value problems of ordinary differential equations in multiple shooting. *Numerische Mathematik*, 27:449–469, 1977.

[80] A. L. Dontchev, W. W. Hager, and K. Malanowski. Error bounds for Euler approximation of a state and control constrained optimal control problem. *Numerical Functional Analysis and Optimization*, 21(5 &(6):653–682, 2000.

[81] A. L. Dontchev, W. W. Hager, and V. M. Veliov. Second-order Runge–Kutta approximations in control constrained optimal control. *SIAM Journal on Numerical Analysis*, 38(1):202–226, 2000.

[82] A. J. Dubovitskii and A. A. Milyutin. Extremum problems in the presence of restrictions. *U.S.S.R. Computational Mathematics and Mathematical Physics*, 5(3):1–80, 1965.

[83] A. J. Dubovitskij and A. A. Milyutin. Extremum problems with constraints. *Soviet Mathematics. Doklady*, 4:452–455, 1963.

[84] I. S. Duff and C. W. Gear. Computing the structural index. *SIAM Journal on Algebraic and Discrete Methods*, 7(4):594–603, 1986.

[85] I. S. Duff, A. M. Erisman, and J. K. Reid. *Direct Methods for Sparse Matrices*. Clarendon Press, Oxford, 1989. Paperback ed. (with corrections).

[86] E. Eich. Convergence results for a coordinate projection method applied to mechanical systems with algebraic constraints. *SIAM Journal on Numerical Analysis*, 30(5):1467–1482, 1993.

[87] M. Emam and M. Gerdts. Sensitivity updates for linear-quadratic optimization problems in multi-step model predictive control. *Journal of Physics. Conference Series*, 2514:012008, 2023. https://doi.org/10.1088/1742-6596/2514/1/012008.

[88] G. Engl, A. Kröner, T. Kronseder, and O. von Stryk. Numerical simulation and optimal control of air separation plants. In H.-J. Bungartz, F. Durst and C. Zenger, editors, *High Performance Scientific and Engineering Computing*, volume 8 of Lecture Notes in Computational Science and Engineering, pages 221–231. Springer, 1999.

[89] D. Estévez Schwarz. *Consistent initialization for index-2 differential-algebraic equations and its application to circuit simulation*. PhD thesis, Mathematisch-Naturwissenschaftlichen Fakultät II, Humboldt-Universität Berlin, Berlin, Germany, 2000.

[90] B. C. Fabien. Parameter optimization using the L_∞ exact penalty function and strictly convex quadratic programming problems. *Applied Mathematics and Computation*, 198(2):833–848, 2008.

[91] B. C. Fabien. dsoa: The implementation of a dynamic system optimization algorithm. *Optimal Control Applications & Methods*, 31(3):231–247, 2010.

[92] W. F. Feehery and P. I. Barton. A differentiation-based approach to dynamic simulation and optimization with high-index differential-algebraic equations. In M. Berz, editor, *Computational Differentiation: Techniques, Applications, and Tools*, volume 25, pages 239–252. SIAM, Philadelphia, 1996.

[93] W. F. Feehery, J. E. Tolsma, and P. I. Barton. Efficient sensitivity analysis of large-scale differential-algebraic systems. *Applied Numerical Mathematics*, 25:41–54, 1997.

[94] A. V. Fiacco. *Introduction to Sensitivity and Stability Analysis in Nonlinear Programming*, volume 165 of Mathematics in Science and Engineering. Academic Press, New York, 1983.

[95] A. V. Fiacco and G. P. McCormick. *Nonlinear Programming: Sequential Unconstrained Minimization Techniques*, volume 4 of Classics In Applied Mathematics. SIAM, Philadelphia, 1990.

[96] A. Fischer. A special Newton-type optimization method. *Optimization*, 24:269–284, 1992.

[97] R. Fletcher. An optimal positive definite update for sparse Hessian matrices. *SIAM Journal on Optimization*, 5(1):192–218, 1995.

[98] R. Fletcher. *Practical Methods of Optimization*, 2nd edition. John Wiley & Sons, Chichester–New York–Brisbane–Toronto–Singapore, 2003.

[99] R. Fletcher and S. Leyffer. Nonlinear programming without a penalty function. *Mathematical Programming*, 91A(2):239–269, 2002.

[100] R. Fletcher, S. Leyffer, and P. Toint. On the global convergence of a filter-SQP algorithm. *SIAM Journal on Optimization*, 13(1):44–59, 2002.

[101] R. Fletcher, A. Grothey, and S. Leyffer. Computing sparse Hessian and Jacobian approximations with optimal hereditary properties. In L. T. Biegler and et al. , editors, *Large-Scale Optimization with Applications. Part 2: Optimal Design and Control. Proceedings of a 3-Week Workshop*, July 10–28, 1995, IMA, University of Minnesota, Minneapolis, MN, USA, volume 93 of IMA Vol. Math. Appl., pages 37–52. Springer, New York, NY, 1997.

[102] C. Führer. *Differential-algebraische Gleichungssysteme in mechanischen Mehrkörpersystemen: Theorie, numerische Ansätze und Anwendungen*. PhD thesis, Fakultät für Mathematik und Informatik, Technische Universität München, München, Germany, 1988.

[103] C. Führer and B. J. Leimkuhler. Numerical solution of differential-algebraic equations for constraint mechanical motion. *Numerische Mathematik*, 59:55–69, 1991.

[104] C. W. Gear. The automatic integration of stiff ordinary differential equations. In *Information Processing 68: Proc. IFIP Congress Edinburgh*, pages 187–193. North-Holland, 1969.

[105] C. W. Gear. Simultaneous numerical solution of differential-algebraic equations. *IEEE Transactions on Circuit Theory*, 18(1):89–95, 1971.

[106] C. W. Gear. Differential-algebraic equation index transformations. *SIAM Journal on Scientific and Statistical Computing*, 9:39–47, 1988.

[107] C. W. Gear. Differential algebraic equations, indices, and integral algebraic equations. *SIAM Journal on Numerical Analysis*, 27(6):1527–1534, 1990.

[108] C. W. Gear, B. Leimkuhler, and G. K. Gupta. Automatic integration of Euler–Lagrange equations with constraints. *Journal of Computational and Applied Mathematics*, 12/13:77–90, 1985.

[109] C. W. Gear and L. R. Petzold. ODE methods for the solution of differential/algebraic systems. *SIAM Journal on Numerical Analysis*, 21(4):716–728, 1984.

[110] C. Geiger and C. Kanzow. *Numerische Verfahren zur Lösung unrestringierter Optimierungsaufgaben*. Springer, Berlin–Heidelberg–New York, 1999.

[111] C. Geiger and C. Kanzow. *Theorie und Numerik restringierter Optimierungsaufgaben*. Springer, Berlin–Heidelberg–New York, 2002.

[112] M. Gerdts. *Numerische Methoden optimaler Steuerprozesse mit differential-algebraischen Gleichungssystemen höheren Indexes und ihre Anwendungen in der Kraftfahrzeugsimulation und Mechanik*, volume 61 of Bayreuther Mathematische Schriften. Bayreuth, 2001.

[113] M. Gerdts. Direct shooting method for the numerical solution of higher index DAE optimal control problems. *Journal of Optimization Theory and Applications*, 117(2):267–294, 2003.

[114] M. Gerdts. A moving horizon technique for the simulation of automobile test-drives. *Zeitschrift für Angewandte Mathematik und Mechanik*, 83(3):147–162, 2003.

[115] M. Gerdts. Optimal control and real-time optimization of mechanical multi-body systems. *Zeitschrift für Angewandte Mathematik und Mechanik*, 83(10):705–719, 2003.

[116] M. Gerdts. Parameter optimization in mechanical multibody systems and linearized Runge–Kutta methods. In A. Buikis, R. Ciegis and A. D. Flitt, editors, *Progress in Industrial Mathematics at ECMI 2002*, volume 5 of Mathematics in Industry, pages 121–126. Springer, 2004.

[117] M. Gerdts. Gradient evaluation in dae optimal control problems by sensitivity equations and adjoint equations. *Proceedings in Applied Mathematics and Mechanics*, 5(1):43–46, 2005.

[118] M. Gerdts. Local minimum principle for optimal control problems subject to index one differential-algebraic equations. Technical report, Department of Mathematics, University of Hamburg, 2005.

[119] M. Gerdts. On the convergence of linearized implicit Runge–Kutta methods and their use in parameter optimization. *Mathematica Balkanica*, 19:75–83, 2005.

[120] M. Gerdts. Solving mixed-integer optimal control problems by branch & bound: A case study from automobile test-driving with gear shift. *Optimal Control Applications & Methods*, 26(1):1–18, 2005.

[121] M. Gerdts. Local minimum principle for optimal control problems subject to index-two differential-algebraic equations. *Journal of Optimization Theory and Applications*, 130(3):443–462, 2006.

[122] M. Gerdts. A variable time transformation method for mixed-integer optimal control problems. *Optimal Control Applications & Methods*, 27(3):169–182, 2006.

[123] M. Gerdts. OCPID-DAE1 – Optimal Control and Parameter Identification with Differential-Algebraic Equations of Index 1: User's guide. Technical report, Institute of Applied Mathematics and Scientific Computing, Department of Aerospace Engineering, Universität der Bundeswehr München, 2018.

[124] C. Büskens and M. Gerdts. Real-time optimization of DAE systems. In M. Grötschel, S. O. Krumke and J. Rambau, editors, *Online Optimization of Large Scale Systems*, pages 117–128. Springer, 2001.

[125] M. Gerdts and C. Büskens. Consistent initialization of sensitivity matrices for a class of parametric DAE systems. *BIT Numerical Mathematics*, 42(4):796–813, 2002.

[126] M. Gerdts and J. Chen. Numerical solution of control-state constrained optimal control problems with an inexact smoothing Newton method. *IMA Journal of Numerical Analysis*, 2011. https://doi.org/10.1093/imanum/drq023.

[127] M. Gerdts, G. Greif, and H. J. Pesch. Numerical optimal control of the wave equation: Optimal boundary control of a string to rest in finite time. *Mathematics and Computers in Simulation*, 79(4):1020–1032, 2008.

[128] M. Gerdts, S. Karrenberg, B. Müller-Beßler, and G. Stock. Generating locally optimal trajectories for an automatically driven car. *Optimization and Engineering*, 10:439–463, 2009.

[129] M. Gerdts and M. Kunkel. A nonsmooth Newton's method for discretized optimal control problems with state and control constraints. *Journal of Industrial and Management Optimization*, 4(2):247–270, 2008.

[130] M. Gerdts and M. Kunkel. A globally convergent semi-smooth Newton method for control-state constrained DAE optimal control problems. *Computational Optimization and Applications*, 48(3):601–633, 2011.
[131] M. Gerdts and F. Lempio. *Mathematische Optimierungsverfahren des Operations Research*. De Gruyter, Berlin, 2011.
[132] E. M. Gertz and S. J. Wright. Object-oriented software for quadratic programming. *ACM Transactions on Mathematical Software*, 29(1):58–81, 2003.
[133] P. E. Gill and W. Murray. Numerically stable methods for quadratic programming. *Mathematical Programming*, 14:349–372, 1978.
[134] P. E. Gill, W. Murray, M. A. Saunders, and M. H. Wright. Inertia-controlling methods for general quadratic programming. *SIAM Review*, 33(1):1–36, 1991.
[135] P. E. Gill, W. Murray, and M. A. Saunders. *Large-scale SQP Methods and their Application in Trajectory Optimization*, volume 115 of International Series of Numerical Mathematics, pages 29–42. Birkhäuser, Basel, 1994.
[136] P. E. Gill, W. Murray, and M. A. S. Snopt. An SQP algorithm for large-scale constrained optimization. *SIAM Journal on Optimization*, 12(4):979–1006, 2002.
[137] P. E. Gill, W. Murray, M. A. Saunders, and M. H. Wright. User's guide for NPSOL 5.0: A FORTRAN package for nonlinear programming. Technical Report NA 98-2, Department of Mathematics, University of California, San Diego, California, 1998.
[138] P. E. Gill, W. Murray, and M. H. Wright. *Practical Optimization*. Academic Press, London, 1981.
[139] I. V. Girsanov. *Lectures on Mathematical Theory of Extremum Problems*, volume 67 of Lecture Notes in Economics and Mathematical Systems. Springer, Berlin–Heidelberg–New York, 1972.
[140] D. Goldfarb and A. Idnani. A numerically stable dual method for solving strictly convex quadratic programs. *Mathematical Programming*, 27:1–33, 1983.
[141] M. Golomb and R. A. Tapia. The metric gradient in normed linear spaces. *Numerische Mathematik*, 20:115–124, 1972.
[142] V. Gopal and L. T. Biegler. A successive linear programming approach for initialization and reinitialization after discontinuities of differential-algebraic equations. *SIAM Journal on Scientific Computing*, 20(2):447–467, 1998.
[143] A. Göpfert and T. Riedrich. *Funktionalanalysis*. Teubner, Leipzig, 1980.
[144] K. Graichen and N. Petit. Solving the Goddard problem with thrust and dynamic pressure constraints using saturation functions. In *Proceedings of the 17th IFAC World Congress, Seoul, Korea*, pages 14301–14306, 2008.
[145] E. Griepentrog. Index reduction methods for differential-algebraic equations. *Seminarberichte Humboldt-Universität Berlin*, 92(1):14–29, 1992.
[146] A. Griewank. *Evaluating Derivatives. Principles and Techniques of Algorithmic Differentiation*, volume 19 of Frontiers in Applied Mathematics. SIAM, Philadelphia, 2000.
[147] A. Griewank. A mathematical view of automatic differentiation. *Acta Numerica*, 12:321–398, 2003.
[148] D. M. Gritsis, C. C. Pantelides, and R. W. H. Sargent. Optimal control of systems described by index two differential-algebraic equations. *SIAM Journal on Scientific Computing*, 16(6):1349–1366, 1995.
[149] L. Grüne. An adaptive grid scheme for the discrete Hamilton–Jacobi–Bellman equation. *Numerische Mathematik*, 75(3):319–337, 1997.
[150] L. Grüne and J. Pannek. *Nonlinear Model Predictive Control. Theory and Algorithms*. Springer, London, 2011.
[151] M. Günther. *Ladungsorientierte Rosenbrock-Wanner-Methoden zur numerischen Simulation digitaler Schaltungen*, volume 168 of VDI Fortschrittberichte Reihe 20: Rechnergestützte Verfahren. VDI-Verlag, 1995.
[152] W. W. Hager. Runge–Kutta methods in optimal control and the transformed adjoint system. *Numerische Mathematik*, 87(2):247–282, 2000.

[153] E. Hairer, C. Lubich, and M. Roche. Error of Rosenbrock methods for stiff problems studied via differential algebraic equations. *BIT*, 29(1):77–90, 1989.

[154] E. Hairer and G. Wanner. *Solving Ordinary Differential Equations II: Stiff and Differential-Algebraic Problems*, 2nd edition, volume 14 of Springer Series in Computational Mathematics, Springer, Berlin–Heidelberg–New York, 1996.

[155] E. Hairer, C. Lubich, and M. Roche. *The Numerical Solution of Differential-Algebraic Systems by Runge–Kutta Methods*, volume 1409 of Lecture Notes in Mathematics. Springer, Berlin–Heidelberg–New York, 1989.

[156] S. P. Han. A globally convergent method for nonlinear programming. *Journal of Optimization Theory and Applications*, 22(3):297–309, 1977.

[157] B. Hansen. Computing consistent initial values for nonlinear index-2 differential-algebraic equations. *Seminarberichte Humboldt-Universität Berlin*, 92(1):142–157, 1992.

[158] C. R. Hargraves and S. W. Paris. Direct trajectory optimization using nonlinear programming and collocation. *Journal of Guidance, Control, and Dynamics*, 10:338–342, 1987.

[159] R. F. Hartl, S. P. Sethi, and G. Vickson. A survey of the maximum principles for optimal control problems with state constraints. *SIAM Review*, 37(2):181–218, 1995.

[160] A. Heim. *Parameteridentifizierung in differential-algebraischen Gleichungssystemen*. Master's thesis, Mathematisches Institut, Technische Universität München, München, Germany, 1992.

[161] H. Hermes and J. P. Lasalle. *Functional Analysis and Time Optimal Control*, volume 56 of Mathematics in Science and Engineering. Academic Press, New York, 1969.

[162] M. R. Hestenes. *Calculus of Variations and Optimal Control Theory*. John Wiley & Sons, New York, 1966.

[163] P. Hiltmann, K. Chudej, and M. H. Breitner. Eine modifizierte Mehrzielmethode zur Lösung von Mehrpunkt-Randwertproblemen. Technical Report 14, Sonderforschungsbereich 255 der Deutschen Forschungsgemeinschaft: Transatmosphärische Flugsysteme, Lehrstuhl für Höhere und Numerische Mathematik, Technische Universität München, 1993.

[164] H. Hinsberger. *Ein direktes Mehrzielverfahren zur Lösung von Optimalsteuerungsproblemen mit großen, differential-algebraischen Gleichungssystemen und Anwendungen aus der Verfahrenstechnik*. PhD thesis, Institut für Mathematik, Technische Universität Clausthal, Clausthal, Germany, 1997.

[165] M. Hintermüller and M. Ulbrich. A mesh-independence result for semismooth Newton methods. *Mathematical Programming Series B*, 101(1):151–184, 2004.

[166] M. Hinze, R. Pinnau, M. Ulbrich, and S. Ulbrich. *Optimization with PDE Constraints*, volume 23 of Mathematical Modelling: Theory and Applications. Springer, Dordrecht, 2009. xi+270 p.

[167] P. Howlett. Optimal strategies for the control of a train. *Automatica*, 32(4):519–532, 1996.

[168] A. Huber, M. Gerdts, and E. Bertolazzi. Structure exploitation in an interior-point method for fully discretized, state constrained optimal control problems. *Vietnam Journal of Mathematics*, 46:1089–1113, 2018.

[169] A. D. Ioffe and V. M. Tihomirov. *Theory of Extremal Problems*, volume 6 of Studies in Mathematics and its Applications. North-Holland Publishing Company, Amsterdam, New York, Oxford, 1979.

[170] K. Ito and K. Kunisch. *Lagrange Multiplier Approach to Variational Problems and Applications*, volume 15 of Advances in Design and Control. Society for Industrial and Applied Mathematics (SIAM), Philadelphia, PA, 2008.

[171] D. H. Jacobson, M. M. Lele, and J. L. Speyer. New necessary conditions of optimality for constrained problems with state-variable inequality constraints. *Journal of Mathematical Analysis and Applications*, 35:255–284, 1971.

[172] L. Jay. Collocation methods for differential-algebraic equations of index 3. *Numerische Mathematik*, 65:407–421, 1993.

[173] L. Jay. Convergence of Runge–Kutta methods for differential-algebraic systems of index 3. *Applied Numerical Mathematics*, 17:97–118, 1995.

[174] H. Jiang. Global convergence analysis of the generalized newton and Gauss–Newton methods of the Fischer–Burmeister equation for the complementarity problem. *Mathematics of Operations Research*, 24:529–543, 1999.

[175] W. Kang, I. M. Ross, and Q. Gong. *Pseudospectral Optimal Control and Its Convergence Theorems*. Springer, Berlin, 2008.

[176] E. Khmelnitsky. A combinatorial, graph-based solution method for a class of continuous-time optimal control problems. *Mathematics of Operations Research*, 27(2):312–325, 2002.

[177] M. Kiehl. Sensitivity Analysis of ODEs and DAEs – Theory and Implementation Guide. TUM-M5004, Technische Universität München, 1998.

[178] C. Kirches, S. Sager, H. G. Bock, and J. P. Schlöder. Time-optimal control of automobile test drives with gear shifts. In *Optimal Control Applications and Methods*, 2010. https://doi.org/10.1002/oca.892.

[179] A. Kirsch, W. Warth, and J. Werner. *Notwendige Optimalitätsbedingungen und ihre Anwendung*, volume 152 of Lecture Notes in Economics and Mathematical Systems. Springer, Berlin–Heidelberg–New York, 1978.

[180] D. Klatte. Nonlinear optimization problems under data perturbations. In *Lecture Notes in Economics and Mathematical Systems*, volume 378, pages 204–235. Springer, Berlin-Heidelberg-New York, 1990.

[181] H. W. Knobloch. Das Pontryaginsche Maximumprinzip für Probleme mit Zustandsbeschränkungen I und II. *Zeitschrift für Angewandte Mathematik und Mechanik*, 55(545–556):621–634, 1975.

[182] H. J. Kowalsky. *Lineare Algebra*, 4th edition. Walter de Gruyter & Co, Berlin, 1969.

[183] D. Kraft. A Software Package for Sequential Quadratic Programming. DFVLR-FB-88-28, Oberpfaffenhofen, 1988.

[184] P. Krämer-Eis. *Ein Mehrzielverfahren zur numerischen Berechnung optimaler Feedback-Steuerungen bei beschränkten nichtlinearen Steuerungsproblemen*, volume 166 of Bonner Mathematische Schriften. Bonn, 1985.

[185] E. Kreindler. Additional necessary conditions for optimal control with state-variable inequality constraints. *Journal of Optimization Theory and Applications*, 38(2):241–250, 1982.

[186] A. Kufner, J. Oldrich, and F. Svatopluk. *Function Spaces*. Noordhoff International Publishing, Leyden, 1977.

[187] B. Kummer. Newton's method for non-differentiable functions. In J. Guddat and et al. , editors, *Advances in Mathematical Optimization*, pages 171–194. Akademie-Verlag, Berlin, 1988.

[188] B. Kummer. Newton's method based on generalized derivatives for nonsmooth functions: Convergence analysis. In W. Oettli and D. Pallaschke, editors, *Advances in Optimization*, pages 171–194. Springer, Berlin, 1991.

[189] P. Kunkel, V. Mehrmann, and R. Stöver. Symmetric collocation for unstructered nonlinear differential-algebraic equations of arbitrary index. *Numerische Mathematik*, 98(2):277–304, 2004.

[190] P. Kunkel, V. Mehrmann, and R. Stöver. Multiple shooting for unstructured nonlinear differential-algebraic equations of arbitrary index. *SIAM Journal on Numerical Analysis*, 42(6):2277–2297, 2005.

[191] P. Kunkel and V. Mehrmann. A new class of discretization methods for the solution of linear differential-algebraic equations with variable coefficients. *SIAM Journal on Numerical Analysis*, 33(5):1941–1961, 1996.

[192] P. Kunkel and V. Mehrmann. The linear quadratic optimal control problem for linear descriptor systems with variable coefficients. *MCSS. Mathematics of Control, Signals and Systems*, 10(3):247–264, 1997.

[193] P. Kunkel and V. Mehrmann. *Differential-Algebraic Equations. Analysis and Numerical Solution*. European Mathematical Society Publishing House, Zürich, 2006.

[194] P. Kunkel and V. Mehrmann. Optimal control for unstructured nonlinear differential-algebraic equations of arbitrary index. *MCSS. Mathematics of Control, Signals and Systems*, 20(3):227–269, 2008.

[195] P. Kunkel and R. Stöver. Symmetric collocation methods for linear differential-algebraic boundary value problems. *Numerische Mathematik*, 91(3):475–501, 2002.

[196] S. Kurcyusz. On the existence and nonexistence of Lagrange multipliers in Banach spaces. *Journal of Optimization Theory and Applications*, 20(1):81–110, 1976.

[197] G. A. Kurina and R. März. On linear-quadratic optimal control problems for time-varying descriptor systems. *SIAM Journal on Control and Optimization*, 42(6):2062–2077, 2004.

[198] G. A. Kurina and R. März. Feedback solutions of optimal control problems with DAE constraints. *SIAM Journal on Control and Optimization*, 46(4):1277–1298, 2007.

[199] R. Lamour. A well-posed shooting method for transferable DAE's. *Numerische Mathematik*, 59(8):815–830, 1991.

[200] R. Lamour. A shooting method for fully implicit index-2 differential algebraic equations. *SIAM Journal on Scientific Computing*, 18(1):94–114, 1997.

[201] J. Laurent-Varin, F. Bonnans, N. Berend, C. Talbot, and M. Haddou. On the refinement of discretization for optimal control problems. In *IFAC Symposium on Automatic Control in Aerospace*, St. Petersburg, 2004.

[202] H. W. J. Lee, K. L. Teo, and X. Q. Cai. An optimal control approach to nonlinear mixed integer programming problems. *Computers & Mathematics with Applications*, 36(3):87–105, 1998.

[203] H. W. J. Lee, K. L. Teo, V. Rehbock, and L. S. Jennings. Control parameterization enhancing technique for time optimal control problems. *Dynamic Systems and Applications*, 6(2):243–262, 1997.

[204] H. W. J. Lee, K. L. Teo, V. Rehbock, and L. S. Jennings. Control parametrization enhancing technique for optimal discrete-valued control problems. *Automatica*, 35(8):1401–1407, 1999.

[205] B. Leimkuhler, L. R. Petzold, and C. W. Gear. Approximation methods for the consistent initialization of differential-algebraic equations. *SIAM Journal on Numerical Analysis*, 28(1):205–226, 1991.

[206] D. B. Leineweber. *Analyse und Restrukturierung eines Verfahrens zur direkten Lösung von Optimal-Steuerungsproblemen*. Master's thesis, Interdisziplinäres Zentrum für Wissenschaftliches Rechnen, Universität, Heidelberg, Heidelberg, Germany, 1995.

[207] F. Lempio. Lineare Optimierung in unendlichdimensionalen Vektorräumen. *Computing*, 8:284–290, 1971.

[208] F. Lempio. *Separation und Optimierung in linearen Räumen*. PhD thesis, Universität Hamburg, Hamburg, Germany, 1971.

[209] F. Lempio. *Tangentialmannigfaltigkeiten und infinite Optimierung*. Habilitationsschrift, Universität Hamburg, Hamburg, 1972.

[210] F. Lempio. Eine Verallgemeinerung des Satzes von Fritz John. *Operations Research-Verfahren*, 17:239–247, 1973.

[211] F. Lempio and H. Maurer. Differential stability in infinite-dimensional nonlinear programming. *Applied Mathematics & Optimization*, 6:139–152, 1980.

[212] L. A. Ljusternik and W. I. Sobolew. *Elemente der Funktionalanalysis*. Verlag Harri Deutsch, Zürich–Frankfurt/Main–Thun, 1976.

[213] A. Locatelli. *Optimal Control: An Introduction*. Birkhäuser, Basel, 2001.

[214] P. Lötstedt and L. R. Petzold. Numerical solution of nonlinear differential equations with algebraic constraints I: Convergence results for backward differentiation formulas. *Mathematics of Computation*, 46:491–516, 1986.

[215] D. G. Luenberger. *Optimization by Vector Space Methods*. John Wiley & Sons, New York–London–Sydney–Toronto, 1969.

[216] K. C. P. Machielsen. *Numerical Solution of Optimal Control Problems with State Constraints by Sequential Quadratic Programming in Function Space*, volume 53 of CWI Tract. Centrum voor Wiskunde en Informatica, Amsterdam, 1988.

[217] K. Malanowski. Sufficient optimality conditions for optimal control subject to state constraints. *SIAM Journal on Control and Optimization*, 35(1):205–227, 1997.

[218] K. Malanowski. On normality of Lagrange multipliers for state constrained optimal control problems. *Optimization*, 52(1):75–91, 2003.

[219] K. Malanowski, H. Maurer, and S. Pickenhain. Second-order sufficient conditions for state-constrained optimal control problems. *Journal of Optimization Theory and Applications*, 123(3):595–617, 2004.

[220] K. Malanowski and H. Maurer. Sensitivity analysis for parametric control problems with control-state constraints. *Computational Optimization and Applications*, 5(3):253–283, 1996.
[221] K. Malanowski and H. Maurer. Sensitivity analysis for state constrained optimal control problems. *Discrete and Continuous Dynamical Systems*, 4(2):3–14, 1998.
[222] K. Malanowski and H. Maurer. Sensitivity analysis for optimal control problems subject to higher order state constraints. *Annals of Operations Research*, 101:43–73, 2001.
[223] K. Malanowski, C. Büskens, and H. Maurer. Convergence of approximations to nonlinear optimal control problems. In A. Fiacco, editor, *Mathematical Programming with Data Perturbations*, volume 195 of Lecture Notes in Pure and Applied Mathematics, pages 253–284. Dekker, 1997.
[224] T. Maly and L. R. Petzold. Numerical methods and software for sensitivity analysis of differential-algebraic systems. *Applied Numerical Mathematics*, 20(1):57–79, 1996.
[225] B. Martens. *Necessary Conditions, Sufficient Conditions, and Convergence Analysis for Optimal Control Problems with Differential-Algebraic Equations*. PhD thesis, Fakultät für Luft- und Raumfahrttechnik, Universität der Bundeswehr München, Neubiberg, Germany, 2019.
[226] B. Martens. Error estimates for Runge–Kutta schemes of optimal control problems with index 1 DAEs. *Computational Optimization and Applications*, 2023. https://doi.org/10.1007/s10589-023-00484-1.
[227] B. Martens and M. Gerdts. Convergence analysis for approximations of optimal control problems subject to higher index differential-algebraic equations and pure state constraints. *SIAM Journal on Control and Optimization*, 59(3):1903–1926, 2021.
[228] B. Martens and M. Gerdts. Convergence analysis for approximations of optimal control problems subject to higher index differential-algebraic equations and mixed control-state constraints. *SIAM Journal on Control and Optimization*, 58(1):1–33, 2020.
[229] B. Martens and M. Gerdts. Convergence analysis of the implicit Euler-discretization and sufficient conditions for optimal control problems subject to index-one differential-algebraic equations. *Set-Valued and Variational Analysis*, 27(2):405–431, 2019.
[230] R. März. Fine decouplings of regular differential algebraic equations. *Results in Mathematics*, 45(1–2):88–105, 2004.
[231] R. März. On linear differential-algebraic equations and linerizations. *Applied Numerical Mathematics*, 18(1):267–292, 1995.
[232] R. März. Criteria of the trivial solution of differential algebraic equations with small nonlinearities to be asymptotically stable. *Journal of Mathematical Analysis and Applications*, 225(2):587–607, 1998.
[233] R. März. EXTRA-ordinary differential equations: Attempts to an analysis of differential-algebraic systems. In A. Balog, editor, *European Congress of Mathematics, vol. 1*, volume 168 of Prog. Math., pages 313–334. Birkhäuser, Basel, 1998.
[234] R. März and C. Tischendorf. Recent results in solving index-2 differential-algebraic equations in circuit simulation. *SIAM Journal on Scientific Computing*, 18(1):139–159, 1997.
[235] H. Maurer and H. J. Oberle. Second order sufficient conditions for optimal control problems with free final time: The Riccati approach. *SIAM Journal on Control and Optimization*, 41(2):380–403, 2002.
[236] H. Maurer and S. Pickenhain. Second-order sufficient conditions for control problems with mixed control-state constraints. *Journal of Optimization Theory and Applications*, 86(3):649–667, 1995.
[237] H. Maurer and J. Zowe. First and second-order necessary and sufficient optimality conditions for infinite-dimensional programming problems. *Mathematical Programming*, 16:98–110, 1979.
[238] H. Maurer. On optimal control problems with boundary state variables and control appearing linearly. *SIAM Journal on Control and Optimization*, 15(3):345–362, 1977.
[239] H. Maurer. On the Minimum Principle for Optimal Control Problems with State Constraints. Schriftenreihe des Rechenzentrums der Universität Münster, **41**, 1979.
[240] H. Maurer. First and second order sufficient optimality conditions in mathematical programming and optimal control. *Mathematical Programming Studies*, 14:163–177, 1981.
[241] H. Maurer and D. Augustin. Sensitivity analysis and real-time control of parametric optimal control problems using boundary value methods. In M. Grötschel, S. O. Krumke and J. Rambau, editors, *Online Optimization of Large Scale Systems*, pages 17–55. Springer, 2001.

[242] H. Maurer and H. J. Pesch. Solution differentiability for nonlinear parametric control problems. *SIAM Journal on Control and Optimization*, 32(6):1542–1554, 1994.

[243] H. Maurer and H. J. Pesch. Solution differentiability for parametric nonlinear control problems with control-state constraints. *Control and Cybernetics*, 23(1–2):201–227, 1994.

[244] H. Maurer and H. J. Pesch. Solution differentiability for parametric nonlinear control problems with control-state constraints. *Journal of Optimization Theory and Applications*, 86(2):285–309, 1995.

[245] R. Mayr. *Verfahren zur Bahnfolgeregelung für ein automatisch geführtes Fahrzeug*. PhD thesis, Fakultät für Elektrotechnik, Universität Dortmund, Dortmund, Germany, 1991.

[246] M. Mayrhofer and G. Sachs. Notflugbahnen eines zweistufigen Hyperschall-Flugsystems ausgehend vom Trennmanöver. In *Seminarbericht des Sonderforschungsbereichs 255: Transatmosphärische Flugsysteme*, pages 109–118. TU, München, 1996.

[247] V. Mehrmann. Existence, uniqueness, and stability of solutions to singular linear quadratic optimal control problems. *Linear Algebra and Its Applications*, 121:291–331, 1989.

[248] V. Mehrmann. *The Autonomous Linear Quadratic Control Problem. Theory and Numerical Solution*, volume 163 of Lecture Notes in Control and Information Sciences. Springer, Berlin, 1991.

[249] M. Mitschke. *Dynamik der Kraftfahrzeuge, Band C: Fahrverhalten*, 2nd edition. Springer, Berlin–Heidelberg–New York, 1990.

[250] T. Moder. Optimale Steuerung eines KFZ im fahrdynamischen Grenzbereich. Master's thesis, Mathematisches Institut, Technische Universität München, München, Germany, 1994.

[251] B. S. Mordukhovich. An approximate maximum principle for finite-difference control systems. *U.S.S.R. Computational Mathematics and Mathematical Physics*, 28(1):106–114, 1988.

[252] P. C. Müller. Optimal control of proper and nonproper descriptor systems. *Archive of Applied Mechanics*, 72:875–884, 2003.

[253] I. P. Natanson. *Theorie der Funktionen einer reellen Veränderlichen*. Verlag Harri Deutsch, Zürich–Frankfurt–Thun, 1975.

[254] M. Neculau. *Modellierung des Fahrverhaltens: Informationsaufnahme, Regel- und Steuerstrategien in Experiment und Simulation*. PhD thesis, Fachbereich 12: Verkehrswesen, Technische Universität Berlin, Berlin, Germany, 1992.

[255] K. Neumann and M. Morlock. *Operations Research*. Carl Hanser Verlag, München, Wien, 2002.

[256] L. W. Neustadt. *Optimization: A Theory of Necessary Conditions*. Princeton University Press, Princeton, New Jersey, 1976.

[257] J. Nocedal and S. J. Wright. *Numerical Optimization*. Springer Series in Operations Research. Springer, New York, 1999.

[258] H. J. Oberle and W. Grimm. BNDSCO – A program for the numerical solution of optimal control problems. Technical Report Reihe B, Bericht 36, Hamburger Beiträge zur Angewandten Mathematik, Department of Mathematics, University of Hamburg, 2001. http://www.math.uni-hamburg.de/home/oberle/software.html.

[259] H. J. Oberle. Numerical solution of minimax optimal control problems by multiple shooting technique. *Journal of Optimization Theory and Applications*, 50:331–357, 1986.

[260] H. B. Pacejka and E. Bakker. The magic formula tyre model. *Vehicle System Dynamics*, 21(sup001):1–18, 1993.

[261] V. G. Palma. *Robust Updated MPC Schemes*. PhD thesis, Fakultät für Mathematik, Physik und Informatik, Universität Bayreuth, Germany, 2015.

[262] C. C. Pantelides. The consistent initialization of differential-algebraic systems. *SIAM Journal on Scientific and Statistical Computing*, 9(2):213–231, 1988.

[263] P. Pedregal and J. Tiago. Existence results for optimal control problems with some special nonlinear dependence on state and control. *SIAM Journal on Control and Optimization*, 48(2):415–437, 2009.

[264] H. J. Pesch. *Schlüsseltechnologie Mathematik: Einblicke in aktuelle Anwendungen der Mathematik*. B. G. Teubner, Stuttgart–Leipzig–Wiesbaden, 2002.

[265] H. J. Pesch. *Numerische Berechnung optimaler Flugbahnkorrekturen in Echtzeit-Rechnung*. PhD thesis, Institut für Mathematik, Technische Universität München, München, Germany, 1978.
[266] H. J. Pesch. Numerical computation of neighboring optimum feedback control schemes in real-time. *Applied Mathematics & Optimization*, 5:231–252, 1979.
[267] H. J. Pesch. Real-time computation of feedback controls for constrained optimal control problems. I: Neighbouring extremals. *Optimal Control Applications & Methods*, 10(2):129–145, 1989.
[268] H. J. Pesch. Real-time computation of feedback controls for constrained optimal control problems. II: A correction method based on multiple shooting. *Optimal Control Applications & Methods*, 10(2):147–171, 1989.
[269] H. J. Pesch and R. Bulirsch. The maximum principle, Bellman's equation and Caratheodorys work. *Journal of Optimization Theory and Applications*, 80(2):199–225, 1994.
[270] L. R. Petzold. A description of DASSL: a differential/algebraic system solver. Rep. Sand 82-8637, Sandia National Laboratory, Livermore, 1982.
[271] L. R. Petzold. Differential/algebraic equations are not ODE's. *SIAM Journal on Scientific and Statistical Computing*, 3(3):367–384, 1982.
[272] L. R. Petzold. Recent developments in the numerical solution of differential/algebraic systems. *Computer Methods in Applied Mechanics and Engineering*, 75:77–89, 1989.
[273] E. Polak. An historical survey of computational methods in optimal control. *SIAM Review*, 15(2):553–584, 1973.
[274] L. S. Pontryagin, V. G. Boltyanskij, R. V. Gamkrelidze, and E. F. Mishchenko. *Mathematische Theorie optimaler Prozesse*. Oldenbourg, München, 1964.
[275] M. J. D. Powell. A fast algorithm for nonlinearily constrained optimization calculation. In G. A. Watson, editor, *Numerical Analysis*, volume 630 of Lecture Notes in Mathematics. Springer, Berlin–Heidelberg–New York, 1978.
[276] R. Pytlak. Runge–Kutta based procedure for the optimal control of differential-algebraic equations. *Journal of Optimization Theory and Applications*, 97(3):675–705, 1998.
[277] L. Qi. Convergence analysis of some algorithms for solving nonsmooth equations. *Mathematics of Operations Research*, 18(1):227–244, 1993.
[278] L. Qi and J. Sun. A nonsmooth version of Newton's method. *Mathematical Programming*, 58(3):353–367, 1993.
[279] J. B. Rawlings and D. Q. Mayne. *Model Predictive Control: Theory and Design*. Nob Hill Publishing, 2009.
[280] P. Rentrop, M. Roche, and G. Steinebach. The application of Rosenbrock–Wanner type methods with stepsize control in differential-algebraic equations. *Numerische Mathematik*, 55(5):545–563, 1989.
[281] F. Riesz and B. Sz.-Nagy. *Functional Analysis*. Dover Publications Inc., New York, 1990. Originally published by Frederick Ungar Publishing Co., 1955.
[282] H.-J. Risse. *Das Fahrverhalten bei normaler Fahrzeugführung*, volume 160 of VDI Fortschrittberichte Reihe 12: Verkehrstechnik/Fahrzeugtechnik. VDI-Verlag, 1991.
[283] S. M. Robinson. Stability theory for systems of inequalities, Part II: Differentiable nonlinear systems. *SIAM Journal on Numerical Analysis*, 13(4):487–513, 1976.
[284] M. Roche. Rosenbrock methods for differential algebraic equations. *Numerische Mathematik*, 52(1):45–63, 1988.
[285] R. T. Rockafellar. *Convex Analysis*. Princeton University Press, New Jersey, 1970.
[286] I. M. Ross and F. Fahroo. *Legendre Pseudospectral Approximations of Optimal Control Problems*. Springer, Berlin, 2003.
[287] T. Roubicek and M. Valásek. Optimal control of causal differential-algebraic systems. *Journal of Mathematical Analysis and Applications*, 269(2):616–641, 2002.
[288] Y. Saad. *Iterative Methods for Sparse Linear Systems*. PWS Publishing Co., Boston, 1996.
[289] S. Sager. MIOCP benchmark site. http://mintoc.de.
[290] S. Sager. Reformulations and algorithms for the optimization of switching decisions in nonlinear optimal control. *Journal of Process Control*, 19(8):1238–1247, 2009.

[291] S. Sager, H. G. Bock, M. Diehl, G. Reinelt, and J. P. Schlöder. Numerical methods for optimal control with binary control functions applied to a Lotka–Volterra type fishing problem. In A. Seeger, editor, *Recent Advances in Optimization (Proceedings of the 12th French–German–Spanish Conference on Optimization)*, volume 563 of Lectures Notes in Economics and Mathematical Systems, pages 269–289. Springer, Heidelberg, 2006.

[292] S. Sager, C. Kirches, and H. G. Bock. Fast solution of periodic optimal control problems in automobile test-driving with gear shifts. In *Proceedings of the 47th IEEE Conference on Decision and Control (CDC 2008)*, pages 1563–1568, Cancun, Mexico, 2008.

[293] S. Sager. *Numerical methods for mixed-integer optimal control problems*. PhD thesis, Naturwissenschaftlich-Mathematische Gesamtfakultät, Universität Heidelberg, Heidelberg, Germany, 2006.

[294] S. Sager, H. G. Bock, and G. Reinelt. Direct methods with maximal lower bound for mixed-integer optimal control problems. *Mathematical Programming Series A*, 118(1):109–149, 2009.

[295] O. Schenk, A. Wächter, and M. Hagemann. Matching-based preprocessing algorithms to the solution of saddle-point problems in large-scale nonconvex interior-point optimization. *Computational Optimization and Applications*, 36(2–3):321–341, 2007.

[296] O. Schenk and K. Gärtner. Solving unsymmetric sparse systems of linear equations with pardiso. *Future Generations Computer Systems*, 20(3):475–487, 2004.

[297] O. Schenk and K. Gärtner. On fast factorization pivoting methods for symmetric indefinite systems. *Electronic Transactions on Numerical Analysis*, 23:158–179, 2006.

[298] O. Schenk, A. Wächter, and M. Weiser. Inertia-revealing preconditioning for large-scale nonconvex constrained optimization. *SIAM Journal on Scientific Computing*, 31(2):939–960, 2008.

[299] K. Schittkowski. The nonlinear programming method of Wilson, Han, and Powell with an augmented Lagrangean type line search function. Part 1: Convergence analysis, Part 2: An efficient implementation with linear least squares subproblems. *Numerische Mathematik*, 383:83–114, 115–127, 1981.

[300] K. Schittkowski. On the convergence of a sequential quadratic programming method with an augmented Lagrangean line search function. *Mathematische Operationsforschung und Statistik. Series Optimization*, 14(2):197–216, 1983.

[301] K. S. NLPQL. A Fortran subroutine for solving constrained nonlinear programming problems. *Annals of Operations Research*, 5:484–500, 1985.

[302] V. H. Schulz. *Reduced SQP Methods for Large-Scale Optimal Control Problems in DAE with Application to Path Planning Problems for Satellite Mounted Robots*. PhD thesis, Interdisziplinäres Zentrum für Wissenschaftliches Rechnen, Universität Heidelberg, Heidelberg, Germany, 1996.

[303] V. H. Schulz, H. G. Bock, and M. C. Steinbach. Exploiting invariants in the numerical solution of multipoint boundary value problems for DAE. *SIAM Journal on Scientific Computing*, 19(2):440–467, 1998.

[304] A. Siburian. *Numerical Methods for Robust, Singular and Discrete Valued Optimal Control Problems*. PhD thesis, Curtin University of Technology, Perth, Australia, 2004.

[305] A. Siburian and V. Rehbock. Numerical procedure for solving a class of singular optimal control problems. *Optimization Methods & Software*, 19(3–4):413–426, 2004.

[306] B. Simeon. *Numerische Integration mechanischer Mehrkörpersysteme: Projizierende Deskriptorformen, Algorithmen und Rechenprogramme*, volume of VDI Fortschrittberichte Reihe 20: Rechnergestützte Verfahren #130. VDI-Verlag, 1994.

[307] E. Sontag. *Mathematical Control Theory – Deterministic Finite Dimensional Systems*, volume 6 of Texts in Applied Mathematics. Springer, New York, 1998.

[308] C. Specht, M. Gerdts, and R. Lampariello. Neighborhood estimation in sensitivity-based update rules for real-time optimal control. In *European Control Conference 2020, ECC 2020*, pages 1999–2006, 2020. art. no. 9143701.

[309] P. Spellucci. *Numerische Verfahren der nichtlinearen Optimierung*. Birkhäuser, Basel, 1993.

[310] M. C. Steinbach. *Fast Recursive SQP Methods for Large-Scale Optimal Control Problems*. PhD thesis, Interdisziplinäres Zentrum für Wissenschaftliches Rechnen, Universität Heidelberg, Heidelberg, Germany, 1995.

[311] H. J. Stetter. *Analysis of Discretization Methods for Ordinary Differential Equations*, volume 23 of Springer Tracts in Natural Philosophy. Springer, Berlin, Heidelberg, New York, 1973.

[312] J. Stoer. Principles of sequential quadratic programming methods for solving nonlinear programs. In K. Schittkowski, editor, *Computational Mathematical Programming*, volume F15 of NATO ASI Series, pages 165–207. Springer, Berlin–Heidelberg–New York, 1985.

[313] J. Stoer and R. Bulirsch. *Numerische Mathematik II*, 3rd edition. Springer, Berlin–Heidelberg–New York, 1990.

[314] R. Stöver. Collocation methods for solving linear differential-algebraic boundary value problems. *Numerische Mathematik*, 88(4):771–795, 2001.

[315] K. Strehmel and R. Weiner. *Numerik gewöhnlicher Differentialgleichungen*. Teubner, Stuttgart, 1995.

[316] K. L. Teo and C. J. Goh. *MISER: An Optimal Control Software. Applied Research Corporation*. National University of Singapore, Kent Ridge, Singapore, 1987.

[317] K. L. Teo, L. S. Jennings, V. Lee, and H. W. J. Rehbock. The control parameterization enhancing transform for constrained optimal control problems. *Journal of the Australian Mathematics Society*, 40(3):314–335, 1999.

[318] F. Tröltzsch. *Optimale Steuerung partieller Differentialgleichungen*. Vieweg, Wiesbaden, 2005.

[319] M. Ulbrich. Semismooth newton methods for operator equations in function spaces. *SIAM Journal on Optimization*, 13(3):805–841, 2003.

[320] M. Ulbrich. *Nonsmooth Newton-like Methods for Variational Inequalities and Constrained Optimization Problems in Function Spaces*. Habilitation, Technical University of Munich, Munich, 2002.

[321] R. J. Vanderbei. *Linear Programming. Foundations and Extensions*, volume 37 of International Series in Operations Research & Management Science. Kluwer Academic Publishers, Dordrecht, 2001.

[322] V. C. Vassiliades, R. W. H. Sargent, and C. C. Pantelides. Solution of a class of multistage dynamic optimization problems. 2. Problems with path constraints. *Industrial & Engineering Chemistry Research*, 33:2123–2133, 1994.

[323] T. von Heydenaber. *Simulation der Fahrdynamik von Kraftfahrzeugen*. Master's thesis, Institut für Mathematik, Technische Universität München, Germany, 1980.

[324] R. von Schwerin and H. G. Bock. A Runge–Kutta starter for a multistep method for differential-algebraic systems with discontinuous effects. *Applied Numerical Mathematics*, 18:337–350, 1995.

[325] O. von Stryk. *Numerische Lösung optimaler Steuerungsprobleme: Diskretisierung, Parameteroptimierung und Berechnung der adjungierten Variablen*, voluem 441 of VDI Fortschrittberichte Reihe 8: Meß-, Steuerungs- und Regeleungstechnik. VDI-Verlag, 1994.

[326] D. Werner. *Funktionalanalysis*. Springer, Berlin–Heidelberg–New York, 1995.

[327] D. V. Widder. *The Laplace Transform*. Princeton University Press, Princeton, 1946.

[328] W. L. Winston. *Operations Research: Applications and Algorithms*, 4th edition. Brooks/Cole–Thomson Learning, Belmont, 2004.

[329] V. M. Zavala and L. T. Biegler. The advanced-step NMPC controller: Optimality, stability and robustness. *Automatica*, 45(1):86–93, 2009.

[330] V. Zeidan. The Riccati equation for optimal control problems with mixed state-control constraints: Necessity and sufficiency. *SIAM Journal on Control and Optimization*, 32(5):1297–1321, 1994.

[331] A. Zomotor. *Fahrwerktechnik: Fahrverhalten*. Vogel Buchverlag, Stuttgart, 1991.

[332] J. Zowe and S. Kurcyusz. Regularity and stability of the mathematical programming problem in Banach spaces. *Applied Mathematics & Optimization*, 5:49–62, 1979.

Index

https://doi.org/10.1515/9783110797893-010

www.ingramcontent.com/pod-product-compliance
Lightning Source LLC
Chambersburg PA
CBHW080125220326
41598CB00032B/4959
* 9 7 8 3 1 1 0 7 9 7 6 9 5 *